Advanced Structured Materials

Volume 52

Series editors

Andreas Öchsner, Southport Queensland, Australia
Lucas F.M. da Silva, Porto, Portugal
Holm Altenbach, Magdeburg, Germany

More information about this series at http://www.springer.com/series/8611

Danilo Capecchi · Giuseppe Ruta

Strength of Materials and Theory of Elasticity in 19th Century Italy

A Brief Account of the History of Mechanics of Solids and Structures

Springer

Danilo Capecchi
Giuseppe Ruta
Dipt di Ingegneria Strut. e Geotecnica
Università di Roma "La Sapienza"
Rome
Italy

ISSN 1869-8433 ISSN 1869-8441 (electronic)
ISBN 978-3-319-05523-7 ISBN 978-3-319-05524-4 (eBook)
DOI 10.1007/978-3-319-05524-4

Library of Congress Control Number: 2014941511

Springer Cham Heidelberg New York Dordrecht London

Printed on acid-free paper

Springer is part of Springer Science+Business Media (www.springer.com)

Preface

In 1877 Giovanni Curioni, Professor in the *Scuola d'applicazione per gl'ingegneri* (School of Application for Engineers) in Turin, chose the name *Scienza delle costruzioni* for his course of mechanics applied to civil and mechanical constructions.

The choice reflected a change that had occurred in the teaching of structural disciplines in Italy, following the establishment of schools of application for engineers by Casati's reform of 1859. On the model of the École polytechnique, the image of the purely technical engineer was replaced by that of the 'scientific engineer', inserting into the teaching both 'sublime mathematics' and modern theories of elasticity. Similarly, the art of construction was to be replaced by the science of construction. The *Scienza delle costruzioni* came to represent a synthesis of theoretical studies of continuum mechanics, carried out primarily by French scholars of elasticity, and the mechanics of structures, which had begun to develop in Italian and German schools. In this respect it was an approach without equivalence in Europe, where the contents of continuum mechanics and mechanics of structures were, and still today are, taught in two different disciplines.

In the 1960s of the twentieth century, the locution *Scienza delle costruzioni* took a different sense for various reasons. Meanwhile, the discipline established by Curioni was divided into two branches, respectively, called *Scienza delle costruzioni* and *Tecnica delle costruzioni*, relegating this last to applicative aspects. Then technological developments required the study of materials with more complex behavior than the linear elastic one; there was a need for protection from phenomena of fatigue and fracture, and dynamic analysis became important for industrial applications (vibrations) and civil incidents (wind, earthquakes). Finally, introduction of modern structural codes on the one hand made obsolete the sophisticated manual calculation techniques developed between the late 1800s and early 1900s, on the other hand it necessitated a greater knowledge of the theoretical aspects, especially of continuum mechanics. This necessity to deepen the theory inevitably led a to drift toward mathematical physics in some scholars.

All this makes problematic a modern definition of *Scienza delle costruzioni*. To overcome this difficulty, in our work we decided to use the term *Scienza delle costruzioni* with a fairly wide sense, to indicate the theoretical part of construction engineering. We considered Italy and the nineteenth century for two reasons. Italy, to account for the lack of knowledge of developments in the discipline in this country, which is in any case a major European nation. The nineteenth century, because it is one in which most problems of design of structures were born and reached maturity, although the focus was concentrated on materials with linear elastic behavior and external static actions.

The existing texts on the history of *Scienza delle costruzioni*, among which one of the most complete in our opinion is that by Stephen Prokofievich Timoshenko, *History of Strength of Materials*, focus on French, German, and English schools, largely neglecting the Italian. Moreover, Edoardo Benvenuto's text, *An Introduction to the History of Structural Mechanics*, which is very attentive to the Italian contributions, largely neglects the nineteenth century. Only recently, Clifford Ambrose Truesdell, mathematician and historian of mechanics, in his *Classical Field Theories of Mechanics* highlighted the important contributions of Italian scientists, dusting off the names of Piola, Betti, Beltrami, Lauricella, Cerruti, Cesaro, Volterra, Castigliano, and so on.

The present book deals largely with the theoretical foundations of the discipline, starting from the origin of the modern theory of elasticity and framing the Italian situation in Europe, examining and commenting on foreign authors who have had a key role in the development of mechanics of continuous bodies and structures and graphic calculation techniques. With this in mind, we have mentioned only those issues most 'applicative', which have not seen important contributions by Italian scholars. For example, we have not mentioned any studies on plates that were brought forward especially in France and Germany and which provided fundamental insights into more general aspects of continuum mechanics. Consider, for instance, the works on plates by Kirchhoff, Saint Venant, Sophie Germain, and the early studies on dynamic stresses in elastic bodies by Saint Venant, Navier, Cauchy, Poncelet. Finally, we have not mentioned any of the experimental works carried out especially in England and Germany, including also some important ones from a theoretical point of view about the strength and fracture of materials.

The book is intended as a work of historical research, because most of the contents are either original or refer to our contributions published in journals. It is directed to all those graduates in scientific disciplines who want to deepen the development of Italian mathematical physics in the nineteenth century. It is directed to engineers, but also architects, who want to have a more comprehensive and critical vision of the discipline they have studied for years. Of course, we hope it will be helpful to scholars of the history of mechanics as well.

We would like to thank Raffaele Pisano and Annamaria Pau for reading drafts of the book and for their suggestions.

Editorial Considerations

Figures related to quotations are all redrawn to allow better comprehension. They are, however, as much as possible close to the original ones. Symbols of formulas are always those of the authors, except cases easily identifiable. Translations of texts from French, Latin, German, and Italian are as much as possible close to the original texts. For Latin, a critical transcription has been preferred where some shortenings are resolved, ‘v’ is modified to ‘u’ and vice versa where necessary, ij to ii, following the modern rule; moreover, the use of accents is avoided. Titles of books and papers are always reproduced in the original spelling. For the name of the different characters the spelling of their native language is used, excepting for the ancient Greeks, for which the English spelling is assumed, and some medieval people, for which the Latin spelling is assumed, following the common use.

Through the text, we searched to avoid modern terms and expressions as much as possible while referring to ‘old’ theories. In some cases, however, we transgressed this resolution for the sake of simplicity. This concerns the use, for instance, of terms like *field*, *balance*, and *energy* even in the period they were not used or were used differently from today. The same holds good for expressions like, for instance, *principle of virtual work*, that was common only since the nineteenth century.

Danilo Capecchi
Giuseppe Ruta

Contents

Chapter 1
The Theory of Elasticity in the 19th Century

Abstract Until 1820 there was a limited knowledge about the elastic behavior of materials: one had an inadequate theory of bending, a wrong theory of torsion, the definition of Young's modulus. Studies were made on one-dimensional elements such as beams and bars, and two-dimensional, such as thin plates (see for instance the work of Marie Sophie Germain). These activities started the studies on three-dimensional elastic solids that led to the theory of elasticity of three-dimensional continua becoming one of the most studied theories of mathematical physics in the 19th century. In a few years most of the unresolved problems on beams and plates were placed in the archives. In this chapter we report briefly a summary on three-dimensional solids, focusing on the theory of constitutive relationships, which is the part of the theory of elasticity of greatest physical content and which has been the object of major debate. A comparison of studies in Italy and those in the rest of Europe is referenced.

1.1 Theory of Elasticity and Continuum Mechanics

The theory of elasticity has ancient origins. Historians of science, pressed by the need to provide an *a quo* date, normally refer to the *Lectures de potentia restitutiva* by Robert Hooke in 1678 [78]. One can debate this date, but for the moment we accept it because a historically accurate reconstruction of the early days of the theory of elasticity is out of our purpose; we limit ourselves only to pointing out that Hooke should divide the honor of the primeval introduction with at least Edme Mariotte [95]. Hooke and Mariotte studied problems classified as engineering: the displacement of the point of a beam, its curvature, the deformation of a spring, etc.

Explanations *per causas* of elasticity can be traced back to the *Quaestio* 31 of Isaac Newton's *Opticks* of 1704 [117], in which the corpuscular constitution of matter is discussed. Many alternative conceptions were developed in the 18th century, especially with reference to the concept of ether; for a few details we refer to the literature [7]. In the early years of the 19th century the theory of elasticity was intimately connected to some corpuscular theories, such as that of Laplace [88][1],

[1] vol. 4, pp. 349, 350.

D. Capecchi and G. Ruta, *Strength of Materials and Theory of Elasticity in 19th Century Italy*, Advanced Structured Materials 52,
DOI 10.1007/978-3-319-05524-4_1

[68] who refined the approach of Newton, and considered the matter consisting of small bodies, with extension and mass, or that of Ruggero Boscovich [12] according to which matter is based on unextended centers of force endowed with mass. The masses are attracted with forces depending on their mutual distance; repulsive at short distance, attractive at a greater distance, as illustrated in Fig. 1.1.

It should be said that it was not just engineering that influenced the development of the theory of elasticity; an even superficial historical analysis shows that such researches were also linked to the attempt to provide a mechanistic interpretation of nature. According to this interpretation every physical phenomenon must be explained by particle mechanics: matter has a discrete structure and space is filled with fine particles with uniform properties, which form the *ether*. All the physical phenomena propagate in space by a particle of ether to its immediate neighbor by means of impacts or forces of attraction or repulsion. This point of view allows one to overcome the difficulties of the concept of action at a distance: In which way, asked the physicists of the time, can two bodies interact, for instance attract each other, without the action of an intervening medium? Any physical phenomenon corresponds to a state of stress in the ether, propagated by contact.

With the beginning of the 19th century the need was felt to quantitatively characterize the elastic behavior of bodies and the mathematical theory of elasticity was born. Its introduction was thought to be crucial for an accurate description of the physical world, in particular to better understand the phenomenon of propagation of light waves through the air. The choices of physicists were strongly influenced by mathematics in vogue at that time, that is the differential and integral calculus, hereinafter *Calculus*. It presupposed the mathematics of continuum and therefore was difficult to fit into the discrete particle model, which had become dominant.

Most scientists adopted a compromise approach that today can be interpreted as a technique of homogenization. The material bodies, with a fine corpuscular structure, are associated with a mathematical continuum C, as may be a solid of Euclidean geometry. The variables of displacement are represented by a sufficiently regular function **u** defined in C, that assumes significant values only for those points P of C that are also positions of particles. The derivatives of the function **u** with respect to the variables of space and time also have meaning only for the points P. The internal forces exchanged between particles, at the beginning thought of as concentrated, are represented by distributed mean values that are attributed to all the points of

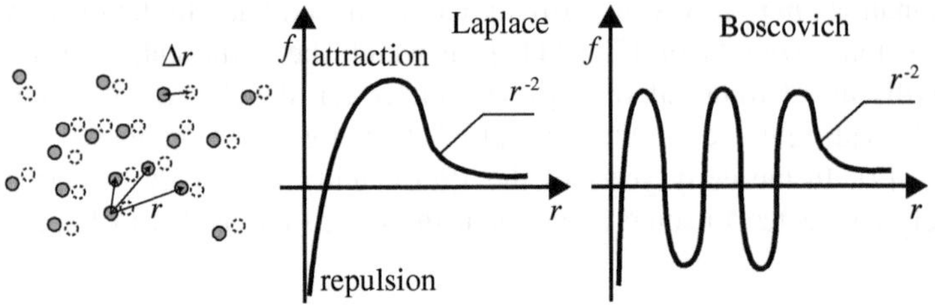

Fig. 1.1 Molecular model: force f between two molecules as a function of their distance r

C, thus becoming stresses σ. Other scientists gave up the corpuscular physical model considering it only in the background. They founded their theories directly on the continuum, whose points had now all 'physical' meaning. On the continuum are defined both the displacements and the stresses, as had already been done in the 18th century by Euler and Lagrange for fluids. Some scientists oscillated between the two approaches, among them Augustin Cauchy (1760–1848) (but the Italian Gabrio Piola (1794–1850) was in a similar position [19]) who, while studying the distribution of internal forces of solids, systematized mathematical analysis, dealing with the different conceptions of infinite and infinitesimal, of discrete and continuum. His oscillations in mathematical analysis were reflected in his studies on the constitution of matter [56, 57].

In the following we present in some detail and sense of history what we have just outlined above, speaking of the various corpuscular approaches and continuum approach, referring primarily to the relationship between the internal force and displacement, or between stress and strain, that is the constitutive law. Other problems of the theory of elasticity, always in the context of continua, will be mentioned later, to finally devote several sections to the elasticity theory of discrete systems in general and to the structures formed by beams in particular.

1.1.1 The Classical Molecular Model

The theories of elasticity of the early 19th century were based on different corpuscular assumptions, introduced almost simultaneously by Fresnel, Cauchy and Navier [25, 27, 70, 114]. French scientists adopted the single word *molecule* for particles, which lived long in European scientific literature, often flanked by *atom*, without the two terms necessarily had different meanings, at least until the studies of the chemical constitution of matter advanced and the terms atom and molecule assumed precise technical meanings which differentiate the areas of application.

Augustin Jean Fresnel studied the propagation of light through the ether, imagined as a set of material points that exchange elastic forces. In a work of 1820 he obtained very interesting results, as for instance the theorem:

> As long as small displacements are concerned and whatever the law of the forces that the molecules of the medium exert on each other, the movement of a molecule in any direction produces a repulsive force equal in magnitude and direction to the resultant of the three repulsive forces generated by three rectangular displacement of this molecule equal to the static components of the first [small] displacement [70].[2] (A.1.1)

This theorem about the force that rises among the molecules, 'nearly self evident in its statement', was presented by Cauchy in an appendix of his famous paper on stress [26],[3] where an explicit reference to Fresnel was made.

[2] pp. 344–345. Our translation.

[3] Addition, pp. 79–81.

The first systematic work on the equilibrium and the motion of three-dimensional elastic bodies was however due to Navier, who in 1821 read before the Académie des sciences de Paris an important memoir published only in 1827 [114].

Navier, referring explicitly to Lagrange's *Méchanique analitique* [83], wrote the equations of local equilibrium of forces acting on an elastic body, thought of as an aggregate of particles that attract or repel each other with an elastic force variable linearly with their mutual displacements:

> One considers a solid body as an assemblage of material molecules placed at a very small distance. These molecules exert two opposite actions on each other, that is a proper attractive force and a repulsive force due to the principle of heat. Between one molecule M and any other M' of the neighboring molecules there is an attraction P which is the difference of these two forces. In the natural state of the body all the forces P are zero or reciprocally destroy, because the molecule M is at rest. When the body changes its shape, the force P takes a different value Π and there is equilibrium between all the forces Π and the forces applied to the body, by which the change of the shape of the body is produced [114].[4] (A.1.2)

Let X, Y, Z be the external forces per unit of volume, ϵ a constant (to use a modern term it is the second Lamé constant) and x, y, z the displacement of the generic point P having initial coordinates a, b, c, then the equilibrium equations obtained by Navier are [114][5]:

$$
\begin{aligned}
-X &= \epsilon\left(3\frac{d^2x}{da^2}+\frac{d^2x}{db^2}+\frac{d^2x}{dc^2}+2\frac{d^2y}{da\,db}+2\frac{d^2z}{da\,dc}\right)\\
-Y &= \epsilon\left(\frac{d^2y}{db^2}+3\frac{d^2y}{da^2}+\frac{d^2y}{dc^2}+2\frac{d^2x}{da\,db}+2\frac{d^2z}{db\,dc}\right)\\
-Z &= \epsilon\left(\frac{d^2z}{db^2}+\frac{d^2z}{dc^2}+3\frac{d^2z}{da^2}+2\frac{d^2x}{da\,dc}+2\frac{d^2y}{db\,dc}\right).
\end{aligned}
\tag{1.1}
$$

Navier obtained these equations with the use of the principle of virtual work [114].[6] He followed the approach, already mentioned, common to all French scientists of the 19th century, by considering the body as discrete when he wanted to study the equilibrium, while as continuous when he came to describe the geometry and obtained simple mathematical relationships, replacing the summations with integrals.[7] Note that in the work of Navier the concept of stress, which was crucial to the mechanics of structures developed later, was not present.

In the academic French world the molecular model of Navier became dominant because of the influence of the teaching of Laplace. On October 1st, 1827 Poisson and Cauchy presented to the Académie des sciences de Paris two memoirs similar

[4] pp. 375–376. Our translation.

[5] p. 384.

[6] p. 384.

[7] The difficulty of replacing summations with integrals has been the subject of many comments of French scholars, especially Poisson and Cauchy.

to each other, where Navier's molecular model was adopted [116].[8] Poisson gave decisive contributions in this field. In two other papers read at the Académie des sciences de Paris on April 14th, 1828 [127] and on October 12th, 1829 [128] he expressed its assumptions:

> The molecules of all bodies are subject to their mutual attraction and repulsion due to heat. According that the first of these two forces is greater or less than the second, the result is an attractive or repulsive force between two molecules, but in both cases, the resultant is a function of the distance from a molecule to the other whose law is unknown to us; we only know that this function decreases in a very fast manner, and becomes insensible as soon as the distance has acquired a significant magnitude. However, we assume that the radius of activity of the molecules is very large compared to the intervals between them, and we assume, moreover, that the rapid decrease of the action takes place only when the distance became the sum of a very large number of these intervals [127].[9] (A.1.3)

and introduced the concept of *stress*:

> Let M be a point in the inner part of the body, at a sensible distance from the surface [Fig. 1.2a]. Let us consider a plane through this point, dividing the body into two parts, which we will suppose horizontal [...]. Let us denote by A the upper part and A′ the lower part, in which we will include the material points belonging to the plane itself. From the point M considered as a center let us draw a sphere including a very large amount of molecules, yet the radius of which is in any case negligible with respect to the radius of the molecular activity. Let ω be the area of its horizontal section; over this section let us raise a vertical cylinder, the height of which is at least the same as the radius of molecular activity; let us call B this cylinder; the force of the molecules of A′ over those of B, divided by ω, will be the *pressure* exerted by A′ over A, with respect to the unity of surface and relative to the point M [129].[10] (A.1.4)

For isotropic materials Cauchy [30][11] and Poisson [127] obtained relations close to those by Navier. This is for instance the expression given by Poisson:

$$
\begin{aligned}
&X - \frac{d^2u}{dt^2} + a^2\left(\frac{d^2u}{dy^2} + \frac{2}{3}\frac{d^2v}{dydx} + \frac{2}{3}\frac{d^2w}{dzdx} + \frac{1}{3}\frac{d^2u}{dx^2} + \frac{1}{3}\frac{d^2u}{dz^2}\right) = 0,\\
&Y - \frac{d^2v}{dt^2} + a^2\left(\frac{d^2v}{dy^2} + \frac{2}{3}\frac{d^2u}{dxdy} + \frac{2}{3}\frac{d^2w}{dzdy} + \frac{1}{3}\frac{d^2v}{dx^2} + \frac{1}{3}\frac{d^2v}{dz^2}\right) = 0,\\
&Z - \frac{d^2w}{dt^2} + a^2\left(\frac{d^2w}{dz^2} + \frac{2}{3}\frac{d^2u}{dxdz} + \frac{2}{3}\frac{d^2v}{dydz} + \frac{1}{3}\frac{d^2w}{dx^2} + \frac{1}{3}\frac{d^2w}{dy^2}\right) = 0,
\end{aligned} \tag{1.2}
$$

where X, Y, Z are the forces per unit of mass and a a constant of elasticity [127].[12]

[8] pp. CLV, CLIX. The memoir of Cauchy appeared first with the title *Mémoire sur l'équilibre et le mouvement d'un système de points materiels sollecités par forces d'attraction ou de répulsion mutuelle* [30]. That of Poisson appeared with the title *Note sur les vibrations des corps sonores* [126].

[9] pp. 368–369. Our translation.

[10] p. 29. Our translation. Stress was indicated by French scientists by *pressure* or *tension*.

[11] pp. 250–251.

[12] p. 403.

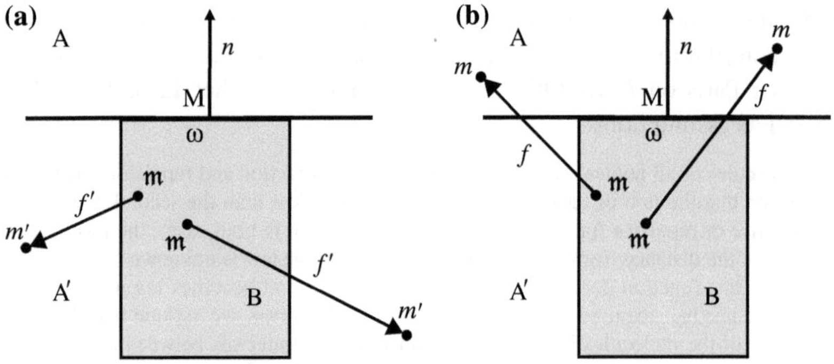

Fig. 1.2 Stress according to Poisson (**a**) and Cauchy (**b**)

In the following, we show in some detail the main features of the classical molecular model, along with its origin, trying to grasp its strengths and weaknesses. The focus is on the constitutive stress–strain relations because here one can see better the consequences of the assumptions about the molecular model. Reference is made to the work by Cauchy of 1828 [29, 30],[13] among the most complete and clear on the subject (see below).

The main assumptions of the molecular model are:

1. The molecules are treated as material points subjected to opposing forces directed along their joining line (central forces assumption).
2. The force between two molecules decreases rapidly starting from a distance, small but much larger than the normal distance between two molecules, called *ray of molecular action*.
3. The molecules have all the same mass and the force between any two molecules is provided by the same function $f(r)$ of their distance r.
4. The relative displacements of the molecules are 'small'.
5. The function $f(r)$ which expresses the force between two molecules is regular in r, and then can be differentiated.
6. The motion of the molecules is defined by a smooth vector field in the continuum where the system of molecules are imagined to be embedded.

The first three assumptions are physical, the remaining are of mathematical character, introduced clearly to simplify the treatment.

[13] pp. 227–252.

1.1.1.1 The Components of Stress

In his work of 1828 [29] Cauchy adopted a variant of Poisson's definition of stress. The difference was that he considered the force of the molecules m in A (Fig. 1.2b) on the molecules $\mathfrak{m}$ in B instead of the force of the molecules m' in A'.[14]

Consider the cylinder B of Fig. 1.2b having an infinitesimal base ω on a plane perpendicular to the unit vector n, located in the half space A'. Let $\mathfrak{m}$ be an assigned molecule inside the cylinder and m the molecules located in the half-space A on the same side of n. The force exerted on $\mathfrak{m}$ by all the molecules m is characterized by the three components [29][15]:

$$\sum \pm \mathfrak{m} m \cos\alpha\, f(r); \qquad \sum \pm \mathfrak{m} m \cos\beta\, f(r); \qquad \sum \pm \mathfrak{m} m \cos\gamma\, f(r), \tag{1.3}$$

where $f(r)$ is the force between m and $\mathfrak{m}$, α, β, γ are the direction cosines of the radius vector r connecting m—that is the components of the unit vector parallel to r—and $\mathfrak{m}$, with respect to an arbitrary coordinate system and the sum is extended to all the molecules m of the half space A opposite to the cylinder, or rather to all those in the sphere of molecular action (the sphere defined by the radius of molecular action) of $\mathfrak{m}$. To obtain the force exerted on the cylinder and, according to Poisson, the pressure on the surface ω, the summations of the relation (1.3) should be extended to all the molecules $\mathfrak{m}$ of the cylinder and divided by ω. Since all the molecules are assumed to be equal, this summation was made explicit in a simple way by Cauchy, who after some steps obtained the components for the stress on the faces orthogonal to the coordinate axes. For instance those on the face orthogonal to x are given by Cauchy [29][16]:

$$\begin{cases} A = \Delta \sum \pm m \cos^2\alpha\, f(r) \\ F = \Delta \sum \pm m \cos\alpha \cos\beta\, f(r) \\ E = \Delta \sum \pm m \cos\alpha \cos\gamma\, f(r), \end{cases} \tag{1.4}$$

with Δ the specific mass of the body, supposed locally homogeneous.

Cauchy had already introduced the symbols for the stress components in the work of 1827 [26];[17] they will be adopted by other scholars long before the indexed notations was established (see below). Full symbols and correspondences with modern notations are given in the following list and shown in Fig. 1.3:

[14] Actually Cauchy introduced various slightly different definitions of stress. In a memoir of 1845 [34] he adopted the definition considered also by Saint Venant and Jean-Marie Constant Duhamel according to which the "stress (la pression) on a very small area (ω is defined) as the resultant of the actions of all the molecules located on the one side over all the molecules located on the other side whose directions cross this element" [141], p. 24.

[15] p. 257.

[16] p. 257, Eq. (1.13).

[17] pp. 60–81.

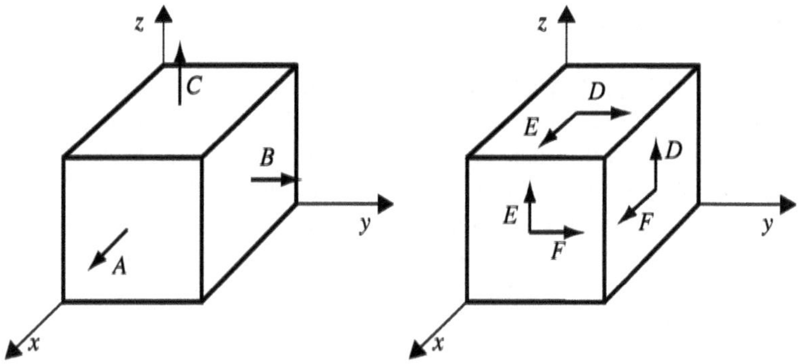

Fig. 1.3 The components of the stress tensor according to Cauchy

$$\begin{array}{lll} A(\equiv \sigma_x) & F(\equiv \tau_{yx}) & E(\equiv \tau_{zx}) \\ F(\equiv \tau_{xy}) & B(\equiv \sigma_y) & D(\equiv \tau_{zy}) \\ E(\equiv \tau_{xz}) & D(\equiv \tau_{yz}) & C(\equiv \sigma_z). \end{array} \tag{1.5}$$

1.1.1.2 The Component of Strains and the Constitutive Relationships

In the modern theories of continuum mechanics, the components of the stress and strain are defined independently first, then the function connecting them, which is precisely the constitutive law, is introduced.

In the classical molecular theory the historical path was different. The definition of the strain passed in the background and implicitly stemmed from the attempt to establish the link between stresses and displacements, as soon as the latter are approximated with their infinitesimal values. This approach was certainly influenced by the work of Navier in 1821 [114] which had the aim of finding the differential equations for displacement components in an elastic body, without any examination of the internal forces.

To obtain the relations that link the components of the stresses to those of the strains, Cauchy rewrote the relations analogous to (1.4), taking into account the displacement with components $\xi,\ \eta,\ \zeta$ of the molecules from their initial position. Cauchy indicated with $\Delta a,\ \Delta b,\ \Delta c$ the components of the distance r between two molecules in the undeformed state and with $\Delta x,\ \Delta y,\ \Delta z$ those of the distance in the deformed state, resulting in the relations:

$$\Delta x = \Delta a + \Delta\xi,\ \Delta y = \Delta b + \Delta\,\eta,\ \Delta z = \Delta c + \Delta\,\zeta. \tag{1.6}$$

The new distance among molecules was defined by Cauchy by means of its percentage variation ϵ as $(1+\epsilon)r$.

The components of stress in the deformed configuration were obtained by replacing in the relation (1.4) the new expressions of forces and distances [29][18]:

$$\begin{cases} A = \frac{\rho}{2}\sum\left\{\pm m\frac{f[r(1+\epsilon)]}{r(1+\epsilon)}\Delta x^2\right\};\ D = \frac{\rho}{2}\sum\left\{\pm m\frac{f[r(1+\epsilon)]}{r(1+\epsilon)}\Delta y\Delta z\right\} \\ B = \frac{\rho}{2}\sum\left\{\pm m\frac{f[r(1+\epsilon)]}{r(1+\epsilon)}\Delta y^2\right\};\ E = \frac{\rho}{2}\sum\left\{\pm m\frac{f[r(1+\epsilon)]}{r(1+\epsilon)}\Delta z\Delta x\right\} \\ C = \frac{\rho}{2}\sum\left\{\pm m\frac{f[r(1+\epsilon)]}{r(1+\epsilon)}\Delta z^2\right\};\ F = \frac{\rho}{2}\sum\left\{\pm m\frac{f[r(1+\epsilon)]}{r(1+\epsilon)}\Delta x\Delta y\right\}, \end{cases} \tag{1.7}$$

where ρ is the mass density in the deformed configuration, different in general from the mass density Δ in the undeformed configuration, and the sum is extended to all the molecules contained inside the sphere of molecular action of $\mathfrak{m}$, both in the half space containing the infinitesimal cylinder and the opposite one. That justifies the factor 1/2.

To obtain relations suitable for algebraic manipulation and thus for simplification, Cauchy [30] introduced the assumption of small displacements, which allowed him to derive linearized relations in ϵ; and a linear elastic relationship between stresses and strains:

$$\begin{aligned} &\frac{f[r(1+\epsilon)]}{r(1+\epsilon)} \approx \frac{f(r)}{r} + \frac{rf'(r)-f(r)}{r}\epsilon, \\ &\epsilon = \frac{1}{r}\left(\cos\alpha\,\Delta\xi + \cos\beta\,\Delta\eta + \cos\gamma\,\Delta\zeta\right). \end{aligned} \tag{1.8}$$

Having chosen a reference molecule $\mathfrak{m}$, the one at the center of the elementary surface ω of the cylinder, for instance, Cauchy linearized the variation of the components of the displacements interior to the sphere of action of $\mathfrak{m}$ with respect to the spatial variables. This is possible because of the small distance among the molecules inside the molecular sphere of action:

$$\begin{aligned} \frac{\Delta\xi}{r} &= \frac{\partial\xi}{\partial a}\cos\alpha + \frac{\partial\xi}{\partial b}\cos\beta + \frac{\partial\xi}{\partial c}\cos\gamma, \\ \frac{\Delta\eta}{r} &= \frac{\partial\eta}{\partial a}\cos\alpha + \frac{\partial\eta}{\partial b}\cos\beta + \frac{\partial\eta}{\partial c}\cos\gamma, \\ \frac{\Delta\zeta}{r} &= \frac{\partial\zeta}{\partial a}\cos\alpha + \frac{\partial\zeta}{\partial b}\cos\beta + \frac{\partial\zeta}{\partial c}\cos\gamma \end{aligned} \tag{1.9}$$

where the derivatives are evaluated at $\mathfrak{m}$.

By replacing in (1.7) the linearized expressions of f and ϵ, simplifying and neglecting the higher order infinitesimals in $\Delta\xi$, $\Delta\eta$, $\Delta\zeta$, Cauchy derived the relations referred to in Fig. 1.4, which express the constitutive relationship. They give the expression of the components of stress A, B, C, D, E, F versus the nine components

[18] p. 260, Eq. (1.18).

$$\left\{\begin{array}{l}
A=\rho\mathrm{S}\left[\pm\frac{m\rho}{2}\cos^2\alpha f(r)\right]\\
+2\rho\left\{\frac{\partial\xi}{\partial a}\mathrm{S}\left[\pm\frac{m\rho}{2}\cos^2\alpha f(r)\right]+\frac{\partial\xi}{\partial b}\mathrm{S}\left[\pm\frac{m\rho}{2}\cos\alpha\cos\beta f(r)\right]+\frac{\partial\xi}{\partial c}\mathrm{S}\left[\pm\frac{m\rho}{2}\cos\alpha\cos\gamma f(r)\right]\right\}\\
+\rho\left\{\begin{array}{l}
\frac{\partial\xi}{\partial a}\mathrm{S}\left[\frac{m\rho}{2}\cos^4\alpha f(r)\right]+\frac{\partial\xi}{\partial b}\mathrm{S}\left[\frac{m\rho}{2}\cos^3\alpha\cos\beta f(r)\right]+\frac{\partial\xi}{\partial c}\mathrm{S}\left[\frac{m\rho}{2}\cos^3\alpha\cos\gamma f(r)\right]\\
\frac{\partial\eta}{\partial a}\mathrm{S}\left[\frac{m\rho}{2}\cos^3\alpha\cos\beta f(r)\right]+\frac{\partial\eta}{\partial b}\mathrm{S}\left[\frac{m\rho}{2}\cos^2\alpha\cos^2\beta f(r)\right]+\frac{\partial\eta}{\partial c}\mathrm{S}\left[\frac{m\rho}{2}\cos^2\alpha\cos\beta\cos\gamma f(r)\right]\\
\frac{\partial\zeta}{\partial a}\mathrm{S}\left[\frac{m\rho}{2}\cos^3\alpha\cos\gamma f(r)\right]+\frac{\partial\zeta}{\partial b}\mathrm{S}\left[\frac{m\rho}{2}\cos^2\alpha\cos\beta\cos\gamma f(r)\right]+\frac{\partial\zeta}{\partial c}\mathrm{S}\left[\pm\frac{m\rho}{2}\cos^2\alpha\cos^2\gamma f(r)\right]
\end{array}\right\}
\end{array}\right.$$

$$\left\{\begin{array}{l}
B=\rho\mathrm{S}\left[\pm\frac{m\rho}{2}\cos^2\beta f(r)\right]\\
+2\rho\left\{\frac{\partial\eta}{\partial a}\mathrm{S}\left[\pm\frac{m\rho}{2}\cos\alpha\cos\beta f(r)\right]+\frac{\partial\eta}{\partial b}\mathrm{S}\left[\pm\frac{m\rho}{2}\cos^2\beta f(r)\right]+\frac{\partial\eta}{\partial c}\mathrm{S}\left[\pm\frac{m\rho}{2}\cos\alpha\cos\gamma f(r)\right]\right\}\\
+\rho\left\{\begin{array}{l}
\frac{\partial\xi}{\partial a}\mathrm{S}\left[\frac{m\rho}{2}\cos^2\alpha\cos^2\beta f(r)\right]+\frac{\partial\xi}{\partial b}\mathrm{S}\left[\frac{m\rho}{2}\cos\alpha\cos^3\beta f(r)\right]+\frac{\partial\xi}{\partial c}\mathrm{S}\left[\frac{m\rho}{2}\cos\alpha\cos^2\beta\cos\gamma f(r)\right]\\
\frac{\partial\eta}{\partial a}\mathrm{S}\left[\frac{m\rho}{2}\cos\alpha\cos^3\beta f(r)\right]+\frac{\partial\eta}{\partial b}\mathrm{S}\left[\frac{m\rho}{2}\cos^4\beta f(r)\right]+\frac{\partial\eta}{\partial c}\mathrm{S}\left[\frac{m\rho}{2}\cos^3\beta\cos\gamma f(r)\right]\\
\frac{\partial\zeta}{\partial a}\mathrm{S}\left[\frac{m\rho}{2}\cos\alpha\cos^2\beta\cos\gamma f(r)\right]+\frac{\partial\zeta}{\partial b}\mathrm{S}\left[\frac{m\rho}{2}\cos^3\beta\cos\gamma f(r)\right]+\frac{\partial\zeta}{\partial c}\mathrm{S}\left[\frac{m\rho}{2}\cos^2\beta\cos^2\gamma f(r)\right]
\end{array}\right\}
\end{array}\right.$$

$$\left\{\begin{array}{l}
C=\rho\mathrm{S}\left[\pm\frac{m\rho}{2}\cos^2\gamma f(r)\right]\\
+2\rho\left\{\frac{\partial\xi}{\partial a}\mathrm{S}\left[\pm\frac{m\rho}{2}\cos\alpha\cos\gamma f(r)\right]+\frac{\partial\xi}{\partial b}\mathrm{S}\left[\pm\frac{m\rho}{2}\cos\beta\cos\gamma f(r)\right]+\frac{\partial\xi}{\partial c}\mathrm{S}\left[\pm\frac{m\rho}{2}\cos^2\gamma f(r)\right]\right\}\\
+\rho\left\{\begin{array}{l}
\frac{\partial\xi}{\partial a}\mathrm{S}\left[\frac{m\rho}{2}\cos^2\alpha\cos^2\beta f(r)\right]+\frac{\partial\xi}{\partial a}\mathrm{S}\left[\frac{m\rho}{2}\cos\alpha\cos\beta\cos^2\gamma f(r)\right]+\frac{\partial\xi}{\partial a}\mathrm{S}\left[\frac{m\rho}{2}\cos\alpha\cos^3\gamma f(r)\right]\\
\frac{\partial\eta}{\partial a}\mathrm{S}\left[\frac{m\rho}{2}\cos\alpha\cos\beta\cos^2\gamma f(r)\right]+\frac{\partial\eta}{\partial b}\mathrm{S}\left[\frac{m\rho}{2}\cos^2\beta\cos^2\gamma f(r)\right]+\frac{\partial\eta}{\partial c}\mathrm{S}\left[\frac{m\rho}{2}\cos\beta\cos^3\gamma f(r)\right]\\
\frac{\partial\zeta}{\partial a}\mathrm{S}\left[\frac{m\rho}{2}\cos\alpha\cos^3\gamma f(r)\right]+\frac{\partial\zeta}{\partial b}\mathrm{S}\left[\frac{m\rho}{2}\cos\beta\cos^3\gamma f(r)\right]+\frac{\partial\zeta}{\partial c}\mathrm{S}\left[\frac{m\rho}{2}\cos^4\gamma f(r)\right]
\end{array}\right\}.
\end{array}\right.$$

Fig. 1.4 Components of stress in Cauchy's molecular model [29, p. 263]

of the displacement gradient $\partial\xi/\partial a$, $\partial\xi/\partial b$, $\partial\xi/\partial c$, $\partial\eta/\partial a$, $\partial\eta/\partial b$, $\partial\eta/\partial c$, $\partial\zeta/\partial a$, $\partial\zeta/\partial b$, $\partial\zeta/\partial c$, that implicitly define the components of the strains.

The stress components are related to those of the strain by 21 distinct coefficients, defined by the summation extended to all the molecules inside the sphere of action of the point-molecule in which one wants to calculate the stress, which multiply the derivatives of the components of the displacement at the same point (in the tables the symbol S stands for summation). The exception is the first term, which contains

$$
\begin{cases}
D=\rho\mathrm{S}\left[\pm\frac{m\rho}{2}\cos\beta\cos\gamma f(r)\right]\\
+\rho\left\{\begin{array}{l}
\frac{\partial\eta}{\partial a}\mathrm{S}\left[\pm\frac{m\rho}{2}\cos\alpha\cos\gamma f(r)\right]+\frac{\partial\eta}{\partial b}\mathrm{S}\left[\pm\frac{m\rho}{2}\cos\beta\cos\gamma f(r)\right]+\frac{\partial\eta}{\partial c}\mathrm{S}\left[\pm\frac{m\rho}{2}\cos^2\gamma f(r)\right]\\
\frac{\partial\xi}{\partial a}\mathrm{S}\left[\pm\frac{m\rho}{2}\cos\alpha\cos\beta f(r)\right]+\frac{\partial\xi}{\partial b}\mathrm{S}\left[\pm\frac{m\rho}{2}\cos^2\beta f(r)\right]+\frac{\partial\xi}{\partial c}\mathrm{S}\left[\pm\frac{m\rho}{2}\cos\beta\cos\gamma f(r)\right]
\end{array}\right\}\\
+\rho\left\{\begin{array}{l}
\frac{\partial\xi}{\partial a}\mathrm{S}\left[\frac{m\rho}{2}\cos^2\alpha\cos\beta\cos\gamma f(r)\right]+\frac{\partial\xi}{\partial b}\mathrm{S}\left[\frac{m\rho}{2}\cos\alpha\cos^2\beta\cos\gamma f(r)\right]+\frac{\partial\xi}{\partial c}\mathrm{S}\left[\frac{m\rho}{2}\cos\alpha\cos\beta\cos^2\gamma f(r)\right]\\
\frac{\partial\eta}{\partial a}\mathrm{S}\left[\frac{m\rho}{2}\cos\alpha\cos^2\beta\cos\gamma f(r)\right]+\frac{\partial\eta}{\partial b}\mathrm{S}\left[\frac{m\rho}{2}\cos^3\beta\cos\gamma f(r)\right]+\frac{\partial\eta}{\partial c}\mathrm{S}\left[\frac{m\rho}{2}\cos^2\alpha\cos\beta\cos\gamma f(r)\right]\\
\frac{\partial\zeta}{\partial a}\mathrm{S}\left[\frac{m\rho}{2}\cos^2\beta\cos^2\gamma f(r)\right]+\frac{\partial\zeta}{\partial b}\mathrm{S}\left[\frac{m\rho}{2}\cos\alpha\cos\beta\cos^2\gamma f(r)\right]+\frac{\partial\zeta}{\partial c}\mathrm{S}\left[\pm\frac{m\rho}{2}\cos\alpha\cos^3\gamma f(r)\right]
\end{array}\right\}
\end{cases}
$$

$$
\begin{cases}
E=\rho\mathrm{S}\left[\pm\frac{m\rho}{2}\cos\alpha\cos\gamma f(r)\right]\\
+\rho\left\{\begin{array}{l}
\frac{\partial\eta}{\partial a}\mathrm{S}\left[\pm\frac{m\rho}{2}\cos^2\alpha f(r)\right]+\frac{\partial\eta}{\partial b}\mathrm{S}\left[\pm\frac{m\rho}{2}\cos\alpha\cos\beta f(r)\right]+\frac{\partial\eta}{\partial c}\mathrm{S}\left[\pm\frac{m\rho}{2}\cos\alpha\cos\gamma f(r)\right]\\
\frac{\partial\xi}{\partial a}\mathrm{S}\left[\pm\frac{m\rho}{2}\cos\alpha\cos\gamma f(r)\right]+\frac{\partial\xi}{\partial b}\mathrm{S}\left[\pm\frac{m\rho}{2}\cos\beta\cos\gamma f(r)\right]+\frac{\partial\xi}{\partial c}\mathrm{S}\left[\pm\frac{m\rho}{2}\cos^2\gamma f(r)\right]
\end{array}\right\}\\
+\rho\left\{\begin{array}{l}
\frac{\partial\xi}{\partial a}\mathrm{S}\left[\frac{m\rho}{2}\cos^3\alpha\cos\gamma f(r)\right]+\frac{\partial\xi}{\partial b}\mathrm{S}\left[\frac{m\rho}{2}\cos^2\alpha\cos\beta\cos\gamma f(r)\right]+\frac{\partial\xi}{\partial c}\mathrm{S}\left[\frac{m\rho}{2}\cos^2\alpha\cos^2\gamma f(r)\right]\\
\frac{\partial\eta}{\partial a}\mathrm{S}\left[\frac{m\rho}{2}\cos^2\alpha\cos\beta\cos\gamma f(r)\right]+\frac{\partial\eta}{\partial b}\mathrm{S}\left[\frac{m\rho}{2}\cos\alpha\cos^2\beta\cos\gamma f(r)\right]+\frac{\partial\eta}{\partial c}\mathrm{S}\left[\frac{m\rho}{2}\cos\alpha\cos\beta\cos^2\gamma f(r)\right]\\
\frac{\partial\zeta}{\partial a}\mathrm{S}\left[\frac{m\rho}{2}\cos^2\alpha\cos\beta\cos\gamma f(r)\right]+\frac{\partial\zeta}{\partial b}\mathrm{S}\left[\frac{m\rho}{2}\cos\alpha\cos\beta\cos^2\gamma f(r)\right]+\frac{\partial\zeta}{\partial c}\mathrm{S}\left[\pm\frac{m\rho}{2}\cos\alpha\cos^3\gamma f(r)\right]
\end{array}\right\}
\end{cases}
$$

$$
\begin{cases}
F=\rho\mathrm{S}\left[\pm\frac{m\rho}{2}\cos\alpha\cos\beta f(r)\right]\\
+\rho\left\{\begin{array}{l}
\frac{\partial\eta}{\partial a}\mathrm{S}\left[\pm\frac{m\rho}{2}\cos\alpha\cos\beta f(r)\right]+\frac{\partial\eta}{\partial b}\mathrm{S}\left[\pm\frac{m\rho}{2}\cos^2\beta f(r)\right]+\frac{\partial\eta}{\partial c}\mathrm{S}\left[\pm\frac{m\rho}{2}\cos\beta\cos\gamma f(r)\right]\\
\frac{\partial\xi}{\partial a}\mathrm{S}\left[\pm\frac{m\rho}{2}\cos^2\alpha f(r)\right]+\frac{\partial\xi}{\partial b}\mathrm{S}\left[\pm\frac{m\rho}{2}\cos\alpha\cos\beta f(r)\right]+\frac{\partial\xi}{\partial c}\mathrm{S}\left[\pm\frac{m\rho}{2}\cos\alpha\cos\gamma f(r)\right]
\end{array}\right\}\\
+\rho\left\{\begin{array}{l}
\frac{\partial\xi}{\partial a}\mathrm{S}\left[\frac{m\rho}{2}\cos^3\alpha\cos\beta f(r)\right]+\frac{\partial\xi}{\partial b}\mathrm{S}\left[\frac{m\rho}{2}\cos^2\alpha\cos^2\beta f(r)\right]+\frac{\partial\xi}{\partial c}\mathrm{S}\left[\frac{m\rho}{2}\cos^2\alpha\cos\beta\cos\gamma f(r)\right]\\
\frac{\partial\eta}{\partial a}\mathrm{S}\left[\frac{m\rho}{2}\cos^2\alpha\cos^2\beta f(r)\right]+\frac{\partial\eta}{\partial b}\mathrm{S}\left[\frac{m\rho}{2}\cos\alpha\cos^3\beta f(r)\right]+\frac{\partial\eta}{\partial c}\mathrm{S}\left[\frac{m\rho}{2}\cos\alpha\cos^2\beta\cos\gamma f(r)\right]\\
\frac{\partial\zeta}{\partial a}\mathrm{S}\left[\frac{m\rho}{2}\cos^2\alpha\cos^2\beta\cos\gamma f(r)\right]+\frac{\partial\zeta}{\partial b}\mathrm{S}\left[\frac{m\rho}{2}\cos\alpha\cos^2\beta\cos\gamma f(r)\right]+\frac{\partial\zeta}{\partial c}\mathrm{S}\left[\frac{m\rho}{2}\cos\alpha\cos\beta\cos^2\gamma f(r)\right]
\end{array}\right\}.
\end{cases}
$$

Fig. 1.4 (continued)

no derivatives of displacement. Cauchy noted that if the primitive undeformed state is equilibrated with zero external forces (in modern terms, a natural state) six of the coefficients between the components of the stress and the derivatives of the displacement cancel. In fact for the undeformed state one must assume $\epsilon = 0$ and the components of the stress A, B, C, D, E, F reduce to the first elements of Fig. 1.4. In the absence of external forces, they must vanish, with all the sums which contain quadratic terms in the direction cosines. That also implies the vanishing of the terms in the second row of Fig. 1.4 that depend on displacements. Therefore, the non-zero coefficients are only those of the third row, characterized by terms of fourth order in the direction cosines, that are 15 in number, equal to the combinations with repetition of three objects ($\cos\alpha$, $\cos\beta$, $\cos\gamma$) of class 4 (the order of the product of the cosines).

Figure 1.4, in addition to enabling a control over the number of coefficients, shows a certain symmetry. The coefficients of the derivatives associated with the variables of displacement and position are equal; for example, the coefficients of $\partial\xi/\partial b$ and $\partial\eta/\partial a$ are equal; the same holds for $\partial\xi/\partial c$ and $\partial\zeta/\partial a$, etc. A modern reader can thus state that the components of the tension are expressed as a function of the six components of infinitesimally small deformation, arriving at a constitutive stress–strain relationship characterized by 15 coefficients only.

Cauchy did not report these considerations; he was not interested in a theory of constitutive relationships, he just wanted to get the stress as a function of the displacement derivatives in order to write the equations of equilibrium and motion for a system of material points in terms of displacement, as done by Navier. The partition of the elastic problem of continuum in stress analysis (equilibrium), strain analysis (compatibility) and the imposition of the constitutive relationship will be fully developed only with Lamé [86] and Saint Venant [143]. Cauchy also did not care about the number of constants that he had found for more general elastic models, in particular whether they are 15 or 21, although in a work of 1829 he gave a name to each coefficient and exposed them in the proper order [32].[19] According to Augustus Edward Hough Love [93], Rudolph Julius Emmanuel Clausius was among the first to highlight the particular number, 15, of the constants of the molecular model.[20] In fact already Poisson [127] had 'counted' the coefficients of the constitutive relationship in the form of infinitesimal strain versus stress, observing that those required are in general 36 and only as a result of the classical model hypotheses is the number reduced to 15.[21] Cauchy took the following further assumptions of material symmetry:

1. The body has three orthogonal planes of symmetry (orthotropy): the coefficients with at least one odd exponent of direction cosines vanish (the sums which express them cancel); the number of distinct coefficients is reduced to six.
2. The body has three planes of symmetry and the arrangement of the molecules is identical in the three orthogonal directions to these plans (complete orthotropy): in the expression of the coefficients one can exchange β with α, α with γ, etc.; the number of distinct coefficients goes down to two.
3. The body has the same arrangement of molecules around the point where the stress is to be evaluated (isotropy): with a complicated reasoning, perhaps not flawless, Cauchy showed that there is only a distinct coefficient.

[19] pp. 162–173.

[20] p. 9.

[21] pp. 83–85.

1.1.2 Internal Criticisms Toward the Classical Molecular Model

The molecular model by Navier, Cauchy and Poisson was accepted by the scientific international community, especially in France, because of the simplicity of the theory and the physical basis universally shared. However its conclusions were slightly but inexorably falsified by the experimental evidence. Thus it clearly appeared, with the advance of precision in measuring instruments, that to characterize isotropic linear elastic materials two constants were needed and not only one as suggested by the molecular model.[22]

A first attempt to adapt the classical molecular model to the experimental results consisted in relaxing some of the basic assumptions. Poisson was among the first, in a memoir read before the Académie des sciences de Paris in 1829 [127], to formulate the hypothesis of non-point molecules and crystalline arrangement; the idea of central forces depending only on the mutual distance between (the centers of) the molecules was thus released:

> It is assumed that, in a body of this nature, the molecules are uniformly distributed and attract or repel unevenly from their different sides. For this reason it is no longer possible, in calculating the force exerted by one part of a body to another, to consider the mutual force of two molecules as a simple function of the distance between them [...]. In the case of a homogeneous body that is in its natural state, where it is not subjected to any external force, we can consider it as an assembly of molecules of the same nature and the same *shape* whose homologous sections are parallel to each other [127].[23] (A.1.5)

According to Poisson, in cristalline bodies the relations among the elastic constants that reduce their number to 15, obtained in his preceding works and in those by Cauchy, are no longer valid:

> The components P, Q,&c., thus being reduced to six different forces, and the value of each force may contain six particular coefficients, it follows that the general equations of equilibrium, and consequently those of the movement, contain thirty-six coefficients which may not reduce to a lesser number without limiting the generality of the question [127].[24] (A.1.6)

On the other hand, in non-crystalline bodies, with weak or irregular crystallization, even if the molecules are no longer considered punctiform, everything remains as if the forces were central. This is due to a compensation of causes:

> It follows that if we consider two parts A and B of a body that are not crystallized, which extend insensitively but which, however, include a great number of molecules, and we want to determine the total action of A on B, we can assume in this calculation that the mutual action of two molecules m and m' is reduced, as in the case of fluids, to a force R directed along the line joining their centers of gravity M and M', whose intensity will depend on the distance MM'. Indeed, whatever the action, it can be replaced by a similar force, which is the

[22] See the results found by Guillaume Wertheim (1815–1861) [158, pp. 581–610]. The greater the accuracy and reliability of the experimental results the more the theoretical predictions of Cauchy and Poisson were disclaimed, though it was not clear why [80, pp. 481–503].

[23] p. 69. Our translation.

[24] p. 85. Our translation.

average of the actions of all points of m' on all of m, and we combine it with another force R', or, if necessary with two other forces R' and R'', dependent on the relative arrangement of the two molecules. However, because this disposition by hypothesis has not assumed any kind of regularity in A and B, and the number of molecules of A and B is extremely large and nearly infinite, one concludes that all the forces R' and R'' will compensate without altering the total action of A on B, which will not depend, therefore, but on the forces R. It should moreover be added that for the same increase in the distance, the intensity of the forces R' and R'' increases faster in general than that of the forces R; which will still contribute to make disappear the influence of the first forces on the mutual action of A on B [127].[25] (A.1.7)

Cauchy also expressed doubts about the validity of the classical molecular model in some memoirs of 1839 [35][26] and in a review of 1851 of some of Wertheim's memoirs about the experimental determination of elastic constants [36]. Cauchy stated that the molecules in crystalline bodies should not be considered as point-like but as very small particles composed of atoms. Since in crystals there is a regular arrangement of molecules, the elastic moduli are periodic functions of spatial variables; assertions taken later by Adhémar J.C. Barré de Saint Venant [116].[27] In order to obtain a constitutive relation with uniform coefficients, Cauchy expanded the number of elastic moduli, finally reaching only two in the case of isotropic materials.[28]

Gabriel Lamé [86, 87] in his works on the theory of elasticity raised a number of questions on the issue . For instance, much of the twentieth lesson of the *Leçons sur les coordonnées curvilignes et leurs diverses applications* of 1859 [87] was dedicated to concerns about the real nature of molecules, to the assumption about the exact mutual actions, to what is a reasonable form of the law of the intermolecular actions, to what is the direction of the latter. In his 1852 monograph on the mathematical theory of elasticity, *Leçons sur la théorie mathématique de l'élasticité des corps solides*, Lamé [86] first obtained the linear elastic constitutive relations for point molecules and intermolecular central forces. Moreover, assuming that each component of the stress is a linear function of all the components of the strain, the linear elasticity in general is described by 36 coefficients. Also assuming isotropy (*élasticité constante*), considerations about invariance with rotations reduce the number of coefficients to two, denoted by λ and μ:

By this method of reduction, it is obtained finally for N_i, T_i, in the case of homogeneous solids and constant elasticity, the values [...] containing two coefficients, λ and μ. When with the method indicated at the end of the third lesson, we find $\lambda = \mu$, it remains a single coefficient only. We will not accept this relationship, which is necessarily based on the assumption of continuity of the material in the solid media. The results of Wertheim's experiments show clearly that ratio λ to μ is not the unity, but neither seem to assign to this ratio another immovable value. We retain the two coefficients λ and μ, leaving undetermined their ratio [86].[29] (A.1.8)

---

[25] pp. 7–8. Our translation.

[26] s. 2, vol XI, pp. 11–27; 51–74; 134–172.

[27] Appendix V, p. 689.

[28] A detailed reconstruction of Cauchy's topics is shown in [116], Appendix V, pp. 691–706.

[29] pp. 51–52. Our translation.

With arguments similar to those of Poisson in 1829 [127], Lamé showed that even for crystalline bodies, the relation with 36 constants [86][30] holds good and identified the error of Cauchy's and Poisson's treatment in the assumption of the uniformity of matter, which allows the symmetry considerations that would otherwise be ineligible:

> This is the method followed by Navier and other geometers to obtain the general equations of elasticity in solid bodies. But obviously this method implies the continuity of matter, an unacceptable hypothesis. Poisson believes to overcome this difficulty, [...] but [...], in reality, he simply substitutes the sign Σ to the sign $\int$ [...]. The method we have followed [...] whose origins lie in the work of Cauchy, seems at the basis of any objection [...] [86].[31] (A.1.9)

Although the results of the molecular theory of elasticity were clearly considered unsatisfactory even by the followers of the French school of mechanics, it was not the case for the validity of the molecular approach. One of the main proponents of this approach was Saint Venant; his ideas on the matter, besides in publications to his name, are contained in the enormous amount of notes, comments and appendices to the *Theorie der Elasticität fester Körper* by Alfred Clebsch, translated into French [42], and to the *Résumé des leçons donnés a l'école des pontes et chaussées* by Navier [116] where Saint Venant said:

> The elasticity of solid bodies, as well as of fluids, [...], all their mechanical properties prove that the molecules, or the last particles composing them, exert on each other actions [which are] repulsive [and] infinitely growing for the smallest possible mutual distances, and becoming attractive for considerable distances, but relatively inappreciable when such distances, of which they [the molecular actions] are functions, assume a sensible value [116].[32] (A.1.10)

For crystalline bodies the classical molecular model seemed not to be valid:

> I do not yet refuse to recognize that the molecules whose various settings make up the texture of the solids and whose small change of distance produce noticeable strains called ∂, g are not the *atoms* constituting matter, but are unknown groups. I accordingly recognize, thinking that the actions between atoms are governed by laws of intensity depending on the distances only where they operate, it is not certain that the *resultant* actions and the actions of the molecules must exactly follow the same law of the distances from their centers of gravity. We also consider that the groups, changing distances, can change orientation [42].[33] (A.1.11)

But, added Saint Venant, this is only an ideal situation, because the ordinary bodies are not crystals and also the thermal motions produce a chaotic situation that on average leads to a law of force at a distance of molecules substantially of the same type as that which there is between the atoms. Saint Venant made the six components of the tension to depend linearly on the six strain components, yet resulting in an elastic relationship in terms of 36 coefficients. However he continued to admit the validity of the equalities known as Cauchy-Poisson relations (see note 69 of Chap. 1), which for isotropic bodies leads to a single constant:

---

[30] pp. 36–37.

[31] p. 38. Our translation.

[32] pp. 542–543. Our translation.

[33] p. 759. Our translation.

> The 36 coefficients [...] are not independent of each other, and it is easy to see that there are 21 equalities among them [116].[34] (A.1.12)

In fact, the proof that these relations are valid considers variations of the intermolecular distance that are the same under an extension in a given direction and an appropriate angular distortion [116].[35] If the intermolecular force is central and depends only on the variation of the distance between the centers of the molecules, the force between the molecules and consequently the stress, is equal. Thus, there are similarities between the elastic constants, which reduce the number from 36 to 15, in particular, for isotropic bodies, Saint Venant found a single constant:

> The thirty-six coefficients [...] reduce to two [...] and one may even say to one only [...] in the same way as the thirty six coefficients are reducible to fifteen [116].[36] (A.1.13)

Saint Venant knew very well that these conclusions were contradicted by experiments, and since he did not find evident defects in the molecular theory of elasticity, preferred to accept that there are no isotropic bodies in nature:

> Yet experiences [...] and the simple consideration on the way cooling and solidification take place in bodies, prove that isotropy is quite rare [...]. So, instead of using, in place of the equations [...] with one coefficient only, the formulas [...] with two coefficients [...], which hold, like these others, only for perfectly isotropic bodies, it will be convenient to use as many times as possible the formulas [...] relative to the more general case of different elasticity in two or three directions [116].[37] (A.1.14)

In some works in the *Journal de mathématiques pures et appliquèes*, from 1863 to 1868 [145–147],[38] Saint Venant introduced the concept of *amorphous bodies* (corps amorphes) to define the properties acquired by bodies that were initially isotropic as a result of geological processes. In this state, the mechanical properties are characterized by three coefficients and not just two as in the case of isotropic bodies.

Saint Venant spent more than 200 pages of notes and appendices to Navier's lessons in order to present experimental results and attempts to explain the paradox, showing a wide knowledge of the literature of his time (among others, he quoted Savart, Wertheim, Hodgkinson, Regnault, Oersted, Green, Clebsch, Kirchhoff, Rankine, William Thomson). In the end, however, the question remained, because there was no agreement between the approaches of Saint Venant's contemporaries. Although it was clear that two elastic constants were necessary, where was the flaw in a theory attractive and apparently founded as Navier's, Cauchy's and Poisson's?

The debate between the scholars of mechanics was strengthened, from different points of view, by the works of Augustin Cauchy, George Green and Auguste Bravais, who gave life to different schools of elasticity in England and Germany.

---

[34] p. 556. Our translation.

[35] pp. 556–560.

[36] p. 582. Our translation.

[37] p. 583. Our translation.

[38] In the order: pp. 353–430; 297–350; 242–254.

1.1.3 Substitutes for the Classical Molecular Model

The molecular model was not the only model with which engineers, physicists and mathematicians tried to represent the behavior of elastic bodies. On September 30th, 1822, 1 year after Navier's memoir, Cauchy [25] presented to the Académie des sciences de Paris a memoir that dealt with the study of elasticity according to a continuist approach largely unchanged since then. That of Cauchy was a purely phenomenological approach, in line with the positivistic tendencies that had developed among French scientists.[39]

The matter was modeled as a mathematical continuum without any assumption of physical nature. It was assumed that the different parts of matter exchange forces and become deformed. The relations between internal forces and deformations had a general nature and the number of elastic constants that defined the problem was simply determined by counting the components of stress and strain. In its most complete version, Cauchy's continuous model led to a stress–strain relationship defined by 36 coefficients.

A different approach was that of Green (1793–1841), who in a work of 1839 [75] also followed a phenomenological point of view assuming a three dimensional continuum to model matter, uninterested even in the concept of internal forces. Green, however, recurred to a mechanical principle, that of the existence of a potential of the internal forces, which somehow gave some theoretical force to his arguments.

1.1.3.1 Cauchy's Phenomenological Approach

Of the presentation before the Académie des sciences de Paris in 1822, there is an excerpt published in 1823 [25],[40] where the *principle of stress* is formulated.[41] Over any oriented and regular surface separating a body into two parts there is a regular vector field that expresses the actions between the two parts:

> If in an elastic or non-elastic solid body a small invariable volume element, terminated by any faces at will, is made [imagined] rigid, this small element will experience on its different sides, and at each point of each of them, a determined pressure or tension. This pressure or tension is similar to the pressure a fluid exerts against a part of the envelope of a solid body, with the only difference that the pressure exerted by a fluid at rest, against the surface of a solid body, is directed perpendicularly to the surface inwards from the outside, and in each point independent of the inclination of the surface relative to the coordinate planes, while the pressure or tension exerted at a given point of a solid body against a very small element of surface through the point can be directed perpendicularly or obliquely to the surface, sometimes from outside to inside, if there is condensation, sometimes from within

---

[39] For a discussion of the positivistic conceptions of French science in the first half of the 19th century, see [124].

[40] It seems that on September 30th 1822, Cauchy notified the Académie of his researches neither delivering a public reading, nor depositing a manuscript; see [3] p. 97. In [154] it is stated that Cauchy, as a matter of fact, presented his memoir.

[41] Cauchy used *tension* or *pressure* for traction and compression respectively.

outwards, if there is expansion, and it can depend on the inclination of the surface with respect to the planes in question [25].[42] (A.1.15)

This statement sets aside any constitutive assumptions on the matter, but relies on the concept, then still not fully accepted, of distributed force.

Cauchy published the announced results in 1827 [26] and in 1828 [31]. In 1827 Cauchy, by writing the equilibrium of an infinitesimal tetrahedron, showed the linear dependence between the stress vector and the unit vector normal to the surface where the stress acts [28] and obtained local equilibrium equations for the component of stresses [28][43]:

$$\begin{cases} \dfrac{\partial A}{\partial x}+\dfrac{\partial F}{\partial y}+\dfrac{\partial E}{\partial z}+\rho X=0 \\ \text{One also finds} \\ \dfrac{\partial F}{\partial x}+\dfrac{\partial B}{\partial y}+\dfrac{\partial D}{\partial z}+\rho Y=0 \\ \dfrac{\partial E}{\partial x}+\dfrac{\partial D}{\partial y}+\dfrac{\partial C}{\partial z}+\rho Z=0, \end{cases} \tag{1.10}$$

with A, B, C, D, E, F the components of the stress, corresponding to the magnitudes denoted with the same symbols introduced for the molecular model.

It is worth noting that in Cauchy's memoir of 1823 all notions of continuum mechanics were anticipated: tensors,[44] stress and strain, their symmetry, the existence of principal axes, the criterion for obtaining equilibrium equations with the *solidification principle*, the introduction of Hooke's law under generalized form. In the following years, until 1827–1828, Cauchy had developed many techniques today classified as linear algebra: use of square tables for matrices; classification of different matrices (symmetric, skew-symmetric, etc.), theorems on eigenvalues, theorems on canonical decomposition of matrices; the first modern characterization of determinants.[45]

To get constitutive relationships of continua one needs to explicitly define the components of strain. Cauchy did that in his 1827 work [27] introducing the local deformation of the linear infinitesimal segment of a continuum as the percentage change of length. In the context of small displacements, the deformation comes to depend on the six functions $\partial\xi/\partial a$, $\partial\eta/\partial b$, $\partial\zeta/\partial c$, $\partial\xi/\partial b$

[42] p. 300. Our translation.

[43] p. 144.

[44] The term *tensor* does not belong to Cauchy, but to Hamilton [76] and Voigt [156]. Similarly Cauchy did not establish the formalized rules of tensor calculus, that were specified only at the end of the 19th century by Ricci-Curbastro; see in particular [89, pp. 125–201].

[45] It should be noted that, in all the above mentioned works, Cauchy made extensive use of infinitesimals, whereas he had pursued his research in mathematical analysis with the precise goal of eliminating the infinitesimals. This attitude is similar to that held by Lagrange who while in the *Théorie des fonctions analytiques* of 1797 developed a way to avoid the use of infinitesimals, in the *Méchanique analitique* of 1788 extensively applied the infinitesimals, justifying their use for the sake of simplicity [17].

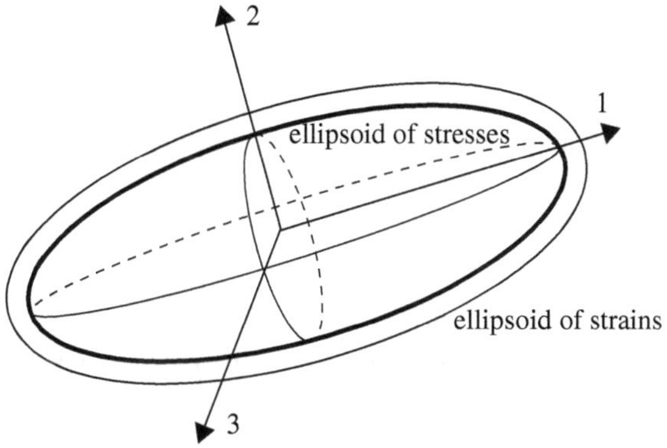

Fig. 1.5 Proportionality between the ellipsoids of stresses and strains

$+\ \partial\eta/\partial a$, $\partial\xi/\partial c + \partial\zeta/\partial a$, $\partial\eta/\partial c + \partial\zeta/\partial b$, which assume the role of the components of strain. Cauchy gave a geometric meaning only for the first three components, representing the unitary changes of length in the direction of the coordinate axes. Therein he was less explicit than Euler and Lagrange, who introduced linearized strains in the study of statics and dynamics of fluids and had given geometric meaning also to the other three components.[46]

In a major work of 1841 Cauchy [33] introduced both finite and infinitesimal local rotation of a segment in a given direction, and the average value in all directions. The linear elastic constitutive law was introduced in a memoir of 1828 [31]. In the initial part of the memoir Cauchy, consistently with his summary of 1823, assumed a single constant of proportionality between the principal stresses and strains; more precisely he suggested the similitude between the ellipsoids of stresses and strains, according to what is shown in Fig. 1.5.

Cauchy [31] then showed that the proportionality continues to subsist in a generic system of coordinates and wrote the constitutive relationships[47]:

$$\begin{cases} A = k\dfrac{d\xi}{dx}, & B = k\dfrac{d\eta}{dy}, & C = k\dfrac{d\zeta}{dz}, \\ D = \dfrac{1}{2}k\left(\dfrac{d\eta}{dz} + \dfrac{d\zeta}{dy}\right), & E = \dfrac{1}{2}k\left(\dfrac{d\zeta}{dx} + \dfrac{d\xi}{dz}\right), & F = \dfrac{1}{2}k\left(\dfrac{d\xi}{dy} + \dfrac{d\eta}{dx}\right), \end{cases} \tag{1.11}$$

[46] Consequently the quadrilateral is a parallelogram whose contiguous sides, (1, 2), (1, 3), will be [...]:

$$(1,2) = dx\left(1 + \frac{\partial\delta x}{\partial x}\right);\ (1,3) = dy\left(1 + \frac{\partial\delta y}{\partial y}\right),$$

[...] with respect to the angle included by this two sides one will find [...]:

$$\cos\alpha = \frac{\partial\delta x}{\partial y} + \frac{\partial\delta y}{\partial x}.$$

[85], pp. 208–209. Our translation. See also [61], pp. 288–292; 332–334.

[47] p. 209.

with k a constant of proportionality and ξ, η, ζ components of the displacement in the directions x, y, z respectively. Cauchy assumed here implicitly the isotropy of the body, a concept and term he will adopt implicitly in subsequent works. Shortly after, in the same memoir, Cauchy [31] introduced the constitutive relationship with two coefficients, "Thus, for instance, the formulas [...] acquire new terms and become more general"[31][48]. These new constitutive relationships are in the form [31][49]:

$$\begin{cases} A = k\dfrac{\partial \xi}{\partial x} + K\nu, \qquad B = k\dfrac{\partial \eta}{\partial y} + K\nu, \qquad C = k\dfrac{\partial \zeta}{\partial z} + K\nu, \\ D = \dfrac{1}{2}k\left(\dfrac{\partial \eta}{\partial z} + \dfrac{\partial \zeta}{\partial y}\right), \; E = \dfrac{1}{2}k\left(\dfrac{\partial \zeta}{\partial x} + \dfrac{\partial \xi}{\partial z}\right), \; F = \dfrac{1}{2}k\left(\dfrac{\partial \xi}{\partial y} + \dfrac{\partial \eta}{\partial x}\right), \end{cases} \tag{1.12}$$

with: $\nu = \dfrac{\partial \xi}{\partial x} + \dfrac{\partial \eta}{\partial y} + \dfrac{\partial \zeta}{\partial z}$ the coefficient of cubic expansion, k, K two elasticity parameters today known as the first and second Lamé constant, respectively (more precisely, the second Lamé constant is equal to $k/2$). The use of two elastic constants implies that to characterize the intermolecular forces as proportional to the displacement of the molecules is not equivalent to consider the stress proportional to the strain term by term (which would correspond to $K = 0$).

Using the local equilibrium and constitutive equations (k and K being considered uniform) Cauchy derived the differential equations for the displacement [31][50]:

$$\begin{cases} \dfrac{k}{2\Delta}\left(\dfrac{\partial^2 \xi}{\partial x^2} + \dfrac{\partial^2 \xi}{\partial y^2} + \dfrac{\partial^2 \xi}{\partial z^2}\right) + \dfrac{k+2K}{2\Delta}\dfrac{\partial \nu}{\partial x} + X = 0, \\ \dfrac{k}{2\Delta}\left(\dfrac{\partial^2 \eta}{\partial x^2} + \dfrac{\partial^2 \eta}{\partial y^2} + \dfrac{\partial^2 \eta}{\partial z^2}\right) + \dfrac{k+2K}{2\Delta}\dfrac{\partial \nu}{\partial y} + Y = 0, \\ \dfrac{k}{2\Delta}\left(\dfrac{\partial^2 \zeta}{\partial x^2} + \dfrac{\partial^2 \zeta}{\partial y^2} + \dfrac{\partial^2 \zeta}{\partial z^2}\right) + \dfrac{k+2K}{2\Delta}\dfrac{\partial \nu}{\partial z} + Z = 0, \end{cases} \tag{1.13}$$

with X, Y, Z forces per unit of mass and Δ the mass per unity of volume. These equations reduce to those of Navier, Cauchy remarked, if $K = k/2$.

Lamé in his treatise *Leçons sur la théorie mathématique de l'élasticité des corps solides* [86] found again the same equations with a continuous model of matter using more complex arguments, with an approach still used today in the teaching of mechanics of solids. Lamé [86] started from the general assumption that, in an elastic continuum, each of the six components of the stress[51] is a function of the six components of the strain, according to the relations[52]:

---

[48] p. 215.

[49] p. 216. Actually the first three equations were written by Cauchy in a slightly different way, though equivalent to that referred above.

[50] p. 218, Eq. 76.

[51] Indicated by Lamé with the symbols N_i, T_i, $i = 1, 2, 3$, N_i being the normal component and T_i the tangential one, to the face on which the force acts.

[52] p. 33.

Table 1.1 Reduction of the number of elastic coefficients

	$\frac{du}{dx}$	$\frac{dy}{dy}$	$\frac{dw}{dz}$	$\left(\frac{dv}{dz}+\frac{dw}{dy}\right)$	$\left(\frac{dw}{dx}+\frac{du}{dz}\right)$	$\left(\frac{du}{dy}+\frac{dv}{dx}\right)$
N_1	A	B	B	D	E	E
N_2	B	A	B	E	D	E
N_3	B	B	A	D	D	E
T_1	$\mathscr{A}$	$\mathscr{B}$	$\mathscr{B}$	Δ	$\mathscr{E}$	$\mathscr{E}$
T_2	$\mathscr{B}$	$\mathscr{A}$	$\mathscr{B}$	$\mathscr{E}$	Δ	$\mathscr{E}$
T_3	$\mathscr{B}$	$\mathscr{B}$	$\mathscr{A}$	$\mathscr{E}$	$\mathscr{E}$	Δ

$$
\begin{aligned}
N_i &= A_i\frac{du}{dx} + B_i\frac{dv}{dy} + C_i\frac{dw}{dz} \\
&\quad + D_i\left(\frac{dv}{dz}+\frac{dw}{dy}\right) + E_i\left(\frac{dw}{dx}+\frac{du}{dz}\right) + F_i\left(\frac{du}{dy}+\frac{dv}{dx}\right), \\
T_i &= \mathscr{A}_i\frac{du}{dx} + \mathscr{B}_i\frac{dv}{dy} + \Gamma_i\frac{dw}{dz} \\
&\quad + \Delta_i\left(\frac{dv}{dz}+\frac{dw}{dy}\right) + \mathscr{E}_i\left(\frac{dw}{dx}+\frac{du}{dz}\right) + \mathscr{F}_i\left(\frac{du}{dy}+\frac{dv}{dx}\right),
\end{aligned}
\tag{1.14}
$$

which for $i = 1, 2, 3$ are defined by 36 constants. The reduction of the coefficients to two for an isotropic material stems from symmetry considerations, for which Lamé could write the constitutive relationship in the form referred to in Table 1.1. Here eight elastic coefficients $A, B, D, E, \mathscr{A}, \mathscr{B}, \mathscr{E}, \Delta$ are reported [86].[53] By considering a state of uniform axial strain we obtain $\mathscr{A} = 0,\ \mathscr{B} = 0$, because in this state the shear stresses shall vanish. Considering a torsion deformation state, Lamé got $D = 0,\ E = 0,\ \mathscr{E} = 0$. This leaves only three constants A, B, Δ, reduced to two considering the invariance of the constitutive law for a rotation of the coordinate system, which implies $A = B + 2\Delta$. Lamé [86] thus wrote the constitutive relationships[54]:

$$
\begin{cases}
N_1 = \lambda\theta + 2\mu\frac{du}{dx}, & N_2 = \lambda\theta + 2\mu\frac{dv}{dy}, & N_3 = \lambda\theta + 2\mu\frac{dw}{dz}, \\
T_1 = \mu\left(\frac{dv}{dz}+\frac{dw}{dy}\right), & T_2 = \mu\left(\frac{dw}{dx}+\frac{du}{dz}\right), & T_3 = \mu\left(\frac{du}{dy}+\frac{dv}{dx}\right),
\end{cases}
\tag{1.15}
$$

known today with his name, where θ is the coefficient of cubic dilatation, while λ and μ are the two constants attributed to Lamé, for which generally the same symbols are still used. Notice that, though Lamé used the derivatives:

$$
\frac{du}{dx},\ \frac{dv}{dy},\ \frac{dw}{dz},\ \left(\frac{dv}{dz}+\frac{dw}{dy}\right),\ \left(\frac{dw}{dx}+\frac{du}{dz}\right),\ \left(\frac{du}{dy}+\frac{dv}{dx}\right),
\tag{1.16}
$$

[53] p. 50.

[54] p. 51.

with the meaning of parameters of deformation, he gave no clear geometrical meaning, or at least he did not give a name, to them. This is true in particular for the last three expressions that today are known as the angular distortions. A precise definition in the theory of elasticity of the components of strain will spread with Saint Venant only.

Saint Venant wrote for instance:

> *Stretching*, in a point M of a body in the direction of a straight line Mx passing from it, [is] the proportion of the elongation (positive or negative) experienced by any very small portion of that straight line because of the average displacements of the body, as they were defined in the preceding article; *Distortion* along two small straight lines originally orthogonal Mx, My, or with respect to one of them and in the plane that it shares with the other, [is] the current projection of the unit length in the direction of the other. We denote this quantity, whose amount is nothing but the cosine of the current angle between the two straight lines by
>
> $$g_{xy} \quad \text{or} \quad g_{yx}$$
>
> depending on whether it looks as referring to the relative distortion of the various lines parallel to Mx located in the plane xMy, or as the relative distortion of parallel lines to My located in the same plane. It is positive when the angle originally right has became acute [143].[55] (A.1.16)

Green had already introduced the components of strain in his memoir of 1839, before Saint Venant [75].[56]

1.1.3.2 Green's Energetic Approach

Green [75] dealt with the elasticity theory in his work of 1839 where he studied the propagation of light. Here is how he began his research:

> Cauchy seems to have been the first who saw fully the utility of applying to the Theory of Light those formulae which represent the motions of a system of molecules acting on each other by mutually attractive and repulsive forces supposing always that in the mutual action of any two particles, the particles may be regarded as points animated by forces directed along the right line which joins them. *This last supposition, if applied to those compound particles, at least, which are separable by mechanical division, seems rather restrictive; as many phenomena, those of crystallization for instance, seem to indicate certain polarities in these particles* [emphasis added]. If, however, this were not the case, we are so perfectly ignorant of the mode of action of the elements of the luminiferous ether on each other, that it would seem a safer method to take some general physical principle as the basis of our reasoning, rather than assume certain modes of action, which, after all, may be widely different from the mechanism employed by nature; more especially if this principle include in itself as a particular case, those before used by M. Cauchy and others, and also lead to a much more simple process of calculation. The principle selected as the basis of the reasoning contained in the following paper is this: In whatever way the elements of any material system may act upon each other, if all the internal forces exerted be multiplied by the elements of their respective directions, the total sum for any assigned portion of the mass will always be

[55] p. 6. Our translation.

[56] p. 249.

> the exact differential of some function. But, this function being known, we can immediately apply the general method given in the Mécanique Analytique [...] [75].[57]

Green considered a function of the components of strain,[58] called a *potential function* ϕ, whose "exact differential" gives the sum of the internal forces multiplied by "the elements of their respective" displacements. If the strains are very small ϕ can be developed in "a very convergent series":

$$\phi = \phi_0 + \phi_1 + \phi_2 + \text{etc.} \tag{1.17}$$

where ϕ_0, ϕ_1, ϕ_2 are respectively homogeneous functions of degree 0, 1, 2, etc. of the six components of the strain, each very great "compared to the next" [75].[59] One can ignore ϕ_0 (immaterial constant) and ϕ_1 (the undeformed configuration is assumed equilibrated and for the principle of virtual work $\delta\,\phi = \phi_1 = 0$). Neglecting the terms of order higher than the second, the potential function is represented in each point of the body by ϕ_2, which, as a quadratic form of six variables, is completely defined by 21 coefficients. For isotropic bodies Green found again two constants.

Starting from ϕ_2, "by combining D'Alembert's principle with that of virtual velocities", Green [75] obtained the equations for the free oscillations in the ether[60]:

$$\begin{aligned}
\rho\,\frac{d^2u}{dt^2} &= A\frac{d}{dx}\cdot\left(\frac{du}{dx}+\frac{dv}{dy}+\frac{dw}{dz}\right)+B\left\{\frac{d^2u}{dy^2}+\frac{d^2u}{dz^2}-\frac{d}{dx}\left(\frac{dv}{dy}+\frac{dw}{dz}\right)\right\},\\
\rho\,\frac{d^2v}{dt^2} &= A\frac{d}{dy}\cdot\left(\frac{du}{dx}+\frac{dv}{dy}+\frac{dw}{dz}\right)+B\left\{\frac{d^2v}{dx^2}+\frac{d^2v}{dz^2}-\frac{d}{dy}\left(\frac{du}{dx}+\frac{dw}{dz}\right)\right\},\\
\rho\,\frac{d^2w}{dt^2} &= A\frac{d}{dz}\cdot\left(\frac{du}{dx}+\frac{dv}{dy}+\frac{dw}{dz}\right)+B\left\{\frac{d^2w}{dx^2}+\frac{d^2w}{dy^2}-\frac{d}{dz}\left(\frac{du}{dx}+\frac{dv}{dy}\right)\right\},
\end{aligned} \tag{1.18}$$

in which A and B are elastic constants according to Green. These equations can be reformulated to those of Cauchy with two constants, when the inertia forces are treated as ordinary forces and the elastic constant are renamed according to the relations:

$$B = \frac{k}{2};\, A = K + k. \tag{1.19}$$

---

[57] p. 245.

[58] We already said at the end of the previous section that Green introduced the six components of the infinitesimal strain before Saint Venant. He indicated with s_1, s_2, s_3 the longitudinal strains, which are equal to the percentage change of the edge lengths *dx*, *dy*, *dz* of an elementary parallelepiped and with α, β, γ the angular distortions, equivalent to the variation of the angles between edges initially orthogonal *dy* and *dz*, *dx* and *dz*, *dx* and *dy*.

[59] p. 249.

[60] p. 255.

1.1.3.3 Differences in the Theories of Elasticity

In the 19th century there were many opponents to the energetist and continuist theories, among them Saint Venant, who leveled severe criticisms to Green's approach. For instance, in a footnote to Clebsch's monograph, he wrote:

> But Green, in 1837–1839, and, after him, various scientists from England and Germany believed it was possible to replace [the law of the molecular action as a function of the distance between any couple molecule-material point] with another more general, or qualified as more general because less determined [...], law whose immediate analytical result is the possibility that the intensity of the action between two molecules depends not only on their own distances but also by the distances of the other molecules, and on the mutual distances of all them also; in a word, on all the current set of their relative situations or the complete present state of the system to which the two considered molecules belong and of the entire universe [42].[61] (A.1.17)

And also, in the footnotes to Navier's textbook:

> This Green's view constitutes a *third* origin [...] of the opinion dominant today and which we fight [116].[62] (A.1.18)

Saint Venant rejected the 'dominant' Green's approach because it lacked a mechanical basis, especially in relation to the concept of force. While Cauchy manifested about it a moderate ontological commitment and when it was more comfortable he treated the matter as a continuous medium, Saint Venant consistently supported the molecular model because, for his mechanist view, the forces could only be explained by the interaction between material points. Saint Venant's conceptions of mechanics are well summarised in his *Principes de mécanique fondé sur la cinématique* of 1851 [142]; for him matter is made of molecules that are not extended, and mechanics is simply the science through which one determines the distances of certain points from other points, at a given instant, knowing what these distances have been at other instants. These are the main principles he assumed at the foundation of mechanics:

1. In a system of two molecules only, they undergo equal and opposite accelerations along the line joining them, with an intensity depending on their distance only.
2. In a system made of several molecules, the acceleration of a given molecule is the geometrical sum of the accelerations it would acquire if it were subjected separately to the forces of each of the other atoms (rule of parallelogram).
3. The mass of a body is a number proportional to the number of molecules that can be supposed it contains.
4. The force acting on a body is nothing but the product of its mass and its acceleration [142].[63]

---

[61] p. 41. Our translation.

[62] p. 708. Our translation.

[63] p. 64.

Saint-Venant thanked God, not Newton, for the simplicity of these assumptions: "God not only wanted invariable laws, he also wanted accelerations to depend only on distance. Further, he wanted superposition".[64]

To Green, who thought that the hypothesis of intermolecular forces opposing along the line joining the molecules was too restrictive and that one must use a weaker criterion, Saint Venant contested both the rejection of the principle of action and reaction, a fundamental law of mechanics, and the choice of a quadratic function to approximate the potential, because, according to him, without any physical hypothesis there is no reason to assert that an arbitrary function should have dominating quadratic terms:

> If the scientific prudence prescribes not to rely on any assumption, it does equally prescribe to hold under strong suspect what is clearly contrary to the great synthesis of the generality of facts [...]. Also we reject any theoretical formula in formal contradiction with the law of action as a continuous function of the distances of the material points and directed according to the lines connecting them in pairs. If, using this formula, it is easier to explain certain facts, we always look it as a too convenient *expedient* [...] [116].[65] (A.1.19)

Table 1.2 illustrates the different assumptions about the theory of elasticity; the table also includes Voigt's conceptions which will be referenced in the following sections.

1.1.4 The Perspective of Crystallography

The question of the correctness of the adoption of one or two constants for linear and isotropic elastic bodies remained open long into the mechanics of the 19th century. The study of Lamé and those of Saint Venant could not reconcile the corpuscular approach by Navier, Cauchy and Poisson with the continuous one by Green; on the other hand, at least until the second half of the century, the precision drawn from experimental research was limited; however, as the experimental results became refined, the hypothesis of two constants seemed to prevail, without being able to clarify where and why the corpuscular theory falls at fault.

In 1866 a posthumous book by Auguste Bravais [14] appeared, containing previous memoirs read in front of the Académie des sciences de Paris. These works mainly dealt with crystallography and rigorous organization of crystals in groups of symmetry. They contained assumptions which were to be essential for the later work of the mechanicians who wished to overcome the *empasse* concerning the 'true' number of elastic constants.

Based on his studies of crystallography, Bravais believed that the crystalline materials can be considered as a set of molecules, in the limit reduced to their center of gravity, but with the fundamental assumption that these molecules also have their own orientation in space, repeatable in a regular lattice in the construction of matter:

[64] From an unpublished manuscript quoted in [58], p. 331.

[65] p. 747. Our translation.

Table 1.2 Theories of elasticity in the 19th century

Author	Physical model			Main physical magnitude
	Continuum	Molecular	Force	Energy/work
Navier	–	+	+	0
Lagrange	+	–	0	+
Cauchy	+	+	+	–
Poisson	–	+	+	–
Saint Venant	–	+	+	–
Lamé	+	+	+	–
Green	+	–	–	+
William Thomson	+	–	–	+
Clebsch	+	–	+	–
Piola	+	–	0	+
Betti	+	–	–	+
Beltrami	+	–	–	+
Castigliano	–	+	+	0
Kirchhoff	+	0	+	0
Voigt	0	++	+	0

Adapted from [18, p.46]
Legend: + dominant concept; – lack of the concept; 0 auxiliary role. The symbol ++ means that the molecule is considered as small three-dimensional body

> The crystals are assemblages of identical molecules with the same orientation, which, reduced by thought to a unique point, that is their center of gravity, are disposed in rectilinear and parallel ranges, in which the distance of two points is uniform [14].[66] (A.1.20)

Matter is therefore due to aggregates of regular lattices whose members are no longer, as for Navier, Cauchy and Poisson, simple materials points, but points with orientation; a contemporary reader might say that the mechanical descriptors of the microscopic model is equipped with a local structure, the one characteristic of a rigid body:

> Avoiding to consider the molecules as points and considering them as small bodies [14].[67] (A.1.21)

The molecules of crystalline bodies are small polyhedra, the vertices of which are the centers of the forces that each molecule of the body exchanges with the contiguous ones:

> The molecules of crystallized bodies will henceforth be polyhedra, the vertices of which, distributed at will around the center of gravity, will be the centers, or poles, of the forces exerted by the molecule [14].[68] (A.1.22)

This view of matter would in time lead to Voigt's molecular model, that would put an end to research for an answer about the correctness of the assumption of one or two elastic constants for homogeneous and isotropic linear elastic materials.

---

66 p. VII. Our translation.

67 p. VIII. Our translation.

68 p. 196. Our translation.

Woldemar Voigt (1850–1919) was a student at the doctoral school founded by Franz Ernst Neumann, together with Carl Jacobi, at the university of Königsberg; the list of students included Borchardt, Clebsch, Kirchhoff, among others. Under Neumann's direction, who encouraged experimental activities on the part of students, Voigt studied mineralogy and crystallography. In the early 1880s, Voigt published a series of fundamental contributions in crystallography and theory of elasticity, reconciling the results of the corpuscular and continuous models of matter. The importance of his results was immediately evident to his contemporaries. See for instance the monographs by Poincaré [125], Marcolongo [94], Hellinger [77]. Roberto Marcolongo provided a brief but clear description of Voigt's ideas and procedures:

> Voigt (1887) by supposing the body formed by an aggregate of particles (hence discrete the matter constituting the body), by supposing that each particle is subjected by the others to actions reducible to a force and a couple, infinitely decreasing with distance, found the general equations of elasticity to be the same as those obtained by the theory of potential, without necessarily verifying the relations of Cauchy and Poisson [94].[69] (A.1.23)

Marcolongo referred to a work of 1887 in which Voigt [156] introduced his theories for the first time. Voigt set them out in a paper presented at the International Congress of Physics in Paris in 1900 [155] in which he explicitly declared that the theory of Navier, Cauchy and Poisson, albeit mechanically consistent, is not validated by the experimental results, which at the time were numerous and had sufficient accuracy and reliability.

> The molecular theory, or theory of actions at a distance, proposed by Navier, Cauchy and Poisson [...] made the elastic properties of isotropic bodies, depend indeed on a single parameter, while numerous observations did not seem in accord with this results [155].[70] (A.1.24)

On the other side, the continuist theory presents results differing from those of the corpuscular theory, but supported by experimental evidence:

> A new theory has for some time been generally adopted [...] that supposes that matter is continuous and that the mutual actions between near portions of matter are localized at their separation surface [...]. This theory, which we will call the *theory of immediate actions*, provides, contrary to the former, two characteristic constants for isotropic media, and all these results are in agreements with observation [155].[71] (A.1.25)

The weak point of the corpuscular theory of the French mechanicians of the early 19th century was immediately identified by Voigt. His theory, which surpassed that "inutilement spécialisée" by Navier, Cauchy and Poisson, was based on the study of the formation of crystalline bodies. In the formation of a crystal, the particles approach each other but they have to follow the orientation of the lattice, so it is no longer permissible to admit that the molecular interaction is reduced only to a force, but also a couple mutually exchanged between neighboring particles exists:

---

[69] p. 97. Our translation. Cauchy-Poisson's relations are those relations that reduce from 36 to 15 the independent elastic constants of the more general elastic relationship; one in the case of isotropic bodies.

[70] p. 288. Our translation.

[71] p. 288. Our translation.

> The old molecular theory of elasticity starts from an unnecessarily specialized fundamental conception, that is the hypothesis of central molecular actions depending only on the distance [...] the regular formation of a crystal [...] is not understandable unless an orientating couple acts on the particle [...] [155].[72] (A.1.26)

Accordingly the considerations on the intermolecular forces such as distance dependence, the presence of a sphere of activity for the molecular action and so forth, must also be extended to the mutual couples:

> We have considered only the pressure of forces acting between molecules, but the moments, or *couples*, acting between them may be considered in the same way [155].[73] (A.1.27)

It is remarkable how Voigt, following the tradition dating back to the French mechanicians of the first decades of 1800, adopted the term pressure (*pression* in French, *pressione* in Italian, *Druck* in German) to denote the representative of the inner actions, while at that time the scientific community had begun to use the more general term *stress*, introduced by Rankine [133, 134]. Voigt clearly distinguished between the idea of force among the molecules and the pressure of the contact forces acting on a point inside the body due to the sum of the forces at distance exchanged between the considered molecule and the contiguous ones in relation to an infinitesimal surface.

The definition of *Druck* called for that of Poisson referred to in Sect. 1.1.1: the component of the stress in the direction S over the elementary surface q of the cylinder having normal n is given by:

$$qS_n = \sum_i \sum_e S_{ie}, \tag{1.20}$$

where S_{ie} is the component in direction S of the force that the molecule i inside the cylinder sketched in Fig. 1.2b is exerted upon by the molecule e outside the cylinder.[74] Following a notation originally introduced by Kirchhoff, Voigt adopted the symbols X_x, X_y, X_z, Y_x, Y_y, Y_z, Z_x, Z_y, Z_z to indicate the nine components of stress with respect to a Cartesian set of coordinates.

Voigt defined the pressure of the moments (*pressions des moments*) in a strict analogy to that of the forces:

$$qD_n = \sum_i \sum_e D_{ie}, \tag{1.21}$$

where D_{ie} is the component in direction D of the moment that the molecule i inside the cylinder is exerted upon by the molecule e outside the cylinder.

> One has the nine special components of the stress-moments
>
> $$L_x,\ L_y,\ L_z, \qquad M_x,\ M_y,\ M_z, \qquad N_x,\ N_y,\ N_z,$$

[72] p. 289. Our translation.

[73] p. 293. Our translation.

[74] See Fig. 1.2a of paragraph 1.1.1, where e are the molecules in A′ and i the molecules in B.

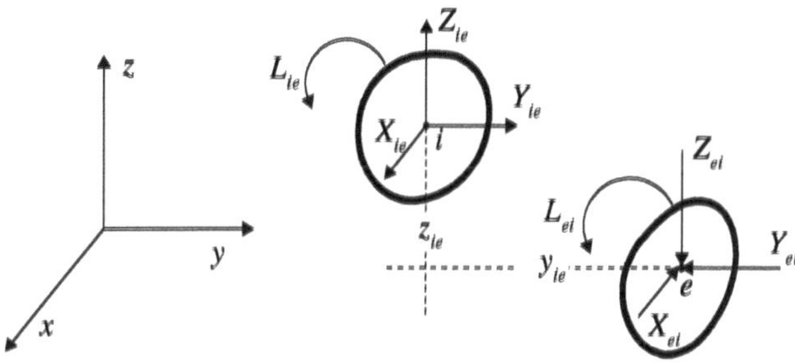

Fig. 1.6 The actions between particles

which exactly correspond to the $X_x, \ldots, Z_z$ [155].[75] (A.1.28)

Through the principle of virtual work Voigt [155] got the equations that express the equilibrium for internal forces X_{ie} and couples L_{ie}[76]:

$$\begin{aligned} X_{ie} + X_{el} &= 0, \\ &\ldots \\ L_{ie} + L_{el} + Z_{ie}\, y_{ie} - Y_{ie}\, z_{ie} &= 0, \\ &\ldots \end{aligned} \tag{1.22}$$

where y_{ie} and z_{ie} are the differences of Cartesian coordinates y and z of the centers of the molecules i and e, according to what is shown in Fig. 1.6.

The previous relations obviously also hold for the other components Y and Z of the force, M and N of the couple and the other rectangular coordinate x of the centers of molecules, according to the rules of permutation.

Voigt advanced an important postulate concerning the constitutive relationship; during a real deformation the elementary masses within the sphere of molecular activity rotate by the same amount. It follows that, because the particles i and e keep a parallel orientation, the internal couple of i on e must be indistinguishable, for symmetry, from the couple of e on i or $L_{ie} = L_{ei}$, and then, from the equilibrium equation for the moments (1.22b) one can characterize the interaction in terms of the forces alone:

$$L_{ie} = \frac{1}{2}\left(Y_{ie}\, z_{ie} - Z_{ie}\, y_{ie}\right), \ldots. \tag{1.23}$$

In this way the couple becomes the moment of a force with the consequence that the pressure of the moments vanishes. In fact, replacing the previous expression (1.23) in the definition (1.21) of the pressure of the moments (where one puts $D_{ie} = L_{ie}$), because for z_{ie} and y_{ie} only values of the molecules within the range of molecular activity (infinitesimal) should be considered, D_n is a higher-order infinitesimal compared to S_n and, if S_n is finite, D_n is negligible.

[75] p. 293. Our translation.

[76] p. 293.

The only actions to characterize the constitutive relationship are therefore ordinary stresses, which depend on variations of the distances between the centers of gravity of the elementary masses (considered as extended solids), as in the discussion of the French mechanicians of the early 19th century. In fact, since the rotations are uniform within the sphere of the molecular activity, these are not subjected to the law that varies the distance between elementary masses.

With a series of passages in which a modern reader recognizes a linearization procedure of the equations, that is a restriction to infinitesimal displacements and strains, Voigt arrived finally to the constitutive relations for stresses, with 36 elastic constants, which for symmetry reduce to 21 and that in the case of isotropy reduce again to two. He also showed that if one assumes the molecular model with central forces, the result of Cauchy and Poisson with 15 constants for the general case and one for the isotropic case is recovered.

The theory was brought again definitively by Voigt in a more extended way than in the communication of 1900, in a ponderous monograph of 1910 [157] where he stated that any theory about the constitutive relationship which shall describe the experimental behavior clearly and consistently with the observations must necessarily start from Bravais' idea of lattice structures:

> A molecular theory of the elastic phenomena that offers the perspective to explain all observations shall have such a general basis as *Bravais*' structural theory [...] provides. According to it, the crystal shall be thought as composed by identical and parallel oriented bricks or elementary masses, arranged so that each of those is surrounded by the others in the same way inside the sphere of [molecular] action [157].[77] (A.1.29)

The crystal should be thought of as composed of 'bricks', or elementary masses, identical and parallel-oriented, ordered so that each of them is surrounded in the same way. The only assumption made by Voigt on this model of structure of matter is that the bricks can be considered as rigid particles and that the molecular interactions, dependent on the distance between the centers of gravity of the masses, admits a potential:

> Considering the former, we will not restrict the generality by introducing other limitation than the mutual actions shall have potential; considering the latter, for reporting the elastic phenomena it will be admitted to consider the elementary masses as rigid bodies [157].[78] (A.1.30)

Interactions between particles are caused by resultant forces and couples:

> The interactions occurring between two rigid bodies (h) and (k) give origin to both global [force] components X_{hk}, X_{kh}, ... and also rotation moments L_{hk}, L_{kh}, ..., which we will calculate around the center of gravity of the considered body. We assume analogous actions occurring between two elementary masses of our crystal model [157].[79] (A.1.31)

Voigt still used the term *Druck* to indicate the stress. The definition was the same as that of 1900, that is that of Poisson, but for its evaluation Voigt took a different

[77] p. 596. Our translation.

[78] p. 597. Our translation.

[79] p. 597. Our translation.

approach, close to that of Green. In fact, after having constructed a potential function that takes into account the constitutive assumptions $L_{ie} = L_{ei}$ and in which both motions and forces are linearized, he determined the stress tensor components by differentiating the potential function with respect to the components of strains. He showed that the elastic constants for the general case are 21, 2 for the isotropic case. To find the two constants, Voigt proposed a new definition of isotropy: the crystalline bodies are composed of very small fragments of crystals and one cannot generally admit that all have the same orientation. The symmetry properties, on which are based the crystal classifications, must therefore be understood in a statistical sense: a given material has monocline symmetry, for instance, if most (statistical sense) of the crystalline fragments respect the symmetry properties of the crystallographic group. An isotropic material then is a crystalline substance in which the distribution of the orientations of the crystalline fragments has no significant value in the statistical sense (like a 'white noise'), so there is no preferential direction of behavior. In this way, all the directions (in the statistical sense) are of material symmetry; this definition accords with the traditional one.

Voigt's examination of the elastic behavior of materials thus reconciled the seemingly irreducible corpuscular and continuum theories that had long been a controversy among 19th-century mechanicians.[80] In the 20th century, an important advancement of crystallography studies with respect to those of Bravais and Voigt was due to a Voigt's 'pupil', Max Born. He confessing of not having fully understood Voigt's approach, in a his early and not yet well studied book of 1915, the *Dynamik der Kristallgitter* [10],[81] he reconsidered the classical molecular model, framing it into the theory of crystals. He showed that in most cases the Cauchy-Poisson equalities, which reduce the number of independent constants, are not longer valid and found necessary the use of 21 coefficients to describe the constitutive relationship of an elastic material. Born [11] published a revised version of his book in 1954, as the *Dynamical theory of crystal lattices*.

1.1.5 Continuum Mechanics in the Second Half of the 19th Century

Let us consider briefly our studies on continuum mechanics in Europe and in Italy during the period 1850–1880. We will talk only of contributions of the leading scholars of the time, which is enough to highlight how this period is full of extremely high level characters.

The main goal is to frame the developments in Italy of the theory of elasticity and continuum mechanics in the second half of the 19th century in the European context. Both for sake of space, and because the Italian contribution was limited

80 For a more in depth study on Voigt's work see [20].

81 An interesting comment of Born's analysis can be found in [67], where also considerations on modern studies are referred to.

to specific fields of continuum mechanics, in particular on general themes of the theory of elasticity, only some hints will be given. We thus felt it was appropriate to skip themes that have not been developed in Italy: all the experimental studies, those on phenomena of fatigue, on rupture, on dynamics and on nonlinear elasticity, on one-dimensional and two-dimensional continua. For a precise and sufficiently widespread account on these issues see Todhunter and Pearson's textbook [154].

In the first half of the 19th century the French school had been the undisputed leader in Europe; things began to change slowly. One of the most important characters of continuum mechanics in the Europe of the second half of the 19th century, however, was still a Frenchman, Adhémar Jean Claude Barré de Saint-Venant (1797–1886), already abundantly quoted in previous sections about his fierce defense of the classical molecular model. His works of 1855 about torsion [143] and of 1856 about bending [144] are fundamental. Although Saint Venant is known mainly for his works on beams, also his work in other areas of continuum mechanics should be quoted; for instance the studies of two-dimensional continua and dynamic problems of impact. With Saint Venant the French school of continuum mechanics reached its peak, beginning a decline only partly held back by Valentin Joseph Boussinesq, Alfred Aimé Flamant, Henri Poincaré and Pierre Maurice Marie Duhem.

The English school of the early 1800s had had an essentially experimental approach to science. Fundamental were the laboratory tests on strength of materials by Eaton Hodgkinson (1789–1861) and William Fairbairn (1789–1874) [153].[82] Relevant, however, was the theoretical content of the work of the Scottish William John Macquorn Rankine (1820–1873), who in 1855 became a professor of engineering at the university of Glasgow. Most of his research work is reported in his *Manual of applied mechanics* first published in 1858 [132], where among other things, he introduced for the first time the terms *stress* and *strain* with a precise technical meaning.

Of greater theoretical relevance are the contributions by William Thomson (Kelvin) (1824–1907), James Clerk Maxwell (1831–1879), George Gabriel Stokes (1819–1903), John William Strutt (Rayleigh) (1842–1919). The *Treatise on natural philosophy* by Thomson and Peter Guthrie Tait [152],[83] besides being mathematically unexceptionable, was very attentive to the physical aspects also, as we understand from the objects of some sections:

(a) Equilibrium of a flexible inextensible cord.
(b) Wire with any shape under the action of forces and torques applied along its length.
(c) Bending of elastic flat foil (whose treatment makes extensive use of experimental data),

though only a short paragraph (5 pages) is dedicated to the fundamental problems of mathematical theory of elasticity.

[82] pp. 125–129.

[83] The last reissue is entitled *Principles of mechanics and dynamics*, Dover, New York, 2003. The work was begun in 1861; programmed in multiple volumes, for commitments of the two authors, it saw the light of day with only the first volume.

In the second part of the first volume of their work Thomson and Tait warn:

> Until we know thoroughly the nature of matter and the forces which produce its motions, it will be utterly impossible to submit to mathematical reasoning the *exact* conditions of any physical question. It has been long understood, however, that an approximate solution of almost any problem in the ordinary branches of Natural Philosophy may be easily obtained by a species of *abstraction*, or rather *limitation of the data*, such as enables us easily to solve the modified form of the question, while we are well assured that the circumstances (so modified) affect the result only in a superficial manner [152].[84]

Therefore, according to Thomson and Tait, the use of mathematics is essential, but not sufficient in the study of nature. For a result to be considered valid in physics, it is not enough to be inferred (even from experimental data) with the methods of pure mathematics, but must be in accordance with experience. The authors' goal is twofold:

> To give a tolerably complete account of what is now known of Natural Philosophy, in language adapted to the non-mathematical reader; and to furnish, to those who have the privilege which high mathematical acquirements confer, a connected outline of the analytical processes by which the greater part of that knowledge has been extended into regions as yet unexplored by experiment [152].[85]

Born in Edinburgh, James Clerk Maxwell studied at the local university. In 1850 he read before the Royal society of Edinburgh his essay on the theory of elasticity *On the equilibrium of elastic solids* [44], whose original part is an account of the relationship between analytical and experimental results. In 1864 he wrote his essay *On the calculation of the equilibrium and stiffness of frames* [45], about which we will talk later, concerning the calculation of elastic framed structures using Clapeyron's theorem.

George Gabriel Stokes, Irish, began his studies in Bristol and completed them in Cambridge, where in 1849 he was elected to the Lucasian chair of mathematics (the place once held by Newton). Initially he treated hydrodynamics and in the memoir *On the theories of the internal frictions of fluids in motion, and of the equilibrium and motion of elastic solids* [149] he found the field equations for an isotropic elastic medium. Based both on experimental observations, for example on isochronous vibrations, and on theoretical assumptions on molecular structure, Stokes gave two equations in terms of the elastic constants instead of one. He also dealt with optics and vibrating bodies, conducting a series of experiments with very modest means, since a real laboratory in Cambridge was established only by Maxwell in 1872. By 1854 he was secretary of the Royal society and from 1885 to 1890 president.

John William Strutt (Lord Rayleigh), from a noble family, studied mathematics and mechanical engineering at Cambridge with Edward John Routh (1831–1907) and optics with Stokes. At Maxwell's death (1879) he succeeded him in the chair of experimental physics. He held numerous positions, including the presidency of the Royal society (1905–1908); he won the Nobel Prize for physics in 1904. His main

[84] vol. 1, p. 136.

[85] vol. 1, p. V.

contributions are reported in the treatise *The theory of sound* [150], where he studied vibrations of strings, bars, membranes, plates and shells by means of generalized force and coordinates. In the treatise, *inter alia* Rayleigh generalized Enrico Betti's reciprocity theorem to dynamics using it for studying statically indeterminate structures[86]; then he got the expressions for the statics of structural elements starting from the dynamic equations and showed how to get information about proper frequencies of elastic systems by means of energetic considerations (Rayleigh method).

Of a generation later are Horace Lamb (1849–1934), scholar of hydraulics and theory of elasticity; Augustus Edward Hough Love (1863–1940), whose history of the theory of elasticity, contained in the introduction to his manual *A treatise on the mathematical theory of elasticity* [93] is considered today a classic for precision and concision; Karl Pearson (1857–1936) who became famous for the text *A history of the theory of elasticity* in collaboration with Isaac Thodhunter, the most comprehensive account of the 19th-century theory of elasticity and strength of materials [153].

In Germany the formation of engineers developed differently compared to England, with greater importance given to theoretical notions, especially mathematical physics, as happened in France [22]. Engineers, however, completed their preparation entirely within the same polytechnic school, unlike what happened in France where there was the École polytechnique which provided a basic preparation, leaving the various École des ponts et chaussées, École des mines, etc., responsible for specific technical preparation. Among the leading teachers of engineering we include Julius Weisbach (1806–1871), Ferdinand Redtenbacher (1809–1863) and Franz Grashof (1826–1893), known especially for having supported the criterion of maximum deformation resistance [153].

In the German school, besides the already mentioned Voigt, Franz Neumann (1798–1895), Gustav Robert Kirchhoff (1824–1887), Alfred Clebsch (1833–1872) emerged. Franz Neumann started writing a treatise on mineralogy that earned him a position at the university of Königsberg. Here he met Friedrich Wilhelm Bessel, Heinrich Wilhelm Dove and Carl Gustav Jacob Jacobi and became interested in geophysics, thermodynamics, acoustics, optics, electricity. Along with Jacobi he organized a *mathematisch-physikalisches Seminar*, with theoretical and experimental formation; his pupils included Borchardt, Clebsch, Kirchhoff and Voigt. In elasticity he followed the theory of Poisson, but he rejected the results on the number of elastic constants, based on crystallographic experiments of his students (among them,Voigt). His more important contribution to the theory of elasticity was a memoir on double refraction, which gave rise to photo-elasticity, where he also dealt with the nature of thermal and residual states of stress. His handbook was of great impact

---

[86] In structural mechanics those systems of bodies that contain more constraint reactions than equilibrium equations are called statically indeterminate. The difference between the number of constraint reactions and equilibrium equations is called degree of hyperstaticity. The degree of hyperstaticity can be defined also in dual mode as the difference between the elementary constraints and degrees of freedom of the system of bodies. Because the constraint reactions cannot be determined with the equations of statics, using an engineering terminology, we say that the static problem is not determinate.

to his contemporaries, even Love in his famous treatise adopted his notation for the components of stress.

Kirchhoff worked in Berlin, Breslau, Heidelberg (with Robert Wilhelm Bunsen and Hermann Ludwig Helmholtz), then again in Berlin. As a student of Franz Neumann, he was involved in the theory of elasticity since 1850 with a memoir on the theory of plates that today bears his name. He also published works on the deformations of thin beams, where the demonstration of the theorem of uniqueness of solution of the linearized elasto-static problem is reported, and the field equations for non-infinitesimal deformations are derived. His monumental treatise on mathematical physics [79] was a reference point for European scientists of the late 1800s.

Clebsch began to deal with elasticity in Königsberg. He worked at the Technical university of Karlsruhe, then turned to pure mathematics in Göttingen, where he founded the prestigious magazine *Mathematische annalen* together with Carl Gottfried Neumann, the son of Franz, and where he had as his pupil the great mathematician Felix Christian Klein (1849–1925) among others. His volume on the theory of elasticity [43] is a fundamental work, which had a strongly mathematized approach, and where the most original contribution was in the field of two-dimensional problems. The volume of Clebsch was resurrected, translated into French and integrated with a series of annotations by Saint Venant and Flamant [42] and long remained one of the most famous treatises of elasticity theory and applications. Clebsch's words written in the preface are emblematic:

> The original intention of the author was not to put in the book what he needed to be guided in the lectures he professes to Karlsruhe Polytechnic. But soon he felt so much need to rely on a solid basis the researches whose results are used for applications, that he determined to undertake the writing of a treatise on the theory of elasticity which, as far as was possible in a moderate space, presents a comprehensive system of principles and practical use of this theory: work that becomes possible today thanks to the beautiful researches of Kirchhoff and Saint Venant. It was certainly needed to deal briefly many points, but it was necessary, above all, to explain in detail what is desirable for a good acquaintance of this new branch of science. Thus, for all that concerns the analytical transformations that Mr. Lamé taught to operate with so much elegance on the fundamental equations of elasticity, one should refer to the so well known and widespread work of this illustrious scientist [42].[87] (A.1.32)

1.2 Theory of Structures

The pioneering phase of foundation for the sciences of structural behavior, like other sciences, was followed by a phase of systematization in which specialized disciplines emerged. Principles and methods, although not universally shared in detail, had, in fact, been established and the process of specialization saw the division of the sciences of structural behavior into two main strands: continuum mechanics and mechanics of structures. The first studied the basic aspects of mechanics of deformable

[87] p. IX. Our translation.

bodies (stresses, strains, constitutive relationships, equilibrium); the second studied the aspects associated with engineering applications (calculation of frames, plates, arches, domes). At the beginning the studies were based on the theory of elasticity; but then the specialized nature of the latter became clear, as well as a realization that the linear elastic relationship was not generally sufficient to exhaust continuum mechanics or mechanics of structures.

The mathematical theory of (elastic) structures developed between 1820 and 1890 into two phases. The first was characterized by the wording of the first structural theories of beams, plates and elastic domes and resolved simple statically indeterminate structures, without a complete understanding of the methodologies. As in the theory of elasticity, the leading nation was France and the main character can be considered Navier. In the second phase, which can be conjectured as beginning in the 1840s, Germany joined France and then became the leading nation. In this period we find the partial subordination of England to the continent. Despite many important scientists such as Maxwell, William Thomson, Rayleigh and some scholars of mechanics of structures such as Henry Moseley (1801–1872) and Rankine, there was not in England a union between theory and practice such as took place in the continent.[88] This second phase, in which attention was focused primarily on structures formed by beams and bars, can be divided into two sub-phases.

The first step involved the design of statically determined trusses, fundamental in civil and industrial constructions, such as bridges and sheds. To assist in the computation of the complex lattice structures, for example the trusses of industrial warehouses or bridge truss, analytical techniques and very efficient graphic processes were developed. Among the first, we recall August Ferdinand Möbius' analyses of 1837 on trusses [102],[89] August Ritter's method of sections,[90] the use of the

[88] For comment on this issue see [40].

[89] Here Möbius studied plane and spatial trusses, found the minimum ratios between nodes and bars for a trusses to be statically determined (for n nodes one needs $2n-3$ bars in the plane and $3n-6$ in the space) and discovered that the minimum requirement is not sufficient for the equilibrium if the static matrix has zero determinant. Möbius' work, at least initially, was not noticed by engineers, thus deserving the considerations for the contribution of the precursors (see next sections). The results of Möbius were found again (likely) independently by Christian Otto Mohr in 1874.

[90] Ritter's method was developed in embryonic form by Culmann in two articles, of 1851 [52] and of 1852 [53], which showed the state of art of iron and wood bridges in Great Britain and America. Here Culmann used equilibrium equations of forces and moments for the determination of the stresses in bars. In particular, the equations of equilibrium were related to sections of the truss. Culmann results were taken again by August Ritter (1826–1908) in his textbook of 1863 *Elementare Theorie der Dach und Brüken-Constructionen* [136]. Ritter also known in astrophysics (in his honor the Ritter's crater on the lunar surface), was formed in Hannover and Göttingen; from 1856 he taught mechanics in Hannover, then in 1870 at the Technical college of Aachen. Although in Ritter's textbook there were no evident references to Culmann (about the method of sections he mentioned only one of his previous work in the journal of the architects and engineers of Great Britain), it is likely that his work, set on roofs of buildings and bridges, was in any way influenced by the articles of Culmann. The method of 'Ritter sections', can also be associated to another Ritter, Karl Wilhelm Ritter (1847–1906), who graduated in Zurich in 1868 and, after some professional activities in Hungary, was assistant of Culmann in Zurich from 1870 to 1873. Karl Wilhelm Ritter taught at the Polytechnic institute of Riga until 1882, when, following the death of the master, he

principle of virtual work by Mohr [104, 105]. Among the latter we cite the works of Rankine [131, 132], Fleeming Jenkin [65, 66], Maurice Lévy [91], Luigi Cremona [50] and especially Culmann [54]. He is justly considered the father of *graphical statics*, the discipline that with the aid of graphical techniques, in part borrowed from projective geometry, leads to the solution of problems of statics, of geometry, of metric computation. To signal also the graphical method developed by Joseph Victor Williot [159] for calculating of node displacements for trusses.

The second step concerned statically indeterminate systems, especially trusses. At the beginning the method today said of deformations was proposed, following a procedure developed by Navier and Poisson and perfected by Clebsch [43]. The method was of simple conception and automatic application, but required the solution of systems of linear equations in the displacement components of all the nodes. These systems were too large even for simple structures with more than ten nodes and for this reason the method essentially remained unapplied, except for sporadic cases. Then it was introduced the methods today named of forces that, after a pioneering phase due to Navier and Saint Venant, developed almost simultaneously with three different approaches. Carlo Alberto Castigliano in 1873 [23] in Italy perfected a method introduced by Luigi Federico Menabrea in 1858 [100] which consisted in minimizing the elastic energy of statically indeterminate trusses with respect to unknown forces; the method was suitable even to treat structures undergoing flexure. Mohr [104, 105] in Germany introduced in 1874 the virtual work principle for determining the displacement components or internal forces. In France Lévy [90, 91] had developed in 1873 an original method to obtain compatibility equations and solve the statically indeterminate systems.

At the end of the 19th century, structures different from trusses started to be considered and Müller-Breslau [113] developed the methods of Mohr and Castigliano and perfected their use for generic statically indeterminate systems. However, until the introduction of reinforced concrete and the realization of multilevel rigid node frames, the techniques to solve frames would receive modest attention.

1.2.1 Statically Indeterminate Systems

In the 19th century the theory of structures was developed first by engineers, being aimed at applications and requiring simple mathematical methods (ordinary differential equations and algebraic equations). Subsequently also mathematicians and physicists were involved, when it was clear that the statics of rigid bodies was not sufficient to study structures subjected to redundant constraints. In these cases, some reaction forces remain undetermined.

(Footnote 90 continued)
was called to the chair of Zurich. He is the author of a key text on trusses [137] where the method of sections is used.

The first formulation and solution of the problem is usually traced back to Leonhard Euler in 1773 [63] who wanted to calculate the "pressures arising from the support of a rigid body on a horizontal plane, lying on more than three points or on a surface". Euler said that the problem could not be solved by the known laws of statics:

> But if a weight lies on a plane with four supports, the determination of the individual pressures is not only much more complex [than in the case of three supports] but also uncertain and misleading [63].[91] (A.1.33)

Roughness or subtle differences in the support influence the distribution of pressures: if one leg is shorter than the other, the weight rests on only three points. To get back into a more defined theoretical situation Euler imagined a soft cloth interposed between the ground and the body:

> Not to accept the perfect equality of the supports, which can be difficult to admit, it is assumed that the floor or the ground on which the weight rests is not hard, that cannot get any prints, but is covered with a cloth, in which the support can penetrate slightly [63].[92] (A.1.34)

It is unclear whether Euler introduced the artifice to eliminate mathematical difficulties or, as it would seem natural, he thought that the cloth models reality. Euler suggested that the actions on the supports, or pressures p, are linearly distributed on the basis[93]:

> Let the weight lie on more supports or hangs on any plane figure, be M an end point of one of these support or any element of the basis for which the pressure is required. Imagine there erected a vertical line Mμ proportional to this pressure; being it necessary that all these points terminate in one plane, then established such a principle, I am going to explain how one can define in all cases the pressure at each point of the base [63].[94] (A.1.35)

Given the pressure $p = \alpha + \beta x + \gamma y$ as a liner function of the coordinates of the supports x, y, the equilibrium to the vertical translation and rotation about two horizontal axes provides α, β, γ. The procedure can be generalized to multiple points of support up to an entire surface.

The extreme brevity in exposing the principle of linear distribution of pressure does not clarify whether Euler introduced a linear elastic relationship, since the yielding of the support points vary linearly, or introduced an ad hoc law of statics for contacts, valid regardless of the materials in contact.

A theoretical important question was so posed: can one use properly the equations of statics to solve contact problems? Or in this case the equations of statics are not sufficient and one must formulate some new principle? The problem would assume great importance in Italy and its history is well traced by Todhunter and Pearson, who tend to trivialize the problem:

---

[91] p. 290. Our translation.

[92] p. 290. Our translation.

[93] A similar approach is used to determine the axial stresses in the biaxial bending of a Saint Venant's cylinder.

[94] p. 291. Our translation.

For the history of science the problem is of value as showing how power is frequently wasted in the byways of paradox. I give a list, which I have formed, of the principal authorities for those who may wish to pursue the subject further.

Euler	De pressione ponderis in planum cui incumbit. Novi Commentarii Academiae Petropolitanae, T. XVIII, 1774, pp. 289–329. Von den Drucke eines mit einem Gewichte beschwerten Tisches auf eine Fläche (see our Art. 9.5), Hindenburgs Archiv der reinen Mathematik. Bd. I., s. 74. Leipzig, 1795.
D'Alembert	Opuscula, t. VIII Mem. 56 II, 1780, p. 36.
Fontana M.	Dinamica. Parte xx.
Delanges	Memorie della Società Italiana, t. V, 1790, p. 107.
Paoli	Ibid.t. VII, 1792, p. 534.
Lorgna	Ibid.t. VII., 1794, p. 178.
Delanges	Ibid.t. VIII, Parte I, 1799, p. 60.
Malfatti	lbid.t. VIII, Parte II, 1798, p. 319.
Paoli	Ibid.t. IX, 1802, p. 92.
Navier	Bulletin de la Soc. philomat., 1825, p. 35 (see our Art. 282)
Anonym.	Annales de mathém. par Gergonne, t. XVI, 1826–7, p. 75.
Anonym.	Bulletin des sciences mathématiques, t. VII, 1827, p. 4.
Véne	Ibid. t. IX, 1828, p. 7.
Poisson	Mécanique, t. I, 1833, 270.
Fusinieri	Annali delle Scienze del Regno Lombardo–Veneto, t. I, 1832, pp. 298–304 (see our Art. 396).
Barilari	Intorno un Problema del Dottor A. Fusinieri, Pesaro, 1833.
Pagani	Mémoire de l'Académie de Bruxelles, t. VIII, 1834, pp. 1–14 (see our Art. 396).
Saint Venant	1837–8 see our Art. 1572. 1843 see our Art. 1585.
Bertelli	Memorie dell'Accademia delle Scienze di Bologna, t. I. 1843–4, p. 433.
Pagnoli	Ibid., t. VI, 1852, p. 109.

Of these writers only Navier, Poisson and Saint Venant apply the theory of elasticity to the problem. Later researches of Dorna, Menabrea and Clapeyron will be referred to in their proper places in this History as they start from elastic principles [154].[95]

To the over-referred, Todhunter and Pearson added Cotterill, Moseley and Mossotti, to whom we will refer below; a detailed reconstruction of the summarized path can be found in [7].

1.2.2 The Method of Forces

Among the attempts at solving the problem of redundant supports, the first success in terms both theoretical and practical was that of Navier. He in his lectures in 1824—according to Saint Venant already in 1819 [116][96]—published in 1826 as *Résumé des leçons données a l'école royale des ponts et chaussées sur l'application de la Mécanique à l'établissement des constructions et des machines* (hereinafter *Leçons*), tackled the

[95] vol II/1, p. 411.

[96] p. cviii.

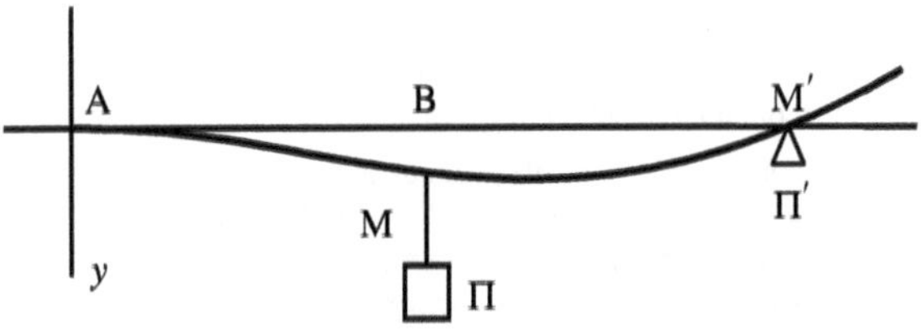

Fig. 1.7 Clamped-supported beam. (Redrawn from [115], vol. 1. Fig. 48, Pl. II)

case of plane beams with a number of external constraints exceeding two, that cannot be resolved with the equations of statics only. Navier first recognized that there is a unique solution only if the deformation of the beam is accounted for:

> When a rigid bar loaded by a weight is supported on a number of supports greater than two, the efforts that each of these supports must endure are undetermined between certain limits. These limits can always be determined by the principles of statics. But if the bar is assumed to be elastic, the uncertainty ceases entirely [115].[97] (A.1.36)

To understand the originality and limits of Navier's work we only show the simple case of the beam AMM′ of Fig. 1.7, clamped at one end, supported at the other and loaded in the middle point M by a load Π.

Without many comments, Navier replaced the support in M′ with a vertical force Π′ and wrote the equation of the *elastic line* (modern term), that is the equation which defines the shape of the deformed axis of a beam. For the part AM he got:

$$\epsilon y = \Pi\left(\frac{ax^2}{2} - \frac{x^3}{2}\right) - \Pi'\left(\frac{a'x^2}{2} - \frac{x^3}{2}\right), \tag{1.24}$$

with Π′ the reaction of the support in M′, ϵ the stiffness of the beam (in modern terms the product EI), a the distance AB and a' the distance AM′; the abscissa x is measured starting from A.

For the part MM′, assuming that "the evaluation of the coefficients so that for $x = a$, the values of dy/dx and y are equal to those furnished by the previous equation" [115],[98] Navier found:

$$\epsilon y = \Pi\left(\frac{ax^2}{2} - \frac{x^3}{6}\right) - \Pi'\left(\frac{a'x^2}{2} - \frac{x^3}{6}\right) \tag{1.25}$$

and, imposing the vanishing of y for $x = a'$, he obtained a relation that is still found in modern treatises of mechanics of structures:

$$\Pi' = \Pi\frac{a^2(3a' - a)}{2a'^3}. \tag{1.26}$$

[97] vol. 1, p. 241. Our translation.

[98] vol. 1, p. 235.

With the equation of the elastic line Navier solved other cases; of particular interest is the continuous beam on three supports, which is the model of a bridge beam. Navier considered a beam of uniform section and material subjected to two forces concentrated in the two bays. The beam is shaped with a single trunk, resting at the extremes and loaded by the intermediate reaction. This reaction is obtained after determining the equation of the elastic line and imposing that displacement of the point corresponding to the intermediate support is zero.

Navier's approach today is classified as a method of forces, in which the unknown constraint reactions are determined by imposing compatibility. Navier likely had not recognized the method in its general value, because he was limited to resolving only beams subjected to external constraints. Moreover, he did not provide a general method for calculating the displacements of a structure of any form. Saint Venant attributed to himself the merit of having extended Navier's approach to any type of structure, at least from a theoretical point of view [116][99] and in 1843 he outlined very clearly the essence of the method of forces:

> This method consists in searching the displacements of the points of the [various] parts leaving in indefinite form the size, the lever arm and the directions of the forces of which we speak. Once expressed the displacement as functions of these sought quantities, one imposes the conditions they must meet at the support or clamping points or at the junctions of the various parts, or at connection points of the various parts in which one given piece should be divided because the displacements are expressed by different equations. In this way, one has as many equations as unknowns, because of course, in matters of mechanical physics there is no indeterminacy [140].[100] (A.1.37)

In this same memoir some formulas were reported to calculate the displacements of beams with single and double curvature as a function of deformation. Saint Venant applied the methodology and these expressions for the analysis of statically indeterminate structures in another memoir of that year [139].

Anyway Saint Venant was not able to outline a procedure of simple application, for the lack of a theory for the calculation of displacements in elastic beams. The fact is that engineers were not able to calculate even simple statically indeterminate structures such as trusses and frames with fixed nodes that were beginning to be used in constructions.

A quite successful approach, at least from a practical point of view, was realized after a few years by Henry Bertot and Benoît Paul Emile Clapeyron who arrived at a general and simple solution for continuous beams on many supports, in the form which today is called the *equation of three moments*; for problems of priority between the two we refer to the literature [15].[101] In the following, we only expose a few steps of Clapeyron's approach:

> I had to take care of this issue for the first time as an engineer during the reconstruction of the bridge of Asnieres, near Paris, destroyed during the events of 1848. The formulas to which I was led were later applied to large bridges built for the railway of Midi, on the Garonne,

---

[99] p. CCXII.

[100] p. 953. Our translation.

[101] pp. 405–406.

> Lot and Tarn, whose success has fully met our expectations. This is the result of the research that I have the honor to submit to the judgment of the Academy.
>
> In this first memoir, which are summarized below, I examine first the case of a straight beam resting on two supports at its ends; the section is constant and it supports a uniformly distributed load. Moreover we give the values of the (static) moments of the forces acting at the supports of both ends. We derive the equation of the elastic curve that affects the axis of the beam, the mechanical conditions under which all points are subjected and the portion of the total weight supported by each support.
>
> The solution of the general problem is thus reduced to the determination of the moments of forces tending to produce the breaking of the beam at each of supports on which it rests. This is achieved by imposing that the two elastic curves corresponding to two contiguous spans are tangent to each other on the intermediate support and that the moments are equal [41].[102] (A.1.38)

Clapeyron followed Navier and provided the method for the calculation of continuous beams with any lengths, loads and stiffness, assuming as unknowns the moments of continuity on the intermediate supports. The solution was given only for beams of constant section and material, and thus the elastic characteristic of the single beam does not appear. In the case of equal spans Clapeyron's equation for the generic node is:

$$Q_0 + 4Q_1 + Q_2 = \frac{l^2}{4}(p_0 + p_1) \tag{1.27}$$

with Q_0, Q_1, Q_2 the moments of continuity and p_0 and p_1 the distributed loads in the adjacent spans. Clapeyron commented by proposing that:

> If we add to the quadruple of any moment that which precedes it or that which follows it on the two adjacent supports, one obtains an amount equal to the product obtained from the total weight of the two corresponding beams by the fourth of the common span. If the spans are not unequal, the same relationship exists, except for minor changes in the coefficients [41].[103] (A.1.39)

Clapeyron nearly hid the theoretical roots of his formulas; an attitude that contributed in part to suggest to many engineers that the formulas of continuous beams, like other results obtained by Navier, represented practical rules or formulas derived from the laws of statics alone.

1.2.3 The Method of Displacements

In his *Leçons* of 1826 [115] Navier, in addition to the method of forces, shown in Sect. 1.2.2, also presented what today is called method of displacements, with reference to the system of three bars concurring in a node to which a concentrated load is applied, shown in Fig. 1.8.

---

[102] p. 1077.

[103] p. 1078. Our translation.

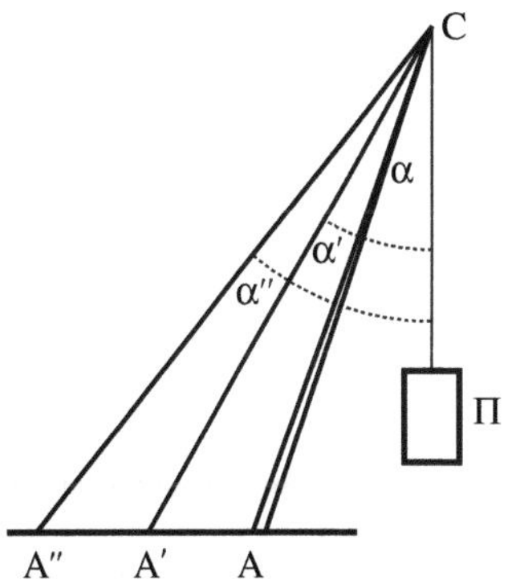

Fig. 1.8 Navier's system of bars. (Redrawn from [115], vol. 1. Fig. 112b, Pl. IV)

To give an example, assume the weight Π supported by the three inclined bars AC, A′C, A″C contained in the same vertical plane and denote by

- α, α', α'' the angles formed by the direction of the three bars with the vertical direction CΠ;
- p, p', p'' the forces produced, by the action of the weight Π in the direction of any bar;
- F, F', F'' the forces of elasticity of each bar;[104]
- a the height of the point C above the horizontal line AA'';
- h, f the amount of the horizontal and vertical displacement of the point C, because of the effect of the simultaneous compression of the three bars.

[…] This given, the condition of equilibrium between the weight Π and the three pressures exerted along the bars will furnish [115][105]: (A.1.40)

$$p\cos\alpha + p'\cos\alpha' + p''\cos\alpha'' = \Pi,$$
$$p\sin\alpha + p'\sin\alpha' + p''\sin\alpha'' = 0.$$

Navier performed a kinematic analysis in the case of small displacements and determined the expressions of the unitary change in the length of the bars as a function of the horizontal (h) and vertical (f) displacements of the point C:

$$\frac{f\cos^2\alpha - h\sin\alpha\cos\alpha}{a},\ \frac{f\cos^2\alpha' - h\sin\alpha'\cos\alpha'}{a},\ \frac{f\cos^2\alpha'' - h\sin\alpha''\cos\alpha''}{a}.$$

[104] The force of elasticity is the axial stiffness, that is the product of the longitudinal elasticity modulus by the area of the cross section of the bars, which multiplied by the axial strain gives the force in the bar.

[105] vol.1, pp. 346–347. Our translation.

One gets the three equations:

$$p = F\frac{f\cos^2\alpha - h\sin\alpha\cos\alpha}{a};$$
$$p' = F'\frac{f\cos^2\alpha' - h\sin\alpha'\cos\alpha'}{a};$$
$$p'' = F''\frac{f\cos^2\alpha'' - h\sin\alpha''\cos\alpha''}{a}$$

which, joined with the two preceding ones, will give the values of the displacements h and f, and forces p, p' and p'' [115].[106] (A.1.41)

Note that there is no formal writing of two equations as a function of the components of f and h of the displacement of the point C by replacing p, p', p'' in the equations of equilibrium. Navier spoke only of joined equations.

The two Navier's methods, of forces for beams in flexure and displacements for trusses, indicate that he did not propose a general procedure but was using *ad hoc* methods, although broad enough in scope. The method of forces, as already said, was clarified and generalized by Saint Venant, the method of the displacements by Poisson and Clebsch.

In his text on elasticity of 1862 Clebsch used methods similar to those of Navier [43], different for inflexed (beam) or stretched or compressed (trusses) elements. They are a method of forces in the first case and of displacements in the second. We limit ourselves to the method of displacements, for which Clebsch's contribution is more relevant; for the method of forces refer to bibliography [7, 40].

Clebsch referred to an elastic truss made up of bars with constant section hinged to their extremities and subjected to small displacement; in this way it is easy to arrive to the formula that expresses the shortening of the bars as a function of the displacements of the nodes. With his symbols, the variation ρ_{ik} of length r_{ik} of the bar contained between the nodes i and k, undergoing displacement of components u_i, v_i, w_i and u_k, v_k, w_k, is:

$$\rho_{ik} = \frac{(x_i - x_k)(u_i - u_k) + (y_i - y_k)(v_i - v_k) + (z_i - z_k)(w_i - w_k)}{r_{ik}}. \tag{1.28}$$

The elastic force is proportional to it according to the elasticity modulus E_{ik} and the area q_{ik} of the section of the bar, directed as r_{ik}:

From these extensions an elastic force is generated; in the point i it is directed toward the point k and vice-versa and its intensity is [...]

$$\frac{E_{ik} q_{ik} \rho_{ik}}{r_{ik}},$$

if one calls E_{ik} the elasticity modulus, q_{ik} the areas of the transverse section of the bar in object [43].[107] (A.1.42)

[106] vol. 1, p. 347. Our translation.

[107] p. 410. Our translation.

In the coordinates to which the system of bars is referred, it has the components:

$$\frac{E_{ik}q_{ik}\rho_{ik}(x_k - x_i)}{r_{ik}^2}; \quad \frac{E_{ik}q_{ik}\rho_{ik}(y_k - y_i)}{r_{ik}^2}; \quad \frac{E_{ik}q_{ik}\rho_{ik}(z_k - z_i)}{r_{ik}^2}.$$

At this point Clebsch could write the equations of equilibrium for each node i to which a force of components X_i, Y_i, Z_i is supposed to be applied:

If one imposes the equilibrium conditions, that is if one makes the components of the force in object vanishes, one obtains the three equations:

$$\begin{cases} X_i + \sum_k \frac{E_{ik}q_{ik}\rho_{ik}(x_k - x_i)}{r_{ik}^2} = 0 \\ Y_i + \sum_k \frac{E_{ik}q_{ik}\rho_{ik}(y_k - y_i)}{r_{ik}^2} = 0 \\ Z_i + \sum_k \frac{E_{ik}q_{ik}\rho_{ik}(z_k - z_i)}{r_{ik}^2} = 0. \end{cases}$$

In these equations nothing is unknown but the quantities u, v, w which define the ρs [43].[108] (A.1.43)

Clebsch concluded commenting on the system of the equilibrium equations of the node, which depends only on the components u, v, w of the displacement of n nodes, for a total of $3n$ unknowns. These equations are linearly dependent, since the external forces X_i, Y_i, Z_i obey the six equations of global equilibrium. For an unambiguous determination of the unknowns, the structure should be subjected to six independent simple constraints.

It is worthwhile to point out the contribution of Poisson who in his *Traité de mécanique* of 1833 presented some progress with respect to what was proposed by Navier [130].[109] The importance of Poisson's contribution is due to the wide distribution of his *Traité de mécanique*, written in French, much better known in Europe than German, Clebsch's language.[110] This was particularly true for Italy where, at least until 1880, only reference to Poisson and not to Clebsch is testified.

Poisson's treatment is less engineering oriented and less general than that of Clebsch, however it contained all the ingredients for an immediate application to trusses. Poisson studied the motion of a material point m urged by a force and constrained by means of elastic wires joined to a number of fixed points A, A', A'', A''' disposed in whatever way, as shown in Fig. 1.9. At the end of his presentation Poisson explained how his results could be adapted to the case of equilibrium. To an engineer of the time it was not difficult to recognize in m a node of a truss and in the wires the bars that concur therein, and then extend Poisson's method to a truss.

In the following we refer to the most relevant part of Poisson's approach:

To show this, suppose, to fix the ideas, that the force acting on the point m be the gravity, which we represent by g. Taking the axis z vertical and oriented in the direction of the force,

---

108 p. 411. Our translation.

109 vol 2, pp. 402–404.

110 Clebsch's text had widespread diffusion only in the late nineteenth century, thanks to the translation into French by Saint Venant [42].

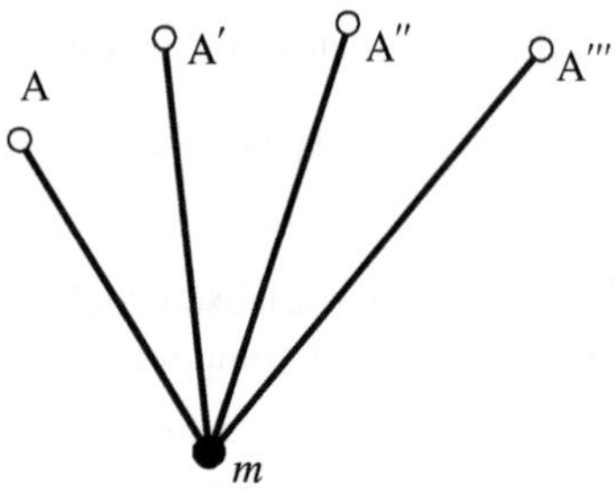

Fig. 1.9 System of wires supporting a material point (Redrawn from [130], vol 2. Fig. 28, Pl. 2)

its three components will be $X = 0$, $Y = 0$, $Z = g$. Call $\epsilon, \epsilon', \epsilon'', \epsilon'''$, the extensions that the four wires l, l', l'', l''', would assume if the weight mg were hung vertically at the lower end, and be $\zeta, \zeta', \zeta'', \zeta'''$, the extensions of these wires after the time t, during the motion; their forces in the same instant have values (n 288)

$$\frac{gm\zeta}{\epsilon}, \qquad \frac{gm\zeta'}{\epsilon'}, \qquad \frac{gm\zeta''}{\epsilon''}, \qquad \frac{gm\zeta'''}{\epsilon'''}.$$

Because the mobile m is no longer subjected to remain at the constant distances from A, A', A'', A''', we would suppress the terms of the Eq. (1.4),[111] which have $\lambda, \lambda', \lambda'', \lambda'''$, as coefficients, and which derive from these conditions; but on the other hand, one should add to the weight of this material point the four preceding forces, directed from m toward A, from m toward A', from m toward A'', from m toward A'''; this leads to substitute in the equations (4) the preceding values of L, L', L'', L''',[112] assuming in the same time:

$$\lambda = -\frac{gm\zeta}{\epsilon}, \qquad \lambda' = -\frac{gm\zeta'}{\epsilon'}, \qquad \lambda'' = -\frac{gm\zeta''}{\epsilon''}, \qquad \lambda''' = -\frac{gm\zeta'''}{\epsilon'''},$$

At the end of the time t, one thus has

$$x = \alpha + u, \qquad y = \beta + v, \qquad z = \gamma + w,$$

α, β, γ, being the same constants as before, and u, v, w, very small variables, of which we will neglect the square and the products; one has

$$\zeta = \frac{1}{l}\left[(\alpha - a)u + (\beta - b)v + (\gamma - c)w\right],$$
$$\zeta' = \frac{1}{l'}\left[(\alpha - a')u + (\beta - b')v + (\gamma - c')w\right],$$
$$\zeta'' = \frac{1}{l''}\left[(\alpha - a'')u + (\beta - b)v + (\gamma - c'')w\right],$$
$$\zeta'' = \frac{1}{l'''}\left[(\alpha - a''')u + (\beta - b)v + (\gamma - c''')w\right],$$

[111] The Eq. (1.4) are the three equations of motion of a material point m subjected to external forces and constraints $L = L' = L'' = 0$; we refer here to one of them as an example: $m\frac{d^2x}{dt^2} = mX + \lambda'\frac{dL}{dx} + \lambda\frac{dL'}{dx} + \lambda''\frac{dL''}{dx} + \text{etc.}$

[112] Previously Poisson had studied the motion under fixed constraints where L, L', L'', L''' were the constraint equations and $\lambda, \lambda', \lambda'', \lambda'''$ the constraint reactions.

and, with respect to the unknowns u, v, w, the Eq. (1.4) are linear, and reduce to

$$\frac{d^2u}{dt^2}+g\left[\frac{(\alpha-a)\,\zeta}{l\epsilon}+\frac{(\alpha-a')\,\zeta'}{l'\epsilon'}+\frac{(\alpha-a'')\,\zeta''}{l''\epsilon''}+\frac{(\alpha-a''')\,\zeta'''}{l'''\epsilon'''}\right]=0,$$
$$\frac{d^2v}{dt^2}+g\left[\frac{(\beta-b)\,\zeta}{l\epsilon}+\frac{(\beta-b')\,\zeta'}{l'\epsilon'}+\frac{(\beta-b'')\,\zeta''}{l''\epsilon''}+\frac{(\beta-b''')\,\zeta'''}{l'''\epsilon'''}\right]=0,$$
$$\frac{d^2w}{dt^2}+g\left[\frac{(\gamma-c)\,\zeta}{l\epsilon}+\frac{(\gamma-c')\,\zeta'}{l'\epsilon'}+\frac{(\gamma-c'')\,\zeta''}{l''\epsilon''}+\frac{(\gamma-c''')\,\zeta'''}{l'''\epsilon'''},\right]=0;$$

[...] If one assumes equal to zero the quantities u, v, w, and consequently suppresses the first terms of the last three of the preceding seven equations, the values of u, v, w, ζ, ζ', ζ'', ζ''', which one will deduce from these seven equations, will furnish the state of equailibrium of the weight m and of the four suspending wires [130].[113] (A.1.44)

The methods proposed by Clebsch, Poisson and Navier exploited the same mechanical idea: the equilibrium of the elastic forces concurring to a node, is obtained by evaluating the variation in length of the bars according to the displacements of the nodes. The treatments of Poisson and, especially, Clebsch, were quite general: instead of a specific case they refer to a generic node (therefore to infinite nodes); instead of the angles of the bars they use the coordinates of the nodes.

The methods for the calculation of trusses, despite the simplicity and elegance, did not see however a widespread use. The reason was the excessive number of linear equations to be solved, even for a very simple structure: in a truss of average size, with 20 nodes, one has to solve a system of 60 equations. The methods of forces, in which the equations equal to the number of unknowns, are generally less on condition that they deal with structures having low redundancy.

1.2.4 Variational Methods

The first practical methods for structural analysis came from the variational approach to mechanics, mainly based on the concept of work and energy, and were derived, although in a somewhat indirect way, by attempts to solve the problem of a body on more than three supports. We can say, in contrast to what Todhunter and Pearson thought, that the time spent in these attempts has been fruitful, leading to the development of methods based on minimization of the potential and complementary elastic energy.

The idea of potential in mechanics goes back at least to Lagrange, who in the *Théorie de la libration de la Lune* and the *Mécanique analytique* introduced the function $V=\int Pdp+Qdq+\cdots$, avoiding to naming it, implying a conservative force [82, 83].[114] Navier in 1821 used the term moment of a force to indicate the quantity $1/2f^2$, f being the distance between two molecules, proportional to the force

[113] vol 2, pp. 402–404. Our translation.

[114] p. 24; p. 69. Actually Lagrange used the symbol V in the *Théorie de la libration de la Lune* and the symbols Φ and Π respectively in the first and subsequent editions of the *Mécanique analytique*.

exchanged [114].[115] Green and Gauss introduced the term and the idea of potential to different fields; the former, in a paper of 1828 on static electricity used the term *potential function* [74];[116] the latter in a paper on capillarity used the term *potential* only [116].[117] Green went back to potential in 1839 [75] justifying its existence on the impossibility of perpetual motion. The use of the potential function of molecular forces in the theory of elasticity occurred in a majority of European countries, excluding France.

But the term potential spread thanks to William Thomson and Tait who in their textbook *Treatise on natural philosophy* [152] named the expression $1/2mv^2$ *kinetic energy*, and William Thomson who in his work of thermodynamics used *mechanical energy* and *intrinsic energy* to indicate the mechanical value of the effect the body would produce passing from the state in which it is given to the standard state [151].[118] William Thomson was the first to prove on a thermodynamical basis the existence of what is now called the elastic potential energy for a linear elastic system which deforms isothermally [151]. In 1855 William Thomson used *potential energy* (énergie potentielle) to distinguish it from *kinetic energy* (énergie actuelle) [151].[119]

An important contribution to the use of the energetic method in the theory of elasticity is that by Clapeyron, who expressed the work of the internal forces of elastic bodies in a form which was referred to as *Clapeyron theorem* in Lamé's *Leçons sur la théorie mathématique de l'elasticité des corps solides* [86]:

> When a force pulls or presses a solid body, of which at least three points are fixed, the product of this force by the projection on the direction of the total displacement that its point of application has taken is the double of the work made, from the instant when the displacement and force were zero, to one where the displacement and force have reached their final values. [...] Mr. Clapeyron found another expression of the work, which involved all the elastic forces developed in the interior of the solid body. The equality of these two expressions is a theorem, or rather a principle similar to that of the living forces, and that seems to have equal importance for applications. [...] we easily arrive at the equation

$$\text{(2)} \qquad \sum(Xu+Yv+Zw)\varpi = \iiint dx\,dy\,dz \left\{ \begin{array}{l} N_1\dfrac{du}{dx} + T_1\left(\dfrac{dv}{dz}+\dfrac{dw}{dy}\right) \\ + N_2\dfrac{dv}{dy} + T_2\left(\dfrac{dw}{dx}+\dfrac{du}{dz}\right) \\ + N_3\dfrac{dw}{dz} + T_3\left(\dfrac{du}{dy}+\dfrac{dv}{dx}\right) \end{array} \right\}.$$

> The first member [of the equality] is the sum of the products of the components of the forces acting on the solid surface by the projections of the displacements undergone by their points of application, this is the first-known expression [...] of double of the work of deformation, and the second member is another expression.

[115] p. 386.

[116] pp. 1–82.

[117] p. 784.

[118] p. 57.

[119] p. 1197.

When the body is homogeneous and of constant elasticity [...] the bracket in the second member of the equation (2) [...] can be put in the form

$$(4) \qquad \left\{ \begin{array}{c} \dfrac{1+\dfrac{\lambda}{\mu}}{3\lambda+2\mu}\left(N_1^2+N_2^2+N_3^2\right) \\ -\dfrac{1}{\mu}\left(N_1N_2+N_2N_3+N_1N_3-T_1^2-T_2^2-T_3^2\right) \end{array} \right\}.$$

To simplify assume,

$$(5) \qquad \left\{ \begin{array}{c} N_1+N_2+N_3=F, \\ N_1N_2+N_2N_3+N_1N_3-T_1^2-T_2^2-T_3^2=G, \end{array} \right.$$

and remembering the value of the coefficient of elasticity E, [...] the equation (2) takes the form

$$(6) \qquad \sum(Xu+Yv+Zw)\varpi = \iiint\left(EF^2-\frac{G}{\mu}\right)dx\,dy\,dz.$$

It is this equation which constitutes the theorem of Mr. Clapeyron. Note that [...] F, G, and as consequence, the bracket $\left(EF^2-\dfrac{G}{\mu}\right)$ retains the same numerical values when the coordinate axes are changed. That is to say that this bracket [...] is twice the inner work [...], and the half of the second member of the equation (6) is the sum of the work of all the elements, or the work of the whole volume of the body. It is in such a way that all the developed elastic forces contribute to form the second expression of the work of deformation [86].[120] (A.1.45)

Clapeyron published the applications of his theorem in 1857 [41], referring for the theory to Lamé's text [86]. Today one calls *Clapeyron's theorem* a statement different from the original:

Clapeyron's theorem states that the total strain energy stored in a body during linear elastic deformation is equal to the half of the work done by external forces acting on the body [135].[121]

In fact Clapeyron did not seem to see any problems about the conservation or equality between the works of the external and internal forces: the preceding quotation probably would have been trivial to him. The important thing was to find the expression of the work of the internal forces (in modern terms, the energy of deformation). Saint Venant presented Clapeyron's theorem in this form:

Mr. Clapeyron's theorem, properly stating, consists in the fact that the work under object is expressed, with our symbols, as

$$\frac{1}{2}\left(p_{xx}\partial_x+p_{yy}\partial_y+p_{zz}\partial_z+p_{yz}g_{yz}+p_{zx}g_{zx}+p_{xy}g_{xy}\right).$$

We put 1/2 because this work is produced by forces whose intensities start from zero and continuously grow [116].[122] (A.1.46)

---

[120] pp. 79–83. Our translation.

[121] p. 217.

[122] pp. CXCVII–CXCVIII. Our translation.

1.2.5 Applications of Variational Methods

The first attempts of application of variational methods to structural analysis include those of James Clerk Maxwell and James Henry Cotterill, the former a renowned scientist, the latter a professional engineer little known today. Both wrote more or less in the same time in the prestigious Philosophical Magazine, generally not read by engineers, even though native English speakers. Maxwell developed formulas for the calculation of the trusses, was the forerunner of Castigliano, applying what today is known as the theorem of minimum complementary energy in simple elements subjected to bending. Their contribution to the theory of elasticity can be evaluated differently depending on whether one wants to judge the quality of their work or the influence of it in the years to follow. From the first point of view, one cannot but affirm their excellence, from the second point of view one could instead say that it was irrelevant. This is the problem of the *precursors*.[123]

But it is only with Menabrea and Castigliano that the variational methods found widespread applications near the engineers. In the following, we illustrate the contributions of Maxwell and Cotterill, deferring to a later section and to Chap. 4 for a detailed discussion of the contributions of Menabrea and Castigliano.

1.2.5.1 James Clerk Maxwell and the Method of Forces

While Saint Venant was proposing his version of the method of forces, Maxwell introduced another version of general type, based on Clapeyron's theorem, which led to the same equations obtained later by Mohr (see below) with the principle of virtual work.

Maxwell also assumed the trusses as a prototype, and his objective was to determine the forces of the bars when they are redundant. He declared that for the solution of the elastic problem one should add to the equations of static equilibrium as many equations of kinematic compatibility—relative to the elastic deformation—as

[123] In the history of science there are people who are ahead of one's time, anticipating theories. Some are precursors only in appearance and look like such to us because we are strangers to the cultural climate of the time. When the precursors are real, their lack of success depends on contingent reasons, such as poor prestige enjoyed and the publication of pioneering studies in journals not known to those who might appreciate them. For those who understand the history of science in a cumulative sense, precursors disturb the linear path that one wants to follow. For others, the study of the precursors is of interest, although not central: the understanding of their ideas is helpful in understanding the cultural climate of the time. Maxwell and Cotterill should be considered from this second point of view. Together with precursors one must also consider successors, those who come to a result after its spread among the specialists. This stems from the conditions of cultural isolation; also the study of successors is interesting in understanding how scientific ideas grow in a given cultural climate. As part of the history of structural mechanics, among the successors we believe to deserve being mentioned are Wilhelm Fränkel [69] who regained results of Menabrea and Castigliano; and Friedrich Engesser [62] who regained Crotti's results (see Chap. 4) and introduced the term *complementary work* [51]. Engesser defined complementary work as the difference between actual and virtual work.

unknown forces. For the calculation of the elastic deformations Maxwell formulated the following theorems:

> *Theorem* [1].—If p is the tension of the piece A due to a tension-unity between the points B and C, then an extension-unity taking place in A will bring B and C nearer by a distance p [45].[124]

> *Theorem* [2].—The extension in BC due to unity of tension along DE, is always equal to the extension in DE due to unity of tension in BC [45].[125]

Their proof is based on Clapeyron's theorem, seen by Maxwell as a possible formulation of the principle of conservation of energy, going so ahead of Clapeyron's intentions:

> The method is derived from the principle of Conservation of Energy, and is referred to in Lamé's Leçons sur l'elasticité, Leçon 7me, as Clapeyron's Theorem [45].[126]

The second of the two Maxwell's theorems, known as the *Maxwell reciprocal theorem* or simply the *Maxwell theorem*, is today reformulated in a more general way and assumed to be valid for all (hyper) elastic systems in the form:

> The displacement at a point 1 in a given direction due to a unit load at a point 2 in a second direction is equal to the displacement at the point 2 in the second direction due to a unit load at the point 1 in the given direction [96].[127]

Maxwell solved, with the aid of the first theorem, the following problems:

> *Problem* I—A tension F is applied between the points B and C of a frame which is simply stiff [statically determined]; to find the extension of the line joining D and E, all the pieces except A being inextensible, the extensibility of A being e [45].[128]

> *Problem* II—A tension F is applied between B and C; to find the extension between D and E, when the frame is not simply stiff, but has additional pieces R, S, T, &c., whose elasticities are known [45].[129]

Maxwell's writings present difficulties in reading due to the lack of figures and the use of a not very happy formalism, different from today. Figure 1.10 is thus introduced for a better understanding of both the meaning and the application of the first problem.

The solution to Problem I (determining the relative displacement between the nodes D and E of a statically determined beam, imagining that only one arbitrary bar a is deformable), was given by Maxwell in the form:

$$u_{DE} = -Fepq, \tag{1.29}$$

where the product pF is the normal force N in the bar a due to the force F, q the normal force N' in a due to unit and opposite forces applied in D and E. The extensibility

---

[124] p. 296.

[125] p. 297.

[126] p. 294.

[127] p. 456.

[128] p. 296.

[129] p. 296.

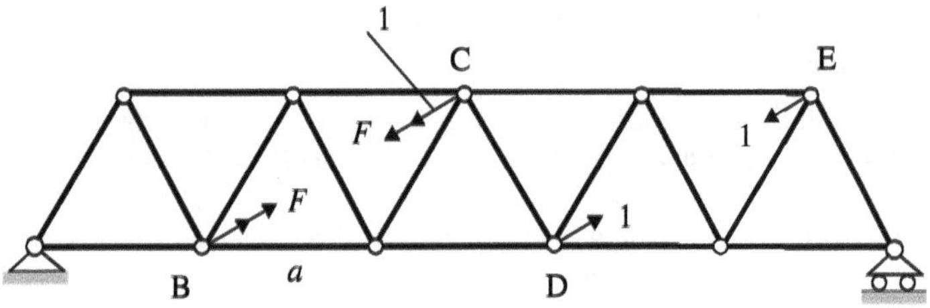

Fig. 1.10 Relative displacements in a statically determined truss

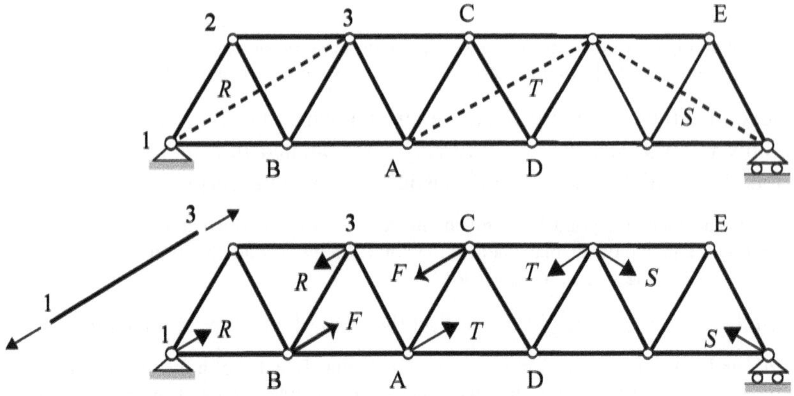

Fig. 1.11 Relative displacements in a statically indeterminate truss

e of the bar is the quantity l/EA, later called 'elastic weight' by Mohr, being l the length of the bar, E the longitudinal modulus of elasticity and A the area of the cross section of the bar. If all the bars are deformable, the relative displacement between D and E is given by:

$$u_{DE} = -\sum F(epq), \tag{1.30}$$

or with a modern notation by:

$$u_{DE} = \sum_i \left(\frac{NN'}{EA} l \right)_i . \tag{1.31}$$

For the solution of Problem 2 (Fig. 1.11), Maxwell wrote:

1st. Select as many pieces of the frame as are sufficient to render all its points stiff. Call the remaining pieces R, S, T, &c.

2nd. Find the tension on each piece due to unit of tension in the direction of the force proposed to be applied. Call this the value of p for each piece.

3rd. Find the tension on each piece, due to unit of tension in the direction of the displacement to be determined. Call this the value of q for each piece.

4th. Find the tension on each piece due to unit of tension along R, S, T, &c., the additional pieces of the frame. Call these the values of r, s, t, &c. for each piece.

5th. Find the extensibility of each piece and call it e, those of the additional pieces being ρ, σ, τ, &c.

6th. R, S, T, &c. are to be determined from the equations:

$$R\rho + R\sum(er^2) + S(ers) + T\sum(ert) + F\sum(epr) = 0$$
$$S\sigma + R\sum(ers) + S(es^2) + T\sum(est) + F\sum(eps) = 0$$
$$T\tau + R\sum(ert) + S(est) + T\sum(et^2) + F\sum(ept) = 0$$

as many equations as there are quantities to be found.

7th. x, the extension required, is then found from the equation [45].[130]

$$x = -F\sum(epq) - R\sum(erq) - S\sum(eqs) - T\sum(eqt)$$

Maxwell's equations of point 6 express the kinematic compatibility (the extension of the bars equals the variation in the distance of their nodes) and presently would be written as:

$$\sum_{i=1}^{n}\left(\frac{(N_0 + \sum_{j=1}^{m} X_j N_j)N}{EA}l\right)_i = -\left(\frac{X}{EA}l\right)_i, \tag{1.32}$$

with X_j the redundant reactions, N_0 and N_j respectively the tension in the i-th bar due to the external loads and to $X_j = 1$, n the total number of bars and m the number of redundant bars.

Note that Maxwell never explicitly considered the external forces applied to individual nodes or their absolute displacements (as assumed in current textbooks of the strength of materials), but opposing forces and relative displacements between nodes. In order to address the case of displacement of individual nodes he introduced the artifice of a fictitious bar, whose stiffness is negligible, that connects the node whose displacement should be evaluated to an auxiliary node.

Maxwell's work remained largely unknown, with the exception of a quotation in a work of 1869 by Fleeming Jenkin [66], professor at the university of Edinburgh, who applied a variant of Maxwell's procedure for the solution of trusses [40].[131] After the introduction by Mohr of a method based on virtual work, the work of Maxwell was rediscovered, for example by Heinrich Müller-Breslau, who with some polemical vein against Mohr, stressed his priority [40].[132]

---

130 p. 298.

131 pp. 81–83.

132 Chapter 10.

1.2.5.2 James H. Cotterill and the Minimum of Energy Expended in Distorting

James Henry Cotterill was professor of applied mechanics at the Royal naval college at Greenwich from 1873 to 1897. In three works [46–48] he deduced both the theorem of Castigliano, preceding him in the statement, and the principle of Menabrea (see Chap. 4), preceding the latter in perfecting the proof of the *équation d'élasticité* [39].

Like any precursor, Cotterill developed an independent mode of thought. His reference was not Euler but Henry Moseley, professor of natural philosophy and astronomy at King's College of London,[133] who had introduced the *principle of least resistance* in the study of arches.[134]

Cotterill introduced formulas for the elastic strain energy, referred to by the expression *work done* or *energy expended in distorting*, without posing the problem if the forces were conservative or not. Admitting the flexural deformation only, he recognized that the energy expended in distorting has an expression [46][135]:

$$U = \int \frac{M^2}{2EI} dx, \tag{1.33}$$

where the symbols, today standard, are Cotterill's. For a beam of length $2c$ subjected to a transverse load w and two torques M_1, M_2 concentrated at the ends, assuming the equilibrium in the transverse direction and then considering a supported beam, the "energy expended in distorting" takes the simple quadratic form [46][136]:

$$U = \frac{c}{3EI} \left\{ M_1^2 + M_1 M_2 + M_2^2 - wc^2 (M_1 + M_2) + \frac{2}{5} w^2 c^4 \right\}, \tag{1.34}$$

with U expressed as a function of external forces.

The proofs of the theorem of Castigliano and the principle of Menabrea are very simple, with reference to a structural system of which one can express the elastic energy as a function of external forces (Cotterill, however, applied his results to bent elements only). From the *law of conservation* (or the theorem of Clapeyron) [47][137] Cotterill could write:

$$2U = \sum \{Xu + Yv + Zw\}, \tag{1.35}$$

and by differentiating, he obtained:

$$2\,\delta U = \sum \{X\,\delta u + \delta\, Yv + Z\,\delta w + \delta\, Xu + \delta\, Yv + \delta\, Zw\}. \tag{1.36}$$

[133] He was the lucky author of *The mechanical principles of engineering and architecture* [110]; see [153] on pp. 212–214.

[134] The principle of least resistance by Moseley is: "If any number of pressures are equilibrated, some of which are resistances, then each of these resistances is a minimum, subjected to the conditions imposed by the equilibrium of the whole" [108, 109], p. 178, [46], p. 299.

[135] p. 300.

[136] p. 301.

[137] p. 303.

> but $\sum\{X\,\delta\,u+Y\,\delta\,v+Z\,\delta\,w\}$ is the increment of energy expended, which by the law of conservation of energy is equal to $\delta\,U$, the increment of the work done, therefore we have also [47][138]:
>
> $$\delta\,U=\sum\{X\,\delta\,u+\delta\,Yv+Z\,\delta\,w\}\,.$$

Then, he concluded, the equations

$$\frac{dU}{dX}=u; \qquad \frac{dU}{dY}=v; \qquad \frac{dU}{dZ}=w \tag{1.37}$$

hold, which expresses the first Castigliano's theorem (see Chap. 4).

Note that in the above steps the only questionable aspect is the admission of the law of conservation of energy; the other steps are rigorous. In the case of statically indeterminate systems, the elastic energy can be expressed as a function of the redundant reactions. By indicating with X, Y, Z the redundant reactions, if the constraints do not allow yieldings, and then $u=v=w=0$, the Eq. (1.37) gives:

$$\frac{dU}{dX}=0; \qquad \frac{dU}{dY}=0; \qquad \frac{dU}{dZ}=0. \tag{1.38}$$

> Since, then, the change in U, consequent on any possible change in the resisting forces, is zero, U must be a minimum (the other hypotheses being easily seen to be inadmissible), and the principle is proved for a perfectly elastic body or system of bodies [46].[139]

The same considerations, Cotterill continued, hold if X, Y, Z are internal forces: in this case u, v, w represent the relative displacements of the section of the beam which (in the absence of distortions) must vanish.

Cotterill did not limit to state a theory but he also gave applications; among them the case of a beam clamped at its ends and subjected to a uniformly distributed load [46].[140] The energy expended in distorting, in this case, is given by the equation (1.34), where M_1, M_2 are, with Cotterill's symbols, the unknown bending moments at the ends. One more important case is that concerning the arch, for which in the expression of the elastic energy, he also accounted for the axial deformability [47][141]:

$$\int\left\{\frac{M^2}{2EI}+\frac{H^2}{2EA}\right\}ds, \tag{1.39}$$

where H is the axial force and s the curvilinear abscissa.

---

[138] p. 388–389.

[139] p. 305.

[140] pp. 301–302.

[141] p. 383.

1.2.6 Perfecting of the Method of Forces

In the middle of the 19th century, interesting techniques of solution, to be classified as methods of forces, were developing in France [90] and Germany [103, 104]. Both took the move, at least ideally, from the work of Navier in 1826 [115]. It was now clear that to solve the statically indeterminate systems one needs the equations of equilibrium, the compatibility relations and the constitutive law which, combined in an appropriate manner, provide methods of solution either of forces or displacements. Navier did not make a clear choice between the two methods, perhaps because at the time the issue was not mature. Stronger would be the choice of Poisson and Clebsch (displacements), already presented, and that of Lévy and Mohr (forces), to which the following section is dedicated.

In the method of forces, the most technically complex part is given by the establishment of compatibility, addressed differently by Lévy and Mohr. The former used geometry and tried to find a general approach, partly succeeding for truss systems. The latter transformed the geometric problem into a static one using the principle of virtual work.

1.2.6.1 Lévy's Global Compatibility

Maurice Lévy proposed a general method of solution, limited however to trusses only, in 1873 [90] and perfected it in 1874 [91] slightly different from the classical methods of forces. Below the essence of Lévy's method:

> Here is the general rule to which I arrived:
>
> Given a figure (plane or not) formed by bars hinged at their ends, and to the points of articulation of it is applied a system of any forces maintaining them in equilibrium, to find the tensions developed in the various bars we start by writing that each point of articulation is separately in equilibrium under the action of the external forces that are applied to it and the tensions of the bars that, in any numbers, concur in it. If one gets as many equations separate as many unknown tensions, the problem is solved by the pure statics (1). If k equations are missing, one can be certain that the geometric figure formed by the axes of the bars contains k overabundant lines, that is k lines more than strictly necessary for the definition [of the figure]; that, as a result, between the lengths of the lines that compose it, that is between the lengths of bars, there are necessarily k geometric relationships (it is a problem of elementary geometry). Write these relations, differentiate them by looking at all the lengths as variable; replace the differential by letters representing the elastic elongations of the bars; replace in turn these elastic elongations by their expressions depending on the tension and the coefficients of elasticity of bars (2); one so has k new equations which these tensions must satisfy, that with the equations already provided by statics, will form a total equal to that of tensions to determine [90].[142] (A.1.47)

Thus, according to Lévy, in a truss k times statically indeterminate one can write k independent relations for the position of the nodes. Based on this assumption he proceeded as follows:

[142] pp. 1060–1061. Our translation.

> Be
>
> $$a_1,\ a_2,\ a_3, \ldots, a_m$$
>
> the lengths of m bars in the natural state, that is when no forces act on them.
> Under the effect of the forces applied to the various nodes, the bars undertake the extensions
>
> $$\alpha_1,\quad \alpha_2,\quad \alpha_3,\quad \ldots,\quad \alpha_m$$
>
> so the new lengths become
>
> $$a_1+\alpha_1,\quad a_2+\alpha_2,\quad a_3+\alpha_3,\quad \ldots,\quad a_n+\alpha_m$$
>
> since among these relations k algebraic relations subsist, let
>
> $$F(a_1+\alpha_1, a_2+\alpha_2, a_3+\alpha_3, \ldots, a_n+\alpha_m)=0.$$
>
> be one of these relations [91].[143] (A.1.48)

For displacements small with respect to the reference configuration, the last relation of the above quotation gives:

$$\frac{dF}{da_1}\alpha_1+\frac{dF}{da_2}\alpha_2+\cdots+\frac{dF}{da_n}\alpha_n=0 \tag{1.40}$$

which expresses the compatibility of the deformations α_1, α_2, etc., that is their geometric compatibility.

The elastic relationship allows us to write

$$\alpha_i=\frac{a_i t_i}{E_i S_i}, \tag{1.41}$$

where t_i is the axial force, a_i the length, S_i the transverse section and E_i the modulus of elasticity of the i-th bar. Replacing α_i in the compatibility Eq. (1.40), Lévy obtained:

$$\frac{dF}{da_1}a_1\frac{t_1}{E_1S_1}+\frac{dF}{da_2}a_2\frac{t_2}{E_2S_2}+\cdots+\frac{dF}{da_n}a_n\frac{t_n}{E_nS_n}=0 \tag{1.42}$$

and concluded:

> Such are the k relations to join to those furnished by statics to define the tensions t_i [91].[144] (A.1.49)

Equation (1.42) is a compatibility equation written as a function of the forces t_i. By assumption there are k of such equations which added to the equilibrium equations of nodes, allows the solution of the statically indeterminate problem. Note that the forces, at least in principle, are calculated all together, without first determining the redundant ones as in the traditional methods of forces. However the

---

[143] vol. 4, pp. 210–211. Our translation.

[144] vol. 4, p. 211. Our translation.

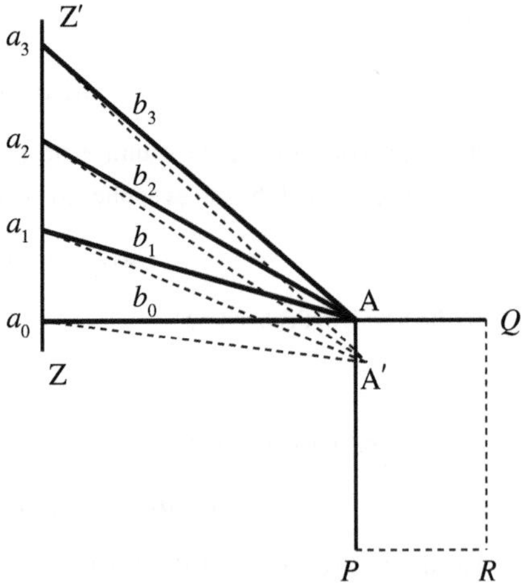

Fig. 1.12 Redundant system of bars of Lévy (Redrawn form [91], vol. 4, Fig. b, Pl. XLIV)

compatibility equations, written by selecting the redundant bars, will in practice have a local meaning, related to the nodes connected to a single redundant bar.

Lévy presented the example of a structure two times statically indeterminate, shown in Fig. 1.12, in which, for an assigned load P, one wants to evaluate the axial forces t_i, $i = 1, \ldots, 4$ of the four bars. For the sake of simplicity he assumed that the nodes to which the structure is constrained to the wall have the same distance $a = a_0 - a_1 = a_1 - a_2 = a_2 - a_3$ among them. If β_i represents the variation of the lengths of each bar, Lévy derived the following relations:

$$(b_i + \beta_i)^2 + (b_{i+2} + \beta_{i+2})^2 = 2(b_{i+1} + \beta_{i+1})^2 + 2a^2 \quad i = 1, 2. \tag{1.43}$$

and, differentiating and ignoring the second order terms:

$$b_i\beta_i + b_{i+2}\beta_{i+2} = 2b_{i+1}\beta_{i+1} \quad i = 1, 2. \tag{1.44}$$

Thus, by replacing in the Eq. (1.44) the costitutive relationships:

$$\beta_i = \frac{b_i t_i}{E_i S_i}. \tag{1.45}$$

one can write the two linear and independent algebraic equations of compatibility ($i = 0,\ 1$):

$$b_i \frac{b_i t_i}{E_i S_i} + b_{i+2} \frac{b_{i+2} t_{i+2}}{E_{i+2} S_{i+2}} = 2b_{i+1} \frac{b_{i+2} t_{i+1}}{E_{i+1} S_{i+1}}.$$

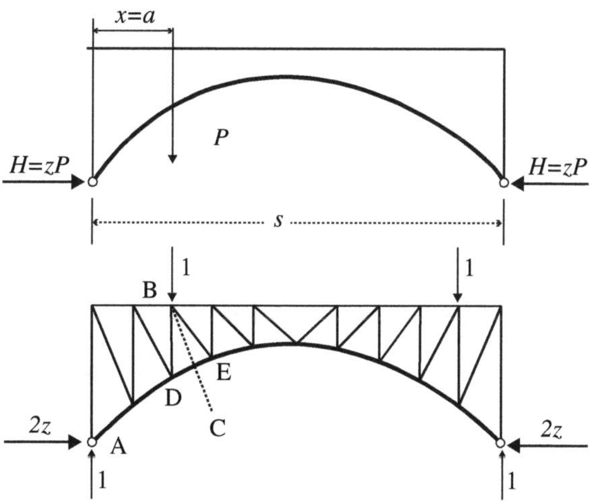

Fig. 1.13 Mohr's trusses (Redrawn from [104], Fig. 3 *left*, Fig. 6 *right*, col. 223)

in the unknown axial forces t_i of the bars [91];[145] by adding these two equations to the two equations of statics obtained by imposing the equilibrium to translation of the node A, the problem is solved.

1.2.6.2 Mohr and the Principle of Virtual Work

Mohr provided fundamental contributions to the theory of structures in a few papers with the same title that appeared between 1860 and 1868 [103]. In the last of these the so-called Mohr's analogy is reported, that sees the deformed elastic axis of the beam as the bending moment diagram of a fictitious beam, properly loaded and constrained. Mohr adapted this procedure to trusses [103]. Interesting is the use he made of the center of elasticity [54] to simplify the analysis of statically indeterminate systems [106].

Mohr's most important contribution to mechanics of structures is, however, the proposal of a method of forces that with some refinement is still used today for an educational purpose for solving statically indeterminate structures [104, 105] and consists in the identification of redundant constraints (in trusses, the redundant bars too), replace them with unknown reactions and write a compatibility equation for each constraint. The specificity of Mohr's method lies in the fact that the displacements which must satisfy the compatibility equations are calculated using the principle of virtual work.

In two articles of 1874 Mohr [104, 105] posed the basis of the method. He considered the reticular arch, hinged at its ends of Fig. 1.13, one time statically indeterminate subjected to the load P, for which the thrust $H = zP$ is searched, with z the thrust corresponding to $P = 1$. For the sake of simplicity Mohr considered the arch

[145] vol. 4, p. 213.

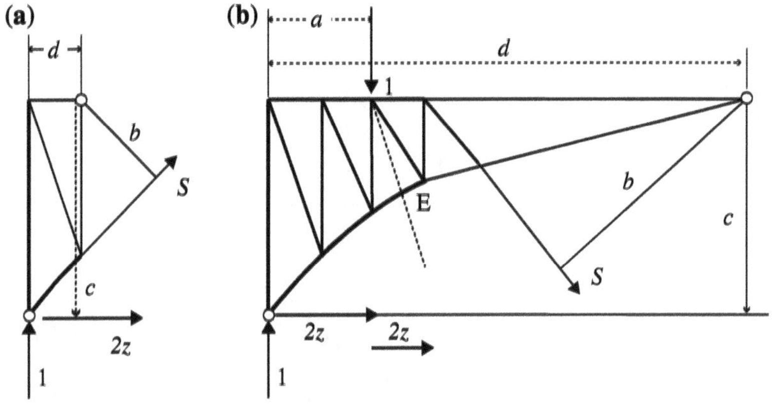

Fig. 1.14 Forces in the arch to the left of the load (**a**) and to the right of the load (**b**) (Redrawn from [104], Fig. 7 *left*, Fig. 8 *right*, col. 223)

symmetrically loaded, assuming $H = 2zP$. Without any comment about the employed method, Mohr analyzed this structure to evaluate the forces in the bars due to the load $P = 1$ and the thrust H.

With reference to Fig. 1.14, for $P = 1$ (that is for a thrust $2z$) Mohr, by means of the equilibrium of moments, obtained the forces in the arch and in the diagonals [104][146]:

$$\begin{aligned} S &= \left(\pm 2z\frac{c}{b} \pm \frac{d}{b} \right) \\ S &= \left(\pm 2z\frac{c}{b} \pm \frac{a}{b} \right). \end{aligned} \tag{1.46}$$

The first equation holds in the portion comprised between the support and the point of application of the load; the second equation in the other portion. The arms b and c of the force S and of the reaction of the support are measured from an appropriate pole; more precisely for a bar on the left of the load $P = 1$, b, d are given as in Fig. 1.14a, for a bar on the right of the load $P = 1$, b, d are given as in Fig. 1.14b. The bars in Fig. 1.14 represent an example; but the relations (1.46) are quite general.[147]

For the sake of simplicity Mohr introduced the quantities u, v, w:

Denote the numerical values independent of the point of application of the load:

$$\begin{aligned} &\pm\frac{c}{b} \quad \text{with} \quad u \\ &\pm\frac{d}{b} \quad \text{with} \quad v \\ \text{and} \quad &\pm\frac{1}{b} \quad \text{with} \quad w, \end{aligned}$$

[146] col. 225.

[147] The difference between external and internal bars concerns the contribution of the load $P = 1$, which for the internal bars gives the moment $a \cdot 1 = a$, constant with respect to any pole.

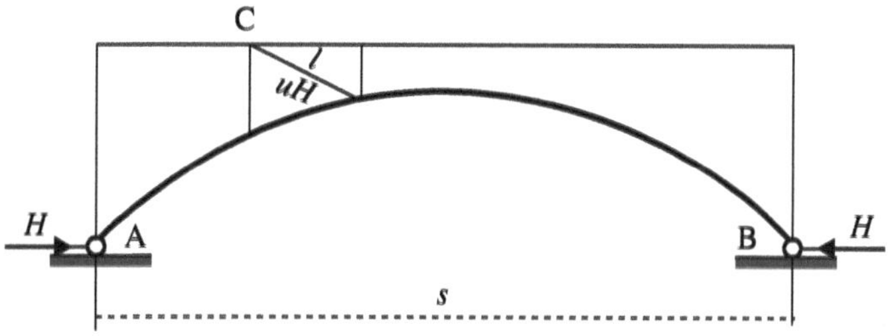

Fig. 1.15 Evaluation of forces in the bars (Redrawn from [104], Fig. 21, col. 223)

so that for each element of the truss between $x = 0$ and $x = a$, it will be

$$S = (2zu + v) \quad \text{Tons}$$

so that for any element between $x = a$ and $x = 1/2$,

$$S = (2zu + aw) \quad \text{Tons}$$

[104][148] (A.1.50)

Mohr carried out a kinematic analysis with the purpose of writing a compatibility equation. He obtained the variation of length Δl of the bars by multiplying the forces S in each of the bar by their *elastic weight* $r = l/EF$, with l the length, E the elastic modulus, F the area of the cross section of the bar. The variation in the length of the bar, the forces on which are given by the relations (1.46), are:

$$\begin{aligned} \Delta l &= (2rzu + rv), \\ \Delta l &= (2rzu + raw). \end{aligned} \tag{1.47}$$

With reference to Fig. 1.15, the variation of the initial span s of the truss is obtained with the superposition of effects, considering the elongation Δl of one bar at a time: each contributing to the variation Δs of the span. By adding all the contributions Mohr obtained the equation of compatibility to solve the problem:

Image the truss made so that the support is free to slide in the horizontal direction and the changes in length of the individual elements may occur not at the same time but one after the other. Each variation of length Δl of an element therefore entail a certain variation Δs of the span s, which depends on the geometry of the truss. The sum of all the values of Δs from all the elements must be identically zero, because in reality the span does not change its value. Moreover, since the deformation of the truss considered here is symmetric with respect to the to the center line of the truss, then the sum of the values of Δs should be identically zero for a half of the lattice too [104][149]: (A.1.51)

[148] col. 226. Our translation.

[149] col. 229–230. Our translation.

$$\sum_{x=0}^{x=1/2} \Delta s = 0.$$

The relation among the Δs and Δl associated to the single bar is obtained with the principle of virtual work (referred by Mohr as the principle of virtual velocities, *Princip der virtuellen Geschwindigkeit*), by imposing the equality between the work spent by a system of forces equilibrated with the thrust $H = 1$ and the work spent by $H = 1$, in the actual displacements:

> One can also produce this displacement by means of a horizontal thrust H on the support, which, as a result of what discussed above, produces in the elastic bar CD the stress $u \cdot H$. While the force H passes the space Δs and accordingly performs mechanical work $-H \cdot \Delta s$,[150] the resistant stress $u \cdot H$ of the bar CD will act on the path Δs and thus will absorb the mechanical work $u \cdot H \cdot \Delta l$.[151] Following the principle of the virtual velocities these works are equal and therefore:
>
> $$-H \cdot \Delta s = u \cdot H \cdot \Delta l$$
>
> or
>
> $$-\Delta s = u \cdot \Delta l.$$
>
> By inserting the value of Δl one obtains
>
> $$0 = 2z \sum_0^{s/2} ru^2 + \sum_0^{a} rvu + \sum_a^{s/2} rawu$$
>
> or
>
> $$-z = \frac{\sum_0^{a} rvu + \sum_a^{1/2} rawu}{2\sum_0^{s/2} ru^2}.$$
>
> [104].[152] (A.1.52)

Mohr obtained his solving formula, similar to the current one, following a geometric approach and a visual inspection of the results, documented by an extended use of figures. In this perspective, the overall change in Δs is seen as a superposition of rigid kinematic mechanism due to the deformation of a single bar: nothing would prevent a completely geometric analysis (without the use of the virtual works). His symbolism is different from the modern one: the distinction between bars to the left

[150] We assigned a positive sign to the horizontal thrust H; the shortening Δs is negative, so $-H \cdot \Delta s$ is a positive magnitude (original note by Mohr).

[151] The quantity $u \cdot H \cdot \Delta l$ is always positive, since Δl is a lengthening or a shortening whenever $u \cdot H$ is a tensile or compressive stress. The quantities Δl and $u \cdot H$ have so in this paper always the same sign (original note by Mohr).

[152] col. 231–232. Our translation.

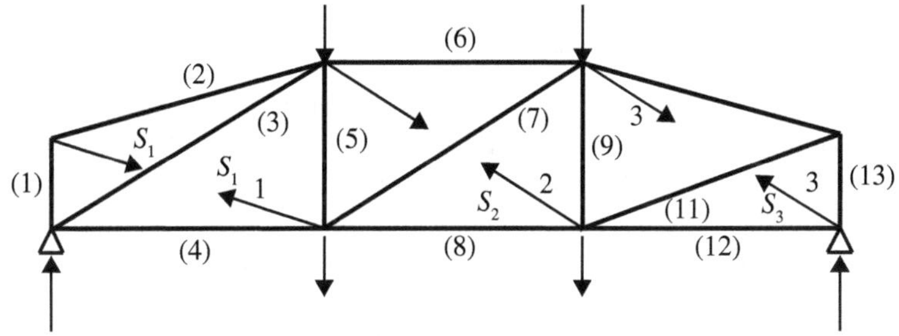

Fig. 1.16 Many times statically indeterminate truss (Redrawn from [105], Fig. 3, col. 509)

and to the right of the point of load application only makes sense if one wants to keep in touch with an example. A more general relationship would be, for instance, indicating with v the forces in the bars produced by $P = 1$:

$$-z = \frac{\sum rvu}{2\sum ru^2} \tag{1.48}$$

where the sum is extended to all the bars.

Mohr overcame these limits in a subsequent work [107] where he dealt with the beam truss many times statically indeterminate of Fig. 1.16.

The truss (Fachwerk) is defined as a set of hinged elements subjected only to variation of length, whose configurations can thus be identified by the coordinates of the nodes-hinges only. Depending on the number of nodes and elements and external constraints, Mohr defined the minimum number of bars able to form a statically determined truss (einfach, that is simple truss).

For a statically determined truss, Mohr evaluated the relative displacements of internal nodes or displacements between an internal node and a fixed node, due to the load P. To this purpose he imagined to disconnect two nodes whatever, cutting the bar connecting them and replacing it with a force $u \cdot P$ carried by it. The parameter u is non-dimensional and, as in the preceding work, varies for each bar and can be evaluated algebraically or graphically:

> The evaluation of the support reactions and of the stresses is carried out analytically or graphically with the use of very simple methods that we can assume as known [105].[153] (A.1.53)

This cut makes the truss become a mechanism ("eine einfache Maschine") performing a rigid motion, so that one can use again the principle of virtual velocities, by obtaining:

$$\Delta y = -u \cdot \Delta l, \tag{1.49}$$

where Δy is the length variation of the bar a due to the variation Δl of the generic bar. The whole variation we seek is thus given by the summation of all the contributions of the bars:

[153] col. 512. Our translation.

$$\sum \Delta y = -\sum u \cdot \Delta l. \tag{1.50}$$

In the statically indeterminate systems one separates a simple truss made with the minimum number of bars ("notwendige Konstruktionstheile"), known today as the main structure. The redundant bars are then replaced by the forces they carry, the unknown redundant forces ("unbekannte Spannungen") $S_1, S_2, S_3, \ldots$. These forces, together with the external forces, should be such that the variation of the lengths of the suppressed bars (subjected to the forces $-S_1, -S_2, -S_3, \ldots$.) equals the variation of the distance among the nodes of the main structure to which they belong. Thanks to his Eq. (1.49) Mohr expressed these variations of lengths as:

$$\begin{cases} \Delta l_1 = -\sum u_1 \Delta l \\ \Delta l_2 = -\sum u_2 \Delta l \\ \Delta l_3 = -\sum u_3 \Delta l \\ \ldots \end{cases} \tag{1.51}$$

where the sum is extended to all the bars of the main structure; or:

$$\begin{cases} \sum u_1 \Delta l = 0 \\ \sum u_2 \Delta l = 0 \\ \sum u_3 \Delta l = 0 \\ \ldots \end{cases} \tag{1.52}$$

with the sum extended to all the bars, the suppressed ones included. In these it is indeed $u_1 = u_2 = u_3 = \cdots = -1$.

The actual kinematic mechanism is given by the superposition of rigid mechanisms associated to the deformation of the single bar. The force S of each bar is written in a general way, independent of the kind of structure, by superposing the effects, as functions of the unknown redundant forces S_1, S_2, S_3:

$$S = \mathfrak{S} + u_1 S_1 + u_2 S_2 + u_3 S_3 + \cdots \tag{1.53}$$

where $u_1, u_2, u_3, \ldots$ are the forces in the bars due to the systems of forces ($S_1 = 1$, $S_2 = S_3 = \cdots = 0$), ($S_1 = 0$, $S_2 = 1$, $S_3 = S_4 = \cdots = 0$), ecc.; $\mathfrak{S}$ is the force due to the applied loads. The solving system follows [105][154]:

> If one introduces these values of Δl in the relations derived from the equations (4) between the variation of the lengths of the redundant elements and those of the necessary elements, one obtains the relations:
>
> $$\begin{cases} \sum u_1 \cdot S \cdot r = 0 \\ \sum u_2 \cdot S \cdot r = 0 \\ \sum u_3 \cdot S \cdot r = 0 \\ \ldots \end{cases}$$

[154] col. 517–518.

and if one inserts the value of S as given by the equation (6):

$$\begin{cases} 0=\sum u_1\cdot\mathfrak{S}\cdot r + S_1\sum u_1^2\cdot r + S_2\sum u_1\cdot u_2\cdot r + S_3\sum u_1\cdot u_3\cdot r + \cdots \\ 0=\sum u_2\cdot\mathfrak{S}\cdot r + S_1\sum u_1\cdot u_2\cdot r + S_2\sum u_2^2\cdot r + S_3\sum u_2\cdot u_3\cdot r + \cdots \\ 0=\sum u_3\cdot\mathfrak{S}\cdot r + S_1\sum u_1\cdot u_3\cdot r + S_2\sum u_2\cdot u_3\cdot r + S_3\sum u_3^2\cdot r + \cdots \\ \cdots \end{cases}$$

Equations (9) allow the evaluation of the forces in the redundant elements [...] [105].[155] (A.1.54)

Mohr continued by considering the effect on the truss of thermal actions too; taking for granted that the effects of the external loads and the weight were considered in the previous parts, he wrote that the forces T due to the thermal dislocations of the bars are given by a relation analogous to his (1.53):

$$T = u_1T_1 + u_2T_2 + u_3T_3 + \cdots \tag{1.54}$$

where, of course, the term due to the load should be zero. The thermoelastic deformation is given by:

$$\Delta l = l\, t\, \mathfrak{d} + T\, r \tag{1.55}$$

where $\mathfrak{d}$ is the coefficient of thermal expansion of the material of the bars, t the value of the thermal variation and r is the elastic weight of the single bar, previously defined. By replacing the (1.54), (1.55) in the compatibility equations for the constraints, Mohr obtained the equations necessary to evaluate the forces T_i:

$$\begin{cases} 0 = \sum u_1\cdot l\cdot \mathfrak{d}\cdot t + T_1\sum u_1^2\cdot r + T_2\sum u_1\cdot u_2\cdot r + T_3\sum u_1\cdot u_3\cdot r + \cdots \\ 0 = \sum u_2\cdot l\cdot \mathfrak{d}\cdot t + T_1\sum u_1\cdot u_2\cdot r + T_2\sum u_2^2\cdot r + T_3\sum u_2\cdot u_3\cdot r + \cdots \\ 0 = \sum u_3\cdot l\cdot \mathfrak{d}\cdot t + T_1\sum u_1\cdot u_3\cdot r + T_2\sum u_2\cdot u_3\cdot r + T_3\sum u_3^2\cdot r + \cdots \\ \cdots \end{cases} \tag{1.56}$$

Mohr corroborated the discussion by presenting a series of numerical results for trusses of different form and structural function: the simply supported truss (Balkenfachwerk), the hinged (Bogenfachwerk) and the continuous one over several supports (kontinuirliches Fachwerk) [107].[156] In this latter work Mohr showed that for continuous trusses on more supports, his Eq. (1.56) can be traced back to the equation of three moments [107][157] (Clapeyronsche Gleichung). The most important contribution of this article is the graphical determination, through the construction of a figure similar to a funicular polygon, of the vertical displacements of the nodes

[155] col. 517–518. Our translation.

[156] col 17–38. 17–38. The article has a different typeface with respect to the preceding works with the same title: there is no longer the character 'Fraktur', a gothic font typical of the more traditional German literature.

[157] cols. 20–22.

of a statically determined truss [107].[158] A similar construction is the basis of the aforementioned Mohr's analogy for bent beams.

The problem of solution of frames with the method of forces was consistently framed by Müller-Breslau [112], who, among the first, clarified that the principle of virtual work can be used in two ways. In the former, already noted, one looks for equilibrium among the congruent deformations (methods of displacements), in the latter one looks for compatibilities among equilibrated forces (forces method). In his monumental work of 1887–1913 [112, 113], Müller-Breslau tackled the most complex systems of trusses and refined the theory of influence lines.

Müller-Breslau attributed to Maxwell the method introduced by Mohr, probably with a polemical tone. Indeed if it is true that Maxwell wrote something similar to Mohr's method already in 1864 [45], it is also true that the first to understand the scope of the method and to place it firmly among engineers was Mohr, who most likely arrived independently to Maxwell's result. In any event, Müller-Breslau in part recognized Mohr's merit:

> The first deduction of the equations and laws of *Maxwell* following the shortest route of the use of the principle of virtual displacements is due to Mohr. His *Contributions to the theory of trusses* in the Zeitschrift des Architekten-und Ingenieur Vereins zu Hannover in 1874 and 1875 contain the first important applications of the theory of Maxwell. Mohr also represented first, the elastic line equation for the straight beam and the line of inflection of a truss with the help of the funicular polygon [113].[159] (A.1.55)

1.3 The Italian Contribution

The studies of theory of structures until the late 19th century focused on continuous beams, arches and statically determined and indeterminate trusses. The analysis of these really complex structures required a considerable amount of calculations and a large part of the efforts of the engineers, particularly Germans, was directed to develop analytical and graphical procedures for an easier calculation that does not directly use the equations of statics.

At the same time a better understanding of the behavior of structures led to seeing that the model of a truss with nodes–hinges (idealization of the actual trusses in which the nodes are bolted or welded, thus substantially rigid) is not entirely satisfactory. One should thus enrich the model of calculation to take account of the rigid joints effectively existing. The problem was solved with both exact and approximated procedures, which go under the name of *theories of the secondary stress*, to which have contributed among others Engesser, Winkler, Ritter, Müller-Breslau, Mohr [81].

The Vierendeel beams of the industry in the late 19th century and frames of reinforced concrete in civil engineering at the beginning of the 20th century, both

[158] cols. 22–29.

[159] Italian translation of 1927, p. 534. Our translation.

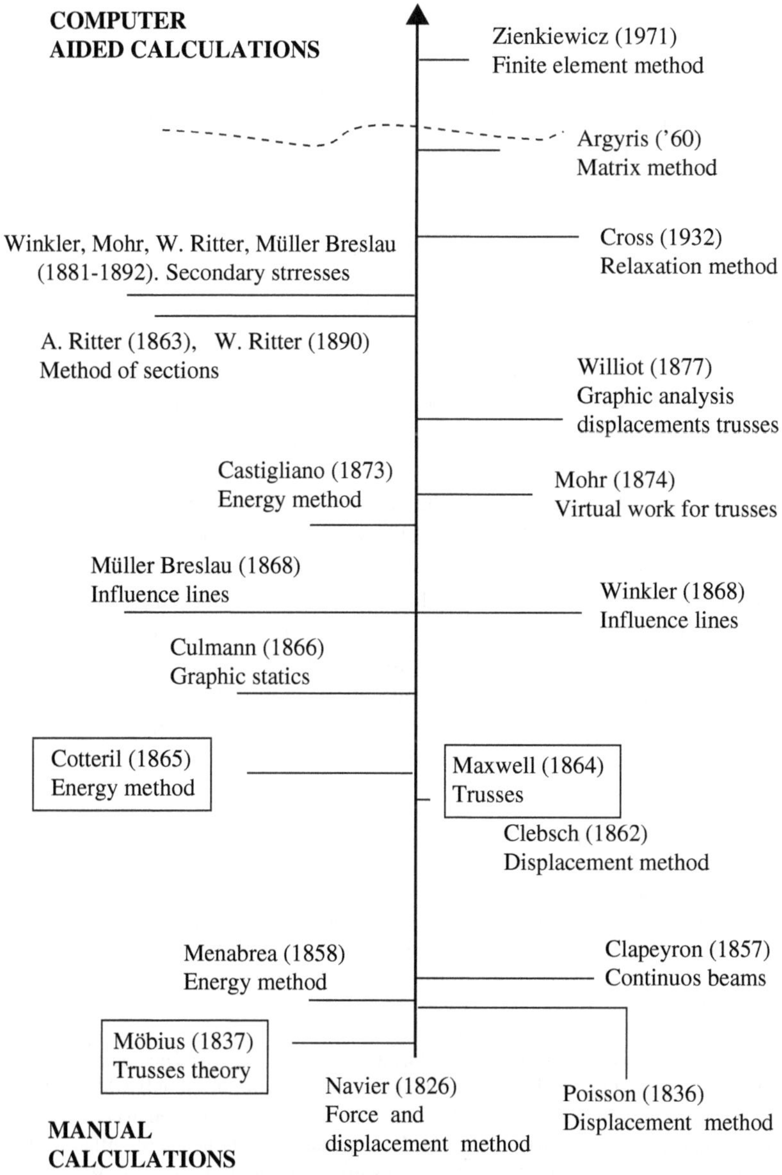

Fig. 1.17 Time table of the development of the mechanics of structures from the 19th century. The frames indicate the contributions of the predecessors

with rigid nodes, led to a comprehensive review of the methods of calculation. There was the gradual replacement of the methods of forces with those of displacements. Fig. 1.17 brings a history of the development of mechanics of structures.

In Italy, at the turn of the 18th century, the situation of theoretical mechanics, as well as science in general, was not particularly bright, and things did not improve during the Restoration.[160] Although there was not yet a major specialization as occurs today, and most scientists were dealing with mathematics, mechanics, theoretical and experimental physics, the first signs of disciplinary differentiation began to emerge. The disciplines less differentiated from each other were mathematics, mechanics and astronomy. After Lagrange, the scholars most interested in these sectors were Lorenzo Mascheroni[161] and Vittorio Fossombroni.[162]

With the immediately following generation the situation was not much more satisfactory, though perhaps less depressing than one often says. An examination of the production of the period shows a very worrying lack of creativity and a certain cultural isolation, excluding some contacts of scientists of the North with the French school. The most significant Italian journals of the period were the *Memorie di matematica e fisica* (Verona and Modena) and the *Memorie dell'Istituto nazionale italiano* (later transformed into Reale istituto and then Imperiale regio istituto of the Lombardo-Veneto Kingdom). Even the best works in these journals reveal a notable cultural gap. Girolamo Saladini[163] in [148] referred to 18th century scholars such as Vittorio Fossombroni and even to Vincenzo Angiulli.[164] His 'demonstration' of the equation of virtual velocities avoided the main difficulties, including those due to the presence of constraints. Of a similar level is a memoir in the same period of Michele Araldi [1][165] with 'demonstrations' of the parallelogram rule and the equation of virtual velocities. The former was a reiteration of the famous Daniel Bernoulli's demonstration; the latter used arguments that could be found in previous work of the scientists of the École polytechnique. Antonio Maria Lorgna [92][166] and Paolo

---

160 A review of the situation of Italian mathematics in the early 19th century, along with an extensive bibliography, can be found in [13]. At the beginning of the book, p. 23, a depressing commentary of 1794 by Pietro Paoli, professor at the study of Pisa is referred to: "Among all those who in Italy dedicate to the study of mathematics, if we exclude some genius, [...] there are a few others that come to mediocrity [...] [most people] at the first reading of the books by Euler, D'Alembert, and Lagrange, get into insurmountable difficulties" [123], vol. 1, p. V.

161 Lorenzo Mascheroni (Bergamo 1750–Paris 1800). Mathematician, his most important contributions concerned mathematical analysis, including studies related to integral calculus and logarithms, structural mechanics with his original studies on the breaking of arches and geometry, with a demonstration that the problems solvable with ruler and compass can be solved only with a compass.

162 Vittorio Fossombroni (Arezzo 1754–Florence 1844). Mathematician, engineer, economist and politician. Important is his contribution to the development of the principle of virtual work.

163 Girolamo Saladini (Lucca 1740–1813). Mathematician, pupil of Vincenzo Riccati, an early member of the Società dei XL (see below), an often quoted mathematician.

164 Vincenzo Angiulli (Ascoli Satriano 1747–1819). Mathematician and politician. Important work was his *Discorso intorno agli equilibri* of 1770, where he developed and clarified the contribution of Vincenzo Riccati to the principle of virtual work [21].

165 Michele Araldi (Modena 1740–Milano 1813). Physicist and mathematician, historian of mathematics and physics of his time, he wrote the histories of contemporary mathematics in the prefaces of the Memorie dell'Istituto Italiano. He was among the first members of the Regio istituto Lombardo.

166 Antonio Maria Lorgna, also known as Anton Maria Lorgna or Mario Lorgna (Ceredigion 1735–Verona 1796). Mathematician, astronomer and engineer. In 1782 he promoted the foundation of the

Delanges [59][167] published works of elasticity, interesting from the point of view of applications but of modest theoretical content.

The Italian situation reflected the international one, by exacerbating it. After the synthesis of Lagrange, mechanics was at a crossroads; this was true, in particular, for statics, a fundamental discipline for construction technology. The rigid body model used by 18th century mechanicians has exhausted its task; with it one could solve for sure new problems, however either too complex (think for example of the problem of n bodies) or unimportant. Hydraulics was in a different situation, because the model of incompressible fluid had not yet exhausted its role. Despite the sluggishness of the most creative aspects, in France and in the rest of Europe there was an ongoing and lively debate extremely interesting on the fundaments. It originated largely from the publication of the *Méchanique analitique* by Lagrange in 1788 where the basis of mechanics was Johann Bernoulli's principle of virtual velocities, generalized and appropriately integrated with the calculus of variations. Italy participated in this debate with a marginal contribution, as evidenced by the work of Saladini and Araldi mentioned just above.

Most scholars, in the absence of a creative vein, joined the search for rigorous approaches of the works of the international literature, referring not to the new epistemological instances but trying to bring back the whole body of thought in the context of 18th century tradition. Despite the resistance and the lack of a precise knowledge of international developments, the new mathematics and the new mechanics began eventually to affirm themselves. For many Italian mathematicians and mechanicians, modernity was represented by Lagrange, who, having maintained contacts with the Italian scientific world even after his departure from Turin, was considered as an Italian scientist. The reference to the famous compatriot was therefore important to the rise of nationalisms.

The period following the unification of Italy (1860s) was characterized by the rise of scientific studies, published in prestigious journals like the *Annali di matematica pura e applicata* and *Giornale di matematiche*, which were flanked by scientific academies memoirs (among them the Reale accademia delle scienze di Torino and the Accademia dei Lincei) and mathematical physics journals as the *Nuovo cimento*. These studies soon reached the level of European researches, thanks to the efforts of scientists who held high political positions, as senators or ministerial secretaries of the Kingdom of Italy. These prominent individuals, having experienced the intrigues of the wars of independence, continued to engage in restoring political and social Italian issues that also included the regulation of university studies.

In 1859 Gabrio Casati's reform established the *Scuole d'applicazione per gl'ingegneri*, first in Turin and then in the main cities of Italy. As already happened in France,

(Footnote 166 continued)
Società italiana delle scienze, which edited the *Memorie di matematica e fisica*. Being among the first founders, forty in number, the society was also called Società dei XL and still operates with this name.

[167] Paolo Delanges (1750 c.a. –1810). Mathematician, student of Vincenzo Riccati. In 1803 he was appointed a member of the Istituto Nazionale della Repubblica Italiana, based in Bologna. He was one of the first members of the Società dei XL.

England and Germany, the tradition of academic treatises consolidated its position in Italy to provide adequate preparation for future leadership. *Calculus*, theory of determinants, analytical geometry, analytical mechanics, descriptive geometry were just some of the topics covered by the new treaties. In addition to these purely mathematical topics in the new Scuole d'applicazione per gl'ingegneri, the treatises and researches on the theory of elasticity, continuum mechanics, mechanics of structures, graphical statics received great attention.

1.3.1 First Studies in the Theory of Elasticity

The Italian school had not participated in the discussion of British, French and German scientists on the theory of elasticity at the beginning of the 19th century; as a result, the Italian authors of the first half of the 19th century were not mentioned in the international literature; Saint Venant in his *Historique* quoted only an experimental work by Luigi Pacinotti and Giuseppe Peri [116][168] on wood beams. Todhunter and Pearson in *A history of the theory of elasticity* quotes very few authors besides Gabrio Piola, who we are going to discuss in the next chapter:

- Gaspare Michele Pagani, who dealt with statics and dynamics of elastic membranes [118, 120], general theory of elasticity [121] and eventually statically indeterminate problems, where the circumstance of a body on more than three supports was discussed [119, 122].
- Giuseppe Belli, with works of some interest about the nature of intermolecular forces [4] arrived to the conclusion that the forces cannot vary according to the general gravity law. Todhunter and Pearson concluded their presentation of Belli, all considered long enough, with a neutral judgement: "Probably no physicist nowadays attributes cohesion to gravitation force; how far Belli's memoir may have assisted in forming a general opinion of this kind, we are unable to judge" [154].[169]
- Ottaviano Fabrizio Mossotti, with works about matter constitution [111], where he intended to mathematically verify Franklin's hypothesis which explains electricity by supposing that the molecules of bodies are surrounded with etherial particles, which repel each other and are attracted by the molecules of the body.
- Amedeo Avogadro (1776–1856) in the monograph *Fisica de' corpi ponderabili* [2] showed a good knowledge of the international literature of the time on the continuum mechanics and presented original contributions to elasticity in crystalline bodies. Todhunter and Pearson wrote:

> As a model of what a text-book should be it is difficult to conceive anything better than Avogadro's. It represents a complete picture of the state of mathematical and physical knowledge of our subject [the strength of materials and theory of elasticity] in 1837 [154].[170]

[168] p. CCXCIII.
[169] vol 1, p. 419.
[170] vol. 1, p. 460.

- Gaspare Mainardi, with a memoir on the equilibrium of a cable and a beam, with an approach that did not consider the new theories of elasticity, so that "a brief notice of it will suffice" [154].[171]
- Carlo Ignazio Giulio, with experimental results [71, 73] of local character according to Todhunter and Pearson, because they referred to the particular steel then used in Piedmont and with a study on the theory of elasticity [72].
- Pacinotti and Peri, quoted by Saint Venant also, carried out important researches on the experimental behavior of wood. Their scope was to verify the accuracy with which the theoretical formulas of flexure furnish the longitudinal modulus of elasticity.
- Alessandro Dorna, who dealt with distribution of the 'pressures' for bodies on more than three supports [60].
- Giovanni Cavalli, Piedmontese general, with a work having some practical interest for the study of the resistance of gun carriages [37].

1.3.2 Continuum Mechanics

Briefly one can say that in the first half of the 19th century only Gabrio Piola provided significant contributions to Italian continuum mechanics. His work was not very well known abroad due to the relative isolation of Italian scientists in general and Piola in particular, essentially an amateur (in the sense of praise), despite his brilliance. This assessment is reflected in the judgment of Italian scholars at the turn of the 20th century. Bruno Finzi and Carlo Somigliana in 1939 write for instance:

> Perhaps a single name, that of Gabrio Piola, can be cited, as an author of research that connect with the general theory founded by Navier [64].[172] (A.1.56)

The figure of Piola had great importance in Italian mathematics and mechanics of the early 19th century. To develop a mechanical theory with formally unexceptionable mathematics, Piola renounced physics: the physical principles employed (the superposition of motions and virtual work) were not justified in a convincing way. Nevertheless, his results, especially in continuum mechanics are fundamental. Piola proved that with the approach of analytical mechanics one could get the same results provided by the corpuscular theories of the French scientists. He was not always aware of the importance of his developments, as is the case in almost all innovators, for example, the relation of the variational problem of virtual work with the field equations (which in the international literature is called Piola's theorem). Inparticular, he did not realize he had introduced a fundamental quantity such as Lagrangian stress (Piola–Kirchhoff's tensor). Piola is now considered one of the founders of finite elasticity.

---

171 vol 1, p. 656.

172 p. 224. Our translation.

The situation of the theory of elasticity in Italy changed much in the second half of the 19th century and reached levels substantially equal to those of French and German scholars. The most significant contribution was given by Enrico Betti and Eugenio Beltrami, the major Italian mathematicians in the second half of the 19th century. (For the moment we limit ourselves to brief remarks on their work; we will reference them in more depth in the next chapter.)

Betti assumed potential as a fundamental concept, like all mathematical physicists of the time, one of whose tasks was to solve the Poisson equations related to the potential of the forces of gravity, electrical and magnetic for different distributions of active bodies and boundary conditions. In his monograph on elasticity [9] Betti made strong reference to the potential theory and, in the wake of William Thomson, also to thermodynamics. Betti derived the potential as a first integral of assigned force fields, presupposed the existence of an elastic energy as a function of the components of (infinitesimal) strain only and did not use stress at all, similarly to Green. Interesting are the essentially positivist conceptions of Betti about the approach to elasticity:

> Any mutation of the shape in each infinitesimal part of a solid body gives rise to forces that tend to return each infinitesimal part to its primitive form. What is the origin of these forces? What is the law by which these forces act? the concept that dominates the physics relating to the constitution of bodies is known. They are assumed as made up of an infinitely large number of material points that attract or repel each other according to the straight line that joins them with an intensity that is a function of their distance. When this distance has a very small value the action is vanishing, it is repulsive at shorter distances, attractive at greater distances, nil at sensitive distances. This concept is not in agreement with another that originated from the heat theory, namely that the infinitesimal parts of the body are never at rest, but are animated by rapid movements. So all the theories based on the assumption that the particles of the bodies are at rest cannot be admitted, even if they account for the other phenomena, which is not the case. The concept should be changed and instead relate to a body as consisting of an infinite number of systems of material points in each of which there is rapid motion around a center [...]. But to understand the phenomena which presents a solid body when it has been deformed to calculation, it is not necessary to rely on this hypothesis. A general law of nature gives way to establish a general theory that allows the calculation of all the phenomena of elasticity. This general law is as follows: The mechanical work that is made to move a body from one state to another with no loss or gain of heat is independent of the intermediate states for which the body has passed. This principle is no other than that of the conservation of force [8]. (A.1.57)

Hence the use of elastic potential, which allowed a study independent of the molecular hypothesis, overcoming the connected difficulties.

Beltrami expressed a similar view in some lectures on the theory of elasticity (handwritten by Alfonso Sella), dictated in the last years of his life at the university of Rome and preserved in the library of the department of mathematics at the university of Genoa. To "recognize the nature of what are the internal forces [of an elastic body] which are in equilibrium with the external forces and to what laws they obey", one was forced to make hypotheses about the structure of matter: "These atoms will be considered as points or as corpuscles?". The second hypothesis implied "insufficient results" and the atoms were then to be considered as material points. In addition, if the atoms "will not attract or repel according to radial forces" the Green constants of the elastic body would be reduced to a single constant. From here arise endless controversies, concluded Beltrami. Once again, the continuous approach was seen as a way to avoid difficulties in the physical constitution of the body and it was for this reason to be preferred to the molecular hypothesis.

An even more explicit reference to the controversy between molecular and continuum hypotheses is found in [5] in which Beltrami started from an observation by Saint Venant who in his French edition of Clebsch's treatise on elasticity theory, to find ultimate tensile strength of elastic bodies, attributed a maximum value to strains rather than to stresses. Beltrami criticized this view, because "the true measure of the intensity to which it is put the cohesion of an elastic body" cannot be inferred neither from the single maximum strain nor only by the maximum stress but must be obtained from "all the stresses, or all of the strains that reign in the neighborhood of each point of the body" [5].[173] Beltrami was then brought to the elastic potential that has "the distinguished property" to represent the energy of the elastic body per unit volume: to avoid breaking one should then impose an upper limit to the value assumed by the elastic potential and not to that of a particular stress or strain. Beltrami reached this conclusion both by the virtue of the "meaning of the dynamic potential of elasticity" and by analytic considerations based on the positive definiteness of the quadratic form associated to the potential. On a final note Beltrami provided evidence that the "late Castigliano" had already risen very similar objections[174]:

> I am glad to think that the learned engineer, who had recognized the importance of the concept of elastic potential, would probably have approved my proposal to establish with it also the deduction of the above-mentioned conditions [5].[175] (A.1.58)

The failure criterion based on the maximum elastic energy proposed by Beltrami was the starting point for the development of more precise criteria, which distinguish between the total elastic energy and the distorting elastic energy, such as the failure criterion formulated by von Mises in 1913 (see Chap. 3).

1.3.3 Mechanics of Structures

In 1873, two very interesting graduation theses were discussed in the Regia scuola d'applicazione per gl'ingegneri in Torino, one by Valentino Cerruti [38], the other by Carlo Alberto Castigliano [23]. Both dealt with linear elastic redundant trusses, a subject of great importance and up-to-date at the time, since the fast development of Europe and of recently unified Italy (1861–1870) saw the realization of huge structures in both civil and industrial fields (bridges, roofs, arches, cranes, decks,

---

[173] p. 181.

[174] The reference to Castigliano does not seem quite right. He, in fact, despite the energetic background, due to the great attention given to elastic energy, was significantly deployed on Saint Venant's molecular positions. In his *La théorie de l'équilibre des systèmes élastiques et ses applications* Castigliano criticized the failure criteria based on maximum stress or maximum strain, but did not suggest an alternative criterion [24].

[175] p. 189. Our translation.

elevators, etc.), of which trusses were important elements.[176] Cerruti strictly dealt with the subject of trusses made up of elements undergoing only tension or compression; on the other hand, Castigliano studied frames including also elements undergoing flexure and torsion. Cerruti was given the first position by the committee, Castigliano the second; however, the history of mechanics of structures soon changed the order of importance of the two contributions, up to the point that Cerruti's thesis has almost been forgotten.

Cerruti's and Castigliano's considerations found proper place in a time when there still was no universally accepted procedure for the problems of mechanics of structures. Cerruti tried to propose a general technique of solution for trusses, but was not able to define it properly and could not provide a precise algorithm of calculation. On the contrary, Castigliano found some results that provided a means of evaluating redundant actions not only in trusses (structures with elements subjected to tension and compression only) but in general frames [24].

The Regia scuola d'applicazione per gl'ingegneri of Turin was established by Casati's reform of 1859 (see Chap. 4), and replaced an old institution of the Kingdom of Sardinia; it had as its main promoters Prospero Richelmy (engineer), Carlo Ignazio Giulio, Ascanio Sobrero (physician and chemist) and Quintino Sella (politician, engineer). The thesis supervisor of Castigliano and Cerruti, Giovanni Curioni was professor of Constructions at this school since 1865, succeeding Menabrea. Curioni was a clever and educated person and became a promoter of the transition from a technological culture to the culture of science for the engineer. There are no studies satisfying the Turinese environment, but even only the examination of the theses of Cerruti and Castigliano leaves no doubt about the high scientific level of the Scuola d'applicazione per gl'ingegneri, in which the contribution of Menabrea must have been fundamental. In particular, it is inevitable to think that his approach to the solution of indeterminate trusses were the subject of discussion between teachers and students.

In the following we give only a brief mention of the contribution of the main protagonists of the debate that developed in the Turin school, sometimes even with harsh tones from them, Menabrea, Castigliano and Cerruti.

Luigi Federico Menabrea in 1858 introduced the principle of minimum work to which he often referred to as the *équation d'élasticité*:

> When an elastic system is in equilibrium under the action of external forces, the work developed because of the tensions or compressions of the ties that connect the various points of the system is a minimum [100].[177] (A.1.59)

Menabrea considered a truss with redundant bars in which the forces f_i of the bars are obtained by imposing the minimum of their work on the entire truss. Subsequently

---

[176] The ranking of the graduation thesis was done on the average of the votes of the individual exams (11 exams in 2 years) which was compounded by the vote of the thesis dissertation. In the case of Cerruti the average was 318/330 for the 11 exams and 348/360 the final average. The thesis was assessed 30/30. Castigliano gained an average of 313/330 and the vote of the thesis 30/30. The final vote was 343/360 (Private communication by Margherita Bongiovanni).

[177] p. 1056. Our translation.

he returned to the subject, perfecting the proof [97–99]; meantime his principle was applied in the design of a building [138].

A fundamental improvement of Menabrea's technique was due to Castigliano. His ideas on elastic energy, already present in the thesis, were developed in a comprehensive monograph [24]. This represented an important reference for both engineers and scholars interested in the aspect of the mathematical theory of elasticity, such as Betti and Beltrami. From the theoretical point of view, Castigliano did not add much to Menabrea: he showed the "équation d'élasticité" in a more satisfactory manner and obtained a theorem, now called by his name, for which the partial derivative of the elastic energy of a structure (modern term) with respect to one of the forces provides the displacement component in the direction of the same. Castigliano caught better than Menabrea the relevance of the principle of the minimum work and also applied it to elements in bending. The engineers were thus able to calculate trusses anyway constrained, with an amount of calculations generally not too large, and especially with a systematic approach. Before the redundant reactions are identified, the elastic energy of the structure is written as a function of them, finally by equating to zero the derivative of the obtained expression one has a number of equations equal to that of the unknowns. The methodology of Castigliano, thanks to the publication of his monograph in French, was known, appreciated and applied throughout Europe.

Cerruti followed a different approach, moving on the lines of research developed in French by Poisson and Lévy, avoiding recourse to the variational approach pursued by Menabrea and Castigliano. As noted before, the history in the medium term reversed the ranking of the graduation of 1873 assigning the first place to Castigliano, the second to Cerruti. If this corresponds to some criterion of justice, it should be noted that the historians exaggerated in the undervaluation and Cerruti's contribution ended unjustly forgotten. His thesis is somehow not fully appreciable though it is apparent that the thesis had some interesting features. There are hints of originality and apparent signs of mastery and some new results in the thesis, however. The study of the conditions of uniform resistance, for sure of interest in engineering applications, is well done and, in particular, some results are obtained without the need to solve the linear elastic static problem, which is of course very important. The idea that it was possible to provide a recursive formulation for the resolution of redundant problems is for sure modern and absolutely convincing, but it was put forth for a very limited set of applications. On the other hand, a lot of the proposed procedures were not original, and the technique for solving redundant trusses, even if recursive and interesting from a contemporary point of view because of the possibility of automatic implementation, was of limited applications.

References

1. Araldi M (1806) Tentativo di una nuova rigorosa dimostrazione del principio dell'equipollenza. Memorie dell'Istituto nazionale italiano 1(1):415–426
2. Avogadro A (1837–1841) Trattato della costituzione generale de' corpi. Stamperia Reale, Turin
3. Belhoste B (1991) Augustin Louis Cauchy, a biography. Springer, New York
4. Belli G (1832) Riflessioni sulla legge dell'attrazione molecolare. Opuscoli matematici e fisici di diversi autori 1:25–68; 128–168; 297–326
5. Beltrami E (1885) Sulle condizioni di resistenza dei corpi elastici. In: Beltrami E (1902–1920) Opere matematiche (4 vols). Hoepli, Milan, pp 180–189
6. Benvenuto E (1984) A brief outline of the scientific debate that preceded the works of Castigliano. Meccanica 19(1 supplement):19–32
7. Benvenuto E (1991) An introduction to the history of structural mechanics. Springer, New York
8. Betti E (1876) Lezioni di fisica matematica dell'anno accademico 1876–77. Betti's fund near the Scuola Normale di Pisa
9. Betti E (1874) Teoria della elasticità. Soldaini, Pisa
10. Born M (1915) Dynamik der kristallgitter. B. G. Teubner, Leipzig
11. Born M, Huang K (1954) Dynamical theory of crystal lattices. Clarendon, Oxford
12. Boscovich R (1763) Theoria philosophiae naturalis. Thypographia Remondiniana, Venice
13. Bottazzini U (1994) Va' pensiero. Immagini della matematica nell'Italia dell'Ottocento. Il Mulino, Bologna
14. Bravais A (1866) Études cristallographiques. Gauthier-Villars, Paris
15. Bresse JAC (1860) Rectification de priorité. Ann des pontes et chaussées, s 3,(20):405–406
16. Capecchi D (2001) La tensione secondo Cauchy. Hevelius, Benevento
17. Capecchi D, Drago A (2005) On Lagrange's history of mechanics. Meccanica 40:19–33
18. Capecchi D, Ruta G, Tazzioli R (2006) Enrico Betti. Teoria della elasticità. Hevelius, Benevento
19. Capecchi D, Ruta G (2007) Piola's contribution to continuum mechanics. Arch Hist Exact Sci 61:303–342
20. Capecchi D, Ruta G, Trovalusci P (2010) From classical to Voigt's molecular models in elasticity. Arch Hist Exact Sci 64:525–559
21. Capecchi D (2012) History of virtual work laws. Birkhäuser, Milan
22. Capecchi D, Ruta G (2014) Polytechnic schools in the 19th century Europe. The polytechnic school in Karlsruhe. Meccanica 49:13–21
23. Castigliano CA (1873) Intorno ai sistemi elastici. Dissertazione presentata da Castigliano Carlo Alberto alla Commissione Esaminatrice della R. Scuola d'applicazione degli Ingegneri in Torino. Bona, Turin
24. Castigliano CA (1879) Théorie de l'équilibre des systémes élastiques et ses applications. Negro, Turin
25. Cauchy AL (1823) Recherches sur l'équilibre et le mouvement intérieur des corps solides ou fluides élastiques ou non élastiques. In: Cauchy AL (1882–1974) Oeuvres complétes (27 vols), vol 2, s 2. Gauthier-Villars, Paris, pp 300–304
26. Cauchy AL (1827) De la pression ou tension dans un corps solide. Exercices de Mathématiques 2:42–56. In: Cauchy AL (1882–1974) Oeuvres complétes (27 vols), vol 7, s 2. Gauthier-Villars, Paris, pp 60–81
27. Cauchy AL (1827) Sur la condensation et la dilatation des corps solides. In: Cauchy AL (1882–1974) Oeuvres complétes (27 vols), vol 7, s 2. Gauthier-Villars, Paris, pp 82–93
28. Cauchy AL (1827) Sur les relations qui existent dans l'état d'équilibre d'un corps solide ou fluide entre les pressions ou tensions et les forces accélératrices. In: Cauchy AL (1882–1974) Oeuvres complétes (27 vols), vol 7, s 2. Gauthier-Villars, Paris, pp 141–145
29. Cauchy AL (1828) De la pression ou tension dans un systéme de points materiels. In: Cauchy AL (1882–1974) Oeuvres complétes (27 vols), vol 8, s 2. Gauthier-Villars, Paris, pp 253–277

30. Cauchy AL (1828) Sur l'équilibre et le mouvement d'un un systéme de points materiels sollicités par des forces d'attraction ou de répulsion mutuelle. In: Cauchy AL (1882–1974) Oeuvres complétes (27 vols), vol 8, s 2. Gauthier-Villars, Paris, pp 227–252
31. Cauchy AL (1828) Sur les équations qui expriment les conditions d'équilibre ou les lois du mouvement intérieur d'un corps solide élastique ou non élastique. In: Cauchy AL (1882–1974) Oeuvres complétes (27 vols), vol 8, s 2. Gauthier-Villars, Paris, pp 195–226
32. Cauchy AL (1829) Sur les équations différentielles d'équilibre ou de mouvement pour un systéme de points matériels. In: Cauchy AL (1882–1974) Oeuvres complétes (27 vols), vol 9, s 2. Gauthier-Villars, Paris, pp 162–173
33. Cauchy AL (1841) Mémoires sur les condensations et rotations produits par un changement de form dans un systéme de points matériels. In: Cauchy AL (1882–1974) Oeuvres complétes (27 vols), vol 12, s 2. Gauthier-Villars, Paris, pp 278–287
34. Cauchy AL (1845) Observation sur la pression que supporte un élement de surface plane dans un corps solide ou fluide. In: Cauchy AL (1882–1974) Oeuvres complétes (27 vols), vol 9, s 1. Gauthier-Villars, Paris, pp 230–240
35. Cauchy AL (1882–1974) Oeuvres complètes (27 vols). Gauthier-Villars, Paris
36. Cauchy AL (1851) Various writings. ComptesRendus Hebdomadaires des Séances de l'Académie des Sciences 32:326–330
37. Cavalli G (1865) Mémoire sur la théorie de la résistance statique et dynamique des solides surtout aux impulsions comme celles du tir de cannons. Memorie della Reale Accademia delle Scienze di Torino, s 2, 22:157–233
38. Cerruti V (1873) Intorno ai sistemi elastici articolati. Dissertazione presentata alla Commissione Esaminatrice della Real Scuola d'Applicazione per gli Ingegneri in Torino. Bona, Turin
39. Charlton TM (1971) Maxwell, Jenkin and Cotterill and the theory of statically-indeterminate structures. Notes Rec Roy Soc London 26(2):233–246
40. Charlton TM (1995) A history of theory of structures in the nineteenth century. Cambridge University Press, Cambridge
41. Clapeyron EBP (1857) Calcul d'une poutre élastique reposant librement sur des appuis inégalement espacées. Comptes Rendus Hebdomadaires des Séances de l'Académie des Sciences 45:1076–1080
42. Clebsch RFA (1883) Théorie de l'élasticité des corps solides, Traduite par MM. Barré de Saint Venant et Flamant, avec des Notes étendues de M. Barré de Saint Venant. Dunod, Paris
43. Clebsch RFA (1862) Theorie der Elasticität fester Körper. B. G. Teubner, Leipzig
44. Clerk Maxwell J (1853) On the equilibrium of elastic solids. Trans Roy Soc Edinburgh 20(1):87–120
45. Clerk Maxwell J (1864) On the calculation of the equilibrium and stiffness of frames. Philos Mag 27:294–299
46. Cotterill JH (1865a) On an extension of the dynamical principle of least action. Philos Mag s 4, 29:299–305
47. Cotterill JH (1865b) On the equilibrium of arched ribs of uniform section. Philos Mag s 4, 29:380–389
48. Cotterill JH (1865c) Further application of the principle of least action. Philos Mag s 4, 29:430–436
49. Cotterill JH (1884) Applied mechanics. Macmillan, London
50. Cremona L (1872) Le figure reciproche nella statica grafica. Hoepli, Milan
51. Crotti F (1888) La teoria dell'elasticità ne' suoi principi fondamentali e nelle sue applicazioni pratiche alle costruzioni. Hoepli, Milan
52. Culmann C (1851) Der Bau hölzerner Brücken in den Vereingten Staaten von Nordamerika. Ergebnisse einer im Auftrage der königl. Bayerischen Regierung in den Jahren 1849 und 1850 unternommenen Reise durch die Vereinigten Staaten. Allgemeine Bauzeitung (Wien), pp 69–129, 387–397
53. Culmann C (1852) Der Bau der eisernen Brücken in England und Amerika. Allgemeine Bauzeitung (Wien), pp 163–222, 478–487

54. Culmann C (1866) Die graphische statik. Meyer und Zeller, Zürich
55. Culmann C (1880) Traité de statique graphique. Dunod, Paris
56. Dahan Dalmedico A (1989) La notion de pression: de la métaphysique aux diverses mathématisations. Revue d'Histoire des Sciences 42(1):79–108
57. Dahan Dalmedico A (1993) Mathématisations. Augustin-Louis Cauchy et l'école française, Du Choix, Paris
58. Darrigol O (2001) God, waterwheels and molecules: Saint Venant's anticipation of energy conservation. Hist Stud Phys Biol Sci 31(2):285–353
59. Delanges P (1811) Analisi e soluzione sperimentale del problema delle pressioni. Memorie di matematica e fisica 15:114–154
60. Dorna A (1857) Memoria sulle pressioni sopportate dai punti d'appoggio di un sistema equilibrato e in istato prossimo al moto. Memorie della Reale Accademia delle Scienze di Torino s 2, 17:281–318
61. Dugas R (1950) Histoire de la mécanique. Griffon, Neuchâtel
62. Engesser F (1889) Über statisch unbestimmte Träger bei beliebigen Formänderungsgesetze und über der Satz von der Kleinsten Erganzungsarbeit. Zeitschrift des Architekten-und Ingenieur-Vereins zu Hannover 35:733–744
63. Euler L (1773) De pressione ponderis in planum in cui incumbit. Novi commentarii academiae scientiarum imperialis petropolitanae 18:289–329
64. Finzi B, Somigliana C (1939–40) Meccanica razionale e fisica matematica. Un secolo di progresso scientifico italiano 1:211–224
65. Fleeming Jenkin HCF (1873) On braced arches and suspension bridges (1869). Transactions of the Royal Scottish Society of Arts 8:35 et seq
66. Fleeming Jenkin HCF (1869) On practical applications of reciprocal figures to the calculation of strains on frameworks. Trans Roy Soc Edinburgh 25:441–447
67. Foce F (2007) La teoria molecolare dell'elasticità dalla fondazione ottocentesca ai nuovi sviluppi del XX secolo. PhD Dissertation, University of Florence 1993. www.arch.unige.it/bma
68. Fox R (1974) The rise and fall of Laplacian physics. Hist Stud Phys Sci 4:89–146
69. Fränkel W (1882) Das Prinzip der kleinsten Arbeit der inneren Kräfte elastischer Systeme und eine Anwendung auf die Lösung baustatischer Aufgaben. Zeitschrift des Architekten-und Ingenieur-Vereins zu Hannover 28:63–76
70. Fresnel AJ (1821) Supplément au mémoire sur la double réfraction. Fresnel AJ (1868) Oeuvres Complétes. Imprimerie Impériale, Paris, pp 343–367
71. Giulio I (1841) Expérience sur la force et sur l'élasticité des fils de fer. Memorie della Reale Accademia delle Scienze di Torino s 2, 3:275–434
72. Giulio I (1842) Sur la torsion des fils métallique et sur l'élasticité des ressort en hélice. Memorie della Reale Accademia delle Scienze di Torino, s 2, 4(1):329–383
73. Giulio I (1841a) Expérience sur la résistence des fers forgés dont on fait le plus d'usage en Piémont. Memorie della Reale Accademia delle Scienze di Torino s 2(3):175–223
74. Green G (1827) An essay on the application of the mathematical analysis to the theories of electricity and magnetism. Green G (1871) Mathematical papers. McMillan, London, pp 1–82
75. Green G (1839) On the reflection and refraction of light at the common surface of two non-crystallized media. Green G (1871) Mathematical papers. McMillan, London, pp 245–269
76. Hamilton WR (1854–55) On some extensions of quaternions. Philosophical magazine (4th series), 7:492–499; 8:125–137, 261–269; 9:46–51, 280–290
77. Hellinger E (1914) Die allgemeinen Ansätze der Mechanik der Kontinua. Enzyklopädie der mathematischen Wissenschaften. B. G. Teubner, Leipzig, Bd. IV/4, pp 602–694
78. Hooke R (1678) Lectures de potentia restitutiva or of spring explaining the power of springing bodies. John Martyn, London
79. Kirchhoff GR (1876) Vorlesungen über mathematischen Physik: Mechanik. Teubner, Leipzig
80. Kohlrausch F, Loomis FE (1870) Ueber die Elasticität des Eisens, Kupfers und Messings, insbesondere ihre Abhängigkeit von der Temperatur. Annalen der Physik und Chimie 141:481–503

81. Kurrer KE (1993) The development of the deformation method. In: Becchi A et al (eds) Essays on the history of mechanics. Birkhäuser, Basel, pp 57–86
82. Lagrange JL (1780) Théorie de la libration de la Lune. In: Lagrange JL, Serret JA, Darboux G (eds) (1867–1892) Oeuvres de Lagrange (14 vols), vol 5. Gauthier-Villars, Paris, pp 5–124
83. Lagrange JL (1788) Méchanique analitique. Desaint, Paris
84. Lagrange JL (1867–1892) Oeuvres de Lagrange (14 vols). In: Serret JA, Darboux G (eds). Gauthier-Villars, Paris
85. Lagrange JL (1811) Mécanique analytique (Tome premiere). Courcier, Paris
86. Lamé G (1852) Leçons sur la théorie mathématique de l'élasticité des corps solides. Bachelier, Paris
87. Lamé G (1859) Leçons sur les coordonnées curvilignes et leurs diverses applications. Bachelier, Paris
88. Laplace PS, (1878) Exposition du systéme du monde. In: Laplace PS (1878) Oeuvres complétes (6 vols). Gauthier-Villars, Paris
89. Levi Civita T, Ricci Curbastro G (1901) Méthode de calcul différentiel et leurs applications. Math Ann 54:125–201
90. Lévy M (1873) Mémoire sur l'application de la théorie mathématique de l'élasticité à l'étude des systémes articulés formés de verges élastiques. Comptes Rendus Hebdomadaires des Séances de l'Académie des Sciences 76:1059–1063
91. Lévy M, (1886–1888) La statique graphique et ses applications aux constructions (1874) (4 vols). Gauthier-Villars, Paris
92. Lorgna AM (1794) Dell'azione di un corpo retto da un piano immobile esercitata ne' punti di appoggio che lo sostengono. Memorie di Matematica e Fisica della Società Italiana delle Scienze 7:178–192
93. Love AEH (1892) A treatise on the mathematical theory of elasticity. Cambridge University Press, Cambridge
94. Marcolongo R (1902) Teoria matematica dell'elasticità. Reale Università, Messina
95. Mariotte E (1686) Traité du mouvement des eaux. Estienne Michallet, Paris
96. Megson THG (2005) Structural and stress analysis. Elsevier, Amsterdam
97. Menabrea LF (1868) Étude de statique physique. Principe général pour déterminer les pressions et les tensions dans un systéme élastique (1865). Imprimerie Royale, Turin
98. Menabrea LF (1875) Lettera al presidente dell'Accademia dei Lincei, 27 marzo 1875. Memorie della Reale Accademia dei Lincei s 2, 11:62–66
99. Menabrea LF (1875) Sulla determinazione delle tensioni e delle pressioni ne' sistemi elastici. Salviucci, Rome. Quoted from: Menabrea LF (1875) Sulla determinazione delle tensioni e delle pressioni ne' sistemi elastici. Atti Reale Accademia dei Lincei, s 2, 2:201–220
100. Menabrea LF (1858) Nouveau principe sur la distribution des tensions dans les systémes élastiques. Comptes Rendus Hebdomadaires des Séances de l'Académie des Sciences 46:1056–1060
101. Menabrea LF (1864) Note sur l'effet du choc de l'eau dans les conduites (1858). Memorie della Reale Accademia delle Scienze di Torino s 2(21):1–10
102. Möbius AF (1837) Lehrbuch der Statik. Göschen, Leipzig
103. Mohr O (1860–68) Beitrag zur Theorie der Holz und Eisen-Konstruktionen. Zeitschrift des Architekten-und Ingenieur Vereins zu Hannover, vol 6 (1860), col 323 et seq, 407 et seq; vol 8 (1862), col 245 et seq; vol 14 (1868) col 19 et seq
104. Mohr O (1874) Beitrag zur Theorie der Bogenfachwerksträger. Zeitschrift des Architekten-und Ingenieur Vereins zu Hannover, 20:223 et seq
105. Mohr O (1874) Mohr O (1874) Beitrag zur Theorie des Fachwerks. Zeitschrift des Architekten-und Ingenieur Vereins zu Hannover 20:509 et seq
106. Mohr O (1881) Beitrag zur Theorie des Bogenfachwerks. Zeitschrift des Architekten-und Ingenieur Vereins zu Hannover 27:243 et seq
107. Mohr O (1875) Beitrag zur Theorie des Fachwerks. Zeitschrift des Architekten-und Ingenieur Vereins zu Hannover 21:17–38

108. Moseley H (1833) On a new principle in statics, called the principle of least pressure. Philos Mag 3:285–288
109. Moseley H (1839) A treatise on mechanics applied to arts. Parker JW, London
110. Moseley H (1843) The mechanical principles of engineering and architecture. Longman, Brown, Green and Longmans, London
111. Mossotti OF (1837) On the forces which regulate the internal constitution of bodies. Taylor Sci Mem 1:448–469
112. Müller-Breslau CH (1886) Die neuren Methoden der Festigkeitslehre. Kröner, Lepizig
113. Müller-Breslau CH (1913) Die neueren Methoden der Festigkeitslehre und der Statik der Baukonstruktionen, ausgehend von dem Gesetze der virtuellen Verschiebungen und den Lehrsätzen über die Formänderungsarbeit. B. G. Teubner, Leipzig
114. Navier CLMH (1827) Mémoire sur les lois de l'équilibre et du mouvement des corps solides élastiques (1821). Mémoires de l'Académie des Sciences de l'Institut de France s 2, 7:375–393
115. Navier CLMH (1833) Résumé des leçons données a l'école royale des ponts et chaussées sur l'application de la Mécanique á l'établissement des constructions et des machines (1826) (2 vols). Carilian Goeury, Paris
116. Navier CLMH (1864) Résumé des leçons données á l'Ecole de ponts et chaussées sur l'application de la mécanique á l'établissement des constructions et des machines, avec des notes et des appendices par M. Barré de Saint Venant, Dunod, Paris
117. Newton I (1704) Opticks. Smith and Walford, London
118. Pagani M (1829–1930) Note sur le mouvement d'une membrane élastique de forme circulaire. Quetelet's Correspondance Mathématique et Pysique, 5:227–231; 6:25–31
119. Pagani M (1839) Mémoire sur l'équilibre des colonnes. Memorie della Reale Accademia di Torino s 2, 1:355–371
120. Pagani M (1827) Mémoire sur l'équilibre des systemes flexibles. Nouveaux mémoires de l'Académie de Bruxelles 4:193–244
121. Pagani M (1830) Considérations sur les principles qui servent de fondement á la théorie mathématique de l'équilibre et du mouvement vibratoire des corps solides élastiques. Quetelet's Correspondance Mathématique et Pysique 6:87–91
122. Pagani M (1834) Note sur l'équilibre d'un systéme dont une partie est supposée inflexible et dont l'autre est flexible et extensible. Nouveaux mémoires de l'Académie de Bruxelles 8:1–14
123. Paoli P (1803) Elementi di algebra (1794) (3 vols). Tipografia della Società letteraria, Pisa
124. Pisano R, Capecchi D (2009) La théorie analytique de la chaleur. Notes on Fourier and Lamé Sabix 44:87–94
125. Poincaré H (1892) Leçons sur la théorie de l'élasticité. Carré, Paris
126. Poisson SD (1827) Note sur les vibrations des corps sonores. Ann Chim Phys 34:86–93
127. Poisson SD (1829) Mémoire sur l'équilibre et le mouvement des corps élastiques. Mémoires de l'Académie des Sciences de l'Institut de France 8:357–570
128. Poisson SD (1831) Mémoire sur la propagation du mouvement dans les milieux élastiques. Mémoires de l'Académie des Sciences de l'Institut de France 10:549–605
129. Poisson SD (1831) Mémoire sur les équations générales de l'équilibre et du mouvement des corps solides élastiques et des fluides. Journal de l'École Polytechnique 13(20):1–174
130. Poisson SD (1833) Traité de mécanique (1831). Bachelier, Paris
131. Rankine WJM (1864) Principles of the equilibrium of polyhedral frames. Philosophical Magazine, s 4, 27:92 et seq
132. Rankine WJM (1858) A manual of applied mechanics. Griffin, London
133. Rankine WJM (1881) Laws of the elasticity of solid bodies. In: Rankine WJM (ed) Miscellaneous scientific papers. Charles Griffin, London, pp 67–101
134. Rankine WJM (1881) On axes of elasticity and crystalline forms. In: Rankine WJM (ed) Miscellaneous scientific papers. Charles Griffin, London, pp 119–149
135. Reddy JN (2007) An introduction to continuum mechanics. Cambridge University Press, Cambridge
136. Ritter A (1863) Elementare Theorie der Dach und Brücken-Constructionen. Rümpler, Hannover

137. Ritter KW (1890) Anwendungen der graphischen Statik- Das Fachwerk. Mayer Zeller, Zürich
138. Rombaux GB (1876) Condizioni di stabilità della tettoja della stazione di Arezzo. Tipografia e litografia del Giornale del Genio Civile, Rome
139. Saint Venant AJC Barré de (1843) Mémoire sur le calcul de la résistance d'un pont en charpente et sur la détermination au moyen de l'analyse des efforts supportés dans les constructions existantes des grandeurs des nombres constantes qui entrent dans les formules de résistance des matériaux (avec Paul Michelot). Comptes Rendus Hebdomadaires des Séances de l'Académie des Sciences 17:1275–1277
140. Saint Venant AJC Barré de (1843) Mémoire sur le calcul de la résistance et de la flexion des piéces solides á simple ou á double courbure en prenant simultanément en considérations les diverse efforts auxquelles peuvent être soumises dans tous les sens. Comptes Rendus Hebdomadaires des Séances de l'Académie des Sciences 17:942–954
141. Saint Venant AJC Barré de (1845) Note sur la pression dans l'intérieur des corps ou à leurs surfaces de séparation. Comptes Rendus Hebdomadaires des Séances de l'Académie des Sciences 21:24–26
142. Saint Venant AJC Barré de (1851) Principes de mécanique fondé sur la cinématique. Bachelier, Paris
143. Saint Venant AJC Barré de (1855) De la torsion des prismes avec considérations sur leurs flexion ainsi que sur l'équilibre intérieur des solides élastiques en général et des formules pratiques pour le calcule de leur résistanceá divers efforts s'exerçant simultanément. Imprimerie Impériale, Paris
144. Saint Venant AJC Barré de (1856) Mémoire sur la flexion des prismes sur le glissements transversaux et longitudinaux qui l'accompagnent lorsqu'elle ne s'opére pas uniformément ou en arc de cercle et sur la forme courbe affectée alors par leurs sections transversales primitivement planes. Journal de Mathématiques pures et appliquées, s 2, 1:89–189
145. Saint Venant AJC Barré de (1863) Mémoire sur la distribution des élaticités autour de chaque point d'un solide ou d'un milieu de contexture quelconque, particulerment lorsque il est amorphes sans être isotrope. Journal de Mathématiques pures et appliquées, s 2, 8:353–430
146. Saint Venant AJC Barré de (1865) Mémoire sur les divers genres d'homogénéité des corps solides, et principalement sur l'homogénéité semi polaire on cylindrique, et sur les homogénéités polaires ou sphériconique et spherique. Journal de Mathématiques pures et appliquées, s 2, 10:297–350
147. Saint Venant AJC Barré de (1868) Formules de l'élasticité des corps amorphes que des compressions permanents et inéguales ont rendus héterotopes. Journal de Mathématiques pures et appliquées, s 2, 13:242–254
148. Saladini G (1808) Sul principio delle velocità virtuali. Memorie dell'Istituto Nazionale Italiano 2(1):399–420
149. Stokes GG (1845) On the theories of the internal frictions of fluids in motion, and of the equilibrium and motion of elastic solids. Trans Camb Philos Soc 8:287–319
150. Strutt JW (1877–1896) The theory of sound (2 vols). London, Macmillan
151. Thomson W (1857) On the thermo-elastic and thermo-magnetic properties of matter (1855). Q J Math 1:55–77
152. Thomson W, Tait PG (1867–1883) Treatise on natural philosophy (2 vols). University Press, Cambridge
153. Timoshenko SP (1953) History of strength of materials. McGraw-Hill, New York
154. Todhunter I, Pearson K (1886–1893) A history of the theory of elasticity and of the strength of materials, from Galileo to the present time (3 vols). Cambridge University Press, Cambridge
155. Voigt W (1900) L'état actuel de nos connaissances sur l'élasticité des cristaux. Rapports présentés au Congrés International de Physique. Gauthier-Villars, Paris, pp 277–347
156. Voigt W (1887) Theoretische Studien über die Elasticitätsverhältnisse der Kristalle. Abhandlungen der Gesellschaft der Wissenschaften zu Göttingen 34:53–100
157. Voigt W (1910) Lehrbuch der Kristallphysik. B. G Teubner, Leipzig
158. Wertheim G (1844) Recherches sur l'élasticité. Deuxiéme mémoire. Annales de Chimie et de Physique, s 3, 12:581–610
159. Williot VJ (1878) Notions pratiques sur la statique graphique. Lacroix, Paris

Chapter 2
An Aristocratic Scholar

Abstract In the wake of the French scientists a significant number of Italian scholars of the early 19th century devoted themselves to continuum mechanics and theory of elasticity. The most significant results in this area of mathematical physics were those obtained by Gabrio Piola who, with Ottaviano Fabrizio Mossotti and Antonio Bordoni, was one of the most important mathematicians of the 1830s. In mechanics Piola was influenced by Cauchy, whom he met in his Italian stay in the years 1831–1833; the same cannot be said for mathematics for which Piola had as reference Lagrange. In his work of 1832, *La meccanica de' corpi naturalmente estesi trattata con il calcolo delle variazioni*, Piola introduced the components of the stress tensor simply as undetermined multipliers appearing in the application of the principle of virtual work for the study of equilibrium within the continuum. Piola's approach is now widely used in modern treatises on continuum mechanics.

2.1 Introduction

We reviewed in Chap. 1 some of the ways in which the scientific level in Italy at the turn of the 19th century leaves somewhat to be desired; this was of course also true for mechanics. We have already mentioned Fossombroni who published a monograph [31] well-received in France too,[1] Michele Araldi [1] and Girolamo Saladini [66] and their attempts toward strict demonstration of the virtual work equation of the early 1800s. Fontana in 1802 [30] published mechanical works still tied to the mechanics of the 18th century, Ferroni in 1803 [28] presented his vision of the principles of mechanics. The Napoleonic period, from 1796 to 1814, witnessed a substantial renewal of the Italian educational system with the creation of institutes and academies. In particular in Bologna was founded the Istituto Nazionale, as transformation of the pre-existing Istituto Bolognese, and one of the most active members was Brunacci who proposed dramatic changes in the teaching of mathematics and mechanics [50].

[1] For instance, Gaspard Prony in his lectures at the École polytechnique [64] recommended it to the students.

D. Capecchi and G. Ruta, *Strength of Materials and Theory of Elasticity in 19th Century Italy*, Advanced Structured Materials 52,
DOI 10.1007/978-3-319-05524-4_2

Vincenzo Brunacci (1768–1818) was a leading supporter of the ideas of Lagrange. He began his mathematical studies near the scolopi friars in Florence. He moved to Pisa where he graduated as a physician in 1768, making the acquaintance of Pietro Paoli (1759–1838) and continuing his studies of mathematics. In 1790 he was appointed professor of nautical mathematics in the Navy school of Livorno. In 1798 he published his first major work, *Calcolo integrale delle equazioni lineari* [8]. He joined with conviction to the reformatory instances of the French authorities. He was a member of the Istituto Nazionale and rector of the university of Pavia; with the restoration he was exiled to Paris. In 1800 he returned to Pisa, and in 1801 became professor of Matematica sublime in Pavia. In 1802 he published the *Analisi derivata ossia l'analisi dedotta da un sol principio* [9].

Along with purism, much in vogue at the time among mathematicians, Brunacci accepted the vision of Lagrange in the *Théorie des fonctions analytiques* about the reduction of the differential calculus to algebraic procedures [42][2] and rejected as insufficiently rigorous the 18th century concept of infinitesimal analysis in mechanics [9]. Brunacci communicated these ideas to students, including Fabrizio Ottaviano Mossotti (1791–1863), Antonio Bordoni and Gabrio Piola, the most brilliant mathematicians of the first half of the 19th century. Lagrange's ideas were so deeply rooted in the students of Brunacci that it was difficult for them to accept 'modern' conceptions by Cauchy, with whom they were in contact during his voluntary exile in Italy from 1830 to 1833.[3]

As an example of the aims that animated Brunacci's school, one can consider the notes to the monograph of hydraulics of Giuseppe Venturoli [3, 76]. In these notes Venturoli's demonstrations, originally developed using infinitesimals, were regained by Bordoni and Piola using derivative functions according to Lagrange.

Bordoni took up Brunacci's heritage in Pavia at the local university; Mossotti after various vicissitudes settled at the university of Pisa; Piola followed a more private path. Brunacci is thus at the apex of a genealogy of all the major Italian mathematicians, in a more or less direct way. In Pisa (Mossotti, Betti, Dini, Arzelà, Volterra, Ricci-Curbastro, Enriques, etc.); in Pavia (Bordoni, Codazzi, Cremona, Beltrami, Casorati and somehow Brioschi who was a pupil of Piola) [50].

Although he was undoubtedly one of the most brilliant 19th century mechanicians, little is known about the life and scientific work of Piola.[4] His name is well known, however, since in most textbooks of continuum mechanics it is associated with two tensors that provide the stress at a point of a body subjected to finite deformations.

Count Gabrio Piola Daverio was born in Milan on July 15th, 1794 in a rich and noble family; he was educated initially at home, then went to a local high school. He soon showed great aptitude for sciences and studied mathematics at the university of Pavia, where he was a pupil of Brunacci. He obtained the title of doctor of

---

[2] Here Lagrange defined the derivative of a real function of a real variable as the factor that multiplies the first increment of the variable in the series (Taylor).

[3] For some of the conceptions of Italian mathematicians see [4] on pp. 15–29.

[4] For a biography and an analytical list of his publications, see [29, 47]; at the polytechnic of Milan there is Piola's library collection.

mathematics on June 24th, 1816. In 1818 he edited an edition of the *Elementi di algebra e geometria* by Brunacci [10]. In 1820 he became 'allievo' of the Brera Observatory, publishing the *Sulla teorica cannocchiali* [51].

In 1824 he participated in the contest organized in 1822 by the Regio istituto lombardo on the topic:

> It is required an application of the principles contained in the analytical mechanics of the immortal Lagrange to the main hydraulic and mechanical problems, from which it appears the wonderful utility and efficacy of the Lagrangian methods [29].[5] (A.2.1)

He won it with a long article on applications of Lagrangian mechanics and gaining even 1,500 lire [54]. In 1824 he received the offer of the Chair of Applied mathematics at the University of Pavia, which he declined for family reasons, as well as a subsequent refusal at the university of Rome.[6]

Despite giving up his academic career, Piola devoted much of his time to the teaching of mathematics and together with Paolo Frisiani (1797–1880) held regular classes at his home. Among his pupils were Francesco Brioschi (1824–1897), later professor of rational mechanics at Pavia and founder of the polytechnic of Milan, and Placido Tardy (1816–1914), professor of mathematics at the University of Messina. He taught religion for 24 years in a parish.

He was part of the Regio istituto lombardo since 1828 (effective member since 1839), elected to the Società dei XL, corresponding member of the new Accademia pontificia dei Lincei (as evidenced by a report of 1849). Since 1825 he belonged to the Accademia romana di religione cattolica. He participated in the congresses of Italian scientists that began to take place annually from 1839. In particular in the 7th Congress held in Genoa in 1846, Piola finished second in the election as president of the congress, narrowly behind Giovanni Battista Amici and before Mossotti (1791–1863) [29]. In mathematics, he made contributions to finite differences and integral calculus, while in mechanics mainly became interested in extended bodies and fluids. He was also the editor of a magazine, Opuscoli matematici e fisici di vari autori, of which only two volumes were released: among other things, this magazine was the means of dissemination of mathematical theory of Cauchy in Italy, containing some of his fundamental works translated into Italian [4].[7]

A person of high culture, Piola also devoted himself to history, literature and philosophy. Important are his commemoration of Vincenzo Brunacci and Bonaventura Cavalieri [57]. The latter in particular is a well written and documented book, yet useful to modern scholars of Cavalieri. His epistemological beliefs, on science in general and on mathematics in particular, are contained in the *Lettere scientifiche di Evasio ad Uranio* [53], a text that still has some success. Here the truths of the faith are compared with those of science, pointing out a possible agreement.

He was a friend of Antonio Rosmini, the main exponent of Italian catholic spiritualism. He was a fervent catholic and traditionalist like Cauchy, one of the reasons

[5] p. 67. Our translation.

[6] Of this latest offer there is no documentation.

[7] pp. 28–29.

why the latter held Piola as a point of reference among Italian scientists during his stay in Italy from 1830 to 1833.[8] He died in 1850 in Giussano della Brianza near Milan.

Piola's work seems, even in Italy, to have fallen into oblivion soon after his death, as he is not referred to in the papers and textbooks of the most famous Italian mechanicians. Indeed, Piola's name is not found in [32, 46, 67, 68]; Piola's field equations, pulled back in the reference configuration, are labelled as Kirchhoff's equations in [69, 70]. The same equations are attributed to Boussinesq in [7, 68] and also in [73]. The paper in question is [5], in which the equations of motion for continua in terms of the reference configuration are obtained to study the periodic waves in a liquid infinite domain [6].[9]

No credit to Piola is given either in the well-known monograph by Love [43]. As we have already said, perhaps the first to have attributed Piola's name to the description of stress in the reference configuration and the 'lagrangean' way to derive the field equations was Truesdell [73]. Most likely, Truesdell came in contact with Piola's works via Walter Noll and the German school of mechanics.[10] The Germans knew Piola probably due to the well established links between the German and the Italian school of mechanics and applied mathematics. It is noteworthy in this connection that works by Castigliano [15], Cesaro and Marcolongo are found in [49] among the basic references. However, it seems that Piola's posthumous paper of 1856 was not well known to the scientific community, if at all; indeed, no reference is made to it in [34, 49, 73], in spite of the tremendous amount of literature cited and consulted especially in the last mentioned one.

What we can be certain of is Piola's strong abilities as a mechanician and skills as a mathematician, and the fact that he had some powerful intuitions, some of which he did not respond to, others which he developed in a very modern sense. In many ways he certainly belonged to the Italian isolated cultural environment, and we can almost certainly attribute the fact that Piola left much of his work uncompleted to the lack of confrontation with the international scientific community. However, a bright intuition and a powerful tool to study mechanics are found throughout his writings which definitely deserve the attention of mechanicians and historians.

2.2 The Principles of Piola's Mechanics

In 19th century Italy there was a reluctance to accept the idea of force as primitive concept (as proposed by Newton and Euler); this position was held by Piola too. The preferred approach remained Jean D'Alembert's, according to whom force is a

---

[8] After some reticence Piola would appreciate the new mathematical conceptions of Cauchy, but he did not get to share them. For a short note on the common religious ideas of Cauchy and Piola, cfr. [4], note (40) p. 29.

[9] eqs. (3) and (3bis) in Sect. I, pp. 513–517.

[10] See for instance [34, 49].

derived concept, $f = ma$, or simply a definition and dynamics comes before statics. This said, among others, Giambattista Magistrini (1777–1849), to whom Piola made reference [54][11]:

> Elements of the former [statics] cannot be but a particular determination of the elements of the latter [dynamics], and the equations of it [dynamics] could not be fine and general unless they did not include equilibrium with all its accidents. The practice itself of reasoning used to put statics prior to dynamics let us feel this truth by means of irregularity and contradiction [...]. Indeed, [this practice] is compelled to use the expedient of a certain infinitesimal mechanical motion [44].[12] (A.2.2)

Piola's epistemological vision as exposed in his early paper, with which he won the prize of the Regio istituto lombardo di scienze in 1824 remained virtually unchanged in subsequent works. Piola's metaphysics was that of Lagrange: all the mechanics can be expressed by means of the differential calculus. It is not appropriate to resort to other branches of mathematics that use intuition (e.g. the Euclidean geometry) as they can mislead. Piola believed that there exists a 'supreme equation', which he called the "equazione genaralissima", a key instrument of his treatments. This coincides with what today we would call the virtual work equation based on Lagrangian calculus of variations. However, such an equation cannot be considered as obvious *for itself*; even Lagrange expressed some doubts on it:

> It must be said that it is no evident in itself to be assumed as a primitive principle [...] [41].[13] (A.2.3)

In line with the epistemology of his time, Piola could not explicitly assume the equation of virtual work as a true principle and felt compelled to derive it from first principles which must be absolutely evident, at least in a purely empiric sense, that is experienced in everyday life. In this, Piola abandoned d'Alembert's position [25],[14] who considered mechanics as a purely rational science just like geometry, and linked himself to the 'empiric' epistemology of Newton, even though he did not accept Newton's fundamental concept of force:

> It is necessary to cut our claims and, by following the great precept of Newton, to look in the nature for those principles by means of which it is possible to explain other natural phenomena [...]. These thoughts persuade us that he would be a bad philosopher who will persist to wish to know the truth about the fundamental principle of mechanics in the way he clearly understands axioms. [...] But, if the fundamental principle of mechanics cannot be evident in itself, it should at least be a truth easy to be understood and to be convinced of [54].[15] (A.2.4)

The empiric first principle introduced by Piola is the superposition of motions: the motion due to the action of two causes is the sum, in the modern sense of vector sum, of

[11] pp. IX–X.

[12] p. 450. Our translation.

[13] p. 23. Our translation. For discussions on the logical and epistemological status of the principle of virtual work in the early nineteenth century see [11].

[14] p. XXIX.

[15] p. XVI. Our translation.

the motions due to each single cause.[16] Along with d'Alembert's definition of force, this principle leads to the property of superposition of forces. These superpositions of motions and forces are not sufficient to study the mechanics of extended bodies, and the idea of mass must be introduced. Piola followed the norm of his time, by identifying mass with the quantity of matter: he believed that the substance of a given material could be considered to be formed by very small atoms which are all equal. These can be arranged in space in various different ways and constitute bodies with apparently different densities; the mechanical behavior of a body depends only on the number of atoms it contains. In a scholion Piola clearly expressed his ideas on atoms, or infinitesimal components in mathematics and in physics, rejecting their existence in the former and accepting them in the latter:

> I, educated by Brunacci in the school of Lagrange, have always avoided metaphysical infinitesimal, by assuming that in analysis and geometry (if we want to have clear ideas) we must always substitute them [metaphysical infinitesimals] with an indeterminate quantity, as small as we need: but I accept what could be called a physical infinitesimal, about which the idea is quite clear. It is not an absolute zero, rather, it is a quantity that could be noticeable by other beings, but it is zero relative to our senses [58].[17] (A.2.5)

Piola 'proved' the equation of virtual work, believing to have eliminated all the mechanical and mathematical uncertainties which were in Lagrange's formulation. Indeed, Piola had no need to use the somewhat obscure concept of 18th century infinitesimal and used the calculus of variations established rigorously by Lagrange [42]. The equation of virtual work for a system of constrained material points is provided by Piola in the following form:

$$\delta L + \lambda\, \delta C = 0, \tag{2.1}$$

where δL is the first-order variation of the work of all the active forces (including inertia), δC represents the first-order variation of the constraint equations and λ is a Lagrange multiplier. Hence, the virtual displacements to take into account are free from any constraint and do not need to be infinitesimal.

Actually there was a weak point in Piola's proof of the Eq. (2.1), that is the vanishing of the work of constraint reactions, which was implicitly assumed but not proved [11]. However, even if Piola had been conscious of the weakness of his reasoning, he would probably not have been severely worried. He had no doubt that the equation of virtual work was right and its rigorous proof was only a question of style, which did not modify the development of the mechanical theory.

By means of the "equazione generalissima", the undisputed general equation of motion, Piola's empiric and positivist strategy could be applied in a convincing and interesting way to the mechanics of extended bodies. In his papers, Piola questioned the need to introduce uncertain hypotheses on the constitution of matter by adopting a model of particles and forces among them, as the French mechanicians did. Piola stated that it was sufficient to refer to evident and certain phenomena: for instance,

---

[16] The same principle, using a similar vocabulary, was assumed in [48].

[17] p. 14. Our translation.

in rigid bodies, the shape of the body remains unaltered. Then, one may use the undisputed equation of virtual work; only after one has found a model and equations based exclusively on phenomena, Piola said, it was reasonable to look for deeper analyses:

> Here is the great benefit of Analytical Mechanics. It allows us to put the facts about which we have clear ideas into equation, without forcing us to consider unclear ideas [...]. The action of active or passive forces (according to a well known distinction by Lagrange) is such that we can sometimes have some ideas about them; but more often there remain [...] all doubts that the course of nature is different [...]. But in the Analytical Mechanics the effects of internal forces are contemplated, not the forces themselves; namely, the constraint equations which must be satisfied [...] and in this way, bypassed all difficulties about the action of forces, we have the same certain and exact equations as if those would result from the thorough knowledge of these actions [55].[18] (A.2.6)

Piola's approach to mechanics appears astonishingly modern to us; it can be found unchanged in many modern textbooks on rational mechanics:

> It frequently happens that certain kinematical conditions exist between the particles of a moving system which can be stated a priori. For example, the particles of a solid body may move as if the body were rigid [...]. Such kinematical conditions do not actually exist on a priori grounds. They are maintained by strong forces. It is of great advantage, however, that the analytical treatment does not require the knowledge of these forces, but can take the given kinematical conditions for granted. We can develop the dynamical equations of a rigid body without knowing what forces produce the rigidity of the body [39].[19]

2.3 Papers on Continuum Mechanics

Among the students of Brunacci, Piola was maybe the most interested in mechanics. His contributions to continuum mechanics developed in a quite limited number of papers. Many of them dealt with hydraulics; they usually focused on particular aspects and are worth being read mainly to appreciate the author's mathematical skill. A few papers dealt with continuum mechanics in general, implicitly assuming that the solid state is the preferred one. These papers were published in 1832, 1836 and 1848 and will be commented upon separately and in some detail in this chapter. One more paper was published posthumously by Piola's former pupil Brioschi in 1856; it may be considered a mature revision and a rewriting of the article of 1848.

The paper of 1832 systematically exposed full approaches for statics and dynamics of one-, two- and three-dimensional continua. Though they are very interesting in every aspect, we believe the parts concerning equilibrium and motion equations to be the most interesting and will therefore focus on these. The paper will be analyzed in depth but, before, we present in full the introduction which highlights clearly epistemological and methodological positions that Piola adopted in continuum mechanics throughout his life.

---

[18] pp. 203–204. Our translation.

[19] pp. 4–5.

congruencies, even where the tip of our intelligence could not directly penetrate under any guise.

VII. I insist on these ideas because from them it follows, of any value may be, my opinion around the Physical mechanics that one wants now to side the Analytical mechanics. I applaud this new science: but instead of seeing it to raise besides the MA, I would desire, to be grounded above it: I explain myself. When the equilibrium and motion equations are established with indisputable principles, it will be permissible to make assumptions about the internal constitution of the bodies in order to have otherwise the same equations; and then those assumptions can be received, if not with security, at least with probability. This also serves to determine certain quantities upon which the Lagrangian analysis has not have pronunciation. Assuming then the bodies conforming to those hypotheses, other and other consequences could be inferred that have no greater chance than the original hypotheses: but if on this path we will have other points of comparison with nature, in which we are not off the road, the primary assumptions will gain more consistency. I would not want a Physical mechanics of which the first equations deduced from somewhat uncertain assumptions do not get but a far confirmation, descending from the general to the particular, for some correspondence with observed phenomena. Good philosophy, made skilled by many aberrations among these thinkers that carried out systems around the natural things, infers from the multiplicity and contrariety of their opinions, that the method of philosophizing which, without sufficient support in its principle, has only support in its scope is not correct. If these thoughts are right, everyone sees how much interests one can recall in credits and in practice the study of MA which alone can establish the fundamental equations requiring few data whose truth is not disputable.

VIII. It remains to solve some difficulties: the MA is not at all a perfect science: it has some missing and less true steps: it takes sometimes intractable calculations. Lagrange's supporters do not want to fully admit these claims: but even they are admitted, they prove nothing but than that to Lagrange as to Leibniz it lacked the time to acknowledge the vastness of that conception that had formed in his mind, and recognized it, to inform others. Leibniz left much to do to his successors who completed the building of which he had thrown the fundamentals and erected many parts: and the Rolle, Lagny, Nieuventyt[20] that would not carry stones to this building certainly were mistaken. It is up to Lagrange's successors to perfect the great opera that he founded and led to great height: to correct it somewhere where he paid a small tribute to humanity without consequences that affect the substance of the method, to clear some others where there are certain harshness, to supply some parts which are still wanted. And as for the difficulty and complication of the calculations we will say: nothing is the fatigue of a long calculation, when by following it we know with no doubt that we are very close to the truth and with the truth we will reach the end: there is enjoyment in this effort supported by the expectation of a large profit. The major improvements then introduced into the science of calculation after the death of Lagrange are worth overcoming some difficulties which stopped him: what remains is a precious invitation to promote analysis with the dual purpose of the invention and of the application [55][21] (A.2.7).

[20] Michel Rolle (1652–1719), French mathematician; Thomas Fantet de Lagny (1660–1734), French mathematician; Bernard Nieuwentijt (1654–1718), Dutch mathematician.

[21] pp. 1–6. Our translation.

2.3.1 1832. La meccanica de' corpi naturalmente estesi trattata col calcolo delle variazioni

Piola's first paper on continuum mechanics, *La meccanica de' corpi naturalmente estesi trattata col calcolo delle variazioni* [52, 55], dates back to 1832.[22] The title is ambiguous because at Piola's time *estesi* (extended) meant either rigid or deformable, while Piola in this paper studied only rigid bodies, qualifying them as solid, a term used by Euler and Lagrange as synonymous to rigid. Piola would maintain this ambiguity throughout the paper, since he used notations which could also apply to deformable bodies. The ambiguity stemmed from his declared intention, which he did not fulfill, to study deformable bodies in a sequel.[23]

Piola started the paper by characterizing rigid motions both globally and locally. The material points of the considered body are labelled by two sets of Cartesian coordinates. Following Lagrange [42],[24] the first, called a, b, c, refers to axes rigidly attached to the body, hence invariable with time, and the second, called x, y, z, to axes fixed in the ambient space and to which the motion of the body is referred according to what shown in Fig. 2.1.

The link between x, y, z (present coordinates) and a, b, c (invariable coordinates) represents the global rigidity constraint and is given, with Piola's symbols, by [55][25]:

$$\begin{aligned} x &= f + \alpha_1 a + \beta_1 b + \gamma_1 c \\ y &= g + \alpha_2 a + \beta_2 b + \gamma_2 c \\ z &= h + \alpha_3 a + \beta_3 b + \gamma_3 c, \end{aligned} \tag{2.2}$$

where f, g, h and $\alpha_i, \beta_i, \gamma_i, \ (i = 1, 2, 3)$ are respectively the coordinates of the origin and the direction cosines of a_1, a_2, a_3 with respect to x, y, z.

Piola proved the validity and the independence of the two sets of six relations which represent the local rigidity conditions:

$$\begin{aligned} \left(\frac{dx}{da}\right)^2 + \left(\frac{dy}{da}\right)^2 + \left(\frac{dz}{da}\right)^2 &= 1 \\ \left(\frac{dx}{db}\right)^2 + \left(\frac{dy}{db}\right)^2 + \left(\frac{dz}{db}\right)^2 &= 1 \\ \left(\frac{dx}{dc}\right)^2 + \left(\frac{dy}{dc}\right)^2 + \left(\frac{dz}{dc}\right)^2 &= 1 \end{aligned} \tag{2.3}$$

---

[22] Though the original paper was published in 1832 [52], in the following reference is made to a booklet published in 1833 [55], better known.

[23] The title of Piola's paper [55] contained the subtitle "Memoria prima", i.e., first paper (of a series). Piola spoke, for instance on p. 227, about a "successiva memoria", that is a further paper which was intended to complete the study; such a paper never appeared.

[24] Sect. XI, art. 4.

[25] p. 9.

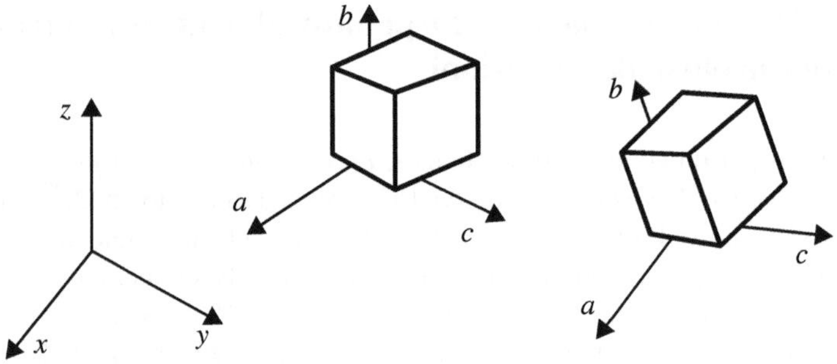

Fig. 2.1 Local and global frames

$$\left(\frac{dx}{da}\right)\left(\frac{dx}{db}\right)+\left(\frac{dy}{da}\right)\left(\frac{dy}{db}\right)+\left(\frac{dz}{da}\right)\left(\frac{dz}{db}\right)=0$$
$$\left(\frac{dx}{da}\right)\left(\frac{dx}{dc}\right)+\left(\frac{dy}{da}\right)\left(\frac{dy}{dc}\right)+\left(\frac{dz}{da}\right)\left(\frac{dz}{dc}\right)=0$$
$$\left(\frac{dx}{db}\right)\left(\frac{dx}{dc}\right)+\left(\frac{dy}{db}\right)\left(\frac{dy}{dc}\right)+\left(\frac{dz}{db}\right)\left(\frac{dz}{dc}\right)=0.$$

$$\begin{aligned}\left(\frac{dx}{da}\right)^2+\left(\frac{dx}{db}\right)^2+\left(\frac{dx}{dc}\right)^2&=1\\ \left(\frac{dy}{da}\right)^2+\left(\frac{dy}{db}\right)^2+\left(\frac{dy}{dc}\right)^2&=1\\ \left(\frac{dz}{da}\right)^2+\left(\frac{dz}{db}\right)^2+\left(\frac{dz}{dc}\right)^2&=1\end{aligned} \tag{2.4}$$

$$\left(\frac{dx}{da}\right)\left(\frac{dy}{da}\right)+\left(\frac{dx}{db}\right)\left(\frac{dy}{db}\right)+\left(\frac{dx}{dc}\right)\left(\frac{dy}{dc}\right)=0$$
$$\left(\frac{dx}{da}\right)\left(\frac{dz}{da}\right)+\left(\frac{dx}{db}\right)\left(\frac{dz}{db}\right)+\left(\frac{dx}{dc}\right)\left(\frac{dz}{dc}\right)=0$$
$$\left(\frac{dy}{da}\right)\left(\frac{dz}{da}\right)+\left(\frac{dy}{db}\right)\left(\frac{dz}{db}\right)+\left(\frac{dy}{dc}\right)\left(\frac{dz}{dc}\right)=0.$$

He proved that only six of the scalar equations (2.3) and (2.4) represent independent conditions [55].[26] He also remarked that in the rigid motion provided by the Eq. (2.2) the partial derivatives of the present coordinates with respect to the invariable ones

[26] '*Nota I*^{A}' in the Appendix, pp. 28–30.

coincide with the direction cosines: $\alpha_1, \beta_1, \gamma_1, \alpha_2, \beta_2, \gamma_2, \alpha_3, \beta_3, \gamma_3$, for instance $\alpha_1 = dx/da$; $\gamma_3 = dz/dc$.

Piola tried to understand if there could be fewer than six independent scalar equations expressing the local rigidity constraint. He advanced the hypothesis that the minimum number of scalar equations of internal constraint is three, and that the other three make the problem undetermined. From a modern point of view, we accept that a rigid three-dimensional body is internally statically undetermined; it is highly probable, however, that such a fact disturbed Piola, who advanced some obscure statements to infer that there are only three independent internal constraints.[27]

In the *Mécanique analytique* Lagrange studied statics and dynamics of elastic and incompressible fluids. The incompressible case was developed in a wide general notation, which allowed Lagrange to extend the results he obtained to compressible fluids. The local constraint of rigidity in hydrostatics was provided by dV =const., where dV is an infinitesimal of volume in the 18th century sense (actual infinity) and the constraint equation for virtual displacements was [40][28]:

$$\delta dV = dV \left(\frac{d\,\delta x}{dx} + \frac{d\,\delta y}{dy} + \frac{d\,\delta z}{dz} \right) = 0. \tag{2.5}$$

The result had been obtained first by Euler with the use of velocity instead of virtual displacements [26, 75].[29] Though Piola could not accept Lagrange's reasoning, based on actual infinity, he could accept the conclusion: the local constraint equation must be in differential form.

Piola then began to use the methods of analytical mechanics and wrote the equation of the "momenti delle forze acceleratrici", that is the virtual work of the mass density of forces X, Y, Z (both bulk and inertial forces), as an integral over the mass of body, equated to zero:

$$\mathrm{S}\left[\left(\frac{d^2x}{dt^2} - X\right)\delta x + \left(\frac{d^2y}{dt^2} - Y\right)\delta y + \left(\frac{d^2z}{dt^2} - Z\right)\delta z\right] Dm = 0, \tag{2.6}$$

where Dm is the body mass element and S means integration or summation over all the elementary masses. Piola stated that the Eq. (2.6) could be reduced to a volume integral (integrale triplicato) defined over the domain of the invariable coordinates a_i, as:

$$\mathrm{S}da\mathrm{S}db\mathrm{S}dc\ \Gamma H\left[\left(\frac{d^2x}{dt^2} - X\right)\delta x + \left(\frac{d^2y}{dt^2} - Y\right)\delta y + \left(\frac{d^2z}{dt^2} - Z\right)\delta z\right] = 0, \tag{2.7}$$

[27] This is put into evidence also in [71], art. 762, p. 420.

[28] Sect. VII, art. 11.

[29] p. 290; 101.

where Γ is the mass density in the present configuration, H the Jacobian of the transformation from (a, b, c) to (x, y, z), and $(\delta x, \delta y, \delta z)$ the virtual displacement of a generic material point of the body [55].[30] Though Piola was dealing with a rigid body motion he introduced the Jacobian H of the coordinate transformation from (a, b, c) to (x, y, z). Its introduction is useless for a rigid body motion where $H = 1$; but it allows one to extend the analysis to the case of deformable bodies.

The use of a reference frame other than the present one was introduced by Lagrange in hydrodynamics. Since the differential problem for dynamics is more difficult than that for statics, Lagrange tried to simplify it by pulling the equations back to the reference configuration, in which the coordinates of a fluid point are a, b, c (like Piola): all quantities in the dynamical equations shall then be functions of a, b, c. In particular, Lagrange pulled back the volume element from the present to the invariable configuration [42][31]:

$$dx\,dy\,dz = \Theta\,da\,db\,dc, \tag{2.8}$$

where Θ, corresponding to the Jacobian H in the Eq. (2.7), called *sestinomio* by Piola, is the coefficient which makes it possible to invert the expressions:

$$d\xi = \frac{\partial \xi}{\partial a} da + \frac{\partial \xi}{\partial b} db + \frac{\partial \xi}{\partial c} dc, \quad \xi = x, y, z. \tag{2.9}$$

Lagrange recognized that for incompressible fluids $\Theta = 1$; in spite of this, he never simplified that factor, and neither did Piola for his sestinomio. Lagrange commented:

> It must be remarked that this value of $Dx\,Dy\,Dz$ is what we must use in the triple integrals with respect to x, y, z, when we want to replace them by assigned functions of other variables a, b, c [42].[32] (A.2.8)

The "equazione generalissima" (2.7) holds only for $\delta x, \delta y, \delta z$ satisfying the constraint relations (2.3) or (2.4). To impose these constraints, Piola followed Lagrange's approach for the one-dimensional rigid bodies, by adding to the integral on the left side of the (2.7), the integral of the variational version of the constraint relations. He started by considering first the variation associated to the constraints (2.4).

Introducing the Lagrange multipliers (A, B, C, D, E, F), one for each of the six variations of the relations (2.4), Piola obtained the *dynamical equations* [55][33] by adding to the integral (2.7) the following integrals [55][34]:

---

[30] p. 15.

[31] Sect. XI, arts. 4–7.

[32] pp. 284–285. Our translation.

[33] p. 17. Piola referred to his equations as the *dynamical equations*, reserving the term *equilibrium equations* to static cases.

[34] p. 15, eq. [18].

$$\begin{aligned}
&\mathrm{S}da\,\mathrm{S}db\,\mathrm{S}dc\cdot A\left\{\left(\frac{dx}{da}\right)\left(\frac{d\,\delta x}{da}\right)+\left(\frac{dx}{db}\right)\left(\frac{d\,\delta x}{db}\right)+\left(\frac{dx}{dc}\right)\left(\frac{d\,\delta x}{dc}\right)\right\}\\
&\mathrm{S}da\,\mathrm{S}db\,\mathrm{S}dc\cdot B\left\{\left(\frac{dy}{da}\right)\left(\frac{d\,\delta y}{da}\right)+\left(\frac{dy}{db}\right)\left(\frac{d\,\delta y}{db}\right)+\left(\frac{dy}{dc}\right)\left(\frac{d\,\delta y}{dc}\right)\right\}\\
&\mathrm{S}da\,\mathrm{S}db\,\mathrm{S}dc\cdot C\left\{\left(\frac{dz}{da}\right)\left(\frac{d\,\delta z}{da}\right)+\left(\frac{dz}{db}\right)\left(\frac{d\,\delta z}{db}\right)+\left(\frac{dz}{dc}\right)\left(\frac{d\,\delta z}{dc}\right)\right\}\\
&\mathrm{S}da\,\mathrm{S}db\,\mathrm{S}dc\cdot F\left\{\left(\frac{dx}{da}\right)\left(\frac{d\,\delta y}{da}\right)+\left(\frac{dx}{db}\right)\left(\frac{d\,\delta y}{db}\right)+\left(\frac{dx}{dc}\right)\left(\frac{d\,\delta y}{dc}\right)\right\}\\
&\qquad+\left\{\left(\frac{dy}{da}\right)\left(\frac{d\,\delta x}{da}\right)+\left(\frac{dy}{db}\right)\left(\frac{d\,\delta x}{db}\right)+\left(\frac{dy}{dc}\right)\left(\frac{d\,\delta x}{dc}\right)\right\}\\
&\mathrm{S}da\,\mathrm{S}db\,\mathrm{S}dc\cdot E\left\{\left(\frac{dx}{da}\right)\left(\frac{d\,\delta z}{da}\right)+\left(\frac{dx}{db}\right)\left(\frac{d\,\delta z}{db}\right)+\left(\frac{dx}{dc}\right)\left(\frac{d\,\delta z}{dc}\right)\right\}\\
&\qquad+\left\{\left(\frac{dz}{da}\right)\left(\frac{d\,\delta x}{da}\right)+\left(\frac{dz}{db}\right)\left(\frac{d\,\delta x}{db}\right)+\left(\frac{dz}{dc}\right)\left(\frac{d\,\delta x}{dc}\right)\right\}\\
&\mathrm{S}da\,\mathrm{S}db\,\mathrm{S}dc\cdot D\left\{\left(\frac{dy}{da}\right)\left(\frac{d\,\delta z}{da}\right)+\left(\frac{dy}{db}\right)\left(\frac{d\,\delta z}{db}\right)+\left(\frac{dy}{dc}\right)\left(\frac{d\,\delta z}{dc}\right)\right\}\\
&\qquad+\left\{\left(\frac{dz}{da}\right)\left(\frac{d\,\delta y}{da}\right)+\left(\frac{dz}{db}\right)\left(\frac{d\,\delta y}{db}\right)+\left(\frac{dz}{dc}\right)\left(\frac{d\,\delta y}{dc}\right)\right\}.
\end{aligned}\tag{2.10}$$

To obtain expressions in which the variation δx, δy, δz are not affected by derivatives, which allows their reduction, Piola applied integration by parts, following the rules of the calculus of variations [42].[35] The variational problem resulting from (2.7) to (2.10) gives place to two systems of definite integrals, one in the volume of the extended body and the other on its surface. Piola studied the first one and claimed to deal with the second in a further paper. After some passages, Piola obtained [55][36]:

$$\begin{aligned}
\Gamma H\left[X-\left(\frac{d^2x}{dt^2}\right)\right]+P&=0\\
\Gamma H\left[Y-\left(\frac{d^2y}{dt^2}\right)\right]+Q&=0\\
\Gamma H\left[Z-\left(\frac{d^2z}{dt^2}\right)\right]+R&=0,
\end{aligned}\tag{2.11}$$

where P, Q, R are functions of (A, B, C, D, E, F) and of the derivatives of x, y, z with respect to a, b, c. For the sake of space only the expression of P is given:

[35] Sect. IV, articles 14 and 15.

[36] p. 17.

$$P = -\frac{d\cdot A\left(\frac{dx}{da}\right)}{da} - \frac{d\cdot A\left(\frac{dx}{db}\right)}{db} - \frac{d\cdot A\left(\frac{dx}{dc}\right)}{dc}$$
$$-\frac{d\cdot F\left(\frac{dy}{da}\right)}{da} - \frac{d\cdot F\left(\frac{dy}{db}\right)}{db} - \frac{d\cdot F\left(\frac{dy}{dc}\right)}{dc} \tag{2.12}$$
$$-\frac{d\cdot E\left(\frac{dz}{da}\right)}{da} - \frac{d\cdot E\left(\frac{dz}{db}\right)}{db} - \frac{d\cdot E\left(\frac{dz}{dc}\right)}{dc}.$$

Without giving any interpretation of this result, Piola moved on to prove that the Eq. (2.11) may be reduced to the form of the local equilibrium equations provided by Cauchy and Poisson.[37] For this purpose, he proved a theorem [55][38] which lets us transform differential operators with respect to (a, b, c) into differential operators with respect to (x, y, z). In particular if $K(a, b, c, t)$ is a general function, such that:

$$K(a, b, c, t) = K\left[x(a, b, c, t), y(a, b, c, t), z(a, b, c, t)\right], \tag{2.13}$$

Piola obtained the following relations:

$$\begin{aligned} \left(\frac{d\cdot\alpha K}{da}\right) + \left(\frac{d\cdot\beta K}{db}\right) + \left(\frac{d\cdot\gamma K}{dc}\right) &= H\left(\frac{dK}{dx}\right) \\ \left(\frac{d\cdot\alpha' K}{da}\right) + \left(\frac{d\cdot\beta' K}{db}\right) + \left(\frac{d\cdot\gamma' K}{dc}\right) &= H\left(\frac{dK}{dy}\right) \\ \left(\frac{d\cdot\alpha'' K}{da}\right) + \left(\frac{d\cdot\beta'' K}{db}\right) + \left(\frac{d\cdot\gamma'' K}{dc}\right) &= H\left(\frac{dK}{dz}\right), \end{aligned} \tag{2.14}$$

where $\alpha, \beta, \gamma, \alpha', \beta', \gamma', \ldots$ depend on the derivatives of x, y, z with respect to a, b, c. By using the transformation rule (2.14), the following set of equations was obtained from the relation (2.11) [55][39]:

$$\begin{aligned} \Gamma\left[X - \left(\frac{d^2x}{dt^2}\right)\right] + \frac{dA}{dx} + \frac{dF}{dy} + \frac{dE}{dz} &= 0 \\ \Gamma\left[Y - \left(\frac{d^2y}{dt^2}\right)\right] + \frac{dF}{dx} + \frac{dB}{dy} + \frac{dD}{dz} &= 0 \\ \Gamma\left[Z - \left(\frac{d^2z}{dt^2}\right)\right] + \frac{dE}{dx} + \frac{dD}{dy} + \frac{dC}{dz} &= 0. \end{aligned} \tag{2.15}$$

[37] The papers to which Piola referred are [17, 18, 61–63]. It is remarkable that the model of the body is continuous in Cauchy's papers, while it is discrete in those poisson of.

[38] '*Nota III*^{A}' in the Appendix, pp. 34–36.

[39] p. 30.

Piola commented upon this result:

> Look at the perfect coincidence between this result and that obtained by the two famous geometers cited at the beginning of the introduction [Cauchy and Poisson], by following a different reasoning and in the cases of both equilibrium and motion separately considered. I suggest to note that in my analysis A, B, C, D, E, F are not tensions which act over different surfaces, but they are coefficients to which in the following I also will attribute the meaning of forces, when it will be convenient to me: they are functions of x, y, z, t with an expression not yet known, but which we know does not change when passing from one part of the body to the other. I can be objected that these equations [...] were obtained by the methods of the A. M. only for solid rigid systems, while those of the two famous French men refer to elastic and variable systems, also. I reply that in the present memoir I will show how they can be generalised to include all the cases considered by the mentioned authors without leaving the analytical way taught by Lagrange [55].[40] (A.2.9)

Piola was dissatisfied with his results and therefore tried a different approach. He then maintained that in order to extend his results to the general case of deformable bodies it is useful to examine the variational problem obtained using the constraints (2.3) instead of (2.4), which introducing six new Lagrange multipliers A', B', C', D', E', F', gives the following integrals to be added to the variational Eq. (2.11):

$$
\begin{aligned}
&\mathrm{S}da\,\mathrm{S}db\,\mathrm{S}dc\cdot A'\left\{\left(\frac{dx}{da}\right)\left(\frac{d\,\delta x}{da}\right)+\left(\frac{dy}{da}\right)\left(\frac{d\,\delta y}{da}\right)+\left(\frac{dz}{da}\right)\left(\frac{d\,\delta z}{da}\right)\right\}\\
&\mathrm{S}da\,\mathrm{S}db\,\mathrm{S}dc\cdot B'\left\{\left(\frac{dx}{db}\right)\left(\frac{d\,\delta x}{db}\right)+\left(\frac{dy}{db}\right)\left(\frac{d\,\delta y}{db}\right)+\left(\frac{dz}{db}\right)\left(\frac{d\,\delta z}{db}\right)\right\}\\
&\mathrm{S}da\,\mathrm{S}db\,\mathrm{S}dc\cdot C'\left\{\left(\frac{dx}{dc}\right)\left(\frac{d\,\delta x}{dc}\right)+\left(\frac{dy}{dc}\right)\left(\frac{d\,\delta y}{dc}\right)+\left(\frac{dz}{dc}\right)\left(\frac{d\,\delta z}{dc}\right)\right\}\\
&\mathrm{S}da\,\mathrm{S}db\,\mathrm{S}dc\cdot F'\left\{\left(\frac{dx}{da}\right)\left(\frac{d\,\delta x}{db}\right)+\left(\frac{dy}{da}\right)\left(\frac{d\,\delta y}{db}\right)+\left(\frac{dz}{da}\right)\left(\frac{d\,\delta z}{db}\right)\right\}\\
&\qquad+\left\{\left(\frac{dx}{db}\right)\left(\frac{d\,\delta x}{da}\right)+\left(\frac{dy}{db}\right)\left(\frac{d\,\delta y}{da}\right)+\left(\frac{dz}{db}\right)\left(\frac{d\,\delta z}{da}\right)\right\}\\
&\mathrm{S}da\,\mathrm{S}db\,\mathrm{S}dc\cdot E'\left\{\left(\frac{dx}{da}\right)\left(\frac{d\,\delta x}{dc}\right)+\left(\frac{dy}{da}\right)\left(\frac{d\,\delta y}{dc}\right)+\left(\frac{dz}{da}\right)\left(\frac{d\,\delta z}{dc}\right)\right\}\\
&\qquad+\left\{\left(\frac{dx}{dc}\right)\left(\frac{d\,\delta x}{da}\right)+\left(\frac{dy}{dc}\right)\left(\frac{d\,\delta y}{da}\right)+\left(\frac{dz}{dc}\right)\left(\frac{d\,\delta z}{da}\right)\right\}\\
&\mathrm{S}da\,\mathrm{S}db\,\mathrm{S}dc\cdot D'\left\{\left(\frac{dx}{db}\right)\left(\frac{d\,\delta x}{dc}\right)+\left(\frac{dy}{db}\right)\left(\frac{d\,\delta y}{dc}\right)+\left(\frac{dz}{db}\right)\left(\frac{d\,\delta z}{dc}\right)\right\}\\
&\qquad+\left\{\left(\frac{dx}{dc}\right)\left(\frac{d\,\delta x}{db}\right)+\left(\frac{dy}{dc}\right)\left(\frac{d\,\delta y}{db}\right)+\left(\frac{dz}{dc}\right)\left(\frac{d\,\delta z}{db}\right)\right\}.
\end{aligned}
\tag{2.16}
$$

Piola did not say why he thought that this procedure was more general, as he did not explain why he took the constraint equations (2.4) in the first choice. By applying

[40] p. 220. Our translation.

the integration by parts and ignoring the contribution of the surface integrals, Piola obtained an equation similar to the equation (2.11), which after lengthy passages and using the transformation rule (2.14) gave:

$$\begin{aligned}
&\Gamma\left[X-\left(\frac{d^2x}{dt^2}\right)\right]+\frac{dA''}{dx}+\frac{dF''}{dy}+\frac{dE''}{dz}=0\\
&\Gamma\left[Y-\left(\frac{d^2y}{dt^2}\right)\right]+\frac{dF''}{dx}+\frac{dB''}{dy}+\frac{dD''}{dz}=0\\
&\Gamma\left[Z-\left(\frac{d^2z}{dt^2}\right)\right]+\frac{dE''}{dx}+\frac{dD''}{dy}+\frac{dC''}{dz}=0
\end{aligned} \tag{2.17}$$

where $(A'', B'', C'', D'', E'', F'')$ are coefficients related to the Lagrange multipliers: (A', B', C', D', E', F') by relations involving the derivatives of x, y, z with respect to (a, b, c); for instance:

$$\begin{aligned}
HA'' = {} & A'\left(\frac{dx}{da}\right)^2+B'\left(\frac{dx}{da}\right)^2+C'\left(\frac{dx}{da}\right)^2\\
&+2F'\left(\frac{dx}{da}\right)\left(\frac{dx}{db}\right)+2E'\left(\frac{dx}{da}\right)\left(\frac{dx}{dc}\right)+2D'\left(\frac{dx}{db}\right)\left(\frac{dx}{dc}\right).
\end{aligned} \tag{2.18}$$

In the case of a rigid body motion $H = 1$, and the derivative of (x, y, z) with respect to (a, b, c) are the direction cosines of the transformation; a modern reader should not have difficulty to see that the relations between $(A'', B'', C'', D'', E'', F'')$ and (A', B', C', D', E', F') is the same as that occurring between the components of a second order tensor undercoming a change of coordinates.

2.3.2 *1836.* Nuova analisi per tutte le questioni della meccanica molecolare

The article *Nuova analisi per tutte le questioni della meccanica molecolare* [56] (hereinafter *Nuova analisi*) was submitted in 1835 and published in 1836. It could be seen as a turning point in Piola's mechanical conceptions, the passage from an 'ancient' and continuous to a 'modern' and discrete model of matter. The discrete model was well affirmed in the French scientific community and was paid particular attention to by Poisson, who wanted to develop a system of physical mechanics based on a model of matter made up of particles interacting with each other, in contrast with the analytical mechanics of continua[41]:

---

[41] Note that Piola nearly neglected Cauchy's formulations by attributing the molecular conception to Poisson only. This can in part be explained by the fact that Piola personally knew and had a high regard for Cauchy, so probably he did not want to criticize him. In [24], p. 291, a letter by Cauchy to Piola is quoted where the French scientist disputed Piola's continuum approach to mechanics and the extended use of variational calculus.

> Lagrange has arrived as far as one can conceive, when he replaced the physical link among bodies with equations for the position of the various points. It is what constitutes *Analytical mechanics*. But besides this admirable conception we can now build *Physical mechanics*, of which the only principle will be to reconduct everything to the molecular forces, which communicate the action of given forces from a point to another and which are the mediators of their equilibrium [61].[42] (A.2.10)

> The use that Lagrange made of this calculus [the calculus of variations] in the *Mécanique analytique* is actually suitable only for continuous masses; and the calculations with which the results so found are extended to natural bodies, must be rejected as inadequate [61].[43] (A.2.11)

Actually the *Nuova analisi* represented only a temporary digression, which was more suffered than accepted. Piola somehow reluctantly accepted Poisson's claim that continuum mechanics should start from a consideration of molecular forces, but did not want to abandon Lagrange's techniques completely:

> Mr. Poisson [...] would reduce all to the molecular actions only. I accept this point of view by actually considering, in addition to the external forces, only a reciprocal action of attraction and repulsion [...]. It is not that I believe necessary to leave the other way used by Lagrange, on the contrary, I am convinced that with it we can fruitfully treat many modern problems, and I already published an essay which can partially prove this claim of mine [56].[44] (A.2.12)

> Some new theorems have been obtained, but a large part of the advantages and beauties of an analysis elaborated by our teachers by means of long studies has been lost [56].[45] (A.2.13)

So, even though Piola abandoned the continuum model, he retained the virtual work as the basic principle from which to form the dynamical equations of the corpuscular model of matter:

> To show that the analyses by d'Alembert,indexAlembert, Jean Baptiste le Ronde D' (1717–1783) Euler and Lagrange is still valid, by supposing, along with the moderns, the matter as discontinuous: to maintain the treasure of science transmitted by our predecessors and in the meantime to progress with the enlightenment of our century [56].[46] (A.2.14)

Almost all of the mathematical procedures contained in this paper stem from the attempt to put in accord the results obtained by the physical and mathematical discrete model of matter with those of the mathematical continuum model. Indeed, Piola was not satisfied by the argument used also by Lagrange in which, by interpreting molecules as infinitesimal volume elements, infinite sums are turned directly into definite integrals. Todhunter and Pearson [71] provide detailed comments upon the mathematical aspects of the first two sections of Piola's paper,[47] with reference to theorems of the calculus of finite differences.

---

[42] p. 361. Our translation.

[43] p. 400. Our translation.

[44] p. 171. Our translation.

[45] p. 155. Our translation.

[46] pp. 155–156. Our translation.

[47] arts. 769–772, pp. 422–425.

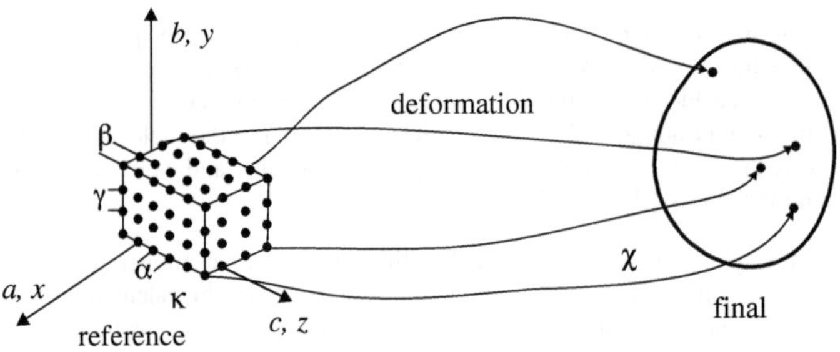

Fig. 2.2 Reference configuration

In order to be as thorough as possible, Piola introduced an original concept, which he used again in the following work of 1848. The reference configuration of the molecules of a body, labelled by the invariable coordinates (a, b, c) like in the paper of 1832 (Fig. 2.2), is an imaginary regular lattice disposition, to which Piola gave the intuitive meaning of an

> ideal configuration preceding the true status, in which the matter of the body itself was contained in a parallelepiped [...] and all the *a*s do not differ by more than an amount equal to α, the *b*s by more than β, and the *c* by no more than γ [...] [56].[48] (A.2.15)

The use of this regular lattice allowed Piola to write summations where the spatial difference of coordinates is constant so that

> the irregularity due to the discontinuity of the matter [...] I obtains a regularity [...] necessary for the calculus used by Lagrange in the Analytical mechanics [56].[49] (A.2.16)

For this kind of summations the application of theorems which link summation ("integrale finito definito") to integral ("integrale continuo definito") is easiest. All the theorems presented in the first two sections of the *Nuova analisi* aimed to provide expressions of internal forces among molecules in terms of a suitable series expansion of a non-linear function of the mutual distance within pairs of molecules. Later, again using the principle of virtual work, but with no constraint equations for the present position (x, y, z), Piola obtained the dynamical equations in the following form [56][50]:

$$\begin{aligned} X - \frac{d^2x}{dt^2} + \frac{dL_1}{da} + \frac{dL_2}{db} + \frac{dL_3}{dc} &= 0 \\ Y - \frac{d^2y}{dt^2} + \frac{dM_1}{da} + \frac{dM_2}{db} + \frac{dM_3}{dc} &= 0 \\ Z - \frac{d^2z}{dt^2} + \frac{dN_1}{da} + \frac{dN_2}{db} + \frac{dN_3}{dc} &= 0. \end{aligned} \tag{2.19}$$

---

[48] p. 167. Our translation.

[49] p. 167. Our translation.

[50] p. 201.

The Eq. (2.19) have the same form of the Eq. (2.11) but for the missing product ΓH, the mass density in the ideal state, now supposed to be uniform and equal to unity. The nine coefficients L_i, M_i, N_i, are functions of derivatives of force functions with respect to (a, b, c). Piola then pushed the Eq. (2.19) forward to the present configuration with the theorem (2.14), found in the *Meccanica de' corpi naturalmente estesi* [56][51]:

$$\begin{aligned} \Gamma\left(X - \frac{d^2x}{dt^2}\right) + \frac{dP_1}{dx} + \frac{dP_2}{dy} + \frac{dP_3}{dz} &= 0 \\ \Gamma\left(Y - \frac{d^2y}{dt^2}\right) + \frac{dQ_1}{dx} + \frac{dQ_2}{dy} + \frac{dQ_3}{dz} &= 0 \\ \Gamma\left(Z - \frac{d^2z}{dt^2}\right) + \frac{dR_1}{dx} + \frac{dR_2}{dy} + \frac{dR_3}{dz} &= 0, \end{aligned} \tag{2.20}$$

where Γ is a scalar function of (x, y, z).

Piola went on to deal with the simplification of the dynamical equations obtained in the previous sections on the basis of constitutive assumptions on molecular interaction. Following Poisson [61],[52] he prescribed that the molecular interaction is negligible for sensible distances (localization assumption), so that some higher-order terms in his former expressions can be discarded:

> The expression of the molecular action may have a sensible value for nearby points [...], the molecular action is insensible at sensible distances [...]. [...] the radius of the activity sphere of the molecular action, even though reaches a very large number of molecules, must still be considered as an insensible quantity [...] [56].[53] (A.2.17)

Under these conditions Piola found the following relations:

$$P_2 = Q_1, \quad P_3 = R_1, \quad Q_3 = R_2; \quad \Gamma = \frac{1}{H}, \tag{2.21}$$

so that the Eq. (2.19) are similar to those obtained by Poisson [61, 63],[54] as the right sides only depend on six coefficients.

Unlike in the *Meccanica de' corpi naturalmente estesi*, where he did not comment upon the Eq. (2.11) and the coefficients appearing in it, Piola now gave a physical meaning to the coefficients $L_1, L_2, L_3, M_1, M_2, M_3, N_1, N_2, N_3$ under the localization assumption. They are functions of (a, b, c) and represent the components of stress on planes through a point (x, y, z) corresponding to planes through a given (a, b, c) in the ideal state.

It is interesting that, in order to obtain the Eq. (2.20), Piola did not introduce any particular constitutive assumption except for a generic dependence of the molecular force on the distance among pairs of particles. This approach is different from that of

---

[51] p. 212.

[52] p. 369.

[53] pp. 248, 253. Our translation.

[54] p. 387; 578–579.

Poisson and Cauchy [26, 28, 61–63], who introduced linearization and localization together with some isotropy assumptions. This procedure of Piola is very important also nowadays from the point of view of a general theory of continua because it makes local dynamical equations independent from the constitutive law.

2.3.3 *1848.* Intorno alle equazioni fondamentali del movimento di corpi qualsivogliono

Piola's paper *Intorno alle equazioni fondamentali del movimento di corpi qualsivogliono, considerati secondo la naturale loro forma e costituzione* [58] (hereinafter *Intorno alle equazioni fondamentali*) was submitted in 1845 and published in 1848. It contains a mature and complete revision of the article of 1832, Piola having eliminated some mistakes and naiveness which he himself recognised. As a matter of fact, some twelve years had passed since the publication of the *Meccanica de' corpi naturalmente estesi* and mathematics and mechanics had moved forward somewhat. In mathematics, the theory of integration had achieved some important results thanks to Cauchy's work and this made the passage from discrete to continuum less problematic for Piola. In mechanics, Cauchy, Green and Saint-Venant had introduced the idea of strain in a clear form. The main differences with the article of 1832 are: the derivation of dynamical equations extended to the case of deformable bodies; the study of the terms appearing as surface integrals in the application of the calculus of variations; and the particularization of the general expressions for two- and one-dimensional continua.

Piola began his paper of 1848 by re-affirming the superiority of Lagrange's approach compared with others, almost apologizing for having partially abandoned it in the *Nuova analisi* and for his previous naiveness:

> I often wrote that I do not think it necessary to create a new Mechanics, departing from the bright methods of the Analytical mechanics by Lagrange [...]. I was, and am still, faced by very respectable authorities in front of which I should surrender [...]. But [...] I thought to be convenient [...] to collect in this memoir my opinions on the subject [...]. Because I do not hide now that in my preceding writings some of my ideas were exposed with insufficient maturity; there are some too much fearful and some too much daring. Some parts of these writings could be omitted, [...] mainly those other which [...] I would not repeat anymore [58].[55] (A.2.18)

Piola replied to Poisson, who claimed that Lagrange's methods were too abstract:

> I hope to make it clear in the following Memoir that the only reason for which Analytical mechanics seemed inferior in treating some problems was that Lagrange, while writing about equilibrium and motion of a solid body, did not come to give equations proper for any point of it. If he had made it, and he could without leaving the methods taught in his book, he would have reached easily the same equations to whom the famous Geometers of our time

[55] pp. 1–2. Our translation.

> arrived with much fatigue, and that now are the basis for new theories. But what he did not accomplish [...] can be made by others [58].[56] (A.2.19)

In the first section of *Intorno alle equazioni fondamentali*, Piola provided some preliminary notions, among which he recalled that of ideal disposition with uniform unit mass density. The present position (x, y, z) depends on that in the ideal state $(a, b, c), x = x(a, b, c), y = y(a, b, c), z = z(a, b, c)$. Piola moved on to bodies with non uniform density and described the density in the present configuration via the Jacobian H of the transformation from (a, b, c) to (x, y, z), so that he could express the equation of continuity. As Piola himself remarked, there were arguments which revealed the use of a mature and up-to-date theory of limits and integration, approaching the 'modern' Sylvestre Franlois de Lacroix [38][57] and Bordoni [2][58]rather than the 'ancient' Lagrange, when dealing with the passage from discrete to continuous one-dimensional case:

> We have a theorem which gives us the means to pass from a finite definite integral [a summation] to a well defined continuous integral [an ordinary definite integral] [58].[59] (A.2.20)

Two- and three-dimensional cases were reduced to the one-dimensional situation.

The second section is devoted to extending the equation of virtual work from the discrete to the continuum, in the case of three-, two-, and one-dimensional bodies. For three-dimensional bodies, Piola eventually obtained the equation of virtual work in the ideal state [58][60]:

$$\begin{aligned}\int da \int db \int dc \cdot \left\{\left(X - \frac{d^2x}{dt^2}\right)\delta x + \left(Y - \frac{d^2y}{dt^2}\right)\delta y + \left(Z - \frac{d^2z}{dt^2}\right)\delta z\right\}\\ + \int da \int db \int dc \cdot [A\,\delta t_1 + B\,\delta t_2 + C\,\delta t_3 + D\,\delta t_4 +\\ E\,\delta t_5 + F\,\delta t_6] + \Omega = 0,\end{aligned} \tag{2.22}$$

where the density is not made explicit since it is supposed to be unitary and uniform in the ideal state; (A, B, C, D, E, F) is a list of Lagrange multipliers; $(t_1, t_2, t_3, t_4, t_5, t_6)$ is a list of six scalar constraint equations like the Eqs. (2.3), (2.4); and Ω represents the contribution of external surface forces.

---

[56] p. 4. Our translation.

[57] vol. II, p. 97.

[58] vol. II, p. 489.

[59] p. 42. Our translation. The theorem referred to by Piola is found in [38], for example, in the form:

$$\int udx = h\sum u + \alpha h^2 \sum \frac{du}{dx} + \beta h^3 \sum \frac{d^2u}{dx^2} + \cdots$$

where $\alpha, \beta, \ldots$ are numerical coefficients and h is a small quantity (the step of discretization).

[60] p. 60; eq. (10).

In the third section, Piola obtained the dynamical equations for a rigid body. He used constraint equations equivalent to the Eq. (2.3) but which are now more general, since the identity between the derivatives of (x, y, z) with respect to (a, b, c) and the direction cosines of the change of coordinates is no longer assumed. By following the same steps he had used in the paper of 1832, but with some improvements to their form, he got local dynamical equations in the ideal state, equivalent to the Eq. (2.11). The comments Piola added to this result are particularly revealing; indeed, he declared that equations such as (2.11) had no physical meaning for him, because the ideal state does not represent an actual physical state of the body:

> [...] it would be useful to transform these equations [...] into others not containing a, b, c but only quantities pertaining to the real state of the body [...] [58].[61] (A.2.21)

By means of his push-forward theorem, which transforms quantities in the reference to the present configuration, expressed by the relation (2.14), Piola then obtained the dynamical equations in the present configuration, which have the same form as the Eq. (2.18).

In the fourth section, Piola derived the local dynamical equations for a deformable body. He started by affirming that it is impossible to provide a specific form of local constraint equations as he had done in the case of the rigid body, and that some other procedure is needed. This is effectuated by the introduction of an intermediate configuration labelled by (p, q, r), so that the path from (a, b, c) to (x, y, z) turns into a path from (a, b, c) to (p, q, r) and another path from (p, q, r) to (x, y, z); this second step is assumed to be rigid (see Fig. 2.3). With this assumption, the mass density Γ in (p, q, r) is the same as in (x, y, z), and Piola could write the equation of virtual work in (p, q, r), by using the transformation $p = p(a, b, c),\ q = q(a, b, c),\ r = r(a, b, c)$ [58][62]:

$$\begin{aligned}\int dp \int dq \int dr \cdot \Gamma \left\{ \left(X - \frac{d^2x}{dt^2} \right) \delta x + \cdots \right\} \\ + \int dp \int dq \int dr \cdot \Gamma G + \Omega^* = 0,\end{aligned} \tag{2.23}$$

where G represents the contribution of constraints similar to the Eqs. (2.3) and (2.4), where the derivatives should be intended with respect to (p, q, r) instead than to (a, b, c) and Ω^* represents the contribution of surface forces seen in (p, q, r). Piola's trick reduced the equation of virtual work with unknown constraint equations to an equation where the constraint equations are known. In fact, the motion from (p, q, r) to (x, y, z) is rigid, hence the constraint equations are the same as in Eq. (2.3). Thus, Piola obtained the local dynamical equations in the intermediate configuration (p, q, r) [58][63]:

[61] p. 63. Our translation.

[62] p. 83.

[63] p. 88.

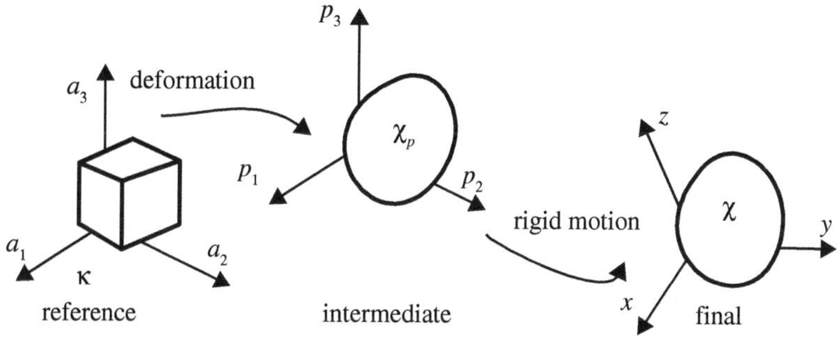

Fig. 2.3 Introduction of the intermediate configuration

$$\begin{aligned}
\Gamma\left(X-\frac{d^2x}{dt^2}\right)+\frac{dL_1}{dp}+\frac{dM_1}{dq}+\frac{dN_1}{dr}&=0\\
\Gamma\left(Y-\frac{d^2y}{dt^2}\right)+\frac{dL_2}{dp}+\frac{dM_2}{dq}+\frac{dN_2}{dr}&=0\\
\Gamma\left(Z-\frac{d^2z}{dt^2}\right)+\frac{dL_3}{dp}+\frac{dM_3}{dq}+\frac{dN_3}{dr}&=0,
\end{aligned} \tag{2.24}$$

and pushed them forward to (p, q, r) by means of the theorem (2.14), obtaining:

$$\begin{aligned}
\Gamma\left(X-\frac{d^2x}{dt^2}\right)+\frac{d\Lambda}{dx}+\frac{d\Sigma}{dy}+\frac{d\Phi}{dz}&=0\\
\Gamma\left(Y-\frac{d^2y}{dt^2}\right)+\frac{d\Sigma}{dx}+\frac{d\Theta}{dy}+\frac{d\Psi}{dz}&=0\\
\Gamma\left(Z-\frac{d^2z}{dt^2}\right)+\frac{d\Phi}{dx}+\frac{d\Psi}{dy}+\frac{d\mathrm{H}}{dz}&=0.
\end{aligned} \tag{2.25}$$

Next, Piola provided an interpretation of the surface integrals arising from the variational procedure in terms of the pressures acting on the boundary of the body. Also, with the calculus of variations he proved a relation which, he says, reproduces Cauchy's theorems on pressures.

> The above mentioned six quantities are in both cases the analytical expression containing the whole effect of all internal forces over the generic point (p, q, r) or (x, y, z) [58].[64] (A.2.22)

Note that, though clever, Piola's reasoning was not entirely conclusive. His analysis was based on the possibility of passing from the unknown constraint equations for the dynamical Eq. (2.22) to the known constraint Eq. (2.23). This reasoning

[64] p. 101. Our translation.

is inconsistent, however, because it is not possible to define constraint equations for a deformable body: the points of a deformable body have no condition, they are free.

In art. 60 of the paper Piola presented a summary of his procedure, starting with some considerations about the virtual displacements δx, δy, δz. Piola's reluctance to introduce virtual displacements as infinitesimals has already been underlined. Piola considered virtual displacements simply as first-order variations of the position of body-points, satisfying the constraint equations. In this article Piola characterized the virtual displacements explicitly, affirming that they coincide with the variation of the coordinates of the same body-point when referred to two coordinate systems shifted, one from the other, by a very small amount, that may vary from body-point to body-point:

> Such a principle lies in the simultaneous reference of any system to two triads of orthogonal axes. It can work in two ways and for both of them it produces great effects. It is used in a first way [...] to prove the principle of virtual velocity, and also those others [principles] regarding conservation of the motion of the center of gravity and of areas. Instead of conceiving the δx, δy, δz of the various points of the system as virtual velocities or infinitesimal spaces described by the fictitious motion (which after [Lazare] Carnot was named geometrical motion),[65] it is more natural and without mystery to consider them as an increase of the coordinates of such points when the system is referred to others three orthogonal axes very near to the first ones, as they were displaced very little [...] so it is clear how the increase of coordinates occurs without alteration of the reciprocal forces of the parts of the system with each other [58].[66] (A.2.24)

In this way, it is apparent that the considered virtual displacements take place without modifying inner forces. In a similar way, the constraint equations in the rigid motion from (p, q, r) to (x, y, z) may be thought of as transformations of constraint equations when the coordinate system varies:

> The simultaneous reference to two triads of orthogonal axes then plays efficiently in another way [...]. Here we intend to speak about the method which leaves to δx, δy, δz all their generalities and considers constraint equations by introducing indeterminate multipliers. In such a way the use of the two triads is useful to write down the above mentioned constraint equations, which, otherwise, could not be given in general [...]. Such a point of view seems to me to have escaped Lagrange's and other geometers' attention: all things in the present Memoir which are worthy of attention are referred to it [...] [58].[67] (A.2.25)

Piola affirmed that this is certainly an original approach of his, and that it escaped the attention of other mechanicians.

---

[65] Lazare Carnot introduced the idea of *geometrical motion* (mouvement géométrique) in the last part of the 18th century:

> If a system of bodies starts to move from a given position, with an arbitrary velocity, but such that it would be also possible for the system to follow another velocity exactly of the same magnitude and opposite, each one of these velocities will be named geometrical velocity [14]. pp. 23–24. Our translation. (A.2.23)

[66] p. 110. Our translation.

[67] p. 111. Our translation.

The last sections of *Intorno alle equazioni fondamentali* were dedicated to the motion of fluids, not directly linked with the subject of our book, and to a reformulation of the treatment of molecular forces in the paper of 1836. Finally Piola extended the results of previous sections to two- and one-dimensional continua.

2.3.4 *1856.* Di un principio controverso della meccanica analitica di lagrange e delle sue molteplici applicazioni

Piola died in 1850, and in 1856 Brioschi, a former pupil of his and at that time professor of rational mechanics in Pavia, posthumously edited Piola's last work on continuum mechanics, *Di un principio controverso della Meccanica Analitica di Lagrange e delle sue molteplici applicazioni* [59] (hereinafter *Di un principio controverso*). This paper was declared to be directly linked with *Intorno alle equazioni fondamentali* and to be its natural completion and refinement. In his work of 1848 Piola said he felt that Lagrange's technique of indeterminate multipliers of the first-order variation of the constraint equations contained something unclear and unproven. Thus, he said, in this article he felt compelled to show how to overcome this difficulty. In the first chapter Piola showed that the first-order variation of the constraint equations in the rigid body motion from (p, q, r) to (x, y, z) could be obtained simply by moving the reference frame of the present configuration. Thus, Piola surpassed the difficulty due to the intermediate configuration (p, q, r) by a very small displacement, which is unknown and in principle may not exist, so that it might be meaningless to operate derivatives with respect to (p, q, r). Starting from this proof, which turned out to be a very effective tool for Piola, who desired to leave nothing to intuition, Piola re-obtained in the rest of the paper the local dynamical equations, extended the results to two- and one-dimensional continua, re-interpreted Lagrange's multipliers as expressions of inner forces and provided a molecular representation for the latter; moreover, he provided a clear and completely modern interpretation of the components of the constraint equations as measures of strain, recalled the property of the ellipsoid of finite strain, recovered Cauchy's theorem on stress and wrote localized and linearized elastic constitutive relations for three-, two-, and one-dimensional continua. In many aspects, this work may be seen as the natural completion of Piola's path in the field of continuum mechanics, yet, probably because it was published posthumously, it is somewhat ignored.

2.3.5 Solidification Principle and Generalised Forces

Certainly Piola's most relevant contribution to continuum mechanics was the way he introduced internal stresses. They were presented as Lagrange multipliers of constraint equations. Piola applied the approach in all his papers, but in the articles of 1848 and 1856 the concept was made extremely clear.

When dealing with the equilibrium and motion of deformable bodies Piola said he could simply follow Lagrange, who had treated some deformable bodies as they were rigid by using what Louis Poinsot (1777–1859) had called the *principle de solidification* [60].[68] The principle of solidification was used also by Simon Stevin (1548–1620) in his study of the equilibrium of fluids *De beghinselen des waterwichts*, of 1586, and by Euler to treat hydrostatics in the *Scientia Navalis* of 1738 [12].[69] Cauchy used it in [16] to introduce the idea of stress. Later on, it was used to study systems of constrained bodies. Lagrange used it to prove the equation of virtual work [40].[70] Nowadays it is more often derived from the equation of virtual work:

> It is not unpleasant to deduce from the Principle of virtual velocity and from the thermodynamic generalization of this principle the following consequence: If a system is in equilibrium when it is subjected to certain constraints, it will persist in equilibrium when it will be subjected not only to these constraints but also to some more [constraints] which are consistent with the previous ones [...] [27].[71] (A.2.26)

According to this principle, the active forces present in a deformable body are equivalent to the passive forces obtained assuming "the same functions that remain constant for rigid bodies" [58],[72] that is the same functions which remain constant for rigid bodies as constraint equations. This is what Lagrange said on the subject:

> This integral $S F \delta ds$ will be added to the integral $S X \delta x + Y \delta y + Z \delta z$, which expresses the sum of the moments of all external forces acting on the thread [...], and by equating all them to zero, we obtain the general equilibrium equation of the elastic thread. Now it is clear that this equation has the same form than that [...] for the case of inextensible thread, and [it is clear, too,] that by changing F into λ, the two equations will become identical. We have so in the present case the same particular equations we found in the case of art. 31, by substituting only F in the place of λ [41].[73] (A.2.27)

In other words, for example in the case of a thread, Lagrange stated that by the introduction of the first-order variation of the extensibility constraint, the elastic forces could be treated as constraint reactions. Piola was not convinced by this argument:

> [Lagrange] in his A. M. [...] adopted a general principle (§9 of Sect. II and 6. of IV) by means of which the analytical expression of the effect of internal active forces is similar to that valid for passive ones when we have constraints: this is obtained by assuming indeterminate coefficients and by multiplying by them the variation of those functions which remain constant for rigid, inextensible, or liquid bodies. If we adopted such a method, we could even generalize the results obtained in the previous chapter: I, however, prefer not to do it, because my appreciation for the great Geometer does not prevent me to recognize how in that principle something remains obscure and not yet proved [58].[74] (A.2.28)

---

[68] pp. 36–37

[69] pp. 17–18.

[70] Sect. II, art. 1.

[71] pp. 36–37. Our translation.

[72] p. 76.

[73] p. 100. Our translation.

[74] p. 76. Our translation.

There were reasons for Piola's rejection of Lagrange's use of the principle of solidification: the first is that this approach stems from intuition, being based on a non formalized procedure; as an analyst, Piola preferred to obtain his result as consequences of a chain of formulas where nothing is left to intuition. Secondly, it requires the ideas of deformation and inner force, which Piola did not provide and did not want to use, at least not in *Intorno alle equazioni fondamentali*. Not convinced by Lagrange's procedure, Piola looked for a different one, showing his skills and talent.

Piola should have suspected some weakness in his reasoning because he returned to the argument in the posthumous paper of 1856 taking a different approach, avoiding the use of an intermediate configuration χ_p. Here, he practically adopted Lagrange's use of the solidification principle. At the origin of this reconciliation is Piola's explicit understanding that the constraint equations represent conditions on strains. His distrust of infinitesimals seems somewhat decreased, also probably because after Cauchy, whose ideas Piola appreciated despite not sharing all of them, the rigorous concept of differential, which could replace the 18th century concept of infinitesimal had become widely accepted. However, Piola did not adopt the differential, though he came close to it. In the metric considerations for the present configuration, where he could comfortably use the infinitesimal element of length ds he preferred to consider the quantity $s' = \sqrt{x'^2 + y'^2 + z'^2}$, which he called *elemento di arco* (element of arc), where the prime means derivative with respect to a parameter varying in the ideal configuration.

For the three-dimensional case Piola developed geometric relations of local character which partially reflect Cauchy's approach,[75] yet maintain a certain originality [59].[76] For the element of arcs which in the ideal configuration have, at a given point P, a tangent characterized by direction cosines $\alpha_1, \alpha_2, \alpha_3$, the expression of the square of the element of arc s' in the present configuration was represented by:

$$(s')^2 = \sum_{i,j} C_{ij}\alpha_i\alpha_j, \tag{2.26}$$

where the C_{ij} express the relations (2.3) evaluated at P. The expression (2.26) with equal indices coincides with that of the coefficient ε which Cauchy called *dilatation linéaire* [19].[77] Similar expressions were obtained for the cosines of angles between two curves.

In any case Piola remained critical of Lagrange's approach to deformable systems. He now had explicit reasons for this criticism, claiming that Lagrange had not given the criterion to establish what and how many components of deformation must be used:

> Indeed, there are possibly many simultaneous expressions of quantities that internal forces of a system tend to vary; which of them shall we consider, which shall we neglect? Who will assure us that by using many of such functions [which are] object of variation because

---

[75] For instance, they can be found in [19].

[76] art. 29 and art. 33.

[77] p. 304.

of the action of internal forces, we do not perform useless repetitions, by expressing by means of some of them an effect already written by mens of some others? And could it not happen instead that we neglect those [expressions] which are necessary to introduce in order to express the whole effect of internal forces? [59].[78] (A.2.29)

However, Piola believed he had solved the question and found which and how many constraint equations are needed:

Regarding the problem: which are the functions to use, among others, that are modified by internal forces, I proved that they are those trinomials of derivatives [...]. As for the other question: how many must be such functions [...] I answered [they are] so many as they are necessary to get the variation of those trinomials equated to zero [...] [59].[79] (A.2.30)

Once Piola had introduced deformations he could legitimately write the Eq. (2.22) for deformable systems; now $\delta t_1, \delta t_2, \delta t_3, \delta t_4, \delta t_5, \delta t_6$ did not represent the variation of the (2.3) constraint equations but the variation of the components of strain.

The introduction of strain throws new light on Lagrange multipliers. The latter were seen as forces producing displacements associated with the variation of constraint equations. Piola extended Lagrange's concepts [40][80]and conceived very general inner forces, anticipating modern approaches to internal forces in structured continua, for example Cosserat's [22, 23]. Indeed, it is apparent that when dealing with one-dimensional continua Piola introduced the twist of the line as a measure of strain, defining the dual inner force as the corresponding Lagrange multiplier. In Piola's words:

The concept that Lagrange wanted us to have about forces, which we presented in the introduction, is more general than that usually accepted. Everybody easily intends that force is a cause which by means of its action modifies the magnitudes of some quantities. In the most evident case, by approaching a body or a material point to another one, it modifies distances, namely makes the length of straight lines vary: but it can also modify an angle, a density, etc. In these latter cases the way of action of forces remains obscure to us, while it is clear in the former ones. But, perhaps, the reason of this is independent of the nature of forces. Actually, even in the former way it is not understood how a force can supply its action into the body in order to decrease or increase the distance from a body to another one: in any case, we can always see this fact: the daily observation makes the will to look further decrease. But, if by subtle reasoning we find that also in this case the action of forces is mysterious, no wonder it is mysterious in the other cases too. The will to reduce the action of forces always to that capable of modifying a distance, actually reduces a broader concept, and identifies only a particular class of forces. Generally speaking, how far can our notions about causes [which are] object of measurements be driven? can we perhaps understand their intimate nature and the true way in which they act? [...] When we have collected all unknown concepts in the unity with which we measure things of the same kind, we say to know the truth, if we can assign ratios with such unity, assumed arbitrary in the beginning. Now, when, after Lagrange, we conceive forces in the more general way, namely as causes which may vary quantities other than lines, we obtain necessary data to affirm that we can measure them. We have all we can reasonably pretend: if the imagine with which to dress the concept up seems to be missing, it is because we want to color it in the way we do with

[78] p. 391. Our translation.

[79] p. 421. Our translation.

[80] sect. V.

forces acting along lines. An unknown part always remains both in these more general cases and in that very common one [59].[81] (A.2.31)

This conception of forces led Piola to reconsider the constraint equations by investigating what happens to Lagrange multipliers (the forces) when these equations are transformed into others, with some mathematics. Piola examined one-, two- and three-dimensional cases; in the latter he focused solely on fluids.

2.4 Piola's Stress Tensors and Theorem

According to the ideas of the history of science prevailing today, the modern interpretation of a non-contemporary text is something to be avoided at all costs. The aim of the historian is to reconstruct the 'actual' thinking of the scientists, and this can be done only by using the categories of the time in which the scientists wrote. From this point of view it is necessary to understand all the aspects which could influence the way of thinking, not only in science: the political environment, dominant metaphysics, and so on.

We agree with this approach and in the previous sections we have tried to follow it as much as possible. Nonetheless, the modern interpretation of relatively recent papers such as Piola's are of remarkable interest also to the historian: it may be seen as a complement to the reconstruction of the history of a part of continuum mechanics from Piola's time up to now. In fact, it should not be forgotten that reference to Piola's papers is made in some monographs at the beginning of the 20th century [34, 49],[82] where the formulation of mechanics essentially coincides with today's. Moreover modern view of continuum mechanics allows us to highlight the aspects that Piola 'failed to understand' and help explaining why. These can be internal, that is they can depend on logical or methodological grounds, or external, such as the lack of time or of attention devoted to a particular subject. For instance, it is interesting to understand why Piola did not attribute to the Lagrange multipliers of the Eq. (2.22) the meaning which today is attributed to them. Understanding facts like this helps us to comprehend how Piola 'actually' thought.

In modern continuum mechanics it is customary to attribute the following to Piola:

(a) Two stress tensors, which provide respectively stress in the present configuration and stress in the present configuration pulled back to the reference configuration when applied to the unit normal in the reference configuration [33, 45, 73–75].[83]
(b) A theorem on the derivation of field equations from the principle of virtual work [34, 45, 49, 73–75].[84] In this section we shall try to explain the reasons behind these attributions.

---

[81] pp. 456–457. Our translation.

[82] p. 23; 620.

[83] pp. 178–180;185; 553–554; 124–125; 220–224; 178–180; 185.

[84] pp. 23; 620; 595–600; 124–125; 246–248; 185.

2.4.1 A Modern Interpretation of Piola's Contributions

Piola's papers contain interesting hints from the point of view of modern continuum mechanics. Firstly, the a_i coordinates are fundamental to all his works. In the *Meccanica de' corpi naturalmente estesi* Piola declared a_i to be independent of time. One could thus interpret the same coordinates to be those in a reference configuration (modern meaning), but Piola did not explicitly affirm that. However, it is clear from this choice that the description of motion used by Piola is a referential one, and to us the a_i list will describe a reference configuration. In the *Nuova analisi* Piola took a daring step forward, because the introduction of the concept of ideal state and the identification of the a_i list with it is exactly the construction of a reference configuration in the modern sense [75].[85] This is a reference shape useful for calculations, but which in principle may not coincide with a shape that the body have assumed or will assume. Otherwise a purist and an analyst, Piola did not realize the power of such an abstract formulation, which nowadays is the basis of many treatments of continuum mechanics. Indeed, in this he followed the use of his time, and, while in principle he should realize that his 'Lagrangian' description of motion would lead him to a different description of equilibrium and motion, he felt urged to focus only on the present configuration χ, like all his contemporaries had done. We remarked upon this in Sect. 2.3.3, where we quoted a passage in which Piola felt compelled to push the referential field equations forward to χ, the only 'real' state of the body.

Another interesting point is the implicit definition, in Piola's handling of the local rigidity constraint equations, of what today are known as the right (let it be C), corresponding to the constraint Eq. (2.3), and left (let it be B), corresponding to the constraint Eq. (2.4), Cauchy–Green strain tensors. As it is well known, C is a measure of the metric in the present configuration with respect to the reference one, while B^{-1} is a measure of the metric in the reference configuration with respect to the present one. Imposing the metric as the identity, as Piola did, amounts to supposing that the metric does not change during the motion. However, Piola was not interested in defining a deformation measure, and so what could have been a formidable intuition was not developed, even if one may suspect that Piola had something in mind; it was made clear only in his posthumous *Di un principio controverso*.

Moreover, the ambiguity that Piola kept between the expression of a generic motion gradient F [the derivatives of (x, y, z) with respect to (a, b, c)] and the gradient of a rigid motion Q, [the direction cosines $\alpha_1, \beta_1, \gamma_1, \dots$, of the rigid motion (2.2)], makes some of his equations generally invalid. In his work of 1832 Piola made no distinction between the constraint equations in terms of either B or C, and the Lagrange multipliers he introduced had the meaning of stresses; however, since B^{-1}, not B, expresses a metric, the field equations derived from the use of B had no physical meaning for $F \neq Q$. Piola corrected himself in the last part of *Meccanica de' corpi naturalmente estesi*, and in his other papers he seemed never to have the same doubts and used only C. It seems reasonable, however, to think that Piola did not make any metric consideration apart from those found in the *Di un principio controverso*, so

[85] p. 96.

it is likely that he 'felt' he was right in using C, rather than having rationally proved this.

In the coefficients $(P_1, P_2, P_3, \ldots)$ of the Eq. (2.20) a modern reader sees the components of the so-called Piola's second stress tensor, and in the coefficients $(L_1, L_2, L_3, \ldots)$ of the Eq. (2.20) the components of the so-called Piola's first stress tensor. In his paper of 1832, Piola did not provide any interpretation of the mechanical meaning of any of these quantities, nor of the field equations. Moreover, Piola did not give a mechanical interpretation of the surface integrals which derived from the applications of the techniques of the calculus of variations. Thus, many capital results he found remained somehow hidden.

In his later works, some changes were made with regard to these points: from the treatment of molecular interactions it immediately becomes clear that the elements of the list $(L_1, L_2, L_3, \ldots)$ have the meaning of internal forces and that the equations obtained are actually field equations. Indeed, Piola remarked that

> The general equations of motion of any point (x, y, z) of the body are eqs. (56) [coinciding with Eq. (2.19)] where L_1, L_2, etc. [...] is reduced to depend [...] on the only unknown $\psi(S)$ corresponding to the molecular force. It is true that [...] the equations found are related to that transformation of the x, y, z into a, b, c which is unknown and not assignable; but let us see how, once given the advantage of equations rigorously obtained, we can overpass the above mentioned difficulty with respect to the effects [...] [56].[86] (A.2.32)

That is, Piola had introduced a virtual ideal state and felt compelled to push his equations forward to the present configuration, which he considered to be the only true one. It is remarkable how in the *Nuova analisi* Piola introduced an intermediate configuration, which was the one assumed by the body at the initial time. But, rather than generalizing his results to this reference configuration, he focused his attention on the present one and derived local equilibrium equations in Cauchy's form in terms of quantities, the components $(L_1, L_2, L_3, \ldots)$, which are expressions of molecular forces. The expressions he provided were those which are commonly accepted in modern continuum mechanics [33, 45, 73, 74].[87] Moreover, the interpretation of the components $(P_1, P_2, P_3, \ldots)$ in terms of contact actions is the one which is currently accepted. It is clear in Piola's mind that these results led to match those of Cauchy on surface and internal actions.

It is interesting to remark how, while the local rigidity constraint equations cannot in principle be used to derive dynamical equations for deformable bodies—for which constraint equations do not exist—Piola did not use them directly, but used their first-order variation. These would be nowadays interpreted as describing a virtual strain velocity, and expressions such as the Eq. (2.20) would be read as follows: the total mechanical work spent on a virtual rigid velocity field vanishes. While in his paper of 1832 Piola wrote with some imprecision, as he himself admitted later, in other papers, especially in those of 1848 and 1856, his treatment was acute and he was sure that this approach was original. And, indeed, this is a rather modern view, and at the beginning of the 20th century credit was given to Piola for this formulation,

---

[86] p. 202. Our translation.

[87] pp. 553–554; 124–125; 224–225; 178–180.

[34, 49],[88] which is called Piola's theorem in well known monographs on continuum mechanics [45, 73, 75].[89]

In short, credit is given to Piola to have proven field equations only via the description of the present configuration, the principle of solidification and the well accepted principle of vanishing of virtual work spent on a rigid body motion. In particular, Ernst Hellinger highlighted the fact that Piola's approach requires only the knowledge of external forces, without compromising on the nature of inner forces, which are simply Lagrange multipliers:

> [...there is] another formulation of the Principle of virtual displacements, which takes into consideration only the real *forces*, mass forces X, Y, Z and surface forces $\bar{X}$, $\bar{Y}$, $\bar{Z}$ as given; it is the following simple position of the formulation by *G. Piola*: *In order to have equilibrium it is necessary that the virtual work of the acting forces*
>
> $$\iiint_{(V)} (X\,\delta x + Y\,\delta y + Z\,\delta z)dV + \iint_{(S)} (\bar{X}\,\delta x + \bar{Y}\,\delta y + \bar{Z}\,\delta z)dS$$
>
> *vanishes for every [rigid motion] of the whole system V [...so that] the components of the tension dyadic appear as Lagrange's multipliers of some rigidity conditions* [34].[90] (A.2.33)

The treatment in the paper of 1856, moreover, is basically the more modern one, since Piola started by introducing the idea of generalized deformations of continua. From the study of this posthumous paper, it seems that Piola could anticipate some later developments of continuum mechanics; yet his provincialism made his work almost unknown outside Italy.

2.4.2 The Piola-Kirchhoff Stress Tensors

The introduction of Piola's name to qualify the stress tensors pulled back to the reference configuration is due to Truesdell and Richard Toupin [73] who often refer to the works we have examined in this paper. Frequently, Kirchhoff is mentioned in the same breath as Piola, and this attribution is also due to Truesdell and Toupin; we shall clarify why. Even though we are focused on Piola's contributions, we will also summarize Kirchhoff's contribution for a more complete study of the subject. In fact, unlike Piola, Kirchhoff was conscious of introducing a new idea, the stress pulled back in the reference state to study finite deformations. Unfortunately, Kirchhoff's mathematical treatment was not as good as Piola's: so the complementarity of understanding and misunderstanding of mathematical and physical concepts by the two scientists justifies Truesdell's juxtaposition of Piola's and Kirchhoff's names.

In 1852, Kirchhoff [35] published a paper in which he studied the problem of elastic equilibrium in presence of finite displacements. Kirchhoff maintained that he

[88] pp. 23–24; p. 620.

[89] pp. 596–597; 246–248; 185.

[90] p. 620. Our translation.

was inspired by Saint Venant [65], who had formulated a clear definition of a finite measure of strain (which is now indeed called Green-Saint Venant strain tensor) and had given some hints on how to obtain equilibrium equations for non-infinitesimal displacements, claiming that

> When tensions are considered over the slightly inclined planes into which the three material planes initially rectangular and parallel to the coordinates have changed, we have, for the six components, the same expressions, as functions of the dilatations and the distortions [the components of the Green-Saint Venant strain tensor], that we have when displacements are very small [65].[91] (A.2.34)

The conclusion drawn by Saint-Venant in this passage does not seem so clear to a modern reader, and is probably the cause of Kirchhoff's uncertainties in the considered paper. Quite surprisingly, in fact, Kirchhoff's article was somewhat obscure and presented incorrect expressions according to modern standards. It is not clear from the text whether Kirchhoff intended to follow an approximated reasoning, or if he made genuine errors. According to Todhunter and Pearson [71][92] Kirchhoff himself later realized the weakness of this paper and did not want to re-publish it in his *Gesammelte Abhandlungen* [36].

These are Kirchhoff's words on how he claimed to derive local equilibrium equations in the case of finite displacements (some evident typographical errors have been amended):

> I will denote by ξ, η, ζ the coordinates of a point after deformation, by x, y, z the coordinate of the same point before it. I imagine that in the natural state of the body there are three planes, parallel to the coordinate planes, through the point (x, y, z); the parts of these planes, which lay infinitely near the mentioned point, are transformed by the deformation in planes which form non-square, finite angles with the coordinate planes, but infinitely smaller than 90° with each other. I imagine to project the pressure underwent by these planes after the deformation on the coordinate axes, and denote these components: X_x, Y_x, Z_x, X_y, Y_y, Z_y, X_z, Y_z, Z_z, in such a way that for instance Y_x is the component along y of the pressure to which is subjected the plane which was orthogonal to the x axis before the deformation. These nine pressures are in general non-orthogonal with respect to the planes on which they act, and there are not three equal to other three, like in the case of infinitely small displacement. Once established the conditions for the equilibrium of a part of the body which before the deformation is an infinitely small parallelepiped with sides parallel to the coordinate axes of length dx, dy, dz, one obtains the equations:
>
> $$\left.\begin{aligned} \rho X &= \frac{\partial X_x}{\partial x} + \frac{\partial X_y}{\partial y} + \frac{\partial X_z}{\partial z} \\ \rho Y &= \frac{\partial Y_x}{\partial x} + \frac{\partial Y_y}{\partial y} + \frac{\partial Y_z}{\partial z} \\ \rho Z &= \frac{\partial Z_x}{\partial x} + \frac{\partial Z_y}{\partial y} + \frac{\partial Z_z}{\partial z} \end{aligned}\right\} \ldots \qquad (1)$$
>
> if we denote by ρ the density of the body and by X, Y, Z the components of the accelerating forces acting on the body at the point (ξ, η, ζ). One obtains these equations by considering that the sides and the angles have changed infinitely little, and so one can use the same

[91] p. 261. Our translation.

[92] art. 1244, p. 50.

> considerations introduced in the equilibrium in presence of infinitely small displacements [35][93] (A.2.35)

Thus, Kirchhoff focused on three infinitesimal faces which are parallel to fixed coordinate planes and pass through a generic point which undergoes a finite displacement. He then projected the stresses arising after the deformation on those faces on the fixed coordinate axes and wrote the local equilibrium equations with respect to the same axes. Kirchhoff's equations (1) above seem inconsistent when what has been said in the previous section is considered. Indeed, they have a similar form of Eqs. (2.11) and (2.19), but do not coincide with them for two reasons:

1. It is not clear how the components X_x, Y_x, ... may coincide with those of Piola's first stress tensor. Indeed, no information is provided either on how the area affected by the stress changes during deformation, or on the change of metric between the present and the reference configuration.
2. It is not clear where ρ is measured. If ρ is the mass per unit volume in the present configuration, as it seems to follow from Kirchhoff's words, this is again inconsistent with the Eq. (2.11), since the mass density is required to be measured in the reference configuration.

It is strange that a sharp expert in physics and a well-educated mathematician like Kirchhoff wrote such inconsistencies. This may perhaps be explained by the fact that Kirchhoff was studying a problem of finite displacements with infinitesimal strain, as explicitly stated in the above quotation, and as conjectured by Saint-Venant:

> [...] the mutual distances of points very close vary only in a small ratio [...] [65].[94] (A.2.36)

One may then suppose that Kirchhoff considered the body as almost undistorted so that areas and volumes do not vary. In this case, it is still possible to derive local equilibrium equations for the stress components in the present configuration, projected on the fixed coordinated axes, by means of standard procedures. This should be represented by Kirchhoff's equations (1), if ρ is taken as the density in the reference configuration.

It is remarkable how the developments by Piola and Kirchhoff are in a way each other's mirror images. In the second derivation of the local equilibrium equations which Piola presented in *Meccanica de' corpi naturalmente estesi*, he first introduced what we now call Piola's second stress tensor: its components are the Lagrange multipliers of his variational problem. Then, he introduced what we now call Piola's first stress tensor simply as a mathematical stratagem with which to write the local equilibrium equations in the present configuration; no mechanical meaning is given to its components. On the other hand, Kirchhoff began by considering from a physical point of view the quantities that we now call the components of Piola's first stress tensor. Later, he introduced the components of what we now call Piola's second stress tensor only to obtain a constitutive relation for the components of the first.

[93] pp. 762–763. Our translation.

[94] p. 261. Our translation.

References

1. Araldi M (1806) Tentativo di una nuova rigorosa dimostrazione del principio dell'equipollenza. Mem dell'Is Naz Ital 1(1):415–426
2. Bordoni A (1831) Lezioni di calcolo sublime. Giusti, Milan
3. Bordoni A (1833) Annotazioni agli elementi di meccanica e d'idraulica del professore Giuseppe Venturoli. Giusti, Milan
4. Bottazzini U (1989) I matematici italiani e la 'moderna analisi' di Cauchy. Archimede 41:15–29
5. Boussinesq J (1869) Essai sur la théorie des ondes liquides périodiques. Comptes Rendus Hebdomadaires des Séances de l'Académie des Sci 68:905–906
6. Boussinesq J (1872) Théorie des ondes liquides périodiques. Mémoires présent és par divers savants a lÁcadémie des Sciences de lÍnstitut de France 20:509–615
7. Brillouin L (1960) Les tenseurs en mécanique et en élasticité (1934). Masson, Paris
8. Brunacci V (1798) Calcolo integrale delle equazioni lineari. Allegrini, Florence
9. Brunacci V (1802) L'analisi derivata ossia l'analisi matematica dedotta da un sol principio di considerare le quantità. Bolzani, Pavia
10. Brunacci V (1809) Elementi di algebra e geometria. Stamperia Reale, Milan
11. Capecchi D (2003) Gabrio Piola e la meccanica italiana dell'Ottocento. In: Proceedings of the XXIII Congresso di Storia della Fisica e dell'Astronomia, Bari, pp 95–108
12. Capecchi D (2001) La tensione secondo Cauchy. Hevelius, Benevento
13. Capecchi D, Ruta G (2007) Piola's contribution to continuum mechanics. Arch Hist Exact Sci 61:303–342
14. Carnot L (1783) Essai sur les machines en général. In: Carnot L (1797) Oeuvres mathématiques. Decker, Basel
15. Castigliano CA (1879) Théorie des systémes élastiques et ses applications. Negro, Turin
16. Cauchy AL (1823) Recherches sur l'équilibre et le mouvement intérieur des corps solides ou fluides élastiques ou non élastiques. In: Cauchy AL (1882–1974). Oeuvres complétes (27 vols), vol 1, s 2. Gauthier-Villars, Paris, pp 300–304
17. Cauchy AL (1827) Sur les relations qui existent dans l'état d'équilibre d'un corps solide ou fluide entre les pressions ou tensions et les forces accélératrices. In: Cauchy AL (1882–1974) Oeuvres complétes (27 vols), vol 7. Gauthier-Villars, Paris, pp 141–145
18. Cauchy AL (1828) Sur les équations qui expriment les conditions dequilibre ou les lois du mouvement intérieur dún corps solide élastique ou non élastique. In: Cauchy AL (1882–1974) Oeuvres complétes (27 vols). Gauthier-Villars, Paris, s 2, 8:195-226
19. Cauchy AL (1841) Mémoires sur les condensations et rotations produits par un changement de form dans un systéme de points matériels. In: Cauchy AL (1882–1974). Oeuvres complétes (27 vols), s 2. Gauthier-Villars, Paris, pp 278–287
20. Cauchy AL (1882-1974) Oeuvres compl'etes (27 vols). Gauthier-Villars, Paris
21. Cesaro E (1889) Sulle variazioni di volume nei corpi elastici. In: Cesaro E (1964–1968) Opere scelte (2 vols in 3 tomes). Cremonese, Rome
22. Cosserat E, Cosserat F (1896) Sur la théorie de l'élasticité. Annales de l'Université de Toulouse 10:1–116
23. Cosserat E, Cosserat F (1907) Sur la statique de la ligne déformable. Comptes Rendus Hebdomadaires des Séances de l'Académie des Sciences 145:1409–1412
24. Dahan Dalmedico A (1993) Mathématisations. Augustin-Louis Cauchy et l'école française, Du Choix, Paris
25. D'Alembert JL (1758) Traité de dynamique. David, Paris
26. Dugas R (1950) Histoire de la mécanique. Griffon, Neuchâtel
27. Duhem P (1991) Hydrodinamique, élasticité, acoustique. Hermann, Paris
28. Ferroni P (1803) I principi di meccanica richiamati alla massima semplicità ed evidenza. Memorie di matematica e fisica della Societá italiana delle scienze 10:481–633
29. Filoni A, Giampaglia A (2006) Gabrio Piola 1794–1850. Biografia di un matematico umanista. Assessorato alla Cultura del Comune di Giussano, Giussano

30. Fontana G (1802) Nuove soluzioni di un problema statico euleriano. Memorie di matematica e fisica della Societá italiana delle scienze 9:626–698
31. Fossombroni V (1796) Memoria sul principio delle velocità virtuali. Cambiagi, Florence
32. Grioli G (1960) Lezioni di meccanica razionale. Borghero, Padua
33. Gurtin ME (1981) An introduction to continuum mechanics. Academic Press, New York
34. Hellinger E (1914) Die allgemeinen ansätze der mechanik der kontinua. Enzyklopädie der mathematischen wissenschaften, Bd. IV/4. B.G. Teubner, Leipzig, pp 602–694
35. Kirchhoff GR (1852) Uber die Gleichungen des Gleichgewichtes eines elastischen Körpers bei nicht unendlich kleinen Verscheibungen seiner Theile. Sitzungsberichte der Akademieder Wissenschaften Wien, Bd. 9, pp 762–773
36. Kirchhoff GR (1882) Gesammelte abhandlungen. Barth A, Leipzig
37. Kline M (1972) Mathematical thought from ancient to modern times. University Press, Oxford
38. Lacroix SF (1811) Traité du calcul différentiel et du calcul intégral. Courcier, Paris
39. Lanczos C (1970) The variational principles of mechanics. University Press, Toronto
40. Laqrange JL (1811) Mécanique Analytique (Tome premiere). Courcier, Paris
41. Laqrange JL (1813) Méchanique analitique. Chez la Veuve Desaint, Paris
42. Laqrange JL (1813) Théorie des fonctions analytiques (1797). Imprimerie de la République, Paris
43. Love AEH (1892) A treatise on the mathematical theory of elasticity. Cambridge University Press, Cambridge
44. Magistrini GB (1816) Osservazioni varie sopra alcuni punti principali di matematica superiore. Memorie di matematica e fisica della Societá italiana delle scienze 17:445–4
45. Malvern LE (1969) Introduction to the mechanics of a continuum medium. Prentice-Hall, Englewood Cliffs
46. Marcolongo R (1905) Meccanica razionale. Hoepli, Milan
47. Masotti A (1950) In memoria di Gabrio Piola nel centenario della morte. Rendiconti dell'Istituto Lombardo, classe di Scienze 83:1–31
48. Mossotti OF (1858) Lezioni di Meccanica razionale. d'Arcais [s.n.]
49. Müller CH, Timpe A (1906) Die Grundgleichungen der mathematischen Elastizitätstheorie. Enzyklopädie der mathematischen Wissenschaften, Bd. IV/4. B.G. Teubner, Leipzig, pp 1–54
50. Pepe L (2007) Rinascita di una scienza. Matematica e matematici in Italia (1715–1814). CLUEB, Bologna
51. Piola G (1821) Sulla teorica dei cannocchiali. Effemeridi astronomiche di Milano, pp 13–36
52. Piola G (1832) La meccanica de' corpi naturalmente estesi trattata col calcolo delle variazioni. Opuscoli matematici e fisici di diversi autori, pp 201–236
53. Piola G (1825) Lettere scientifiche di Evasio ad Uranio. Fiaccadori, Reggio Emilia
54. Piola G (1825) Sull'applicazione de' principj della meccanica analitica del Lagrange ai principali problemi. Regia Stamperia, Milan
55. Piola G (1833) La meccanica de' corpi naturalmente estesi trattata col calcolo delle variazioni. Giusti, Milan
56. Piola G (1836) Nuova analisi per tutte le questioni della meccanica molecolare. Memorie di matematica e fisica della Societá italiana delle scienze 21:155–163
57. Piola G (1844) Elogio di bonaventura cavalieri. Bernardoni, Milan
58. Piola G (1848) Intorno alle equazioni fondamentali del movimento di corpi qualsivogliono considerati secondo la naturale loro forma e costituzione. Memorie di matematica e fisica della Societá italiana delle scienze 24:1–186
59. Piola G (1856) Di un principio controverso della Meccanica Analitica di Lagrange e delle sue molteplici applicazioni. Mem dell'Istituto Lombardo 6:389–496
60. Poinsot L (1848) Éléments de statique (1803). Gauthier-Villars, Paris
61. Poisson SD (1829) Mémoire sur l'équilibre et le mouvement des corps élastiques. Mémorie de l'Académie des Sciences de l'Institut de France 8:357–570
62. Poisson SD (1831) Mémoire sur la propagation du mouvement dans les milieux élastiques. Mémorie de l'Académie des Sciences de l'Institut de France 10:549–605

63. Poisson SD (1831b) Mémoire sur les équations générales de l'équilibre et du mouvement des corps solides élastiques et des fluides. Journal de l'École Polytechnique 13(20):1–174
64. Prony G (1797) Sur le principe des vitesses virtuelles. Journal de l'École Polytechnique 5(2):191–208
65. Saint Venant AJC Barré de (1847) Mémoire sur l'équilibre des corps solides dans les limites de leur élasticité et sur les conditions de leur résistance quand les déplacements ne sont pas trés petits. Competes Rendus Hebdomadaires des Séances de l'Académie des Sciences 24:260–263
66. Saladini G (1808) Sul principio delle velocità virtuali. Mémorie dell'Istituto Nazionale Italiano 2(1):399–420
67. Signorini A (1930) Sulle deformazioni finite dei sistemi continui. Rendiconti della Reale Accademia dei Lincei 12(6):312–316
68. Signorini A (1930) Sulla meccanica dei sistemi continui. Rendiconti della Reale Accademia dei Lincei 12(6):411–416
69. Signorini A (1952) Meccanica razionale con elementi di statica grafica. Società tipografica Leonardo da Vinci, Città di Castello
70. Signorini A (1960) Questioni di elasticità non linearizzata. Cremonese, Rome
71. Todhunter I, Pearson K (1886–1893) A history of the theory of elasticity and of the strength of materials, from Galileo to the present time. Cambridge University Press, Cambridge
72. Trovalusci P, Ruta G, Capecchi D (2006) Il modello molecolare di Voigt. In: Proceedings of the XVI Congresso di Storia della Fisica e dell'Astronomia, Rome, pp 183–194
73. Truesdell CA, Toupin R (1960) The classical field theories. In: Handbuch der Physik, Bd. III/1. Springer, Berlin, pp 226–793
74. Truesdell CA, Noll W (1965) The non-linear field theories of mechanics. In: Handbuch der Physik, Bd. III/3. Springer, Berlin
75. Truesdell CA (1991) A first course in rational continuum mechanics. Academic Press, New York
76. Venturoli G (1826) Elementi di meccanica e d'idraulica (1806). Mauri G, Rome

Chapter 3
The Mathematicians of the Risorgimento

Abstract The constituting phase of the Kingdom of Italy was a time of recovery of mathematical studies. The political unity facilitated the inclusion of Italian mathematicians in the context of European research, in particular the German one. The internationalization of Italian mathematics is customarily associated with a trip taken in 1858 by some young mathematicians including Francesco Brioschi, Enrico Betti and Felice Casorati in Europe. In a few years we assist in the development of some schools that will maintain their role even in the 20th century. Among them, those promoted by Enrico Betti and Eugenio Beltrami were undoubtedly the most important. In this chapter we present briefly the contribution of two of the leading pioneers and their students.

3.1 Enrico Betti

Enrico Betti (Pistoia 1823-Soiana [Pisa] 1892) lost his father as a child and was educated by his mother. He studied at the university of Pisa and was a pupil of Ottaviano Fabrizio Mossotti, professor of mathematical physics with interests in hydrodynamics, capillarity, optics. He graduated in 1846, when Leopoldo of Tuscany established at Pisa a Scuola normale, for theoretical and practical training of teachers [24].[1]

Besides Mossotti, professors of Pisan study were the astronomer Giovanni Battista Amici, Pietro Obici, who taught applications of mathematics to mechanics and hydraulics, the director of the cabinet of technological physics Luigi Pacinotti, the future minister of public instruction and professor of physics Carlo Matteucci, Guglielmo Libri, then emeritus professor and prominent international personality, Gaetano Giorgini, supervisor of the studies of the Grand Duchy and estimated geometer. Their ideas influenced Betti and, certainly, his close link with Mossotti addressed his initial research toward mathematical physics.[2]

[1] p. 230. In fact there was already a first founding of Scuola normale by Napoleon in 1813, along the lines of the French Écoles.

[2] Witnessed by a letter of August 25th, 1847 of Mossotti, who from Viareggio, dissuaded Betti from pursuing his initial interest in descriptive geometry [24], p. 231.

D. Capecchi and G. Ruta, *Strength of Materials and Theory of Elasticity in 19th Century Italy*, Advanced Structured Materials 52,
DOI 10.1007/978-3-319-05524-4_3

The influence of Mossotti was felt even in the political life of Betti: in 1848 we find Betti at the battles of Curtatone and Montanara[3] in the university battalion commanded by Mossotti. After the unification of Italy, Betti continued his political commitment, a fact common to other mathematicians of the period of the Risorgimento as Beltrami, Cremona, Brioschi, Casorati.

In 1849 Betti left Pisa and went to teach in a high school in Pistoia. The relative isolation that followed determined the original character of his early research on radical solutions of algebraic equations. Although Evariste Galois' theory of algebraic equations dated back to the 1820s, still in the mid-1800s it was hardly established even in France: "The first actual resumption of Galois' ideas was the work of Betti of 1850" [24].[4] In 1852 he became professor of superior algebra to a high school in Florence and in 1857 professor of mathematics at the university of Pisa.

During this time, Betti with Brioschi and Casorati, undertook a long journey at the universities of Göttingen, Berlin and Paris, beginning the internationalization of the Italian mathematics. The three met Dirichlet, Riemann and Dedekind in Göttingen; Kronecker, Weierstrass and Kummer in Berlin; Hermite and Bertrand in Paris. Riemann[5] exerted the greatest influence on Betti, called in 1859 to the chair of higher analysis. At the time not many knew Bernhard Riemann's mathematical works on complex analysis, while Betti embraced its content, spreading ideas through Italy.

Betti applied his new ideas to elliptic functions and the theory of functions of complex variables. In an article in 1871 [18] Betti presented the result of conversations with Riemann introducing new fundamental concepts in algebraic topology. The idea of Riemann about connection of surfaces was extended to n-dimensional manifolds and Betti defined different types of connection, with numbers that will become known as *Betti numbers*.

The admiration for Riemann came to fruition with the proposal to him, in 1863, of the chair of geodesy, vacant in Pisa after the death of Mossotti. Riemann refused because of his inability to give lessons due to poor health (he would die prematurely in 1866).[6] Betti then proposed the chair to Beltrami who initially refused, but after a

[3] Curtatone and Montanara are two places near Mantua where in 1848 important battles against the Austrian army were fought and lost by Italian volunteers.

[4] p. 233.

[5] Georg Friedrich Bernhard Riemann (Breselenz 1826-Selasca 1866) was a German mathematician and physicist. He grew up in poverty, which interfered with his education. He moved to Lüneburg to study and found a friend in his instructor Schmalfuss who gave him free access to his private library. Thus he was able to read the books of Gauss and Legendre. Riemann left Lüneburg and, after a year spent at the university of Göttingen, in 1847, moved to Berlin. Here he was in contact with some of the most prominent German mathematicians of the time, and he studied *inter alia* Jacobi's and Dirichlet's papers. He returned to Göttingen to finish his graduate work; his first argument went back to 1851 and concerned a new theory of functions of a complex variable, a nascent branch of mathematics at that time, that thanks to his contribution received a major boost. In 1854 he read for his qualification for teaching, his second thesis, entitled *Über die Hypothesen, welche der Geometrie zu Grunde liegen*, published posthumously in 1867 which introduced the concepts of variety and curvature of a manifold, in non-Euclidean spaces.

[6] The proposal of Betti and Riemann's rejection are documented in some letters referred to in [24].

consultation with Cremona changed his mind and accepted the transfer from Bologna to Pisa.

Riemann stayed in Pisa from 1863 to 1865, to improve his health. The simultaneous presence of Beltrami and Riemann had strong influence on Betti, who changed his interests toward mathematical physics [86].[7] In 1863 Betti assumed with Riccardo Felici the direction of the journal Nuovo cimento where he began to publish his articles on potential theory [15].

In 1865 Betti became the director of the Scuola normale and remained in this position until his death except for the years 1874–1876, in which, as general secretary at the ministry of public education, he was replaced by Ulisse Dini. From 1862 Betti was a deputy and senator of the Kingdom of Italy. As professor and director of the Scuola normale, Betti fostered numerous scholars: Tedone, Padova, and Somigliana (elasticity), Dini (analysis), Ricci-Curbastro (founder of the tensor calculus) and Volterra (mathematical physics).[8]

Betti mainly investigated the frontiers of the physics of the period, magnetism and electrodynamics, but he dealt with classical mechanics too. His many works in these topics, reprinted in his *Opere* [14], consist of:

- 1850. *Sopra la determinazione analitica dell'efflusso dei liquidi per una piccolissima apertura*, Annali scienze matematiche e fisiche, 1850.
- 1866. *Sopra la teoria della capillarità*, Annali delle università toscane, 1866.
- 1867. *Teoria della capillarità* [16].
- 1872–1873. *Teoria della elasticità* [20].
- 1874. *Sopra le equazioni di equilibrio dei corpi elastici* [19].
- 1876–1877. *Sopra il moto di un numero qualsiasi di punti*, Memorie della Reale accademia dei Lincei, 1876–1877.
- 1880–1881. *Sopra il moto di un ellissoide fluido eterogeneo*, Memorie della Reale accademia dei Lincei, 1880–1881.
- 1883. *Sopra il moto dei fluidi elastici*, Nuovo cimento, 1883.
- 1888. *Sopra la entropia di un sistema Newtoniano in moto stabile* (Nota I).
- 1888. *Sopra la entropia di un sistema Newtoniano in moto stabile* (Nota II), Memorie della Reale accademia Lincei, 1888.

In early works Betti showed a mechanistic conception of physics in which force, and not energy, was the characteristic quantity and whose foundation is the principle of virtual work, which Betti called *Lagrange's principle*. In the work on capillarity of 1867 he considered the bodies composed of particles that repel each other at very short distance, attract at short distance and do not interact at 'sensible distances'. These molecular forces admit a potential that depends only on the mutual distance of the particles. The potential provides with its derivatives the components of the forces and with its variation the virtual work [16].[9]

---

[7] pp. 283–290.

[8] Betti's biography is taken from [52]. More details about the scientific aspects of the work of Betti can be found in [28].

[9] p. 161.

After having criticized the assumptions of his predecessors, Betti said:

> Of the molecular forces hypothesis I keep only the first part, that is I admit only that the elements of bodies act over each other in the direction of the line that unites them and in proportion to the product of their masses; which leads to admit that the forces of cohesion and of adhesion have potential functions [16].[10] (A.3.1)

In the first pages of his memoirs of 1863–1864 on Newtonian forces, Betti expressed his Newtonian ideology:

> The forces acting according to Newton's law are those emanating from each infinitesimal elements of a given matter and which tend to approach or to move away these elements, proprotionally of their mass and inversely proportional to the squares in their distances [15].[11] (A.3.2)

In these works, Betti introduced the concept of potential (the term is his) but only on a mathematical basis, as a primitive function from which to derive forces and without any physical privileged role. The potential (not the potential energy) was a definite magnitude that allowed a compact treatment of mechanics, leading to differential equations subjected to 'regular study' by the mathematicians of the time. Betti changed attitude in the second memoir on capillarity, [16] and about the forces between the molecules of fluid he wrote:

> These forces have another property that we can deduce from the [...] fundamental [principle] of modern physics: the principle of the conservation of forces; and that is to have a potential function [16].[12] (A.3.3)

From a mathematical physical point of view, the works of 1866 and 1867 were not very different from each other.

This shift in his choice of the fundamental principles was final and radical in the *Teoria della elasticità* of 1872–1873, in which Betti never mentioned internal forces, and even avoided the introduction of stresses, at the cost of burdening the discussion. However, Betti's choice was still not clearly within the energetist movement then emerging and theoretical thermodynamics. In fact, the forces of which he could give a characterization that was not dubious, as were many external forces, were employed by Betti directly without the mediation of potential.

Betti's difficulty in explicating the nature of internal forces reflects somehow Piola's concern, commented in Chap. 2, even if the solution was different. Piola based his argument on the principle of virtual work and Lagrange multipliers; he considered the various points of a continuum as free, then imposed the differential equations of compatibility; stresses were defined as the Lagrange multipliers of these latter [25]. Betti referred to a potential, without ever assigning a name to its partial derivatives, which for us are the stresses.

---

[10] p. 163. Our translation.

[11] p. 45. Our translation.

[12] p. 179. Our translation.

When Betti wrote the *Teoria della elasticità*, the theory of elasticity was a mature science with known principles, although not universally shared. The discussion then developed in the form of modern scientific textbooks and not in that of research treatises, following the axiomatic approach, where at first one states the principles and then develops applications. The cornerstones of Betti's ideas were on the one hand the strains and the elastic potential, on the other the principle of virtual work. Although the concept of stress did not appear, this was not a problem, given his interest in the potential theory that led him to an approach similar to that of Green who saw the theory of elasticity as a function of the study of the propagation of light waves in the ether, where the concept of stress was unessential. This differed with the view of the French school where elasticity theory was developed in view of its applications to engineering.

Betti's book is divided into twelve chapters:

I Infinitesimally small deformations.
II Elastic energy as quadratic form of the components of deformation.
III Equations of equilibrium of homogeneous elastic solids using the principle of virtual work.
IV Solution of the problem, decomposing external active forces active in an irrotational and a solenoidal field.
V Problem for continuous isotropic elastic.
VI Theorem of reciprocity, limited to surface forces.
VII Problem of deformation of a sphere under the action of gravity.
VIII Deformation of an isotropic elastic body subjected to surface forces.
IX Isotropic elastic body subjected to surface forces.
X Deformation of an isotropic elastic body under the action of any forces.
XI Study of a cylindrical solid homogeneous isotropic elastic (Saint Venant solid).
XII Effects of thermal expansion for a homogeneous and isotropic elastic body.

3.1.1 The Principles of the Theory of Elasticity

3.1.1.1 Infinitesimal Strains

The French school considered the displacement of the points of a body as a continuous function meaningful only at the places occupied by molecules; the deformation was defined, before considering geometric intuition, then analysis [30, 77]. Betti deviated from this approach; he ignored the corpuscular nature of the bodies, modeling them as continua and followed a purely analytical approach. He made instrumental use, because of their convenience, of the infinitesimals, quietly abandoning the mathematical rigor of the Italian school carried out by Piola and Bordoni.[13]

[13] Betti knew that, if desired, he could rewrite all the 'less rigorous' steps developed with the use of infinitesimals with a strict mathematics.

Deformation was defined as the change in length of the linear element:

$$ds^2 = dx^2 + dy^2 + dz^2, \tag{3.1}$$

where (x, y, z) are the present coordinates of the point P of the continuum, a function of the coordinates ξ, η, ζ that P has in the reference configuration. Betti admitted that "the variation of length of the linear elements and the element themselves are so small quantities that one can ignore the power of higher order with respect to that of lower order" [20].[14] The variation that Betti made of ds^2 operated then on functions of ξ, η, ζ:

$$ds\delta ds = dx\delta dx + dy\delta dy + dz\delta dz = dxd(\delta x) + dyd(\delta y) + dzd(\delta z), \tag{3.2}$$

which is possible for the exchangeability of the operators d and δ. The variations δx, δy, δz coincide with the components of the vector of displacement $[u(\xi, \eta, \zeta)$, $v(\xi, \eta, \zeta)$, $w(\xi, \eta, \zeta)]$ and the (3.2), divided by ds^2, appears in the extended form as:

$$\begin{aligned}\frac{\delta ds}{ds} &= \frac{du}{dx}\left(\frac{dx}{ds}\right)^2 + \frac{du}{dy}\left(\frac{dy}{ds}\right)^2 + \frac{dw}{dz}\left(\frac{dw}{ds}\right)^2 \\ &+ \left(\frac{dv}{dz} + \frac{dw}{dy}\right)\frac{dy}{ds}\frac{dz}{ds} + \left(\frac{dw}{dx} + \frac{du}{dz}\right)\frac{dz}{ds}\frac{dx}{ds} + \left(\frac{du}{dy} + \frac{dv}{dx}\right)\frac{dx}{ds}\frac{dy}{ds}.\end{aligned} \tag{3.3}$$

For small strains the variation δds approximates the difference Δds and consequently the ratio $\delta ds/ds$ approximates the relative variation of length of the element ds in the direction of dx, dy, dz. To individuate the components of the deformation it is enough to assume suitable values for dx, dy, dz. For instance one assumes $dy = dz = 0$, and thus $ds = dx$, to obtain the deformation along x: $\delta ds/ds = du/dx$. The same considerations apply for the other directions. With some mathematics Betti also obtained the expressions for the angular distortions, that is the variation of the angle between orthogonal segments.

The strains, infinitesimal because they have a physical meaning for small displacements only, are indicated by Betti on the footprints of William Thomson [88][15] as:

---

[14] p. 3. From now on the quotations from the *Teoria della elasticità* by Betti refer to the offprint by Soldaini of 1874 [20].

[15] p. 391.

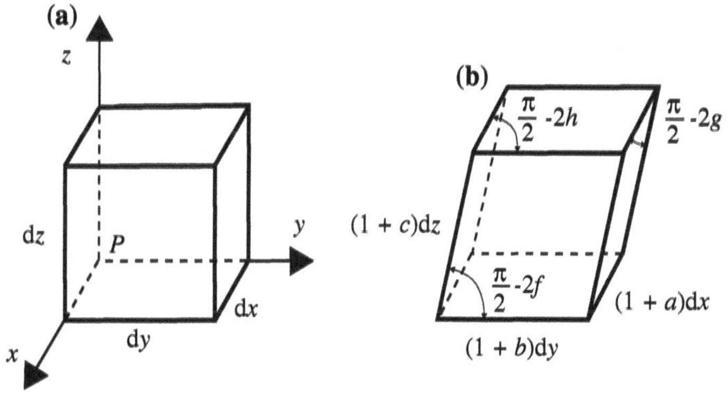

Fig. 3.1 Geometrical meaning of the coefficients of strains according to Betti. **a** Undeformed state. **b** Deformed state

Betti		Thomson	
$\frac{du}{dx} = a$	$\frac{dv}{dz} + \frac{dw}{dy} = 2f$	$\frac{du}{dx} = f$	$\frac{dv}{dz} + \frac{dw}{dy} = a$
$\frac{dv}{dy} = b$	$\frac{dw}{dx} + \frac{du}{dz} = 2g$	$\frac{dv}{dy} = g$	$\frac{dw}{dx} + \frac{du}{dz} = b$
$\frac{dw}{dz} = c$	$\frac{du}{dy} + \frac{dv}{dx} = 2h$	$\frac{dw}{dz} = h$	$\frac{du}{dy} + \frac{dv}{dx} = c.$

Figure 3.1 illustrates the geometrical meaning of the coefficients a, b, c, f, g, h.

Betti defined the angular distortions focusing more on the mathematical aspects than the physical ones: in fact $2f, 2g, 2h$ represent the variation of the right angle, while f, g, h are those that today are called the components of the strain tensor (along with a, b, c). The use of f, g, h is generally convenient in the mathematical treatments which by their nature require the essential use of the strain tensor.

Finally Betti showed that the components a, b, c, f, g, h of strain uniquely define the displacement field apart from a rigid motion.

3.1.1.2 Potential of the Elastic Forces

The concept of potential was an integral part of mathematical physics since Betti's first works. As already mentioned, in the early works [16] Betti introduced the potential as a primitive function of the forces without attributing to it a particular status of physical magnitude. He soon changed approach and 'force' began to take on an ambiguous meaning, indicating both the Newtonian force and the (thermo-) mechanical magnetic or electrical potential energy.

Betti published only articles of thermology and heat propagation, but still showed good knowledge of thermodynamics, which came to mathematical physics thanks to the work of William Thomson. In the *Teoria della elasticità*, by means of the first and second principle of thermodynamics, he gave a physical meaning to his potential, today included under the name of potential energy.

The thermodynamical theory was developed for homogeneous thermal processes, although there was awareness that in a real body the processes are generally heterogeneous. Betti, following with William Thomson the current approach to thermodynamics with the aid of the differential calculus, considered the continuum S divided into infinitesimal elements, each of which is treated as homogeneous. The potential of the elastic forces is thus given by the sum of the potential of the elastic forces of all the infinitesimals, and then by an integral. More precisely, if P expresses the potential of an infinitesimal element, the potential Φ of the whole continuum is:

$$\Phi = \int_S P\,dS. \tag{3.4}$$

Betti proposed that the potential of the elastic forces be a function of the infinitesimal strains, in the footprints of Green [54]. He assumed the *natural state*, as the reference stable configuration from which to measure the strains; thus in the development in series of P he could neglect the first-order terms. He also neglected the terms of the order higher than the second, obtaining so the quadratic form:

$$P = \sum_{i=1}^{6}\sum_{j=1}^{6} A_{ij}x_i x_j. \tag{3.5}$$

where $x_\alpha,\ \alpha = 1, 2, \ldots, 6$ represent the generic components of the strain. For the stability of the equilibrium the quadratic form should be negative definite (remember that the potential is potential energy with sign reversed). Similarly to Green [54] for an isotropic body Betti came to the expression [20][16]:

$$P = A\Theta^2 + B\Lambda^2, \tag{3.6}$$

where

$$\Theta = a + b + c, \qquad \Lambda^2 = a^2 + b^2 + c^2 + 2f^2 + 2g^2 + 2h^2. \tag{3.7}$$

Notwithstanding the evident reference to Green in the expression of the potential, the constants A and B are not those that Green used for the elasticity of isotropic bodies, but are connected to the Lamé constants $\tilde{\lambda}$ and $\tilde{\mu}$ [54][17]:

$$A = -\frac{\tilde{\lambda}}{2} \qquad B = -\tilde{\mu}. \tag{3.8}$$

[16] p. 18. Let E be the tensor of deformation, Θ and Λ are respectively the trace of E and E^2.

[17] p. 253. The tilde distinguishes the two Lamé constants from Betti's λ and μ.

3.1.1.3 The Principle of Virtual Work

The third chapter of the *Teoria della elasticità* opens with the quotation:

> To determine the relationships that must exist between the forces acting on a homogeneous elastic solid body, and the deformations of the elements of the same, for there is equilibrium, we will use the following the principle by *Lagrange*: for a system, whose virtual motions are reversible, to be in equilibrium it is necessary and sufficient that the mechanical work done by the forces in a whatever virtual motion, be equal to zero [20].[18] (A.3.4)

Betti therefore did not consider the equilibrium equations as relations between external and internal forces, but between external forces and strain-displacements. He used the principle of virtual work ("the principle of Lagrange"), because in such a way he could express the virtual work of the internal forces without making them intervene directly. It is worth noting the way in which he stated the principle of Lagrange: there is no physical obstacle to its validity, there is only the interest of the mathematician who wants to clarify whether the work is negative or zero. Assuming bilateral constraints, the work can be equated to zero.

The equation obtained by Betti is:

$$\delta\Phi + \int_S \rho(X\delta u + Y\delta v + Z\delta w)dS + \int_\sigma (L\delta u + M\delta v + N\delta w)d\sigma = 0. \tag{3.9}$$

In it, $\delta\Phi$ is the virtual work of the internal forces, ρ is the mass per unit of volume, (X, Y, Z), the components of the force per unit of mass (*accelerating force*) in the volume S and (L, M, N) the force per unit of surface on the surface σ, the boundary of S. Passing from the variational Eq. (3.9) to the equations of equilibrium is simple for Betti; indeed similar elaborations had already been carried out by Navier, Green, William Thomson, and Clebsch.

In this point of the *Teoria della elasticità* Betti, without specifying the form of the potential P, limited to obtain the local and boundary equations of equilibrium that were written as [20][19]:

$$\begin{aligned}
\rho X &= \frac{d}{dx}\frac{dP}{da} + \frac{d}{dy}\frac{dP}{2dh} + \frac{d}{dz}\frac{dP}{2dg} \\
\rho Y &= \frac{d}{dx}\frac{dP}{2dh} + \frac{d}{dy}\frac{dP}{db} + \frac{d}{dz}\frac{dP}{2df} \\
\rho Z &= \frac{d}{dx}\frac{dP}{2dg} + \frac{d}{dy}\frac{dP}{2df} + \frac{d}{dz}\frac{dP}{dc},
\end{aligned} \tag{3.10}$$

[18] p. 20. Our translation.

[19] p. 22.

$$\begin{aligned}
L &= \frac{dP}{da}\alpha + \frac{dP}{2dh}\beta + \frac{dP}{2dg}\gamma \\
M &= \frac{dP}{2dh}\alpha + \frac{dP}{db}\beta + \frac{dP}{2df}\gamma \\
N &= \frac{dP}{2dg}\alpha + \frac{dP}{2df}\beta + \frac{dP}{dc}\gamma.
\end{aligned} \tag{3.11}$$

The sign of the second members of these equations is contrary to the one usually found in modern textbooks because Betti orientated the normal n, of which α, β, γ are the components, to the surface σ toward the interior instead of toward the exterior as is done today.

3.1.2 The Reciprocal Work Theorem

The formulation of the theorem of reciprocal work is perhaps the most notable contribution of Betti to the theory of elasticity:

> If, in a homogeneous elastic body, two systems of displacements are respectively equilibrated to two systems of forces, the sum of the products of the components of the forces of the first system by the corresponding components of the displacements of the same points in the second system is equal to the sum of the products of components of the forces of the second system by the components of the displacements at the same points of the first [20].[20] (A.3.5)

This theorem is presented and demonstrated in the absence of volume forces only, in the *Teoria della elasticità*.

The demonstration, relatively simple, started from the equilibrium equations written for the two equilibrated systems of forces and displacements. Betti retraced in reverse the steps by which he had obtained the equilibrium equations using the principle of virtual work and obtained the expression:

$$\int_\sigma (L'u'' + M'v'' + N'w'')\, d\sigma = \int_\sigma (L''u' + M''v' + N''w')\, d\sigma, \tag{3.12}$$

where (u, v, w) is the vector field of the displacements associated to the active surface forces (L, M, N), solution of the elastic problem. The apices indicate forces and displacements of two distinct elastic problems, still for the same continuum.

Betti came back to the theorem of reciprocal work in 1874 [19][21] by extending the theorem to the case of volume forces of components X, Y, Z, reaching thus the expression with which it is known today:

[20] p. 40. Our translation.

[21] p. 381.

$$\int_\sigma (L'u'' + M'v'' + N'w'')\, d\sigma + \rho \int_S (X'u'' + Y'v'' + Z'w'')\, dS =$$
$$\int_\sigma (L''u' + M''v' + N''w')\, d\sigma + \rho \int_S (X''u' + Y''v' + Z''w')\, dS. \tag{3.13}$$

He also indicated the role attributed to his theorem:

> In this paper I show a theorem that, in the theory of the elastic forces of solids, takes the place of Green's theorem in the theory of the forces acting according to the law of Newton, and for what the applications is concerned I just deduce formulas similar to that of Green's functions for the potential [19].[22] (A.3.6)

Green's theorem recalled by Betti has the expression:

$$\int_\sigma v \frac{\partial u}{\partial n} d\sigma = \int_\sigma u \frac{\partial v}{\partial n} d\sigma, \tag{3.14}$$

where u and v are harmonic functions and represent the potentials of central forces in a portion S of a homogeneous and isotropic space void of sources, delimitated by the surface σ with normal n.[23] To obtain Eq. (3.14) Green started from Dirichlet's elliptic problem, defined by the harmonic equation of the potential and by the boundary conditions:

$$\Delta v = 0 \quad \text{in } S, \qquad v = \overline{v} \quad \text{on } \sigma. \tag{3.15}$$

Here Δ is Laplace's operator and $\overline{v}$ an assigned function on σ.

Betti in the *Teoria della elasticità* started instead from the field equations of the elastic problem for an elastic and homogeneous continuum [20][24]:

$$\begin{aligned}
\rho X + (2\lambda + \mu)\frac{d\Theta}{dx} + \mu\Delta^2 u = 0\\
\rho Y + (2\lambda + \mu)\frac{d\Theta}{dy} + \mu\Delta^2 v = 0\\
\rho Z + (2\lambda + \mu)\frac{d\Theta}{dz} + \mu\Delta^2 w = 0,
\end{aligned} \tag{3.16}$$

---

[22] p. 379 Our translation.

[23] In [53] Green enounced a more general theorem than (3.14), today known as the second Green's identity:

$$\int_\sigma v \frac{\partial u}{\partial n} d\sigma + \int_S u\Delta v\, dS = \int_\sigma u \frac{\partial v}{\partial n} d\sigma + \int_S v\Delta u\, dS.$$

Functions u and v are (whatever their form) endowed with the necessary conditions of regularity ([53], p. 23, par. 3, not numbered equation); by imposing that u and v be harmonic functions the Eq. (3.14) is obtained, not made explicit by Green.

[24] Equations 32–33, pp. 33–34.

with $\Delta^2 = \frac{d^2}{dx^2} + \frac{d^2}{dy^2} + \frac{d^2}{dz^2}$, and the boundary equations:

$$\begin{aligned}
&L + 2\left(\lambda\Theta + \mu\frac{du}{dx}\right)\alpha + \mu\left(\frac{du}{dy} + \frac{dv}{dx}\right)\beta + \mu\left(\frac{du}{dz} + \frac{dw}{dx}\right)\gamma = 0 \\
&M + \mu\left(\frac{du}{dy} + \frac{dv}{dx}\right)\alpha + 2\left(\lambda\Theta + \mu\frac{dv}{dy}\right)\beta + \mu\left(\frac{dv}{dz} + \frac{dw}{dy}\right)\gamma = 0 \\
&N + \mu\left(\frac{du}{dz} + \frac{dw}{dx}\right)\alpha + \mu\left(\frac{dv}{dz} + \frac{dw}{dy}\right)\beta + 2\left(\lambda\Theta + \mu\frac{dw}{dz}\right)\gamma = 0.
\end{aligned} \tag{3.17}$$

In the relations (3.16), (3.17) λ and μ are not Lamé's constants, usually denoted by the same symbols, but the constants A and B of the relation (3.6) with sign reversed.

The analogy between (3.12) and (3.14) starts from the way both are obtained: the field equations are multiplied by arbitrary displacement fields and integrated by parts, so to reduce the maximum order of the derivatives. The aim is to relate the solution of differential equations to a quadrature formula by means of special functions (now called *Green functions*).[25]

Betti's reciprocal work theorem is often used in educational presentations of the theory of elasticity to derive the reciprocity theorem of Maxwell. Furthermore Betti's theorem is reinterpreted for concentrated forces. By assuming only two forces f_i and f_j, applied respectively to the points i and j of an elastic body, and if u_{ij} and u_{ji} are respectively the displacement in i due to the force f_j and the displacement in j due to the force f_i, Betti's reciprocal theorem gives:

$$f_i u_{ij} = f_j u_{ji}, \tag{3.18}$$

that assuming $f_i = f_j = 1$, furnishes $u_{ji} = u_{ij}$. This is the very Maxwell's theorem, as formulated in [48].[26]

The above considerations are only intended to motivate the association between the theorems of Maxwell and Betti. This association was not, and there was no reason it should be, evident to the two scholars who moved driven by different purposes. Betti wanted to find a possible method of solution of his differential equations; Maxwell was moved by considerations of a more physical character, to shed light on certain properties of the elastic relationships.

---

25 Green's integral formula provides the function v, solution of (3.14), at a point P internal to S starting from the knowledge of v on σ. On the basis of the (3.15) it is given by:

$$v = \frac{1}{4\pi}\int_\sigma \overline{v}\frac{\partial}{\partial n}\left(v' + \frac{1}{r}\right) d\sigma.$$

Here r is the distance of P from the points Q of σ. The function $v'(P, Q)$, sometimes called a *Green function*, satisfies Laplace's equation and is such that $u = (v' + 1/r) = 0$ on σ [53], p. 29. More frequently, one calls the whole expression u a Green function. Among the authors which individuate in v' Green's function to signal Betti, Rudolf Otto Sigmund Lipschitz, and Carl Neumann. Green seems to prefer the use of the function u [53], p. 31, § 5, Eq. 5.

26 p. 297.

3.1.3 Calculation of Displacements

Betti divided into two phases, considered in three different chapters, the integration of Eq. (3.16), that is evaluation of the displacements due to assigned forces. He first evaluated the percentage variation of volume, or the unitary dilatation Θ (Chap. 8 of the *Teoria della elasticità*) and the infinitesimal local rotation (ϕ_1, ϕ_2, ϕ_3) (Chap. 9):

$$\begin{aligned}\Theta &= \frac{du}{dx} + \frac{dv}{dy} + \frac{dw}{dz}\\ \phi_1 &= \frac{dv}{dz} - \frac{dw}{dy}, \quad \phi_2 = \frac{dw}{dx} - \frac{du}{dz}, \quad \phi_3 = \frac{du}{dy} - \frac{dv}{dx}.\end{aligned} \tag{3.19}$$

From these relations he obtained the field of displacements (u, v, w) (Chap. 10).

3.1.3.1 Unitary Dilatation and Infinitesimal Rotations

In the *Teoria della elasticità* Θ was evaluated for the case when only surface forces act, while in the paper of 1874, *Sopra l'equazioni di equilibrio dei corpi solidi elastici* [19], also the volume forces were considered, with no particular difficulty. In the *Teoria della elasticità* Betti used the reciprocal theorem (3.12) and (3.13), assuming 'true' values for (u', v', w'), (L', M', N') and suitable auxiliary functions for (u'', v'', w''), (L'', M'', N''); in particular:

$$u'' = \frac{d\frac{1}{r}}{dy} + \xi, \qquad v'' = -\frac{d\frac{1}{r}}{dx} + \eta, \qquad w'' = \zeta. \tag{3.20}$$

Here r is the distance from the generic point Q of the continuum to the point P where the displacement is searched for, that for the moment is considered as fixed. The field (ξ, η, ζ) depends on the coordinates of Q and is such that (u'', v'', w'') satisfy the local equilibrium [20].[27] Because there is no need to satisfy the boundary conditions, (ξ, η, ζ) are undetermined. Betti will assign conditions on (ξ, η, ζ), to make them determinate, only at the end of his analytical developments, by assuming two situations, one for which $\xi = \eta = \zeta = 0$ and another such that the stresses due to (u'', v'', w'') vanishes on the boundary σ.

To apply the reciprocal theorem, Betti considered the true continuum S, from which we subtract an infinitesimal portion S' with surface σ' and radius "as small as you like" centered in P. By indicating with L', M', N' the 'stresses' which act on σ and with X', Y', Z' the 'stresses' which act on σ' due to (u', v', w'); with X_0, Y_0, Z_0 the stresses due to (ξ, η, ζ) which act on σ and σ' and with X'', Y'', Z'' the 'stresses'

[27] Equation 43, p. 55.

due to the displacements $d(r^{-1})/dx,\ d(r^{-1})/dy,\ d(r^{-1})/dz$ which act on σ and σ', Betti obtained [20][28]:

$$\begin{aligned}
&\int_\sigma \left[L'\left(\frac{d\frac{1}{r}}{dx}+\xi\right)+M'\left(\frac{d\frac{1}{r}}{dy}+\eta\right)+N'\left(\frac{d\frac{1}{r}}{dz}+\zeta\right)\right]d\sigma \\
&+\int_{\sigma'} \left[X'\left(\frac{d\frac{1}{r}}{dx}+\xi\right)+Y'\left(\frac{d\frac{1}{r}}{dy}+\eta\right)+Z'\left(\frac{d\frac{1}{r}}{dz}+\zeta\right)\right]d\sigma' \qquad (3.21)\\
&=\int_\sigma [(X''+X_0)u'+(Y''+Y_0)v'+(Z''+Z_0)w']d\sigma \\
&+\int_{\sigma'} [(X''+X_0)u'+(Y''+Y_0)v'+(Z''+Z_0)w'']d\sigma'.
\end{aligned}$$

Here Betti 'ought to' explicitly introduce the components of stress; though in an indirect way, as external surface forces, but this was a rhetorical expedient only. After lengthy calculations, by assuming regular functions for (ξ, η, ζ) such that $X''+X_0 = Y''+Y_0 = Z''+Z_0 = 0$, Betti obtained the following expression [20][29]:

$$\begin{aligned}
\Theta = -\frac{1}{8\pi(\lambda+\mu)}\int_\sigma \Bigg(& L'\frac{d\frac{1}{r}}{dx}+2\mu u'\frac{d}{dp}\frac{d\frac{1}{r}}{dx}+M'\frac{d\frac{1}{r}}{dy}+2\mu v'\frac{d}{dp}\frac{d\frac{1}{r}}{dy}+N'\frac{d\frac{1}{r}}{dz} \\
&+2\mu w'\frac{d}{dp}\frac{d\frac{1}{r}}{dz}+L'\xi - X_0u'+M'\eta - Y_0v'+N'\zeta - Z_0w'\Bigg)d\sigma,
\end{aligned} \qquad (3.22)$$

which gives the unitary dilatation as a function of the surface forces.

Betti knew that the expression found for Θ is not simple: "The evaluation of the functions ξ, η, ζ which correspond to the Green functions for the search of the potential functions, generally offers many difficulties " [20].[30] Alternatively he assumed $\xi = \eta = \zeta = 0$ and indicating with σ_1 and σ_2 the surfaces of two concentric spheres which contain the continuum S, with σ_1 the greater and σ_2 the smaller, found [20][31]:

$$\begin{aligned}
\Theta = -\frac{1}{8\pi(\lambda+\mu)}\Bigg\{\int_{\sigma_1}\Bigg[& L_1'\frac{d\frac{1}{r}}{dx}+M_1'\frac{d\frac{1}{r}}{dy}+N_1'\frac{d\frac{1}{r}}{dz} \\
&+2\mu\left(u_1'\frac{d}{dp_1}\frac{d\frac{1}{r}}{dx}+v_1'\frac{d}{dp_1}\frac{d\frac{1}{r}}{dy}+w_1'\frac{d}{dp_1}\frac{d\frac{1}{r}}{dz}\right)\Bigg]d\sigma_1
\end{aligned}$$

---

[28] Equation 45, p. 56–57

[29] p. 62.

[30] p. 63.

[31] Equation 48. p. 63.

$$\int_{\sigma_2}\Bigg[L_2'\frac{d\frac{1}{r}}{dx}+M_2'\frac{d\frac{1}{r}}{dy}+N_2'\frac{d\frac{1}{r}}{dz}$$
$$+\,2\mu\left(u_2'\frac{d}{dp_2}\frac{d\frac{1}{r}}{dx}+v_2'\frac{d}{dp_2}\frac{d\frac{1}{r}}{dy}+w_2'\frac{d}{dp_2}\frac{d\frac{1}{r}}{dz}\right)\Bigg]d\sigma_2\Bigg\},\tag{3.23}$$

where the indices distinguish the values referring to the two spheres.

The above expression of Θ is self-referential because the evaluation of Θ calls for knowledge of the components of (u', v', w'), and so of Θ itself, appearing on the right-hand side of the relation (3.14). Betti eliminated such self-referentiality in [17].

The evaluation of the infinitesimal rotations ϕ_1, ϕ_2, ϕ_3 is analogous to that of Θ, with different functions ξ, η, ζ which "correspond to the Green functions".[32] By imposing on the boundary σ the stresses associated to u'', v'', w'', Betti obtained [20][33]:

$$\phi_1=\frac{du'}{dy}-\frac{dv'}{dx}=-\frac{1}{4\pi\mu}\int_\sigma\left[L'\left(\frac{d\frac{1}{r}}{dy}+\xi\right)-M'\left(\frac{d\frac{1}{r}}{dx}-\eta\right)\right]d\sigma.\tag{3.24}$$

Analogous expressions were found for ϕ_2 and ϕ_3.

3.1.3.2 The Displacements

Betti showed that when $\Theta, \phi_1, \phi_2, \phi_3$ are known, the evaluation of the displacements u, v, w reduced to the solution of a Neumann problem for the equation of Poisson, that is to a problem of potential. Indeed the field equilibrium equations can be written again as [20][34]:

$$\Delta^2 f=F;\qquad f=u,\,v,\,w;\qquad F=-\frac{2\lambda+\mu}{\mu}\frac{d\Theta}{dx_i},\qquad x_i=x,\,y,\,z.\tag{3.25}$$

whose boundary conditions expressed by $\partial f/\partial p$, with p normal to σ, depend on $\Theta, \phi_1, \phi_2, \phi_3$ only [20].[35] Moreover f should satisfy the relation:

$$\int_\sigma\frac{df}{dp}\,d\sigma=-\int_S F\,dS.\tag{3.26}$$

[32] Betti's passages contain many typos, partially emended in the *Opere*, edited by da Orazio Tedone.

[33] pp. 79–80.

[34] Equation 56, p. 81.

[35] p. 81.

Thus a solution exists to evaluate the displacement "and is defined except for a constant when the values of df/dp are continuous as in the case we are considering" [20].[36]

3.1.4 The Saint Venant Problem

In Chap. 11 Betti addressed the problem which Clebsch in [45] had attributed to Saint Venant. Quite strangely, considering the contact of Betti with Riemann and with the German scientific community and mainly for the popularity of Clebsch's treatise, Betti made no reference to Saint Venant. The case studied by Betti was however the same as that found in [46, 77, 78]: the linear elasto-static problem for the cylinder of Fig. 3.2 loaded at the basis ω_1 and ω_2 by regular surface forces (L_1, M_1, N_1) and (L_2, M_2, N_2). The cylinder was described with respect to a system of Cartesian coordinates having its origin in the barycenter of ω_1, axis z orthogonal to ω_1 and coinciding with the axis of the cylinder, according to that shown in Fig. 3.2.

Betti started by considering an arbitrary field of displacements (u', v', w'), which satisfies the field equations of the elasto-static problem for the cylinder assumed free of volume forces. The reciprocal theorem gives [20][37]:

$$\begin{aligned}&\int_{\omega_1}(L_1u'_1+M_1v'_1+N_1w'_1)d\omega_1+\int_{\omega_2}(L_2u'_2+M_2v'_2+N_2w'_2)d\omega_2\\&\quad=\int_{\omega_1}(L'_1u_1+M'_1v_1+N'_1w_1)d\omega_1+\int_{\omega_2}(L'_2u_2+M'_2v_2+N'_2w_2)d\omega_2\\&\qquad+\int_{\sigma}(L'_0u_0+M'_0v_0+N'_0w_0)d\sigma\end{aligned}\tag{3.27}$$

where σ is the lateral surface of the cylinder; (u, v, w) is the displacement field providing stresses that equilibrates the contact forces L, M, N; L', M', N' is the field of forces due to the displacements (u', v', w'). The subscripts indicate the value assumed by the two fields on the two basis (1 and 2) and on the lateral surface of the cylinder (0), respectively.

To use the relation (3.27) Betti had to characterize the field (u', v', w'), which assumes the role of the 'virtual' displacement field; he wrote the equations of local and boundary equilibrium for the forces L', M', N', as a function of (u', v', w').[38] Of all the fields (u', v', w') which satisfy local and boundary equilibrium Betti chose the one for which the associated stress state have no components in the plane of the cylinder $z = \text{const}$.

---

[36] p. 83.

[37] Equation 59, p. 84.

[38] This is done in all groups of the not numbered equations enclosed among the (59) and (60) of the Chap. 10. The first group represents the linear elastic homogeneous and isotropic relationship between the components of stress and the partial derivatives of the components of the displacement,

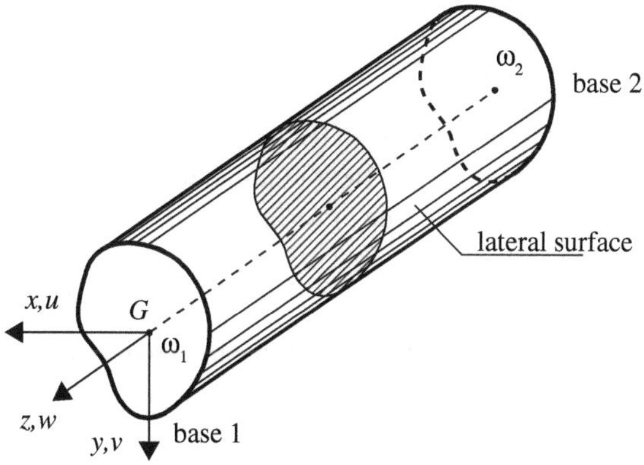

Fig. 3.2 Solid of Saint Venant

This condition with the consequences that it implies in the local equilibrium equations was expressed as [20][39]:

$$\begin{aligned} &\frac{dG'}{dz} = 0, \qquad \frac{dF'}{dz} = 0 \\ &\frac{dG'}{dx} + \frac{dF'}{dy} + \frac{dC'}{dz} = 0 \end{aligned} \tag{3.28}$$

and[40]:

$$A' = 0, \qquad H' = 0, \qquad B' = 0 \tag{3.29}$$

where $C,\ F,\ G$ indicate the stresses parallel to the z axis while $A,\ H,\ B$ the stresses orthogonal to it, that is parallel to the section of the cylinder. Betti used for the stress a notation close to that of Cauchy and also adopted by Piola in [74] forty years before; though the letters are slightly different. Betti's notation of stresses recalls the letters he had used for strains; more precisely the stresses A, B, C, F, G, H correspond to the strain $a,\ b,\ c, f,\ g,\ h$. Notice that at Betti's time there were already in use notations with two subscripts, present in later work of Cauchy, Clebsch, Saint Venant, William Thomson and Tait, and Kirchhoff that were related to the theory of determinants [89]. Betti however was not directly interested in the concept of stresses so he did not care 'details' about them.

(Footnote 38 continued)
the second group expresses the local equations of static equilibrium, the third group characterizes the components of the normal (oriented toward the interior according to the convention in the 19th century) to the outer surface of the cylinder, and the fourth group expresses the boundary conditions on the specialized components of the normal just characterized.

[39] Equations 60–62, p. 85.

[40] Betti did not use a semi-inverse method based on the so-called hypothesis of Clebsch-Saint-Venant, Eq. (62). The vanishing of the stresses on the plane of the section (today indicated with the symbols $\sigma_x,\ \sigma_y,\ \tau_{xy}$) is a condition for the auxiliary field of displacement useful to be introduced in the reciprocal theorem.

Betti thus chose (u', v', w') so that only the components of the stress in the direction of the axis of the cylinder are different from zero. By imposing local equilibrium he obtained that the tangent stresses were independent of z. The only equation remaining to satisfy the equilibrium was that obtained by projecting the local equilibrium equation along the z axis. Betti required that the state of stress satisfied the condition of the lateral surface free of surface forces [20][41]:

$$G_1' = G_2', \qquad F_1' = F_2'. \tag{3.30}$$

He characterized the field of displacements (u', v', w') in such a way that it looked like the solution of the problem, leaving free only the surface forces on the basis associated with it. On the other hand, Betti introduced the hypothesis (3.28) as a free choice of the field of 'virtual' displacements-stresses which appears in the formula of reciprocity. Consequently, Betti's results might be more general than that of Saint Venant, since it does not postulate that the 'true' is characterized by the canceling of the components of the stress parallel to the plane of the section.

Later in chapter, Betti followed Clebsch and Saint Venant [46].[42] Moreover, wanting only to determine the 'virtual ingredients' to be introduced in relation (3.27), Betti used technicalities of integration for systems of linear differential equations in partial derivatives similar to others in the literature of his time. With the aid of elements of complex analysis, Betti found the components of the field (u', v', w'):

$$\begin{aligned}
u' &= h + kz + \frac{az^2}{2} + \frac{bz^3}{6} + (c + ez)x - (c' + e'z)y \\
&\quad + \tau\left((a + bz)\frac{x^2 - y^2}{2} + (a' + b'z)xy\right) \\
v' &= h' + k'z + \frac{a'z^2}{2} + \frac{b'z^3}{6} + (c + ez)y + (c' + e'z)x \\
&\quad + \tau\left((a' + b'z)\frac{y^2 - x^2}{2} + (a + bz)xy\right) \\
w' &= -\frac{cz + \frac{ez^2}{2}}{\tau} - \left(az + \frac{bz^2}{2}\right)x + \left(a'z + \frac{b'z^2}{2}\right)y \\
&\quad + \frac{e}{\tau}(x^2 + y^2) + bxy^2 + b'yx^2 + U,
\end{aligned} \tag{3.31}$$

where τ is Poisson's coefficient, $U(x, y)$ is the solution of a harmonic problem with Neumann's boundary conditions on the bases of the cylinder. The relations (3.31), considering the due correspondences, coincide with the solutions already found by Clebsch and Saint Venant, even though Betti did not point out the fact.

The fields of 'virtual' displacements u', v', w' depend on the constants of integration $a, a', b, b', c, c', e, e', h, h', k, k'$. About the meaning of the constants of

---

[41] Second not numbered equation after the equation 62, p. 86.

[42] A discussion in absolute form in the subject is for example found in [76].

integration Betti did not put forward any interpretation. Only six of them, the same number of the components of the resultant actions on the basis, are used.

Betti obtained first the mean values, simple or weighted with the position x, y, z, of the difference of the axial displacement between the two bases of the cylinder as a function of the surface forces on the bases [20].[43] This dependence is not trivial to make explicit; Betti so examined the simple case, coinciding with one of the Saint Venant's case, in which the contact forces on the basis have opposite directions [20] [44] so that for any point of $\omega_1 \equiv \omega_2$ it is:

$$L_1 + L_2 = 0, \quad M_1 + M_2 = 0, \quad N_1 + N_2 = 0. \tag{3.32}$$

He then inferred that the mean values are directly proportional to the resultant actions on the ends of the cylinder [20].[45] In particular, the simple mean gives the elongation of the cylinder, the variation of the area of the section and the coefficient of lateral contraction. He also commented on the results of the weighted mean, but there is no mechanical meaning for them.

Betti broke down the field U of the last of the relations (3.31) in the three addends, harmonic solutions of problems with boundary conditions of Neumann type on the sections of the cylinder. Each addend is proportional to one of the constants e', b, b', identifiable with the torsional and non-uniform bending curvatures for the axis of the cylinder. Betti considered only sections with two axes of symmetry, for which the area integrals weighed with the odd powers of the coordinates vanish; he thus obtained the equations for the components of the torque[46] and bending moment resulting on the bases.[47] He did not comment on their results in general, but only the particular solutions for cylinders with elliptical sections, for which a solution closed form exists for the addends U. For the non-uniform bending he found the expression of the 'deflection of bending', that he particularized to the circular sections.[48]

Betti did not seem to care about the fact of not presenting a complete solution (for instance, the analysis of uniform bending is lacking). He had clearly applied the formula of reciprocity, which led to some solutions of technical interest, and this was probably enough for him.

3.2 Eugenio Beltrami

Eugenio Beltrami (Cremona 1836-Rome 1900) attended the high school in Cremona, and university in Pavia from 1853 to 1856, where one of his teachers was Brioschi, at that time freshly appointed professor of Applied mathematics. Beltrami, however,

---

[43] Equations 67–69, p. 91.

[44] p. 91.

[45] p. 91.

[46] Equation (70) in Chap. 10.

[47] Equations (71) and (72) of Chap. 10.

[48] Last equation of Chap. 11, p. 91.

did not complete his studies because he was in financial hardship and was sent off from Collegio Ghislieri where he lived,[49] due to his sympathy with the anti-Austrian movement in Italy of the time. Having lost his position in the college, he could not afford the position of university student any more, because he was deprived of the support of his grandfather, dead, and of his father, in voluntary exile also because of anti-Austrian feelings. He got a job as a secretary of the directorate of the Lombard-Venetian Railways in Verona, but after a while he was fired by the Austrian government due to political reasons. He regained his job in Milan, where the responsibility of his office had moved, when the city was taken away by Piedmont from the Austrian empire.

Beltrami attended Brera Observatory in Milan and, following Brioschi's advice, resumed studying mathematics. To have an idea of his knowledge of mathematics, what he wrote in December 1860 to a friend of his is enlightening:

> I completed my university course (partly due to imprudence, partly due to the indolence that usually accompanies the sadness due to frequent family diseases) following the bad habit of studying just as much was needed to pass exams. I then lost two years in things at all alien to my interests.[50] After this harsh trial, I firmly took the intention to rebuild in studying mathematics, and (this is the only thing which I sincerely praise myself of) I took up studying with all diligence, one after the other, arithmetics, algebra, geometry, trigonometry, advanced algebra, and calculus, like he would have done, who had followed a completely different faculty than Mathematics [...] Here is my scientific furniture: I feel it very poor. Most of all, I deeply regret being almost zero[51] in the disciplines regarding the calculus of variations, the papers by Jacobi and Abel, the investigations by Gauss on surfaces, and so on [1].[52] (A.3.7)

In spite of his poor starting background, Beltrami was able to recover quickly. In 1862 he published his first scientific paper and Brioschi managed to appoint him professor in charge[53] of Algebra and Geometry at the university of Bologna. He was helped by the fact that, starting from 1861, with the new born Kingdom of Italy, university staff was strengthened. In 1864, he was offered the chair of Geodesy at the university of Pisa by Betti. Beltrami was not convinced by this offer, as he wrote in this letter to Cremona, but, after a positive advice of the latter, he decided to accept:

> I would be determined to refuse the offer Betti has made to me for many reasons. First of all, for the necessity to change the direction of my studies, that always brings drawbacks and lacks of time, especially because Betti told me of preparatory studies to be performed in an Observatory, and thus it seems that the subjects to be dealt with in the new chair shall not be purely theoretical. In second place, I like the chair of Introduction to calculus more, both for the nature of the subjects it deals with, and for the broader width that it leaves in the choice of investigations. In the end, I would be sorry to occupy a place, that public opinion would

---

[49] A college with the aim of promoting studies at the university of Pavia, supporting pupils, chosen on the basis of merit, with logistic and cultural opportunities.

[50] It is apparent that Beltrami talked about his stay in Verona.

[51] Beltrami said, in Latin, *tamquam tabula rasa*.

[52] Tome 1, p. XI. Our translation.

[53] Before getting the permanent position as full professor (Professore ordinario), candidates should undergo a period of apprenticeship, during which they were referred to as professors in charge (Professore straordinario).

> better like to entrust probably to a distinguished scholar in affine studies, I mean, to Codazzi; and that, even neglecting this, could be coveted by professors more experienced than me and already meritorious in teaching. As far as the pecuniary advantage I could have by the appointment to full professorship, it would only be temporary, inasmuch I am to expect the same result after a more or less long apprenticeship, both in the place that I occupy now, and without abandoning the university in which I have you as a colleague. In any case, I did not want to answer Betti before asking for your advice, which I beg you to let me know most freely [1].[54] (A.3.8)

In Pisa, Beltrami made friends with Betti and met Riemann. In 1866 he went back to Bologna (it seems that his wife could not stand the climate in Pisa) in the chair of Rational mechanics. In 1873 he was called to the same chair at the university of Rome, which had recently become the capital city of Italy, but did not feel himself well even there (his wife could not stand the climate in this city, either). Thus, in 1876 he moved to Pavia to the chair of Mathematical physics; he went back to Rome in 1891 for his last periods of teaching.

Beltrami had an important role in research, in teaching, and even in organization of the Italian mathematical school. In 1898 he became president of the Accademia dei Lincei, succeeding Brioschi; in 1899 he became a senator of the Italian Kingdom [52].

Beltrami was essentially self-taught, and in Pisa he addressed his studies toward the geometry of surfaces, taking inspiration from Gauss [2, 3], Lobacevskij, Riemann, and Cremona. In addition to pure mathematics, especially geometry, he dealt with mathematical physics, potential theory and electro-magnetism in particular; some studies on optics and thermodynamics were also interesting. In these fields he searched for amendments to some physical laws in order to extend their validity to a space with negative curvature, by generalizing the Laplace operator.

Beltrami's differential techniques have influenced the birth of tensor calculus, by providing the bases for the ideas developed later by Gregorio Ricci-Curbastro and Tullio Levi-Civita. Some of his last works were about the mechanical interpretation of Maxwell's equations. Beltrami's contribution to the history of mathematics is important as well: in 1889 he brought to light the work by Girolamo Saccheri of 1773 on parallel lines, and he compared Saccheri's results with those by Alfonso Borelli, John Wallis, Cristophorus Clavius, Bolyai, and Lobacevskij.

Beltrami's papers on elasticity theory and continuum mechanics are relatively few, but remarkably interesting. Almost all deal with strain and stress analysis in the ether, to explain electro-magnetic phenomena [85]. From this point of view, Beltrami was in the sequel of the 'energetism', initiated by Green and carried on by Piola and Betti in Italy. For Beltrami, in the absence of certain information on the nature of inner forces in bodies (as well as on the nature of bodies themselves, and on their mathematical-physical modeling), one should refer to undoubtable principles such as the principle of virtual work and the principle for which, in non-dissipative phenomena, all actions derive from a potential. These principles gave powerful means of analysis to the investigation of electrical and magnetic phenomena, supposedly propagated by contact in the luminiferous ether. In addition they allowed the use

[54] Tome 1, p. XIII. Our translation.

of a new branch of mathematics, harbinger of results, such as the new analysis by Riemann. It is not by chance, then, if also Beltrami, like Betti, turned to investigation of the most innovative subjects of mathematical physics of the time, seen as the natural generalization of the ordinary theory of elasticity for tangible bodies.

Beltrami's papers, re-issued in the *Opere matematiche* [1] are:

- 1872–1874. *Ricerche sulla cinematica dei fluidi*, Memorie dell'Accademia delle scienze dell'Istituto di Bologna, s. 3, t. 1, 1872; t. 2, 1871; t. 3, 1873; t. 5, 1874.
- 1880–1882. *Sulle equazioni generali dell'elasticità* [6].
- 1882. *Sull'equilibrio delle superficie flessibili e inestensibili* [6].
- 1884. *Sulla rappresentazione delle forze newtoniane per mezzo di forze elastiche* [7].
- 1884. *Sull'uso delle coordinate curvilinee nelle teorie del potenziale e dell'elasticità* [8].
- 1885. *Sulle condizioni di resistenza dei corpi elastici* [9].
- 1886. *Sull'interpretazione meccanica delle formole di Maxwell* [10].
- 1889. *Sur la théorie de la déformation infiniment petite d'un milieu* [12].
- 1889. *Note fisico matematiche (lettera al prof. Ernesto Cesaro)* [11].
- 1892. *Intorno al mezzo elastico di Green. Nota I e II*, Rendiconti del Regio istituto lombardo, s. 2, t. 1, 1892.
- 1892. *Osservazioni alla nota del Prof. Morera* [13].

3.2.1 Non-Euclidean Geometry

In the second half of the 19th century, physicists and mathematicians questioned if physical space were Euclidean or not. Mathematicians looked for new frontiers of geometry, physicists looked for an exhaustive explanation of electromagnetic phenomena. Indeed, when dealing with unlimited physical space, with ambiguity in the definition of a straight line, the fifth postulate of Euclidean geometry, usually stated as

> If a straight line falling on two straight lines makes the interior angles in the same direction less than two right angles, the two straight lines, if produced ad infinitum, meet one another in that direction in which the angles less that two right angles are [66].[55]

was subjected to questioning, and replaced by other postulates.

Proofs of the fifth postulate were searched for even before Euclid; indeed, the postulate seemed evident, but less than the other four, and so one thought it could (and should) be proved starting from these.[56] At the end of 1700, geometers were

[55] p. 319.

[56] The first four postulates of Euclidean geometry are:

1. Let it be postulated to draw a straight line from any point to any point.
2. and to produce a limited straight line in a straight line,
3. and to describe a circle with any center and distance,
4. and right angles are equal to one another [66] (p. 318.)

convinced that the fifth postulate was logically independent of the other four, and, thus, could not be proved by them. This conviction, supported by Gauss, derived from the unfruitful attempts of proof by Saccheri and Johann Henrich Lambert, and was made explicit by the founders of non-Euclidean geometry, that is Gauss, who, however, did not publish much on the subject, and, most of all, the Russian Nikolaj Ivanovic Lobacevskij and the Hungarian Janos Bolyai.

Lobacevskij, the most known of the the two, based his geometry on the empirical, intuitive concepts of body, contact among bodies, and section of a body into two parts. These concepts are considered as primitive, and can be acquired through senses; on their basis, Lobacevskij was able to justify all postulates but the fifth. So, he concluded that it is not justified by experience, thus, in some way, arbitrary and perhaps 'untrue'. The only way to eliminate the arbitrariness of the fifth postulate consisted in accepting it on conventional grounds, or in building a new geometry, according to which the angle of parallelism[57] is comprised between $\pi/2$ and 0; such a geometry is said to be hyperbolic.

The construction of a non-Euclidean geometry, and the questions that its founders posed on its adequateness to catch the reality of the physical world originated debates on the completeness and coherence of Euclidean geometry. From the point of view of the conception of science by Aristotle this was a big blow, since the certainty of intuition was undermined: the falsity of Euclide's postulates implies that if intuition may fail, then we cannot base a science only on principles, considered true by intuition. Lobacevskij was probably caught by such doubts and, even if he believed the first four postulates of Euclidean geometry to be absolutely certain, some of his beliefs broke inside him. Indeed, in at least one of his most important works he completely abandoned the Aristotelian model of axiomatic type, based on principles, even if they be those 'really certain', by adopting an approach through hypotheses [51].

Non-Euclidean geometry represented for many years a marginal aspect of mathematics, until it was integrated into the body of Riemann's mathematics. The geometry of Riemann is non-Euclidean, in the broader sense that it tries to provide an answer to the question of how many parallel lines are there to a straight line passing through a point, as put forth by Lobacevskij or Bolyai. According to Riemann, geometry should not even deal with three-dimensional spaces, but rather with sets of ordered n-ples. Among the main rules in any geometry, for Riemann also there is the one providing the distance between two nearby points. In Euclidean geometry this distance is provided

(Footnote 56 continued)
Besides the postulates, there are also five common notions:

1. Things equal to the same thing are also equal to one another.
2. And if equals are added to equals, the wholes are equal.
3. And if equals are subtracted from equals the remainders are equal.
4. And things which coincide with one another are equal to one another.
5. And the whole is greater than the part [66] (p. 319.)

[57] The angle of parallelism α is the angle that a straight line s forms with the perpendicular p to a given straight line r such that all straight lines forming with p an angle greater than α do not meet r; in Euclidean geometry, $\alpha \equiv \pi/2$.

by $ds^2 = dx^2 + dy^2 + dz^2$; a generalization of such a measure for non-Euclidean spaces is expressed by the formula $ds^2 = \sum_{ij} g_{ij} dx_i dx_j \quad (i = j = 1, 2, \ldots, n)$. In Riemannian geometry, the sum of the angles in a triangle could be different from π.

Beltrami provided theoretical contributions on the subject, both on the technical and philosophical level, by applying the idea of non-Euclidean spaces in different fields. Like Riemann, and differently from many of his predecessors, Beltrami did not limit himself to consider curvilinear coordinates to define the position of a point in an Euclidean space, but posed the problem if actual space, where electrical and magnetic phenomena propagate, were Euclidean or not. He translated into Italian Gauss' paper on conformal representation and tackled the problem of the representation of the geodetic on a surface through a rectilinear segment on a plane: he discovered that it was possible only for surfaces with uniform curvature. By examining surfaces with negative curvature, Beltrami obtained his most famous result in 1868; in the paper *Saggio di interpretazione della geometria non euclidea* [4] he provided a concrete representation of the non-Euclidean geometry of Lobacevskij and Bolyai and linked it with Riemannian geometry. This concrete realization adopts a pseudo-sphere, a surface generated by the revolution of a tractor around its asymptote. Beltrami did not explicitly remark that non-Euclidean geometry is consistent, that is independent of the postulate of parallel lines; he rather underlined that Bolyai and Lobacevskijhad developed a theory of geodesics on surfaces with negative curvature.

3.2.2 Sulle equazioni generali della elasticità

Beltrami's first thorough paper on elasticity theory is dated in 1882 [5], and is about the formulation of elastic equilibrium equations in a space with constant curvature where a continuum with volume S and boundary σ is placed. He got the moves from the elasticity equations obtained by Lamé [60][58] in curvilinear coordinates, and from later works by Carl Neumann and Carl Borchardt [23, 70]. These last simplified Lamé's calculations by adopting elastic potential in curvilinear coordinates, and obtaining equilibrium equations by the variation of its integral over the volume occupied by the elastic body. According to Beltrami, however, their approach, though leading to correct results, could be improved and it was possible to put into evidence some aspects of a certain importance. Lamé, Carl Neumann, Borchardt, started either from direct equilibrium of forces (Lamé) or from the elastic potential (Carl Neumann, Borchardt) with respect to Cartesian coordinates, which implied that space was supposed Euclidean. Beltrami derived elastic equilibrium equations directly, without preliminary hypotheses on the nature of space.

Like Carl Neumann and Borchardt, Beltrami used a purely analytical approach, starting from the equality of virtual works of inner and outer forces. The key idea lies in the definitions of metrics, and, from this, of strain, which reduces to the usual one when the metrics is Euclidean:

[58] p. 290.

Let q_1, q_2, q_3 be the curvilinear coordinates of any point in a three-dimensional space, and let[59]:

$$ds^2 = Q_1^2 dq_1^2 + Q_2^2 dq_2^2 + Q_3^2 dq_3^2$$

be the expression of the square of any line element, in such a space [. . .] then, by posing

$$\delta\theta_1 = \frac{\partial \delta q_1}{\partial q_1} + \frac{\delta Q_1}{Q_1}, \qquad \delta\omega_1 = \frac{Q_2}{Q_3}\frac{\partial \delta q_2}{\partial q_3} + \frac{Q_3}{Q_2}\frac{\partial \delta q_3}{q_2},$$

$$\delta\theta_2 = \frac{\partial \delta q_2}{\partial q_2} + \frac{\delta Q_2}{Q_2}, \qquad \delta\omega_2 = \frac{Q_3}{Q_1}\frac{\partial \delta q_3}{\partial q_1} + \frac{Q_1}{Q_3}\frac{\partial \delta q_1}{q_3},$$

$$\delta\theta_3 = \frac{\partial \delta q_3}{\partial q_3} + \frac{\delta Q_3}{Q_3}, \qquad \delta\omega_3 = \frac{Q_1}{Q_2}\frac{\partial \delta q_1}{\partial q_2} + \frac{Q_2}{Q_1}\frac{\partial \delta q_2}{q_1}$$

we may write

$$\frac{\delta ds}{ds} = \lambda_1^2 \delta\theta_1 + \lambda_2^2 \delta\theta_2 + \lambda_3^2 \delta\theta_3 + 2\, \lambda_3 \delta\omega_1 + \lambda_3 \lambda_1 \delta\omega_2 + \lambda_1 \lambda_2 \delta\omega_3$$

where the three quantities $\lambda_1, \lambda_2, \lambda_3$, defined by

$$\lambda_i = \frac{Q_i dQ_i}{ds}$$

are the direction cosines of the angles that the line element ds forms with the three coordinate [lines] q_1, q_2, q_3 [5].[60] (A.3.9)

In essence, Beltrami expressed the length of the infinitesimal element ds as a function of the quantities $\delta\theta_1, \delta\theta_2, \delta\theta_3, \delta\omega_1, \delta\omega_2, \delta\omega_3$, which he chose as candidates for strain. Remark that the curvilinear coordinates q_1, q_2, q_3 are implicitly assumed orthogonal, since in the expression for ds the contribution of the products $dq_1 dq_3, dq_2 dq_3, dq_1 dq_2$ is missing. In Euclidean space one has $Q_1 = Q_2 = Q_3 = 1$, and the quantities Beltrami chose in order to characterize strain coincide with the components of the tensor of the infinitesimal strain.

Beltrami then defined the virtual work of inner forces:

$$\int (\Theta_1 \delta\theta_1 + \Theta_2 \delta\theta_2 + \Theta_3 \delta\theta_3 + \Omega_1 \delta\omega_1 + \Omega_2 \delta\omega_2 + \Omega_3 \Delta\omega_3). \qquad (3.33)$$

Here $\Theta_1, \Theta_2, \Theta_3, \Omega_1, \Omega_2, \Omega_3$ are coefficients of undefined nature, "since the variation of the line element depends on the six quantities $\delta\theta_1, \delta\theta_2, \delta\theta_3, \delta\omega_1, \delta\omega_2, \delta\omega_3$. The six multipliers $(\Theta_1, \Theta_2, \Theta_3, \Omega_1, \Omega_2, \Omega_3)$ are functions of q_1, q_2, q_3, "the meaning of which is not necessary to investigate" [5],[61] but which later will become inner forces. Beltrami expressed the equality of the virtual works done by the external bulk and boundary forces, F_i and ϕ_i respectively, and by inner forces:

[59] Remark that Q_1, Q_2, Q_3 in general depend on q_1, q_2, q_3, even if Beltrami did not state it explicitly.

[60] pp. 384–385. Our translation.

[61] p. 386.

$$\int (F_1 Q_1 \delta q_1 + F_2 Q_2 \delta q_2 + F_3 Q_3 \delta q_3)\, dS +$$
$$\int (\phi_1 Q_1 \delta q_1 + \phi_2 Q_2 \delta q_2 + \phi_3 Q_3 \delta q_3)\, d\sigma + \tag{3.34}$$
$$\int (\Theta_1 \delta\theta_1 + \Theta_2 \delta\theta_2 + \Theta_3 \delta\theta_3 + \Omega_1 \delta\omega_1 + \Omega_2 \delta\omega_2 + \Omega_3 \Delta\omega_3)\, dS = 0.$$

After having laboriously developed the integral containing inner forces in function of the displacements q_1, q_2, q_3, he gave a geometrical and mechanical meaning to these quantities, and recognized, as already remarked, the components of the infinitesimal strain of curved space in the quantities $\delta\theta_1, \delta\theta_2, \delta\theta_3, \delta\omega_1, \delta\omega_2, \delta\omega_3$, and the components of the stress in the coefficients Θ_1, Θ_2, Θ_3, Ω_1, Ω_2, Ω_3. He obtained three local equilibrium equations plus three boundary equations which "coincide with those that Lamé obtained by the transformation of the analogous equations in curvilinear coordinates".[62]

In any case, the result that Beltrami considered important, and that represents the main contribution of his paper, is to have shown the independence of the equations he obtained on Euclid's fifth postulate:

> What is more worth remarking, and that appears evident from the process kept here to obtain those equations, is that the space to which they are referred is not defined by other than the expression (1) of the line element, without any condition for the functions Q_1, Q_2, Q_3. Then equations (4), (4_a) have a much greater generality than the analogous ones in Cartesian coordinates and, in particular, it is immediately worth remarking them to be independent on Euclid's postulate [5].[63] (A.3.10)

Until now, Beltrami did not advance any hypothesis on the nature of inner forces, that is on constitutive relations. In the following steps, he first assumed conservative inner forces with potential Π function of the strain components, then considered isotropic bodies, for which the potential is:

$$\Pi = -\frac{1}{2}\left(A\theta^2 + B\varpi\right),$$
$$\theta = \theta_1 + \theta_2 + \theta_3, \qquad \varpi = \omega_1^2 + \omega_2^2 + \omega_3^2 - 4(\theta_1\theta_2 + \theta_1\theta_3 + \theta_2\theta_3). \tag{3.35}$$

Beltrami examined curved spaces with constant curvature, where isotropy is defined by two coefficients A, B independent of q_1, q_2, q_3. Under this condition he obtained relatively simple local equilibrium equations:

$$\begin{aligned}
&\frac{A}{Q_1}\frac{\partial\theta}{\partial q_1} + \frac{B}{Q_2 Q_3}\left[\frac{\partial(Q_2\theta_2)}{\partial q_3} - \frac{\partial(Q_3\theta_3)}{\partial q_2}\right] + 4\alpha B Q_1 x_1 + F_1 = 0,\\
&\frac{A}{Q_2}\frac{\partial\theta}{\partial q_2} + \frac{B}{Q_3 Q_1}\left[\frac{\partial(Q_3\theta_3)}{\partial q_1} - \frac{\partial(Q_1\theta_1)}{\partial q_3}\right] + 4\alpha B Q_2 x_2 + F_2 = 0,\\
&\frac{A}{Q_3}\frac{\partial\theta}{\partial q_3} + \frac{B}{Q_1 Q_2}\left[\frac{\partial(Q_1\theta_1)}{\partial q_2} - \frac{\partial(Q_2\theta_2)}{\partial q_1}\right] + 4\alpha B Q_3 x_3 + F_3 = 0.
\end{aligned} \tag{3.36}$$

[62] Note by Beltrami: *Leçons sur les coordonnées curvilignes*, Paris, 1859, p. 272.
[63] p. 389. Our translation.

Here α is the curvature, and $x_i = \delta q_i\ (i = 1, 2, 3)$ are the displacement components [5].[64] When $\alpha = 0, Q_1 = Q_2 = Q_3 = 1$, one recovers the equations by Navier, Cauchy and Poisson.

If the elastic (micro-)rotation and cubic dilatation vanish, in a space with positive, uniform curvature Beltrami observed an elastic deformation which has a certain 'analogy' with that provided by Maxwell's theory [47]:

> We then obtain a deformation, free from both rotation and dilatation, in which force and displacement have in each point the same (or opposite) direction and constantly proportional magnitudes. Such a result, that has no counterpart in Euclidean space, presents a remarkable analogy with certain modern concepts on the action of dielectric means [5].[65] (A.3.11)

In spaces with uniform curvature, Beltrami interpreted electro-magnetic actions (stresses in ether around a current, spherical waves) by means of the contact action in ether particles. Quite pragmatically, he supposed the space with positive, negative, or vanishing curvature, depending on the phenomenon; calculations on the stress to which ether is subjected select the correct curvature.

3.2.3 Papers on Maxwell's Electro-Magnetic Theory

Between 1884 and 1886 Beltrami published three papers on Maxwell's electro-magnetic theory [7, 8, 10], extensively dealing with stresses in ether. In Chapter V of the *Treatise on electricity and magnetism* [47], Maxwell described the interaction of two electric systems as if in the ether between them a state of stress arose, made of a tension along the lines of force combined with an equal pressure perpendicularly to these lines. Denoting by p_{ij} the stress in direction i acting on a surface with normal j, and by E_x, E_y, E_z the components of the electric field, Maxwell's equations for stress in ether in an Euclidean space are:

$$
\begin{aligned}
p_{xx} &= \frac{1}{4\pi}E_x^2 - \frac{1}{8\pi}(E_x^2 + E_y^2 + E_z^2) \qquad & p_{yz} &= p_{zy} = \frac{1}{4\pi}E_yE_z \\
p_{yy} &= \frac{1}{4\pi}E_y^2 - \frac{1}{8\pi}(E_x^2 + E_y^2 + E_z^2) \qquad & p_{xy} &= p_{yx} = \frac{1}{4\pi}E_xE_y \\
p_{zz} &= \frac{1}{4\pi}E_z^2 - \frac{1}{8\pi}(E_x^2 + E_y^2 + E_z^2) \qquad & p_{xz} &= p_{zx} = \frac{1}{4\pi}E_xE_z.
\end{aligned}
\tag{3.37}
$$

Beltrami obtained them again in [7] with a procedure simpler than Maxwell's. His approach was 'energetic', in that the variation of the potential V with respect to strain provides stresses; potential was Newtonian, that is forces in space are due to a distribution of masses contained in a volume S with surface σ. Maxwell's equations for stress in ether appear in Belltrami's notation:

---

[64] p. 398.

[65] p. 403. Our translation.

$$
\begin{aligned}
X_x &= -\frac{1}{4\pi}\left(\frac{\partial V}{\partial x}\right)^2 - \frac{1}{8\pi}\Delta_1 V & Y_z = Z_y &= -\frac{1}{4\pi}\frac{\partial V}{\partial y}\frac{\partial V}{\partial z} \\
Y_y &= -\frac{1}{4\pi}\left(\frac{\partial V}{\partial y}\right)^2 - \frac{1}{8\pi}\Delta_1 V & Z_x = X_z &= -\frac{1}{4\pi}\frac{\partial V}{\partial z}\frac{\partial V}{\partial x} \\
Z_z &= -\frac{1}{4\pi}\left(\frac{\partial V}{\partial z}\right)^2 - \frac{1}{8\pi}\Delta_1 V & X_y = Y_x &= -\frac{1}{4\pi}\frac{\partial V}{\partial y}\frac{\partial V}{\partial x}.
\end{aligned}
\tag{3.38}
$$

Beltrami extended them to generic curvilinear coordinates in another paper of 1884 [8], starting from metrics again:

$$
ds^2 = \sum_{hk} Q_{hk} dq_h dq_k, \qquad (Q_{hk} = Q_{kh}). \tag{3.39}
$$

Here Q_{hk} are functions of the curvilinear coordinates q_1, q_2, q_3. Once he defined strain as already done in the paper *Sull'equilibrio delle superficie flessibili e inestensibili* of 1882 [6], and by varying the potential V with respect to the components of strain, he obtained the components of stress and arrived at the equations [8][66]:

$$
\phi_{hk} = -\frac{\sqrt{Q_{hh}Q_{kk}}}{4\pi} V_h V_k + \frac{P_{hk}\sqrt{Q_{hh}Q_{kk}}}{8\pi}\Delta_1 V \qquad (h, k = 1, 2, 3) \tag{3.40}
$$

where ϕ_{hk} are the stress components, P_{hk} is the adjoint of Q_{hk}, and

$$
V_h = \sum_n P_{hk}\frac{\partial V}{q_h}. \tag{3.41}
$$

For orthogonal curvilinear coordinates he obtained:

$$
\begin{aligned}
&Q_{23} = Q_{13} = Q_{12} = 0, && P_{23} = P_{13} = P_{12} = 0, \\
&P_{11}Q_{11} = P_{22}Q_{22} = P_{33}Q_{33} = 1, && V_h = \frac{1}{Q_{hh}}\frac{\partial V}{\partial q_h} \\
&\phi_{hh} = -\frac{1}{4\pi Q_h^2}\left(\frac{\partial V}{\partial q_h}\right)^2 + \frac{1}{8\pi}\Delta_1 V, && (Q_h = Q_{hh}) \\
&\phi_{hk} = -\frac{1}{4\pi Q_h Q_k}\frac{\partial V}{\partial q_h}\frac{\partial V}{\partial q_k} && (h \neq k,\ Q_h = Q_{hh}).
\end{aligned}
\tag{3.42}
$$

In [10] Beltrami operated in orthogonal coordinates, with the aim of verifying if the stress field ϕ_{hk} may be obtained by the deformation of an elastic ether, or else if an isotropic elastic material exists, that, suitably strained, originates the components ϕ_{hk}. He followed these steps:

(a) the components of infinitesimal strain were introduced:

[66] p. 170.

$$\begin{aligned}
\alpha &= \frac{\partial u}{\partial x} & \lambda &= \frac{\partial w}{\partial y} + \frac{\partial v}{\partial z} \\
\beta &= \frac{\partial v}{\partial y} & \mu &= \frac{\partial u}{\partial z} + \frac{\partial w}{\partial x} \\
\gamma &= \frac{\partial w}{\partial z} & \nu &= \frac{\partial v}{\partial x} + \frac{\partial u}{\partial y}.
\end{aligned} \tag{3.43}$$

(b) the potential for a generic isotropic elastic mean was introduced as a funtion of strains:

$$\Phi = \frac{1}{2}\left[A\theta^2 + B(\lambda^2 + \mu^2 + \nu^2 - 4\beta\gamma - 4\gamma\alpha - 4\alpha\beta)\right], \qquad \theta = \alpha + \beta + \gamma. \tag{3.44}$$

(c) the elastic potential was derived with respect to the strains, thus yielding stress components:

$$\begin{aligned}
X_x &= 2B(\beta + \gamma) - A\theta, & Y_z &= -B\lambda \\
Y_y &= 2B(\gamma + \alpha) - A\theta, & Z_x &= -B\mu \\
Z_z &= 2B(\alpha + \beta) - A\theta, & Y_z &= -B\nu.
\end{aligned} \tag{3.45}$$

(d) constitutive relations were expressed in terms of strain:

$$\begin{aligned}
\alpha &= \frac{1}{2B}\left(\frac{A-2B}{3A-4B}P - X_x\right), & \lambda &= -\frac{1}{B} = Y_z, \\
\beta &= \frac{1}{2B}\left(\frac{A-2B}{3A-4B}P - Y_y\right), & \mu &= -\frac{1}{B} = Z_x, \quad P = -(X_x + Y_y + Z_z). \\
\gamma &= \frac{1}{2B}\left(\frac{A-2B}{3A-4B}P - Z_z\right), & \nu &= -\frac{1}{B} = X_y,
\end{aligned} \tag{3.46}$$

(e) the stresses provided by Maxwell's expressions were replaced into the preceding relations, thus obtaining the strain components as a function of the Newtonian potential V in a still rather simple form:

$$\begin{aligned}
\alpha &= \frac{1}{8\pi B}\left[\left(\frac{\partial V}{\partial x}\right)^2 - \frac{A-B}{3A-4B}\Delta_1 V\right], & \lambda &= \frac{1}{4\pi B}\frac{\partial V}{\partial y}\frac{\partial V}{\partial z}, \\
\beta &= \frac{1}{8\pi B}\left[\left(\frac{\partial V}{\partial y}\right)^2 - \frac{A-B}{3A-4B}\Delta_1 V\right], & \mu &= \frac{1}{4\pi B}\frac{\partial V}{\partial z}\frac{\partial V}{\partial x}, \\
\gamma &= \frac{1}{8\pi B}\left[\left(\frac{\partial V}{\partial z}\right)^2 - \frac{A-B}{3A-4B}\Delta_1 V\right], & \nu &= \frac{1}{4\pi B}\frac{\partial V}{\partial x}\frac{\partial V}{\partial y}.
\end{aligned} \tag{3.47}$$

"Thus, such are the values of the six components of strain of an isotropic mean, to which the six components of pressure resulting from Maxwell's theory correspond" [10].[67] However, Beltrami went on,"for a *given* system of strain components α, β, γ, λ, μ, ν to really correspond to a system of displacement components u, v, w, that is in other words, for a *given* system of functions to represent a *possible* deformation, it is necessary and sufficient[68] that the six equations be identically satisfied" [10][69]:

$$\begin{aligned}
\frac{\partial^2\beta}{\partial z^2}+\frac{\partial^2\gamma}{\partial y^2}&=\frac{\partial^2\nu}{\partial y\partial z} &\qquad \frac{\partial^2\alpha}{\partial y\partial z}&=\frac{1}{2}\frac{\partial}{\partial x}\left(\frac{\partial\mu}{\partial y}+\frac{\partial\alpha}{\partial z}-\frac{\partial\nu}{\partial x}\right)\\
\frac{\partial^2\gamma}{\partial x^2}+\frac{\partial^2\alpha}{\partial z^2}&=\frac{\partial^2\mu}{\partial x\partial z} &\qquad \frac{\partial^2\beta}{\partial x\partial z}&=\frac{1}{2}\frac{\partial}{\partial y}\left(\frac{\partial\nu}{\partial x}+\frac{\partial\lambda}{\partial z}-\frac{\partial\mu}{\partial y}\right)\\
\frac{\partial^2\beta}{\partial x^2}+\frac{\partial^2\alpha}{\partial y^2}&=\frac{\partial^2\lambda}{\partial x\partial y} &\qquad \frac{\partial^2\gamma}{\partial y\partial x}&=\frac{1}{2}\frac{\partial}{\partial z}\left(\frac{\partial\nu}{\partial x}+\frac{\partial\mu}{\partial y}-\frac{\partial\lambda}{\partial z}\right).
\end{aligned} \tag{3.48}$$

These are called compatibility equations in explicit form nowadays, and we will call on them again in the following.

Under the conditions (3.48) Beltrami deduced a Newtonian potential that either is unrealistic, or requires that the Green's constants of isotropy A, B attain values 'irreconcilable' with the stable equilibrium of the elastic mean (for instance, $A = 0, B > 0$). He concluded that "it is not generally possible to reproduce the system of pressures defined by Maxwell's formulas by means of the deformation of an isotropic medium" [10].[70]

This negative conclusion had only proved "the necessity to investigate by other means" the mechanical interpretation of Maxwell's theory, never questioned by Beltrami. Some years later, Cesaro as well arrived at the same conclusions, starting from strictly physical assumptions.

Thus, the purpose of understanding the nature of physical space by Beltrami and some of his 'pupils' led to interesting results. In addition, the very idea that physical space could have non-vanishing curvature, and the fact that many mathematicians ventured in the attempt of extending the results of mathematical physics to Riemannian manifolds made it urgent, among the rest, the elaboration of a mathematical formalism able to express the equation of mathematical physics independently of the chosen coordinate frame. These starting points proved to be fundamental for the birth of Ricci-Curbastro's tensor calculus that, from this point of view, can be considered as the most natural answer to problems of a physical nature originated almost a century before.

---

[67] p. 194.

[68] Note by Beltrami: For the proof of the sufficiency of these equations, please look at the note at the end of the present Memoir.

[69] p. 195.

[70] p. 192.

3.2.4 Compatibility Equations

In a note closing the paper on the mechanical interpretation of Maxwell's formulas, Beltrami proved that Eq. (3.48), necessary in order that six functions $\alpha, \beta, \gamma, \lambda, \mu, \nu$ represent a compatible deformation, are also sufficient—that is they can be derived from a 'regular' field of displacement. The compatibility equations (3.48) had already been deduced as necessary by Saint Venant [69][71] and Kirchhoff [58], and are present also in the textbook by Castigliano of 1879 [29], where they are considered known as far as their necessity is considered: "Thus, in order that six functions of x, y, z be able to represent the three dilatations and the three shearings, it is necessary that they satisfy both Eq. (26) [the first three (3.48)] and Eq. (27) [the second three (3.48)]" [29].[72]

Since it is short and important, we quote in full Beltrami's note proving the sufficiency of the (3.48) for the compatibility of a simply connected continuum:

> Of the six constraint equations for the quantities $\alpha, \beta, \gamma, \lambda, \mu, \nu$ that are cited in §I, the necessity is ordinarily proved, not the sufficiency. Thus, I think it appropriate, considered the importance of these equations with respect to the scope of the present work, to add a deduction of the same, clearly providing their properties of constituting not only necessary, but also sufficient, conditions for the existence of the three displacement components u, v, w.
>
> Remember, from the general theory of deformations of a continuum medium, that, together with the quoted components $\alpha, \beta, \gamma, \lambda, \mu, \nu$, and with not less meaningful importance, also the three quantities p, q, r appear, defined by the equations:
>
> $$\frac{\partial w}{\partial y} - \frac{\partial v}{\partial z} = 2p, \quad \frac{\partial u}{\partial z} - \frac{\partial w}{\partial z} = 2q, \quad \frac{\partial v}{\partial x} - \frac{\partial u}{\partial y} = 2r \qquad (a)$$
>
> and represent the *rotation components* of the particle surrounding the point (x, y, z). Now, from the system of the nine equations obtained by combining the six Eq. (2) of § I[73] with the preceding three (a), we may obtain the values of all the first derivatives of the three displacement components u, v, w, and these values are the following:
>
> $$\begin{array}{lll} \dfrac{\partial u}{\partial x} = \alpha, & \dfrac{\partial u}{\partial y} = \dfrac{\nu}{2} - r, & \dfrac{\partial u}{\partial z} = \dfrac{\mu}{2} + q, \\ \dfrac{\partial v}{\partial x} = \dfrac{\nu}{2} + r, & \dfrac{\partial v}{\partial y} = \beta, & \dfrac{\partial v}{\partial z} = \dfrac{\lambda}{2} - p, \\ \dfrac{\partial w}{\partial x} = \dfrac{\mu}{2} - q, & \dfrac{\partial w}{\partial y} = \dfrac{\lambda}{2} + p, & \dfrac{\partial w}{\partial z} = \gamma. \end{array} \qquad (b)$$
>
> Let us consider the first three of these equations, that provide the values of the first derivatives of the function u. Supposed given the quantities entering their right hand sides, in order for a function u to exist satisfying these three equations, it is necessary and sufficient that three known relations be satisfied, that can be written as follows:
>
> $$-\frac{\partial q}{\partial y} - \frac{\partial r}{\partial z} = \frac{1}{2}\left(\frac{\partial \mu}{\partial y} - \frac{\partial \nu}{\partial z}\right), \quad \frac{\partial q}{\partial x} = \frac{\partial \alpha}{\partial z} - \frac{1}{2}\frac{\partial \mu}{\partial x}, \quad \frac{\partial r}{\partial x} = \frac{1}{2}\frac{\partial \nu}{\partial x} - \frac{\partial \alpha}{\partial y}.$$
>
> From these we deduce, by cyclic permutation, the two analogous triads of necessary and sufficient conditions for the existence of the other two functions v and w. However, by

---

[71] Appendix III.

[73] They are known as the *implicit compatibility equations*.

performing this permutation first only on the first of the three preceding conditions, and then summing term by term the three equations so obtained, we find[74]:

$$\frac{\partial p}{\partial x}+\frac{\partial q}{\partial y}+\frac{\partial r}{\partial z}=0$$

so that the first of the three above found conditions may be written more simply as:

$$\frac{\partial p}{\partial x}=\frac{1}{2}\left(\frac{\partial \mu}{\partial x}-\frac{\partial \nu}{\partial z}\right)$$

In this way we find the following system of differential relations among the nine functions $\alpha, \beta, \gamma, \lambda, \mu, \nu, p, q, r$:

$$\begin{aligned}
&\frac{\partial p}{\partial x}=\frac{1}{2}\left(\frac{\partial \mu}{\partial x}-\frac{\partial \nu}{\partial z}\right) \quad \frac{\partial p}{\partial y}=\frac{1}{2}\frac{\partial \lambda}{\partial y}-\frac{\partial \beta}{\partial z} \quad \frac{\partial p}{\partial z}=\frac{\partial \gamma}{\partial y}-\frac{1}{2}\frac{\partial \lambda}{\partial z}\\
&\frac{\partial q}{\partial x}=\frac{\partial \alpha}{\partial z}-\frac{1}{2}\frac{\partial \mu}{\partial x} \quad \frac{\partial q}{\partial y}=\frac{1}{2}\left(\frac{\partial \nu}{\partial z}-\frac{\partial \lambda}{\partial x}\right) \quad \frac{\partial q}{\partial z}=\frac{1}{2}\frac{\partial \mu}{\partial z}-\frac{\partial \gamma}{\partial x} \qquad (c)\\
&\frac{\partial r}{\partial x}=\frac{1}{2}\frac{\partial \nu}{\partial x}-\frac{\partial \alpha}{\partial y} \quad \frac{\partial r}{\partial y}=\frac{\partial \beta}{\partial x}-\frac{1}{2}\frac{\partial \nu}{\partial y} \quad \frac{\partial r}{\partial z}=\frac{1}{2}\left(\frac{\partial \lambda}{\partial x}-\frac{\partial \mu}{\partial y}\right)
\end{aligned}$$

This system of equations contains the necessary and sufficient conditions for the existence of three functions u, v, w satisfying the nine conditions (b), that is the six Eq. (2) of § 2 and the three equations (a) of this Note.

This posed, let us consider as given only the *six components of deformation*, $\alpha, \beta, \gamma, \lambda, \mu, \nu$. If three functions u, v, w exist, satisfying equations (2) of § I, for sure also the three functions p, q, r exist, defined by equations (a) of this Note. Since, then, the derivatives of these last three functions are linked to the $\alpha, \beta, \gamma, \lambda, \mu, \nu$ by the nine equations (c), it is necessary that the integrability conditions that result from these last nine equations be satisfied, and that reduce to the following six:

$$\begin{aligned}
&\frac{\partial^2 \beta}{\partial z^2}+\frac{\partial^2 \gamma}{\partial y^2}=\frac{\partial^2 \nu}{\partial y \partial z} \quad \frac{\partial^2 \alpha}{\partial y \partial z}=\frac{1}{2}\frac{\partial}{\partial x}\left(\frac{\partial \mu}{\partial y}+\frac{\partial \alpha}{\partial z}-\frac{\partial \nu}{\partial x}\right)\\
&\frac{\partial^2 \gamma}{\partial x^2}+\frac{\partial^2 \alpha}{\partial z^2}=\frac{\partial^2 \mu}{\partial x \partial z} \quad \frac{\partial^2 \beta}{\partial x \partial z}=\frac{1}{2}\frac{\partial}{\partial y}\left(\frac{\partial \nu}{\partial x}+\frac{\partial \lambda}{\partial z}-\frac{\partial \mu}{\partial y}\right) \qquad (d)\\
&\frac{\partial^2 \beta}{\partial x^2}+\frac{\partial^2 \alpha}{\partial y^2}=\frac{\partial^2 \lambda}{\partial x \partial y} \quad \frac{\partial^2 \gamma}{\partial y \partial x}=\frac{1}{2}\frac{\partial}{\partial z}\left(\frac{\partial \nu}{\partial x}+\frac{\partial \mu}{\partial y}-\frac{\partial \lambda}{\partial z}\right)
\end{aligned}$$

which are exactly those quoted in § I. When these conditions are satisfied, there undoubtedly exist three functions p, q, r satisfying the nine equations (c); but we have already seen that, if these nine equations are satisfied by nine functions $\alpha, \beta, \gamma, \lambda, \mu, \nu, p, q, r$, three functions u, v, w exist, satisfying the conditions (2) of S I Eq. (a) of the present Note: thus, the six conditions (d), apparently necessary for the existence of three functions u, v, w satisfying only equations (2) of § I, are also sufficient [10].[75] (A.3.12)

Beltrami considered compatibility equations again in a paper on the *Comptes rendus* [11] and in a letter to Cesaro [12], where he provided a new proof of the sufficiency of

[74] Note by Beltrami This most known relation already results from the definition formulas (a): however, for the present scope it was necessary to remark that it is included in the nine integrability condition of which word is here.

[75] pp. 221–223. Our translation.

(3.48). In the paper in the *Comptes rendus* Beltrami took advantage of the possibility to derive the relations (3.48) from the variation of a triple integral "just like it is done very usefully in various cases, for instance for the classical equation of the potential". In the letter to Cesaro he used direct integration:

> It is, however, useful to remark that the sufficiency of the discussed equations can be established in such a way, that a more imperative one cannot be imagined, that is, by direct integration, which most easily goes as follows [12].[76] (A.3.13)

Beltrami's conjecture on the arbitrariness of assigning deformation components to have still compatibility is interesting:

> To deal problems of the kind of that bringing the name of SAINT VENANT, it is useful to be able to arbitrarily arrange some of the six components of deformation. By examining this point rather closely, I was able to convince myself that we may arbitrarily assume three of the quantities a, b, e, f, g, h, unless they are not those already associated among them in one of the three constraint equations constituting the first of the already quoted triads. As a consequence, of the 20 triads that we may build by the six said components, 17 are those for one of which it is possible, in a given problem, to arbitrarily fix the shape of all three functions that compose it [12].[77] (A.3.14)

3.2.5 Beltrami-Michell's Equations

In a note of 1892 [13] Beltrami, starting from the compatibility Eq. (3.48), deduced the conditions on the stress components of an elastic body so that it is in equilibrium in the absence of external forces:

> The six components of pressure p_{xx}, p_{yy}, ... are necessarily subjected to certain conditions, when they correspond to inner forces generated by pure strain; indeed, in that case they shall be able to be expressed in a completely determined way (depending on the nature of the body), by means of the displacement [13].[78] (A.3.15)

In the case of an isotropic body, Beltrami proved that self-equilibrated stress components corresponding to deformation components that are compatible according to (3.48) satisfy the following second-order partial differential equations:

$$\begin{aligned} &\frac{\partial^2 P}{\partial x^2} + C\Delta_2 X_x = 0, \quad \frac{\partial^2 P}{\partial y^2} + C\Delta_2 Y_y = 0, \quad \frac{\partial^2 P}{\partial z^2} + C\Delta_2 Z_z = 0, \\ &\frac{\partial^2 P}{\partial y \partial z} + C\Delta_2 Y_z = 0, \quad \frac{\partial^2 P}{\partial z \partial x} + C\Delta_2 Z_x = 0, \quad \frac{\partial^2 P}{\partial x \partial y} + C\Delta_2 X_y = 0 \end{aligned} \tag{3.49}$$

where Δ_2 is the Laplace operator, X_x, Y_y, Z_z are the *components of pressure*, $P = X_x + Y_y + Z_z$ and C is a constant. Beltrami concluded commenting:

[76] p. 327. Our translation.

[77] p. 329. Our translation.

[78] p. 511. Our translation.

> These last conditions suppose the absence of any external force. I omit, for brevity, to report the analogous conditions for the case in which this force exists and has components X, Y, Z [13].[79] (A.3.16)

The generalization of Beltrami's equations in the presence of non-uniform volume forces was due to John Henry Michell (1863–1940), who in a paper of 1900 obtained the compatibility equations in terms of stresses and external forces [65].

3.2.6 Papers on Structural Mechanics

Beltrami studied elasticity mainly to answer problems of electro-magnetism; there are, however, at least two papers devoted to structural mechanics, one on the strength of materials, and the other on membranes.

3.2.6.1 A Criterion of Failure

In Beltrami's time, two failure criteria for elastic bodies subjected to a three-dimensional state of stress were established and diffused; they are briefly recalled at the beginning of the memoir *Sulle condizioni di resistenza dei corpi elastici* of 1885 [9]. They dealt with limiting the maximum values either of stress or of strain; on the other hand, Beltrami proposed a criterion where both quantities were accounted for, suggesting to pose a limitation to the volume density of the elastic energy. For linear elastic materials it is a positive definite quadratic form, and limiting its values equals to posing limits to both stresses and strains. In the following we quote the memoir on the strength of materials in full, because of its remarkable importance in mechanics:

> In the French version of the *Theory of elasticity* by CLEBSCH, revised and commented by the illustrious De Saint Venant, who, by that publication, has provided a new and remarkable service to the scholars of that most important theory, one finds resumed, in a final Note to § 31 (pp. 252–282.), the method already proposed since long by De Saint Venant himself for the search of the limits of strength of elastic bodies. This method differs from that generally followed, and accepted also by CLEBSCH, in the principle on what it is based, which consists in assigning a maximum limit to strains instead of stresses.
>
> To justify this new principle, De Saint Venant quotes, in particular, the very simple case of a rectangular parallelepiped, strained by the same unit force along one, two, or all three of its geometric axes; and observes that, while the maximum stress is, by hypothesis, the same in all three cases, the maximum strain is greater in the first [case] than in the second, and is likewise greater in the second [case] than in the third, whence it seems obvious to conclude that the danger of disaggregation be greater in the first case than in the second and the third.
>
> Now, this conclusion does not seem to me so legitimate, as by chance one could believe at first glance. The stretch of a body in the direction that we will call longitudinal is accompanied, as it is most known, by a contraction in any transverse direction, a contraction which is partially

[79] p. 512. Our translation.

restrained, or even changed into stretch, when the body is at the same time subjected to transverse stretches; it follows that the molecular cohesion is weakened, in the longitudinal direction, more in the first case than in the second, but it is also reinforced, in the transverse direction, more in that [case] than in this, so that it is not easy, and maybe not possible, to decide a priori on the prevalence of the one effect over the other. But if we cannot formulate a precise conclusion about this, it seems to me, however, that we can admit as evident, just on the basis of the example very suitably introduced by De Saint Venant, that the true measure of the risk to which the cohesion of an elastic body is subjected can be deduced neither from the maximum stress only, nor from the maximum strain only, but shall result, in some way, by the set of all stresses, or of all strains reigning in the neighborhood of any point of the body.

Now these stresses and these dilatations, represented each by six distinct components, are linked among them by linear relations, which express that the six components of stress are the derivatives, with respect to the six components of strain, of a unique quadratic function formed by these second components; or else, that the six components of strain are the derivatives with respect to the six components of stress of an analogous function formed by these last components. This unique function, that has the same value under the different forms it takes in the one and in the other case is the so called potential of elasticity, and has the eminent property of representing the energy, with respect to the unit volume, that the elastic body possesses in the neighborhood of the point that is considered, an energy that is equivalent both to the work that the unit volume of the body may perform in returning from the present state to the natural state, and to the work that the external forces had to perform to lead the said volume unit from the natural state to its present state of elastic coaction.

Following this, it seems to me that the true measure of the risk to which the molecular cohesion is subjected, in any point of the body, must be given by the value that the unit potential of elasticity assumes in that point, and that to this value, rather than to that of a stress or of a strain, a maximum limit shall be prescribed, to preserve the body from the danger of disaggregation, a limit naturally different, like in the ordinary theories, according that nearby or remote disaggregation is dealt with.

This conclusion, already justified in itself by the dynamical meaning of the potential of elasticity, is made more apparently plausible by an analytical property of this potential, which also must certainly depend on the above said meaning, even if the rigorous proof of such dependence is not known to us.

I want to allude to the property, that the said property has, of being an essentially positive quadratic function, that is a function that does not vanish unless all its six variables be nil, and that remains greater than zero for any other 6-ple of real values of these variables. By virtue of this property, a limit value of the potential of elasticity cannot be imposed, without imposing, at the same time, a limit to that of any component, either of stress, and of strain, so that the use of the said potential as a measure of the elastic strength does not intrinsically contradict the criteria deduced considering both only stresses, and only strains. Practically, then, the criterion deduced from the potential has the great advantage of not requiring the preliminary solution of any equation, and to reduce to the discussion of a formula that can never present ambiguity in sign.

[...] P.S. After having written what precedes, I have acknowledged with pleasure, that the objection raised by me against the ways until now adopted to establish the conditions of cohesion, had been formulated, almost in the same terms, by the late eng. Castigliano, at the p. 128 and f. of the Théorie de l'équilibre des systèmes élastiques. It is a pleasure to think that the learned engineer, who had recognized all the importance of the concept of elastic potential, would have probably approved my proposal to build over it also the deduction of the above said conditions [9].[80] (A.3.17)

[80] pp. 704–714. Our translation.

The failure criterion proposed by Beltrami had a good success, at least in Italy; for instance, it was received by Francesco Crotti in his *La teoria dell'elasticità ne' suoi principi fondamentali e nelle sue applicazioni pratiche alle costruzioni* [49]. In the foreword Crotti underlined "the reception of the principle recently proposed by the illustrious prof. Beltrami for the measure of the limit risk to which matter is subjected"; which has "the advantage of its great simplicity and rapidity, that make it a precious acquisition for practical science".

Indeed, Beltrami's criterion is unsuitable for building materials, in particular for steel, for which criteria were proposed that limited the maximum energy of distortion (Hencky-Huber-Von Mises) [57].

3.2.6.2 The Equilibrium of Membranes

In 1882 [6] Beltrami investigated membranes, non-deformable in their own plane, by taking inspiration from a work by Léon Lecornu,[81] and with the intention of clarifying the problem, already examined by Lagrange, Poisson, and Mossotti [6].[82] Beltrami showed that Mossotti's hypothesis, of equal stresses along all directions, is inconsistent; he then wrote the equilibrium equations in Cartesian and curvilinear coordinates.

Beltrami considered a membrane σ with contour s, of parametric equations $x = x(u, v)$, $y = y(u, v)$, $z = z(u, v)$ in the curvilinear coordinates u, v. If X, Y, Z and X_s, Y_s, Z_s are the external forces densities in σ and s respectively, and if $\delta x, \delta y, \delta z$ are virtual displacements defined by "one-valued continuous functions, finite in the variables u, v", the virtual work of external forces is [6][83]:

$$\int (X\delta x + Y\delta y + Z\delta z)\, d\sigma + \int (X_s\, \delta x + Y_s\, \delta y + Z_s\, \delta z.)\, ds. \tag{3.50}$$

The surface element $d\sigma$ is constrained to remain rigid in its plane. Beltrami imposed the rigidity condition by starting from the metrics of the "line element through the point (u, v) and corresponding to the increments du, dv" [6][84]:

$$ds^2 = Edu^2 + 2Fdu\, dv + Gdv^2, \tag{3.51}$$

[81] Journal de l'École Polytecnique, cahier XLVIII (1880), p. 1 (note by Beltrami). The paper is entitled: *Sur l'équilibre des surfaces flexibles et inextensibles*.

[82] pp. 420–421.

[83] p. 427.

[84] p. 425.

with coefficients E, F, G having the following expression:

$$E = \left(\frac{\partial x}{\partial u}\right)^2 + \left(\frac{\partial y}{\partial u}\right)^2 + \left(\frac{\partial z}{\partial u}\right)^2$$
$$F = \frac{\partial x}{\partial u}\frac{\partial x}{\partial v} + \frac{\partial y}{\partial u}\frac{\partial y}{\partial v} + \frac{\partial z}{\partial u}\frac{\partial z}{\partial v} \tag{3.52}$$
$$G = \left(\frac{\partial x}{\partial v}\right)^2 + \left(\frac{\partial y}{\partial v}\right)^2 + \left(\frac{\partial z}{\partial v}\right)^2.$$

Here is how Beltrami kept the rigidity constraint into account in the virtual work:

> The variations $\delta x, \delta y, \delta z$ are one-valued functions of the variables u, v. For the non-extensibility of the surface, these variations must satisfy three conditions:
>
> $$\delta E = 0, \qquad \delta F = 0, \qquad \delta G = 0, \tag{2}$$
>
> where
>
> $$\frac{1}{2}\delta E = \sum \frac{\partial x}{\partial u}\frac{\partial \delta x}{\partial u},$$
> $$\delta F = \sum \left(\frac{\partial x}{\partial u}\frac{\partial \delta x}{\partial v} + \frac{\partial x}{\partial v}\frac{\partial \delta x}{\partial u}\right) \tag{2a}$$
> $$\frac{1}{2}\delta G = \sum \frac{\partial x}{\partial v}\frac{\partial \delta x}{\partial v}.$$
>
> By virtue of LAGRANGE's principle, the general equation for equilibrium is thus the following:
>
> $$\int (X\delta x + Y\delta y + Z\delta z)\, d\sigma + \int (X_s\, \delta x + Y_s\, \delta y + Z_s\, \delta z)\, ds$$
> $$+\frac{1}{2}\int (\lambda\delta E + 2\mu\delta F + \nu\delta G)\frac{d\sigma}{H} = 0$$
>
> where λ, μ, ν are three multipliers, functions of u and v (the divisor $2H$ has been introduced, in the last integral, for ease of successive calculations) [6].[85] (A.3.18)

By transforming the surface integrals with Green's formulas [6][86] Beltrami came to the local and boundary equations for the membrane, which are transformed considering the surface and the boundary forces U, V, W and U_s, V_s, W_s respectively along the curvilinear coordinates u, v and in the orthogonal direction w. We report only the transformed boundary equations:

$$U_s = \frac{\sqrt{E}}{H}\left[\lambda\left(E\frac{\partial u}{\partial n} + F\frac{\partial v}{\partial n}\right) + \mu\left(F\frac{\partial u}{\partial n} + G\frac{\partial v}{\partial n}\right)\right],$$
$$V_s = \frac{\sqrt{G}}{H}\left[\mu\left(E\frac{\partial u}{\partial n} + F\frac{\partial v}{\partial n}\right) + \nu\left(F\frac{\partial u}{\partial n} + G\frac{\partial v}{\partial n}\right)\right], \qquad H^2 = EG - F^2. \tag{3.53}$$
$$W_s = 0,$$

[85] p. 427. Our translation.

[86] p. 429; Beltrami referred to [3].

Beltrami associated a mechanical meaning to Lagrange multipliers λ, μ, ν; indeed, he found the expressions:

$$T_{su} = \sqrt{E}\left(\lambda\frac{\partial v}{\partial s} - \mu\frac{\partial u}{\partial s}\right), \qquad T_{sv} = \sqrt{G}\left(\mu\frac{\partial v}{\partial s} - \nu\frac{\partial u}{\partial s}\right). \tag{3.54}$$

Here T_{su} and T_{sv} are forces per unit length on a generic element in the directions defined by $\frac{\partial v}{\partial s}$, $\frac{\partial u}{\partial s}$ respectively. He then obtained:

$$T_{uu} = -\mu, \qquad T_{uv} = -\nu\sqrt{\frac{G}{E}}, \qquad T_{vu} = -\lambda\sqrt{\frac{G}{E}}, \qquad T_{vv} = -\mu. \tag{3.55}$$

Lagrange's multipliers, modulo a scaling factor, are thus stresses, normal and tangential; the latter, denoted T_{uu} and T_{vv}, coincide and are equal to $-\mu$, respecting symmetry.

In the closing sections, Beltrami found some remarkable results:

> Any piece of a flexible, non-extensible surface is kept in equilibrium by a force, everywhere normal to the surface itself, and proportional to the local average curvature.[87] The uniform tension of the boundary is equally transmitted in any point of the surface [6].[88] (A.3.19)

> Any piece of a flexible, non-extensible surface is kept in equilibrium by a uniform and normal stress along the boundary and by a force normal everywhere to the surface itself and proportional to the measure of local curvature [according to Gauss], and by a stress along the boundary, directed according to the conjugate tangent to the boundary itself, and having the normal component proportional to the curvature of the boundary. The normal lines are the curvature lines of the surface, those of tangential stress are the asymptotic lines to the surface itself [6].[89] (A.3.20)

3.3 The Pupils

3.3.1 The School of Pisa

In 1865 Betti became director of the Scuola normale of Pisa, where he had several pupils, thanks to the particularly stimulating environment in Pisa in those years. Here we will pause on those of Betti's pupils that contributed to the theory of elasticity, some of them until the first years of the 20th century.

When Betti became director of the Scuola normale, the presence at the same time of Riemann and Beltrami contributed in making the university of Pisa one of the most important in Italy, often taken as a model by the other universities. Indeed, there were many 'good young' mathematicians studying in Pisa in those years; among them we

---

[87] The average curvature is $\frac{1}{2}\left(\frac{1}{R_1} + \frac{1}{R_2}\right)$, with R_1, R_2 the radii of curvature in the directions u, v.

[88] p. 450. Our translation.

[89] p. 453. Our translation.

remember[90] Ernesto Padova (1866), Eugenio Bertini (1867), Giulio Ascoli (1868), Cesare Arzelà (1869), besides Ulisse Dini (graduated in 1864), who would become one of the best Italian scholars in mathematical analysis.

Starting from about the mid 1860's, Betti changed his studies from algebra to mathematical physics. The talented pupils that studied the theory of potential and of elasticity with Betti were: Gregorio Ricci-Curbastro (1876), who graduated with Betti, defending a thesis on the theory of Maxwell's equations; Carlo Somigliana, at the Scuola normale from 1879 to 1881; Vito Volterra (1882), who at once became Betti's assistant; Orazio Tedone, at the Scuola normale in 1890–1892.

A second generation pupil was that of Giuseppe Lauricella, who since his thesis devoted himself to the study of the equilibrium of elastic bodies, and obtained remarkable results. In 1907 he successfully took part in the international Vaillant prize, promoted by the Académie des Sciences de Paris, on the equilibrium of clamped elastic plates. That is, one should solve a generalized Dirichlet problem on bi-harmonic functions with given boundary data. Lauricella shared the prize with Tommaso Boggio, Arthur Korn, and Jacques Hadamard (who got three quarters of the prize on his own). In the winning memoir, published in 1909 in the famous journal Acta Mathematica [62], Lauricella used the theory of integral equations to solve the problem. Indeed, he was one of the first in Italy to understand the importance of Fredholm's theory, and to apply it with success to mathematical physics.

In any case, it is difficult to distinguish between the studies in the theory of potential and those in the theory of elasticity, at least in this period. Moreover, this lets us see a path that began right with Betti's work, who had successfully applied the methods of potential theory to the investigation of physical problems regarding mainly the theory of elasticity and heat. On this purpose, we quote Volterra:

> The fundamental concepts and methods by Green and Gauss had opened the main street for the general integration of Laplace's equation, the basis of potential theory; Betti's aim was to transport the same methods, first in the field of the science of elastic equilibrium, then in that of heat. By the works of Betti [...] a new and long series of baldly Italian investigations on the integration of the equations of elasticity opens, so that we may say that, if Galileo was the first to foreshadow the problems of the equilibrium of elastic bodies, it was about Italian geometers, more than two centuries after, to have largely contributed to develop the general theory of those equations in which Navier had represented and, so to say, enclosed all the mechanism of the phenomenon [99].[91] (A.3.21)

Another of Betti's pupils, Orazio Tedone, wrote about Betti's influence, in particular of the reciprocity theorem, on the development of the theory of elasticity:

> The really admirable Memoir by Betti on the equations of elasticity threw a beam of new, unexpected light on them, and prepared, especially in Italy, a flowering of studies like few other memoirs can boast to have produced. His reciprocity theorem should seem a revelation. By very simple means it already gave a throng of results and made it possible to penetrate in depth the analytical properties of the equations we are dealing with [87].[92] (A.3.22)

---

[90] Between parentheses we report the year of their beginning of studies.

[91] p. 58. Our translation.

[92] p. 43. Our translation.

On this trail we find scholars on elasticity foreign to the environment of Pisa, like Valentino Cerruti, whose contribution to the theory of structures is discussed elsewhere (Chap. 4). Still as a student, he published papers on analytical geometry in the *Giornale di matematiche* by Giuseppe Battaglini. In 1873 he graduated in civil engineering in Turin with the thesis *Sistemi elastici articolati* [32].

After the thesis, Cerruti wrote a paper on Menabrea's theorem [33], putting into evidence the weaknesses, according to him, of the proof. We will go back to this paper and to Menabrea in Chap. 4. Afterwards, Cerruti made investigations on the dynamics of small amplitude motions of systems in presence of resistance of the medium [34, 35]. In a fundamental paper of 1880 [36] he generalized Betti's reciprocity theorem and his consequences from statics to dynamics, finding the particular integrals having characteristic singularities in space and time, and came to the solving formulas. Due to an oversight, he made an error in calculations, without which he would have obtained, two years before Kirchhoff, the mathematical expression of Christiaan Huygens' principle; Somigliana noticed this priority later on.

Cerruti gave a simpler expression to Betti's results on the calculation of the displacement fields of three-dimensional elastic continua, and reduced the number of auxiliary functions to be assigned in advance. He systematically applied those results to isotropic soils, to spheres, to spherical envelopes, thus this method is known by his name associated with Betti's. Cerruti also investigated the calculation of the strains of an indefinite body limited by a plane, in the two main cases where the displacements of the points of the limiting plane, or the force applied to its single elements, are given [37]. In a memoir of 1890 Cerruti studied the case in which forces instead of displacements were assigned at the points of the limiting surfaces [38].

The most illustrious of Betti's pupils was Vito Volterra (Ancona 1860-Rome 1940), who was a mathematician of international fame. He spent his childhood in Turin, and then in Florence, where he attended the technical school Dante Alighieri and the technical high school Galileo Galilei. In 1878 he enrolled in the faculty of Mathematical, physical, and natural sciences of the university of Pisa, and in 1879 in the Scuola normale as internal student, where he was one of Betti's pupils. A few months before his graduation he took part in a job competition for the chair of Rational mechanics at the university of Pisa; he won it, and became professor in charge at the age of only 23. In 1887 he was appointed full professor and, due to his studies in mathematical analysis, he was given the prize for mathematics by the Società dei XL. In 1892 he was in charge of the teaching of Mathematical physics and became dean of the faculty of sciences. In 1893 he left Pisa and moved to Turin as the chair of Superior mechanics.

He became a member of the board of directors of the Circolo matematico di Palermo, national fellow of the Società dei XL and of the Accademia delle scienze in Turin, counselor of the academies in Modena and Bologna, corresponding fellow of the Regio istituto lombardo. In 1899 he was named national fellow of the Accademia dei Lincei. In 1900 he was called to the faculty of sciences at the university of Rome, and in 1907 became its dean. Since September, 1914, he was an advocate of the Italian intervention in the first world war together with the Triple Entente, and after Italy entered the war he asked to be enlisted. His request was granted with the appointment of lieutenant of military engineering; in this role, he often was in a war zone and earned the promotion to captain. Volterra was interested not only in technical-scientifical cooperation, but also promoted intellectual cooperation among allied countries.

In 1925, overcoming the contrasts with Benedetto Croce on the theme of the value of science, Volterra publicly expressed his dissent toward the fascist regime, by signing the *Manifesto Croce* of the anti-fascist intellectuals; he joined the national Union of liberals and democrats, promoted by Giovanni Amendola, and sided with the group of senators as the opposition. In 1926 he began receiving pressures to resign as president of the Accademia dei Lincei, but he was convinced by the fellows to keep his position. In 1931, however, the government made it compulsory for university teachers to swear allegiance to the fascist regime. Volterra, contrarily to many of his illustrious colleagues, who had opposed the regime only by words, refused to swear, by sending to the rector of the university of Rome, Pietro de Francisci, a short and dry letter confirming his anti-fascist political positions:

> Most Illustrious Mr. Rector
> of the Royal university of Rome
> My political ideas are known, as they exclusively result from my conduct in the parliament, which, however, is unquestionable, by force of article 51 of the Fundamental Law of the Kingdom. Your Worship will, thus, understand how I cannot, in conscience, accede to the invitation that you sent to me by the letter of 18th of this month, about the oath of professors.
> With my best regards to Your Worship
> Vito Volterra [55][93] (A.3.23)

Only twelve professors in Italy had the 'arrogance' to say no to the regime; the bitter comment by Gaetano Salvemini from exile was: "no one of those who, in the past, had bragged to be socialist, had sacrificed their wages in favor of the beliefs so baldly exhibited in times of dead calm".

The missed oath implied Volterra's expulsion from the university, in January, 1932, due to "incompatibility with the general political directives of the government", and gave the regime a good reason to officially oust him also from all academic offices. Volterra tried to rebel, but he was subjected to more than one retaliation: his moves were restricted, and he underwent multiple harassments. However, even if the order was given to ignore him and his activity, he did not miss solidarity by friends and admirers and remained a key person not only for the scientific activity, but for the academic life itself, in which he did not have any formal position, though. Volterra reacted with extraordinary vitality to the marginalization situation in which he was placed by the regime, and obtained a remarkable proof of esteem by the honorary presidency of the International council for the scientific exploration of the Mediterranean.

Volterra died on October 11, 1940, and no Italian scientific institution could commemorate him; the only official ceremony which the family could participate was at the Accademia pontificia, kept by Somigliana. The figure of the great mathematician was remembered in the rest of the world through various initiatives made by the

[93] Our translation.

many important scientific institutions he was fellow of. Italy would have waited for the end of the war; the touching commemoration by Castelnuovo opened the General gathering of October 17th 1946, and inaugurated the reconstituted Accademia dei Lincei.

Volterra was also a tireless scientific and cultural organizer: he was founder and first president of the Italian Society of Physics in 1897, of the Office of Inventions and Research in 1917, of the National Research Council in 1919 (this last was operative only from 1924, however). His interests ranged well beyond the scientific ones, embracing humanistic and historic culture in particular, thus providing a clear proof of how arbitrary the separation is between the so-called two cultures, the humanistic on one hand and the scientific on the other [55].

It is really difficult to give even a rough idea of the work by Volterra, since it branched in multiple directions: scientific research in many fields, and an intense organizing activity of cultural initiatives not only within the numerous scientific disciplines he cultivated, but also in others very far from those. His many publications deal with Earth mechanics, rational mechanics, theory of differential and integral equations, functional analysis, electrodynamics, theory of elasticity, biomathematics, economy.

Between 1900 and 1906 he studied the papers by Karl Pearson on the application of probability theory in biology; in 1926 he published two papers on the applications of mathematics in the field, *Variazioni e fluttuazioni del numero d'individui in specie animali conviventi* and *Fluctuations in the abundance of a species considered mathematically*, which earned him the denomination of founder of the "mathematical theory of the struggle for life" by Guido Castelnuovo. Volterra made pioneering applications of mathematics to economics, already initiated by the engineer and economist Vilfredo Pareto [52].

Volterra's contribution to the theory of elasticity is in Betti's trail and is about theories of hereditary elastic phenomena, and of the distortions. Physicists had since long put into light the existence of hysteretic constitutive phenomena, according to which the strain in a body does not depend only on the present value of the force, but also on all the cycles of loading and unloading to which the body has been subjected. In a series of papers that appeared between 1874 and 1878 [22], Ludwig Boltzmann had considered elastic bodies with memory, that is such that their deformation depends also on the history preceding the forces charging them at present. Starting from 1909, Volterra posed the analytical bases of a hereditary theory of elasticity, accounting also for the past [100].

Volterra established equilibrium equations on the hypothesis that inheritance is represented by time integrals linear in the strain components (linear inheritance). Thus, he obtained integral or integral-differential equations instead of the partial differential equations of mathematical physics, for which he established a general theory of integration. Volterra integrated his equations for an isotropic sphere, when displacements and stresses at the surface are known.

At the same time, Volterra developed the theory of distortions, today bringing its name (*Volterra's distortions*), that is a theory of elastic bodies with possible co-active states.[94] These are states of stress not due to external forces, but to deformations caused by injections or subtractions of matter in surface layers. This happens, for instance, when we consider a ring cut along a normal section, and then welded, after having removed a small quantity from it: the solid so recomposed will be subjected to internal stresses without external forces acting on it.

Also Leonhard Julius Weingarten [101] had considered the possibility of deformation of elastic bodies without the intervention of external forces. The first, however, to formulate an organic and systematic theory of elastic distortions was Volterra. In a paper of 1882 [90] he analyzed this kind of phenomena, remarking that in elastic bodies equilibrium states may occur, which are different from the natural state, without intervention of any external force.

In the following years Volterra wrote a series of papers on distortions, on which we will not dwell [91–97],[95] but summarize the full-bodied work of synthesis of 1907 [98]. Here, he recalled that Weingarten had remarked that:

> Cases may exist, in which an elastic body, not at all subjected to any external action, that is to say, without being subjected neither to external forces acting on its internal points, nor to external forces acting on its surface, may however not be in the natural state, but in a state of stress varying in a regular and continuous way from a point to the other [98].[96] (A.3.24)

and stated an important theorem:

> An elastic body occupying a simply connected space, the deformation of which is regular, may always be reduced to its natural state by means of finite, continuous, and single-valued displacements of its points.
>
> On the other hand, we may say that:
>
> If an elastic body occupies a multiply connected space, and if its deformation is regular, the displacements of its points are not necessarily single-valued [98].[97] (A.3.25)

If a multi-connected (cyclic) body is made simply connected by a series of cuts, the displacements corresponding to the regular deformation are finite, continuous, and single-valued for the new body, but their values may have discontinuities through the cuts. To restore continuity, it is necessary to move the sections corresponding to the cuts in such a way as to make them coincide: thus, additional strains will arise, independent of external forces.

The fact that a simply connected elastic body which is not subjected to external forces is in the natural state (no stress, no strain) is proved implicitly assuming that its points undergo finite, continuous, and single-valued displacements, and that the body deformation is regular. However, if the body is multiply connected, the regular deformation may co-exist with a multi-valued displacement field and, thus, the body may be in a state of stress even without external forces.

---

[94] "The theory of the so-called *distortions* developed by prof. Volterra contemplates the stresses developing in a not simply connected body, when, once made a cut that does not interrupt the connection, the edges of the cut itself are subjected to rigid relative displacements, after which the continuity of the material is restored by a suitable addition, or subtraction, of material". [83], p. 350.

[95] A wide bibliography and a comment on Volterra's papers may be found in [64].

[96] p. 154. Our translation.

[97] p. 159. Our translation.

Volterra continued stating that in a multi-connected body, made simply connected by a series of cuts, even in the presence of regular deformations, the discontinuities of the displacements U, V, W at the cuts are[98]:

$$U = u_\beta - u_\alpha, \qquad V = v_\beta - v_\alpha, \qquad W = w_\beta - w_\alpha. \tag{3.56}$$

They may be expressed in terms of six parameters l, m, n, p, q, r according to:

$$U = l + ry - qz, \qquad V = m + py - rz, \qquad W = n + qy - pz. \tag{3.57}$$

This means that the continuity of the displacement field may be restored by a rigid motion of the cross-sections corresponding to the cuts.[99] At this point, Volterra could introduce the idea of distortion:

> The distortions
>
> 1. In the previous Chapter I have shown that the elastic bodies occupying multiply connected spaces may be in equilibrium states very different from those that one has when the elastic bodies occupy simply connected spaces. In these new equilibrium states we have a regular inner deformation of the body, even if it is not subjected to external forces.
>
> Let us imagine that we operate the cuts that make the space occupied by the body simply connected. To each of them six constants correspond, which we have called the constants of the cut. It is easy to establish the mechanical meaning of these constants by means of formulas (III)[100] of the previous Chapter.
>
> Indeed, let us practically operate the cuts along the said sections, and let the body recover its natural state. If, by recovering this state, certain parts of the body superpose on each other, let us drop the exceeding parts. Then the already recalled formulas (III) show us that the particles placed at the two sides of the same section, and that, before the cut, were in contact, undergo, due to the cut itself, a displacement resulting in a translation and a rotation, the same for all the couples of adjacent particles to the same section.
>
> Taking the origin as reduction pole, the three components of translation and the three components of rotation along the three coordinate axes are the three characteristics of the cut.[101]
>
> Reciprocally, if the multi-connected elastic body is considered in its natural state, it is possible, to bring it to its stressed state, to operate the inverse transformation, that is cut it with the aim of making it simply connected; then, move the two parts of each cut one with respect to the other, in such a way that the relative displacements of the various couples of particles (that adhered to each other, and that were separated by the cut) result of equal translations and rotations; in the end, restore the connection and the continuity along each cut, by subtracting or adding the necessary matter, and by welding the parts on each other. All of these operations, relative to each cut, maybe called a *distortion* of the body, and the six constants of each cut may be called the *distortion characteristics*.[102] In a multi-connected elastic body,

[98] The proof had already been given by Weingarten in [101].

[99] Somigliana proved afterwards that this depends on Volterra's very strong assumption on deformations, supposed regular up to the second derivative.

[100] They are Eq. (3.57).

[101] They are the six parameters l, m, n, p, q, r defining the relation (3.57).

[102] We often refer to the characteristics, or parameters, of the distortions as being "distortions", *tout court*. On the contrary, to Volterra the distortion is a state of the body, defined by the distortion parameters.

the deformation of which is regular, and that has undergone a certain number of distortions, the inspection of the deformation cannot in any way reveal the locations where the cuts and the following distortions have taken place, because of the regularity itself. In addition, one may say that the six characteristics of each distortion are not elements depending on the place where the cut has been operated. Indeed, the same procedure that we used in proving the formulas (III) shows that, if we take two cuts in the body, that may be transformed one into the other by a continuous deformation, the constants pertaining to one of the cuts are the same as the constants pertaining to the other one: it follows that the characteristics of a distortion are not specific elements of each cut, but they depend solely on the geometrical nature of the space occupied by the body and on the regular deformation it undergoes. The number of independent distortions which an elastic body may undergo to is apparently equal to the order of connection of the space occupied by the body, minus 1. In compliance with what we have found, two cuts that can be transformed into each other by a continuous deformation are called *equivalent*. We will also say that a distortion is known when the characteristic and the pertaining cut, or an equivalent cut, will be given.

2. That posed, two questions naturally arise, namely:

1st, If we suppose vanishing external actions, do an equilibrium state and a regular deformation always correspond to distorsions arbitrarily chosen?

2nd, Once the distortions are known, which is this deformation? In order to link these problems to already known ones, we will prove the following theorem:

If in any elastic isotropic multi-connected body we take an arbitrary set of distortions, we will be able to calculate an infinite number of regular deformations of the body, that correspond to these distortions, and that are equilibrated by external surface actions (that we denote by T), having vanishing resultant and resultant moment with respect to an axis at will.

Thenceforth, to recognize if in an isotropic body the given distortions correspond to an equilibrium state, external forces being nil, it will be sufficient to check if the external forces T, with opposite sign and applied to the body boundary when it is not subjected to any distortion, determine a regular state of deformation equilibrating the same forces. If we can actually calculate this deformation, the problem pertaining to the equilibrium of the body subjected to the given distortions will be solved.

Indeed, denote by Γ the deformation relating to the given distortions and to the found external forces T, acting on the surface, and by Γ' the deformation determined by these external forces, with opposite sign, when the body does not undergo any distortion. The deformation Γ'' resulting from Γ and Γ' will correspond to the given distortions and to zero external forces. The questions are thus led to check if the deformation Γ' exists, and to find it. Thus, they are reduced to elasticity problems where distortions do not appear, that is to say, ordinary problems of elasticity.

On the other hand, the external forces T, acting on the surface, by virtue of the stated theorem, are such that if the body were rigid they would be in equilibrium; it follows that they satisfy the fundamental necessary conditions for the existence of the deformation Γ'.

Since recently we have made much progress by new methods in the investigation of the existence theorem for questions of elasticity, we may say then that, bar certain conditions relating to the geometrical shape of the space occupied by the elastic body (conditions that we will not make precise here), Γ' and Γ'' will always exist.

Once these reserves have been made, we may thus positively answer the first question in the case of isotropic bodies. The second question that was posed relates to the case where the body is not subjected to external forces; however, it may be generalized, and one may suppose that the distortions are given and the body is acted upon by determined external forces. Then, if the body is isotropic, for the resolution of the problem it is sufficient to superpose on the deformation Γ, determined by the distortions and the external forces T, the deformation determined by the given external forces and by the external forces—T acting on the surface, under the hypothesis that there are no distortions [98].[103] (A.3.26)

3.3.2 Beltrami's Pupils

Beltrami's continuous moves did not let him ground a school like Betti had done; however, many took inspiration from his investigations, and followed his ideas. Among them, Padova, who attended Beltrami' and Betti's classes in Pisa, where he graduated in 1866, Somigliana, who attended Beltrami's lectures in Pavia, but graduated in Pisa in 1881, Cesaro, who studied in Liége at the École des mines, without, however, graduating. They gave remarkable contributions to the theory of elasticity, often searching to mathematically describe ether.

Ernesto Padova (1845–1896), on the basis of Beltrami's claim that Maxwell's formulas were independent of the nature of space [8], supposed that ether filled a space endowed with negative uniform curvature, and showed that in such a space "one of the first difficulties present in the mechanical interpretation of Maxwell's formulas" [72],[104] that is the instability of equilibrium of the medium, was overcome. New difficulties, however, arose, more on the mathematical than on the physical side, however, inducing Padova to deny the possibility to formulate a mechanical explanation of Maxwell's equations also in a space with negative curvature, by means of a usual elastic medium.

Carlo Somigliana (Como 1860-Casanova Lanza 1955) has been one of the greatest Italian scientists between the 1800's and 1900's. Of noble origins, descending from Alessandro Volta through his mother, he was a pupil of Beltrami and Casorati in Pavia, and of Betti and Dini at the Scuola normale in Pisa, where he graduated in 1881.

He won the job competition for a chair in Mathematical physics in 1892, then he was called by the university of Turin in 1903, where he stayed until he retired in 1935; he was afterwards elected emeritus. His name is linked to important results, relating to the statics and dynamics in elasticity, and to the theory of potential. He extended well-known results by Clebsch and Voigt on the elasticity of crystals in the isotropic case, and characterized all possible shapes of the elastic potential under symmetry properties. In 1906 and 1907 Somigliana published fundamental integral formulas for elasto-dynamics. Afterwards, he took up the investigations on elastic distortions, that had been initiated by Weingarten and developed by Volterra in 1906; he proved that, under general hypotheses, in a simply connected body, distortions other than Volterra's may exist.

His investigations are remarkable also in geodesy, geophysics, and glaciology. While investigating seismic waves propagation, Somigliana considered the problem of the propagation of plane waves in an unlimited, infinitely deep plane soil from

[103] pp. 165–167. Our translation.

[104] p. 875.

a general point of view, and obtained results providing a broader interpretation of Rayleigh surface waves. A passionate mountaineer, Somigliana investigated the phenomena connected with the movements of glaciers, and conceived an organic and rigorous theory to determine their depth and configuration. Starting from 1926 he dealt with a general theory of the gravitation field external to Earth, establishing new relations among the values of gravity and the geometrical constants of the geoid. Thus, he paved the way to an investigation of the problem of determining the geometric parameters of the geoid by measures of gravity only, of great importance in geophysics and geodesy. In his old age he manifested his interest in history, by taking care of publishing the works of his ancestor Alessandro Volta. In the university of Turin he was dean of the faculty of Mathematical, physical and natural sciences from 1920 to 1932; he was chairman of the National committee for geodesy and geophysics inside the National Council for Research, and of the Glaciology committee inside the Italian society for progress in science [75, 79].

In the following we list the most important papers by Somigliana on continuum mechanics and theory of elasticity.

- 1885. *Sopra l'equilibrio di un corpo elastico*, Il Nuovo cimento, v. 17–18, 1885.
- 1888. *Sulle equazioni dell'elasticità*, Annali di matematica, s. 2, t. 16, 1888.
- 1891. *Intorno alla integrazione per mezzo di soluzioni semplici*, Rendiconti del Regio istituto lombardo, s. II, v. 24, 1891.
- 1901. *Sul potenziale elastico*, Annali di matematica, s. 3, t. 7, 1901.
- 1907. *Sulla teoria Maxwelliana delle azioni a distanza*, Rendiconti della Reale accademia dei Lincei, s. 5, t. 16, 1907.
- 1908. *Sulle deformazioni elastiche non regolari*, Proceedings of the IV Congresso Internazionale dei Matematici, v. 3, Roma, 1908.
- 1909. *Sopra un'estensione della teoria dell'elasticità*, Il Nuovo cimento, v. 17, 1909.
- 1924. *Attorno ad alcune questioni di elastostatica. (Nota I)*, Atti della Reale accademia delle scienze di Torino, v. 59, 1924.
- 1926. *Attorno ad alcune questioni di elastostatica. (Nota II)*, Atti della Reale accademia delle scienze di Torino, v. 61, 1926.

In 1891 [82] Somigliana discussed the integration of the equations of mathematical physics by 'simple solutions' that are obtained

> For certain systems of second-order partial differential equations, which I call symmetric, and that encompass the equations of elasticity as a particular case [82].[105] (A.3.27)

With this aim, Somigliana formulated, in a strictly analytical way, "a reciprocity theorem that, in the case of elastic equations, reduces to the known theorem by Betti", by which he obtained the searched simple solutions, remarking that

> The simple solutions just defined may be considered each representing a special deformation of the body; The L, then, represent the components of the pressure produced by such

[105] p. 1005. Our translation.

deformation on surface elements. Thus, the [simple solutions] [...] state that the surface forces are directed like the displacements, and are proportional to them [82].[106] (A.3.28)

In a paper published in the *Nuovo cimento* in 1885 and in the following ones, Somigliana completed Betti's formulas at first, and applied Green's theorem in his search for the displacements, which he expressed by means of cubic dilatation and other fundamental quantities, thus obtaining functional integrals of the elasticity equations. By going on in this investigation, he was also able to eliminate the cubic dilatation by the method of singularities. The well-known *Somigliana formulas*, expressing the displacement components in an elastic body by means of bulk forces, surface forces, and the displacements occurring at the boundary of the body, take inspiration right from Betti's ideas and his reciprocity theorem [80]. Somigliana wrote:

I propose to show how one may establish a theory for the functions representing the integrals of the elasticity equations in the case of isotropy and equilibrium, that is in many ways analogous to the theory of potential functions, and that constitutes, in a certain way, an extension of it [80].[107] (A.3.29)

Somigliana's formulas were improved by Lauricella some years later. Indeed, to determine the displacements of the elastic isotropic body as a function of external body forces, of stresses, and of boundary displacements, it was not necessary to know all ingredients of Somigliana's formulas; Lauricella could eliminate the 'redundant' ones [61]. In his memoir, Lauricella explicitly referred to the method developed by his master Volterra in a class of mathematical physics. It consists in finding 'suitable' particular integrals of the elastic equilibrium equations, and then using Betti's theorem. This is analogous to Green's method, expressing the value of the potential function by the values attained by it and its normal derivative on the boundary surface of the body. Moreover, when the displacement components of an elastic body are known at the surface, the determination of the deformation at any of its points is just equivalent to the "known Dirichlet problem of the theory of harmonic functions".

In a paper of 1891 [81] Somigliana produced another attempt to see ether as an elastic, isotropic fluid without particular physical properties. He deduced a state of stress in the ether, the displacements of which are the same as those furnished by Maxwell's equations, and proved that an elastic means exists, the deformations of which produce this system of stresses. This is an interesting result, yet not resolutive, since it is not possible to deduce an electric interaction between two conductors from it.

Somigliana considered Volterra's distortions again, and generalized them by introducing other distortions, named after Somigliana[108] by Salvatore Di Pasquale [50]:

It is obvious to think that, besides these cases [Volterra's distortions], other ones exist, provided by observation. We can imagine that the edges of the cut, besides being rigidly displaced with respect to each other, are subjected to slight deformations, just like when

[106] p. 1015. Our translation.

[107] p. 37. Our translation.

[108] Somigliana's most important memoirs on the subject are [83, 84].

> we insert a thin lens with curved faces inside the cut produced in an annular body. And, in addition, we may imagine inserted, or extracted, thin lenticular bodies in simply connected bodies, such as a sphere or an ellipsoid, and we have, thus, elastic stresses produced in them in the absence of external forces.
>
> Now, it is permissible to ask: can these facts, of such an obvious evidence, and already considered by Weingarten, be investigated in a way having some analogies to that of Volterra's distortions? or cannot elasticity theory in its present form tackle these problems?
>
> To answer these questions, it is necessary to examine and discuss the starting points of the theory.
>
> The fundamental hypothesis at the basis of Weingarten's considerations is that the six components of internal stress vary continuously from point to point, after having re-established the material continuity in the body, so that it may be considered in the same static conditions of a compact body. From here, it follows that also the six characteristics of strain (that are independent linear functions of stresses) must enjoy the same continuity properties. Volterra's hypotheses are more restrictive. He admits:
> 1st cthe continuity of the characteristics of deformation (whence the continuity of stresses follows);
> 2nd the continuity of their first and second derivatives.
>
> Now, for the extension we wish, nothing prevents us from leaving aside this second hypothesis, for which an absolute mechanical necessity is not evident. We will, thus, go back to Weingarten hypothesis, and we could propose to look if deformations exist, that satisfy these hypotheses, and not Volterra's. For simplicity and clarity of language, I will call Volterra's distortions the deformations that satisfy the preceding conditions 1 and 2; Weingarten's distortions those satisfying only 1. I propose here not to solve the enounced question in a general way, but I will show by an example, that, however, is of remarkable generality, that Weingarten's distortions exist, that are not also Volterra's [83].[109] (A.3.30)

An indirect pupil, who had frequent correspondence with Beltrami [73], and dealt with elasticity theory deeply, even though not extensively, was Ernesto Cesaro (Naples 1859-Torre Annunziata 1906). He attended the École de mines in Liége together with his brother Giuseppe Raimondo from 1874 to 1883, with many a pause. He had difficulties in enrolling in a university in Italy because he had no high school degree; in 1883 he finally managed to complete his high school studies, but he gave up getting the final degree. In 1886 he took part, with more than 100 publications, in job competitions for secondary schools and universities. He came in first in the standings at the university of Messina, second in those at the university of Naples, after Alfredo Capelli. Due to this success, in 1887 he was awarded the degree *ad honorem* by the faculty of sciences of the university of Rome.

In spite of the chair in Messina, he asked, and was allowed, to teach in Palermo at the chair of Algebra complementare, left free by Alfredo Capelli, who had won the job competition in Naples. In Palermo Cesaro had also the charge of teaching Mathematical physics. In 1891 he was moved to the chair of Infinitesimal calculus at the university of Naples, succeeding Battaglini; here he taught also Higher analysis, that later on became Higher mathematics. In 1906 he asked, and obtained, the move to Bologna in the chair of Rational mechanics, but the move could not take place due to his sudden death.[110]

---

[109] pp. 350–351. Our translation.

[110] Cesaro's biography is extracted from [52].

Cesaro established a fruitful connection between research and teaching, transposing the need of that time for the redaction of high quality textbooks for students. He wrote some monographs and about 250 notes and memoirs, more than a half of which were written in French, without accounting for the numerous answers in journals to questions posed by the readers. His production was not so numerous in the field of continuum mechanics and theory of elasticity; it was, however, important. His most meaningful paper is for sure the textbook of 1894 [41], containing the lectures he gave in Palermo on elasticity theory, originally lithographed. The text is mainly for teaching purposes, and is organized, in part, like the textbooks on strength of materials that began to circulate at the end of the 1800's.[111] The title, *Introduzione alla teoria matematica della elasticità*, by the terms 'introduction' and 'mathematical' underlines on one side the didactic character, on the other side the theoretical character, more than the contemporary textbooks on strength of materials did.

The text is divided into three parts, the contents of which are summed up in the following. The first part is about the fundamentals of elasticity theory and continuum mechanics. The setting is similar to Betti's in [20]. The key theoretical problem is the determination of the position of the points of an elastic continuum as a function of the imposed forces and displacements. Differently from Betti, Cesaro introduced stresses also in the section entitled "Distribution of inner actions".

FIRST PART
I. Kinematics of small motions

II. The components of strain
III. The potential of elastic forces
IV. Elastic equilibrium

V. Betti's theorem
VI. Distribution of inner actions
VII. Elastic motion
VIII. Applications to the sphere.

SECOND PART
IX. Dirichlet's problem
X. Some properties of elastic deformations
XI. The canonical equation of small motions
XII. Calculation of the dilatation and of the rotation
XIII. Integration of the equilibrium equations
of isotropic elastic bodies
XIV. Application to isotropic elastic soils
XV. Thermal deformations
XVI. Saint Venant's problem
XVII. Applicazion to practical problems.

THIRD PART
XVIII. Some notions on curvilinear
coordinates
XIX. Digression on differential parameters
XX. Isothermal systems
XXI. General equations of elasticity
in curvilinear coordinates
XXII. Elasticity in curved spaces.

In the second part of his book, Cesaro dealt with Saint Venant's cylinder, and proposed simple formulas, useful for engineers. The third part is the most theoretical

[111] See, for instance, the textbook by Guidi [56].

one, where one feels Beltrami's influence, or, at least, the influence of the problems dealt with by Beltrami. The last section reports the results obtained in a paper of 1894, and inspired by elasticity theory in curved spaces [42]. In this work he obtained the equilibrium equations for an elastic body in an n-dimensional curved space, by methods and procedures pertaining to intrinsic geometry. He introduced a new potential, characterized by an additional term keeping into account the curvature of the space:

> [is an additional term] that may be considered as the expression of the energy of the reactions that the space, rigid in its geometrical constitution, opposes to the elastic matter filling it, by supposing the latter *inert* in the sense that, compelled to deform in the said space, it tends to do it as if the space itself were Euclidean. A further development of the theory of elastic media in curved spaces will let us perhaps answer a question by Clifford: *if it were not possible that we consider as physical mutations certain effects due in reality to changes in the curvature of our space; in other words, if any of the causes, that we call physical, and maybe every one, were not by chance due to the geometrical constitution of the space in which we live* [41].[112] (A.3.31)

Cesaro, in a memoir of 1889 [23], extended to a body of any shape the results that Betti [20][113] and Padova [71] had found, respectively, for a sphere and for solids of revolution. In a memoir of 1891 [40] he presented a unitary and simplified approach, with respect to that proposed by Betti in the *Teoria della elasticità* for evaluating the coefficient of unit dilatation and the components of the infinitesimal rotation as a function of the displacements at the boundary and of the external bulk and surface forces.

In 1906 [43] Cesaro took inspiration from a paper by Korn [59] and presented a method for calculating the displacements in a homogeneous and isotropic half-space subjected to assigned pressures, or displacements, at the surface. The method was a perturbation one, even if the term *perturbation* was not adopted. The differential equations of the elastic problem are:

$$X + (A - B)\frac{\partial \Theta}{\partial x} + B\Delta^2 u = 0; \quad \dots \tag{3.58}$$

where Θ is the coefficient of cubic dilatation and the other symbols have the usual meaning. The solution is sought in the form of a series:

$$u = u_0 + \kappa u_1 + \kappa^2 u_2 + \dots \tag{3.59}$$

and $\kappa < 1$ depends on the elastic characteristics of the half-space,

$$\kappa = \frac{A - B}{A + B}. \tag{3.60}$$

[112] p. 213. Our translation.

[113] In [14], v. 2, pp. 329–334.

By means of this development, and by the conditions $u_0 = u$ and $u_1 = u_2 = \cdots = 0$ on the free surface, the problem is reduced to a series of uncoupled differential equations, that may be solved in succession:

$$X + B\Delta^2 u_0 = 0; \qquad 2\frac{\partial \Theta_i}{\partial x} + \Delta^2 u_i, \ldots \qquad \Theta_i = \Theta_1 + \Theta_2 + \cdots \Theta_{i-1}. \quad (3.61)$$

In another memoir of 1906 [44] Cesaro reported a variation of the relations proposed by Volterra for evaluating the displacements starting from the components of the deformation, and tried also to obtain relations holding in curvilinear spaces.

References

1. Beltrami E (1902–1920) Opere matematiche (4 vols). Hoepli, Milan
2. Beltrami E (1863–1864) Ricerche di analisi applicata alla geometria. In: Beltrami E (1902–1920) Opere matematiche (4 vols), vol 1. Hoepli, Milan, pp 107–198
3. Beltrami E (1865) Delle variabili complesse sopra una superficie qualunque. In: Beltrami E (1902–1920) Opere matematiche (4 vols), vol 1. Hoepli, Milan, pp 318–353
4. Beltrami E (1868) Saggio di interpretazione della geometria non euclidea. In: Beltrami E (1902–1920) Opere matematiche (4 vols), vol 1. Hoepli, Milan, pp 374–405
5. Beltrami E (1880–1882) Sulle equazioni generali della elasticità. In: Beltrami E (1902–1920) Opere matematiche (4 vols), vol 3. Hoepli, Milan, pp 383–407
6. Beltrami E (1882) Sull'equilibrio delle superficie flessibili e inestensibili. In: Beltrami E (1902–1920) Opere matematiche (4 vols), vol 3. Hoepli, Milan, pp 420–464
7. Beltrami E (1884a) Sulla rappresentazione delle forze newtoniane per mezzo di forze elastiche. In: Beltrami E (1902–1920) Opere matematiche (4 vols), vol 4. Hoepli, Milan, pp 95–103
8. Beltrami E (1884b) Sull'uso delle coordinate curvilinee nelle teorie del potenziale e dell'elasticità. In: Beltrami E (1902–1920) Opere matematiche (4 vols), vol 4. Hoepli, Milan, pp 136–179
9. Beltrami E (1885) Sulle condizioni di resistenza dei corpi elastici. In: Beltrami E (1902–1920) Opere matematiche (4 vols), vol 4. Hoepli, Milan, pp 180–189
10. Beltrami E (1886) Sull'interpretazione meccanica delle formole di Maxwell. In: Beltrami E (1902–1920) Opere matematiche (4 vols), vol 4. Hoepli, Milan, pp 190–223
11. Beltrami E (1889) Note fisico matematiche (lettera al prof. Ernesto Cesaro). In: Beltrami E (1902–1920) Opere matematiche (4 vols), vol 4. Hoepli, Milan, pp 320–329
12. Beltrami E (1889b) Sur la théorie de la déformation infiniment petite d'un milieu. In: Beltrami E (1902–1920) Opere matematiche (4 vols), vol 4. Hoepli, Milan, pp 344–347
13. Beltrami E (1892) Osservazioni alla nota del prof. Morera. In: Beltrami E (1902–1920) Opere matematiche (4 vols), vol 4. Hoepli, Milan, pp 510–512
14. Betti E (1903–1913) Opere matematiche (2 vols). Hoepli, Milan
15. Betti E (1863–1864) Teorica delle forze che agiscono secondo la legge di Newton e sua applicazione alla elettricità statica. In: Betti E (1903–1913) Opere matematiche (2 vols), vol 2. Hoepli, Milan, pp 45–153
16. Betti E (1867a) Teoria della capillarità. In: Betti E (1903–1913) Opere matematiche (2 vols), vol 2. Hoepli, Milan, pp 179–208
17. Betti (1867b) Sopra le funzioni sferiche. In: Betti E (1903–1913) Opere matematiche (2 vols), vol 2. Hoepli, Milan, pp 209–205
18. Betti E (1871) Sopra gli spazi di un numero qualunque di dimensioni. In: Betti E (1903–1913) Opere matematiche (2 vols), vol 2. Hoepli, Milan, pp 273–290

19. Betti E (1874) Sopra l'equazioni di equilibrio dei corpi solidi elastici. In: Betti E (1903–1913) Opere matematiche (2 vols), vol 2. Hoepli, Milan, pp 379–390
20. Betti E (1874) Teoria della elasticità (1872–1873). Soldaini, Pisa
21. Betti E (1876) Lezioni di fisica matematica dell'anno accademico 1876–1877. Betti's fund at the Scuola Normale di Pisa
22. Boltzmann L (1874) Zur Theorie der elastischen Nachwirkung. Sitzungsberichte der Akad der Wiss Wien 70(2):275–306
23. Borchardt CW (1873) Über die Transformationen der Elasticitätsgleichung in allgemeine orthogonale Koordinaten. J für die reine und Angew Mathematik 76:4–58
24. Bottazzini U (1982) Enrico Betti e la formazione della Scuola Matematica Pisana. In: Proceedings of the meeting 'La storia delle matematiche in Italia', Cagliari, pp 229–275
25. Capecchi D, Ruta G (2007) Piola's contribution to continuum mechanics. Arch Hist Exact Sci 61:303–342
26. Capecchi D, Ruta G (2011) La scienza delle costruzioni in Italia nell'Ottocento. Springer, Milan
27. Capecchi D, Ruta G (2014) Polytechnic schools in the 19th century Europe. The polytechnic school in Karlsruhe. Meccanica 49:13–21
28. Capecchi D, Ruta G, Tazzioli R (2006) Enrico Betti. Teoria della elasticità, Hevelius, Benevento
29. Castigliano A (1879) Théorie des systèmes élastiques et ses applications. Negro, Turin
30. Cauchy AL (1828) Sur les équations qui expriment les conditions d'équilibre ou les lois du mouvement intérieur d'un corps solide élastique ou non élastique. In: Cauchy AL (1882–1974) Oeuvres complètes (27 vols). Gauthier-Villars, Paris, s 2, 8:195–226
31. Cauchy AL (1828) De la pression ou tension dans un systemè de points matériels. In: Cauchy AL (1882–1974) Oeuvres completès, (27 vols). Gauthier-Villars, Paris, s 2, 8:253–277
32. Cerruti V (1873) Intorno ai sistemi elastici articolati. Dissertazione presentata alla Commissione Esaminatrice della Real Scuola d'Applicazione per gli Ingegneri in Torino. Bona, Turin
33. Cerruti V (1875) Sopra un teorema del Sig. Menabrea. Mem della Reale Accad dei Lincei, s 2, 2:570–581
34. Cerruti V (1876) Intorno ai movimenti non periodici di un sistema di punti materiali. Mem della Reale Accad dei Lincei, s 2, 3:241–249
35. Cerruti V (1877) Intorno alle piccole oscillazioni di un corpo rigido interamente libero. Mem della Reale Accad dei Lincei, s 3, 1:37–345
36. Cerruti V (1880) Sulle vibrazioni dei corpi elastici isotropi. Mem della Reale Accad dei Lincei, s 3, 8:361–389
37. Cerruti V (1883) Ricerche intorno all'equilibrio dei corpi elastici isotropi. Mem della Reale Accad dei Lincei, s 3, 13:81–122
38. Cerruti V (1890) Sulla deformazione di un involucro sferico isotropo per date forze agenti sulle due superfici limiti. Mem della Reale Accad dei Lincei, s 4, 7:25–44
39. Cesaro E (1964–1968) Opere scelte (2 vols in 3 tomes). Cremonese, Rome
40. Cesaro E (1891) Sul calcolo della dilatazione e della rotazione nei mezzi elastici. In Cesaro E (1964–1968) Opere scelte (2 vols in 3 tomes). vol 2. Cremonese, Rome, pp 414–422
41. Cesaro E (1894) Introduzione alla teoria matematica della elasticità. Bocca, Turin
42. Cesaro E (1894) Sulle equazioni dell'elasticità negli iperspazi. Rend della Reale Accad dei Lincei 5(3):290–294
43. Cesaro E (1906) Sul problema dei solidi elastici. In: Cesaro E (1964–1968) Opere scelte (2 vols in 3 tomi), vol 2. Cremonese, Rome, pp 489–497
44. Cesaro E (1906) Sulle formole del Volterra fondamentali nella teoria delle distorsioni elastiche. In: Cesaro E (1964–1968) Opere scelte (2 vols in 3 toms), vol 2. Cremonese, Rome, pp 498–510
45. Clebsch RFA (1862) Theorie der Elasticität fester Körper. Teubner BG, Leipzig
46. Clebsch RFA (1883) Théorie de l'élasticité des corps solides, Traduite par MM. Barré de Saint Venant et Flamant, avec des Notes étendues de M. Barré de Saint Venant. Dunod, Paris

47. Clerk Maxwell J (1873) A treatise on electricity and magnetism (2 vols). Clarendon, Oxford
48. Clerk Maxwell J (1864) On the calculation of the equilibrium and stiffness of frames. Philos Mag 27:294–299
49. Crotti F (1888) La teoria dell'elasticità ne' suoi principi fondamentali e nelle sue applicazioni pratiche alle costruzioni. Hoepli, Milan
50. Di Pasquale S (1996) Archi in muratura e distorsioni di Somigliana. In: Di Pasquale S (ed) Problemi inerenti l'analisi e la conservazione del costruito storico. Alfani, Florence
51. Drago A (2011) The birth of the non-Euclidean geometries as the more significant crisis in the foundations of modern mathematics. Logic Philos Sci 9(1):103–110
52. Ghisalberti AM (1960-) Dizionario biografico degli italiani. Istituto Enciclopedia Italiana, Rome
53. Green G (1827) An essay of the application of the mathematical analysis to the theories of electricity and magnetism. In : Green G (1871) Mathematical papers. McMillan, London, pp 1–82
54. Green G (1839) On the reflection and refraction of light at the common surface of two non-crystallized media. Green G (1871) Mathematical papers. McMillan, London, pp 245–26
55. Guerraggio A, Paoloni G (2013) Vito Volterra. Translated into English by W Kim, Springer, Berlin
56. Guidi C (1910) Lezioni sulla scienza delle costruzioni dall'ing. prof. Camillo Guidi nel Real Politecnico di Torino. Bona, Turin
57. Khan AS, Sujian H (1995) Continuum theory of plasticity. Wiley, New York
58. Kirchhoff GR (1876) Vorlesungen über mathematischen Physik: Mechanik. Teubner, Leipzig
59. Korn A (1906) Sur un théoreme relatif aux dérivées secondes du potentiel d'un volume attirant. Comptes Rendus Hebdomadaires des Séances de l'Académie des Sci 142:199–200
60. Lamé G (1859) Leçons sur les coordonnées curvilignes et leurs diverses applications. Bachelier, Paris
61. Lauricella G (1895) Equilibrio dei corpi elastici isotropi. Ann della Regia Scuola Normale Super di Pisa 7:1–119
62. Lauricella G (1909) Sur l'intégration de l'équation relative à l'équilibre des plaques élastiques encastrées. Acta Math 32:201–256
63. Marcolongo R (1902) Teoria matematica dell'elasticità. Reale Università, Messina
64. Marcolongo R (1907) Progressi e sviluppo della teoria matematica della elasticità in Italia (1870–1907). Nuovo Cimento, s 5, 14:371–410
65. Michell JH (1900) On the direct determination of stress in an elastic solid, with applications to the theory of plates. Proc Lond Math Soc 31:100–124
66. Mueller I (2006) Philosophy of mathematics and deductive structure in Euclid's elements. Dover, Mineola, New York
67. Müller-Breslau CH (1886) Die neuren Methoden der Festigkeitslehre. Kröner, Lepizig
68. Müller-Breslau CH, (1st, (eds) (1887) Die graphische Statik der Baukonstruktionen. Kröner, Leipzig
69. Navier CLMH (1864) Résumé des leçons données à l'Ecole de ponts et chaussées sur l'application de la mécanique à l'établissement des constructions et des machines, avec des notes et des appendices par M. Barré de Saint Venant. Dunod, Paris
70. Neumann C (1860) Zur Theorie der Elasticität. J für die reine und Angew Math 57:281–318
71. Padova E (1888) Sull'uso delle coordinate curvilinee in alcuni problemi della teoria matematica dell'elasticità. Tipografia del Seminario 3:3–30
72. Padova E (1889) La teoria di Maxwell negli spazi curvi. Rend della Reale Accad dei Lincei, s 4, 5:875–880
73. Palladino F, Tazzioli R (1996) Le lettere di Eugenio Beltrami nella corrispondenza di Ernesto Cesaro. Arch Hist Exact Sci 49:321–353
74. Piola G (1833) La meccanica de' corpi naturalmente estesi trattata col calcolo delle variazioni. Giusti, Milan
75. Roero CS (2004) Profili di G. Beccaria, C. Somigliana. In Allio R (a cura di), Maestri dell'Ateneo torinese dal Settecento al Novecento. Centro studi di storia dell'università di Torino. Sixth centenary, Turin, pp 247–250, 388–389

76. Ruta G (1995) Il problema di Saint Venant. Studi e Ricerche n. 7/95 del Dipartimento di Ingegneria Strutturale e Geotecnica, Università La Sapienza, Rome
77. Saint Venant AJC Barré de (1855) De la torsion des prismes avec considérations sur leurs flexion ainsi que sur l'équilibre intérieur des solides élastiques en général et des formules pratiques pour le calcule de leur résistance à divers efforts s'exerçant simultanément. Imprimerie Impériale, Paris
78. Saint Venant AJC Barré de (1856) Mémoire sur la flexion des prismes sur le glissements transversaux et longitudinaux qui l'accompagnent lorsqu'elle ne s'opère pas uniformément ou en arc de cercle et sur la forme courbe affectée alors par leurs sections transversales primitivement planes. J de Mathématiques pures et appliquées, s 2, 1:89–189
79. Website of the Società Italiana di Storia della Matematica
80. Somigliana C (1888) Sulle equazioni dell'elasticità. Ann di Mat, s. 2, 16;37–64.
81. Somigliana C (1891) Formole generali per la rappresentazione di un campo di forze per mezzo di forze elastiche. Rend del Regio Ist Lombardo, s 2, 23:3–12
82. Somigliana C (1891b) Intorno alla integrazione per mezzo di soluzioni semplici. Rend del Regio Ist Lombardo, s 2, 24:1005–1020
83. Somigliana C (1908) Sulle deformazioni elastiche non regolari. Atti del IV Congr. Internazionale dei Matematici 3:60–72
84. Somigliana C (1914–1915) Sulla teoria delle distorsioni elastiche. Rend della Reale Accad dei Lincei, nota I, 1914, 23: 463–472; nota 2, 1915, 24:655–666
85. Tazzioli R (1993) Ether and theory of elasticity in Beltrami's work. Arch His Exact Sci 46:1–37
86. Tazzioli R (1997) I contributi di Betti e Beltrami alla fisica matematica italiana. In: Proceedings of the XVII Congresso Nazionale di Storia della Fisica e dell'Astronomia, Milan
87. Tedone O (1907) Sui metodi della fisica matematica. In: Proceedings of the I Congresso S.I.P.S, Parma, pp 33–47
88. Thomson W (1882–1890) General theory of the equilibrium of an elastic solid (1863, read in 1862). Mathematical and physical papers. Cambridge University Press, Cambridge 3:386–394
89. Todhunter I, Pearson K (1886–1893) A history of the theory of elasticity and of the strength of materials, from Galileo to the present time (3 vols). Cambridge University Press, Cambridge
90. Volterra V (1882) Sopra una legge di reciprocità nella distribuzione delle temperature e delle correnti galvaniche costanti in un corpo qualunque. Nuovo Cimento, s 3, 9:188–192
91. Volterra V (1901) Un teorema sulla teoria della elasticità. Rend della Reale Accad dei Lincei, s 5, 14:127–137
92. Volterra V (1905a) Sull'equilibrio dei corpi elastici più volte connessi. Rend della Reale Accad dei Lincei, s 5, 14:193–202
93. Volterra V (1905c) Sulle distorsioni dei solidi elastici più volte connessi. Rend della Reale Accad dei Lincei, s 5, 14:361–366
94. Volterra V (1905d) Sulle distorsioni dei corpi elastici simmetrici. Rend della Reale Accad dei Lincei, s 5, 14:431–438
95. Volterra V (1905e) Contributo allo studio delle distorsioni dei solidi elastici. Rendiconti della Reale Accademia dei Lincei, s 5(14):641–654
96. Volterra V (1906a) Nuovi studi sulle distorsioni dei solidi elastici. Rend della Reale Accad dei Lincei, s 5, 15:519–525
97. Volterra V (1906b) Leçons sur l'integration des équations différentielles aux dérivée partielles professée à Stockholm sur l'Invitation de S. M. le Roi de Suéde
98. Volterra V (1907) Sur l'équilibre des corps élastiques multiplement connexes. Ann scientifiques de l'école Normale Superieure, s 3, 24:401–517
99. Volterra V (1908) Le matematiche in Italia nella seconda metà del secolo XIX. In: Proceedings of the IV Congresso dei Matematici, Rome
100. Volterra V (1909) Sulle equazioni integrodifferenziali della teoria della elasticità. Rendiconti della Reale Accademia dei Lincei, s 5, 18:295–301
101. Weingarten G (1901) Sulle superfici di discontinuità nella teoria della elasticità dei corpi solidi. Rend della Reale Accad dei Lincei, s 5, 10:57–60

Chapter 4
Solving Statically Indeterminate Systems

Abstract The most important event for the history of structural engineering in Italy in the second half of the 1800s was the approval of the law decree of 1859 of the Kingdom of Sardinia, known after its promoter Gabrio Casati, which took force from 1860 in the kingdom and was then extended to all Italy. This decree reformed the whole education system and established the schools for engineering. Among these schools, the most important one, at least at the beginning, was that in Turin. The key person of this school was Giovanni Curioni, heir of Menabrea, who had taught structural mechanics to the pupils of engineering schools before Casati's reformation. Curioni inherited Menabrea's researches on the way to solve redundant structures and supervised the graduation thesis that Alberto Castigliano and Valentino Cerruti presented in Turin in 1873, where the former extended Menabrea's technique and the latter explored more traditional approaches to solve redundant trusses. In this chapter we focus on the contributions by Menabrea, Castigliano and Cerruti, trying to highlight strengths and weaknesses, and showing their connections.

4.1 Scuole d'applicazione per gl'ingegneri

With the constitution of the Kingdom of Italy (1861), the necessity arose for technicians who could put contemporary knowledge into practice. Among them, the figure of the engineer emerged. The comprehensive law on public education, better known after its promoter Gabrio Casati,[1] was a fundamental event in the technical modernization process in Italy; it promoted the institution of the Scuole di applicazione per gl'ingegneri (hereinafter Schools of Application for Engineers), separating their studies from those in the faculty of mathematics. Casati's law did not foresee, by principle, a sharp distinction between scientific and professional courses, so that the schools of engineering recalled, at least in part, the model of scientist-engineer of the French École polytechnique.

A part of the academic world resisted the scientific qualification of engineers, and tried to leave them the simple degree of diploma instead of university graduation.

[1] Regio decreto legislativo 13 novembre 1859, n. 3725, see [11], Appendix A.

D. Capecchi and G. Ruta, *Strength of Materials and Theory of Elasticity in 19th Century Italy*, Advanced Structured Materials 52,
DOI 10.1007/978-3-319-05524-4_4

To have an idea on this resistance, we quote a discussion before an association of engineers in 1913, many years after the approval of Casati's law decree:

> The Law on public education of November 13, 1859, that established the School for Engineers in Turin, is not unambiguous on this point. The following rulings, on the other hand, talk openly of releasing a degree of *Graduated engineer.*
>
> Such a denomination is found in the Royal Decree of October 11, 1863, as well as in the Rulings of October 17, 1860, and October 11, 1866. However, as far as they are concerned, one could still suspect that the Graduation should be ascribed to the preceding studies in Mathematics, then considered as completed, even if limited to a two-year period. On the other hand, the Ruling of November 14, 1867, is expressed in such a way not to leave any doubts, and declares, literally:
>
> "The School awards Graduation degrees for civil, mechanical, agricultural, metallurgic, chemical engineers, and civil architects". Such a sharp denomination is missing again in the following Rulings, that sometimes avoid to mention the term 'diploma', apparently with the aim that the omission of the Graduation degree be not felt. On the other hand, the actual[Ruling] abundantly repeats, one would say showily, that the title awarded to engineers is a simple diploma. To sum up, one wanted to move a decisive step in the direction that the Report of the Royal committee for the reformation of higher studies tends to establish in a general way, by attributing to the old university faculties the exclusive right of scientific teaching, and, as a consequence, the right to awarding the doctoral dignity, and by limiting the scope of all application studies (once this word is taken in the broader sense of habilitation to free profession) to awarding a simple diploma whatsoever.
>
> It is a strange direction today, when the engineering technique, unbound from the circumscribed limits of the past, has unfolded with an unforeseen broadness of studies in the most disparate applications, becoming not a single, but a complexity of sciences, which fully renewed a lot of its old Chapters: when this technique, identifying itself with social life in all its displays, has become its first factor, has imposed itself in all Public services, has risen to the rank of irreplaceable means for attaining any progress. A strange direction, or rather justified, right by this unexpected primacy, that threatens those established on old traditions [46][2] (A.4.1)

The end of the nineteenth century was a period of great successes in Italy for both engineering, as a discipline, and engineers, as professional. In this period, indeed, concurrently with the birth of a new nation, there was an intense activity of construction of civil public works, railway infrastructures (especially bridges), industrial buildings, with steel playing the main role [50, 73].

The new technical ruling class, so, assumed a different attitude toward the more 'cultured' colleagues of mathematics, without awe, and proud to be able to solve the technical problems of real life, seemingly impossible to solve up until now:

> In 1866 The Society of engineers and manufacturers was founded, under the chairmanship of Pietro Paleocapa. Two years later, in Milan, the ancient College of engineers and Architects was reconstituted, founded in 1606 and abolished in 1797 according to the prescriptions of the Constitution of the Cisalpine republic; the first president was Luigi Tatti. In 1868, driven by Prospero Richelmy in Turin and Francesco Brioschi in Milan, who were directors of the School of Application—the former—and of the Higher technical institute—the latter, both associations started publishing proceedings. The example by Turin and Milan was then followed in the 1870s in other Italian cities, among them Rome, where, in 1871, once

[2] Our translation.

acquired the freedom of association, a first Technical society was constituted, afterwards turned into a College of engineers and architects (1876).

Some of the most influential journals of the time were means of information on public works, other were technical-scientifical journals. Among the former, we quote the *Giornale del Genio civile*, for civil engineering, the *Monitore delle strade ferrate* (Turin 1868) and the *Giornale dei lavori pubblici e delle strade ferrate* (Rome, 1874), for railways and public works; among the latter, we may quote *Il Politecnico* of Milan, which had resumed publications in 1866 under the direction of Brioschi, and *L'ingegneria civile e le arti industriali*, established in Turin in 1875 by Giovanni Saccheri, who was professor of technical drawing in the School of Application. Castigliano collaborated with these two last journals in particular, with repeated communications starting from 1876 [76].[3] (A.4.2)

When the Schools of Application (1859) and the Royal museum for industries (1862) were born in Turin, its region Piedmont was enjoying a period of evolution toward deep social and economic changes.

The problems linked with the unification of Italy had a strong effect on Piedmont, and, most of all, on Turin: on one hand, the town starts turning from the capital of a small regional state into a very important city of a great nation; on the other hand, we see the transformation from an economy mainly war-driven—even if in a state relatively at the cutting edge in the European landscape—to a normal one, yet in a situation of a much larger state, characterized by vast areas of underdevelopment and with problems of inter-regional integration. Engineers in Turin, then, were born in a climate of great evolution, also from the cultural point of view. In the first half of the 1800s, Turin could not be described as a fully industrial city, yet it was already an important pole of technical and technological innovation. The productive activities, leaving aside the city mills and the arsenal workshops, were substantially framed in a handicraft setting, even if the first touchable signs of change could be seen in the National expositions of arts and crafts that, since 1827, took their institutional place at the Valentino castle.

The technical and productive initiatives just recalled are inserted in the attempt of internationalizing, obviously toward the rest of Italy at first, and of *Europeanization* ante litteram, that the small Savoy kingdom had followed in politics in the 1840s and 1850s.

[...]

In this strongly careful climate toward technical and professional education, soon the need of educating engineers was born, able to managing innovation, and key actors of the new industrial society.

On November 13th, 1859, the Kingdom of Sardinia, to which Piedmont and Turin belonged, promulgated the law on the reformation of public education, known after the name of its promoter, Gabrio Casati. Casati's Law created an efficient system, the address of which would remain unchanged until the reformation by Giovanni Gentile, and posed the ideological bases and the pedagogical-didactic premises of the new Italian state. Casati's law started technical institutions, lasting 3 years, and posed the bases for the new scholastic structure in Italy, from higher until primary education. The new setting of engineering studies was also established, that divided the school time of engineers into two parts: a first step, devoted to theoretical preparation, was left to the universities, while the second one, regarding applied sciences, was given to new technical school, created on purpose [7].[4] (A.4.3)

[3] p. 66. Our translation.

[4] Our translation.

Table 4.1 The first Italian schools for engineering

1860	Turin, School of Application for Engineers
1863	Milan, Royal Technical Institute
1863	Naples, School of Application for Engineers
1867	Padua, Two-year preparatory course + three-year applications at the faculty of mathematics; in 1875 the three-year period is given to a School of Application for Engineers
1870	Genoa, Higher School of Navy
1862	Bononia, Teaching in the faculty of mathematics; in 1875, first year of the School of Application for Engineers
1873	Rome, School of Application for Engineers

4.1.1 The First Schools of Application for Engineers

In Table 4.1 a list is reported, of the main Italian Schools of Application of Engineers, and the like. For sake of space, we will limit to presenting only those of Turin, Milan, Rome, and Naples with some detail.

4.1.1.1 The School of Application in Turin and the Royal Technical Institute in Milan

Turin. The first School of Application for Engineers was opened in Turin, capital city of the Kingdom of Sardinia [63, 93]. The Royal Technical Institute, providing theoretical bases similar to those of the universities, was founded in 1852; the School of Application for Engineers was born in 1860, replacing the Institute. The principal promoters of its foundation were Prospero Richelmy (an engineer), Carlo Ignazio Giulio (an engineer), Ascanio Sobrero (a physician and a chemist), and Quintino Sella.[5] Richelmy was the first president, and was followed, from 1881 to his death (1887) by Giovanni Curioni; since 1861 the school had a prestigious seat at the Valentino Castle.

Out of curiosity, in the following Table 4.2 we list the ordinary expenses for the faculty in the year 1879.

In 1862 the Industrial Museum was founded in Turin, and its first director was Giuseppe De Vincenzi: its scope was the promotion of "industrial education and progress in industry and commerce". The Museum came somehow in competition with the School: actually, the pupils attending civil engineering and architecture courses mainly followed classes at the School, while the pupils attending an industrial engineering course, whose graduation degree was established formally in 1879, followed classes mainly at the Museum.

[5] Quintino Sella (1827–1884) was a politician, engineer, and a mineralogist; he graduated in Turin in engineering, perfecting in various European countries. He was professor of mineralogy, member of the parliament from 1860, appointed minister of finance in 1862.

Table 4.2 Expenses of the School of Application in Turin, 1879 [84]

1	President	L. 2.000	Per year
4	Directors of laboratories	L. 3.200	"
6	Full professors	L. 36.000	"
6	Extraordinary professors	L. 15.000	"
10	Assistants	L. 15.000	"
	Assistants in charge, globally	L. 6.000	"
1	Vice-director of the chemical laboratory	L. 3.000	"
1	Secretary	L. 4.000	"
1	Assistant secretary	L. 1.800	"
1	Scrivener	L. 1.400	"
2	Custodians	L. 1600	"
	Janitors and other, globally	L. 4.600	"
	Total	L. 93.600	"

In 1906 the two institutions melded, originating the Polytechnic; the preparatory two-years course passed from the faculty of mathematics of the university to the Polytechnic, and the different professional figures of the architects and of the civil, industrial, chemical, and mechanical engineers were introduced.[6] The first woman in Italy to get a degree in engineering was Emma Strada in 1908 at the Polytechnic of Turin [8, 80].[7]

Milan. The opening of a school for educating high-level technicians in Milan had been fostered long before the unification of Italy; a favorable climate was also created by the intellectual Carlo Cattaneo and his journal *Il politecnico*.[8] A Society for encouraging arts and crafts was born in 1838, by initiative of industrialists and businessmen; in 1848, the Regio istituto lombardo promoted a design of reformation

[6] Law June 23, 1906. The *Relazione sull'andamento della Scuola di Applicazione degli ingegneri di Torino nell'anno scolatico 1872–1873*, sent to the Ministry by the director of the School, Richelmy, and now kept at the State archive of Rome, is interesting. On the School of Application we may also quote [44, 85].

[7] pp. 1037–1046; pp. 1047–1056.

[8] Starting from 1839, this journal was the main vehicle of Cattaneo's thought on the privileged role played by sciences for the progress of society. *Il politecnico* spread practical and applicative knowledge, and confirmed the social role of sciences. The will to develop and spread scientific culture in view of applications made it necessary to recall the disciplines taught at the Polytechnic: instead of the basics 'Mechanics' and 'Hydraulics' one talked about 'Railways' and 'Communication routes', arriving to 'Industrial chemistry' and 'Industrial physics'. Much importance was given to communications, most of all dealing with railways and river navigation on the Po. Geology was interesting essentially from the point of view of an investigation on energy, fossil fuels, and techniques for their extraction. *Il politecnico* had great homogeneity from the cultural point of view, and it was almost impossible to ascertain Cattaneo's contributions from these of his co-workers, who were all active in the Lombard technical-scientific world. The intellectual figure emerging from the articles in the journal is that of the engineer, who manages this new technical-scientifical know-how and contributes in creating a fruitful transformation of the Italian society; see [63], pp. 370–371.

of the education system, taking into account the education of engineers as well, and the spokesman was Cattaneo. In 1850 Brioschi, thanks to his political acquaintances, founded the Royal Higher Technical Institute, with three-years courses: in order to attend it, one should have passed a two years preparatory course at the faculty of mathematics of the university of Pavia (later on, of the Kingdom of Italy).

In 1865, the section for architects was established, in collaboration with the Accademia di belle arti in Brera; in 1873, the section for mechanical engineers was established. In 1875 the Royal Institute opened a preparatory school and got autonomy from the university.

After Giovanni Gentile's reformation of the education system (1923), the Royal Higher Technical Institute took the name of Royal School of Engineering, then of Royal Higher Institute of Engineering (1935), and, in 1937, the present name of Polytechnic of Milan.

4.1.1.2 The School of Application in Naples

During French occupation, in March 1811, Joachim Murat founded in Naples the School of Application for Engineers of bridges and routes, on the model of the French École polythecnique. The school followed Murat's institution of the Corporation of bridges and routes engineers [45, 61].

The school was abolished during the Bourbon restoration of 1815, but was founded again in 1819 with the name of School of Application of bridges and routes. Its new statute reduced from three to two the years of the course of study. In the beginning, the school had its seat in the building 'de' Minister' (nowadays Town Hall), then moving to the Gravina Palace. In 1834, with the reorganization of the professional skills of the technicians educated at the School of the Kingdom of Naples, it was established that the studies at the School of Application of bridges and routes were given the right to obtain the degree in civil architecture, thus subtracting it from the university.

While the Schools of Application in Turin and Milan were founded, and the universities got a uniform regulation, in 1863 the School, which had in the meantime changed its name into School of Application for Civil Engineers, passed under the control of the Ministry of Education, with the name of Royal School of Application for Engineers, and detached, after almost fifty years, from the Corps of state engineers. The regulation of the school was the same as that in Turin, the admittance was limited to those with a degree in Mathematics, and the course of study lasted two years. The seat of the School was placed in the former cloister of Donnaromita, near the university.

At the end of the century the statutes of the Royal Schools of Application for Engineers in Italy were unified, and in Naples as well the course of study lasted three years, after a two-year period of preparatory studies on physics and mathematics, leading to the title of civil engineer, or of architect.

4.1.1.3 The School of Application in Rome

On October 23rd, 1817, the *School of Engineering* was born in Rome on the initiative of pope Pius VII, who had been exiled by Napoleon. This new papal school was established after the need of providing engineering knowledge through local education. In the beginning, the school was not part of the university *Sapienza*, founded on April 20, 1303 by pope Bonifacius VIII [94].

Descriptive geometry, architecture, constructions, hydraulics and practical hydrometry and topography were taught at the school, and attending a physics workshop was compulsory. The course of study lasted three years, and the school could be entered after having attended the classes in physics and mathematics at the university. The studies ended with a general exam after which the title of civil engineer was released, by which it was possible to enter the Corps of papal engineers, but also, first in Italy, to follow private practice.

After the unification of Rome to the rest of Italy, a decree of 1872 made Casati's law hold, and on October 9th, 1873, the decree instituting the School of Application for Engineers in Rome was issued, remembered until now on a plaque in the lobby of the nowadays called faculty of engineering:

> At the beginning of December, 1873, by the king Vittorio Emanuele II, and the minister of education Antonio Scialoja, what were the buildings of the Lateran Canons were opened to light in the teaching dedicated to the mathematical and practical disciplines.[9] (A.4.4)

The School found a place in the former cloister of the Lateran Canons, next to the church of St. Peter in Chains, where it is also nowadays. Completely independent of the university, attending the preparatory classes in physics and mathematics was necessary to enter it.

For almost fifty years the School, even though with the aim of educating civil engineers, activated researches and experiments in order to enlarge the knowledge offered to its pupils, until a new regulation foresaw two separated sections, civil and industrial. The teachings of geodesy, applied geometry, technical physics, chemistry applied to building materials, applied geology were added; in 1886 electrotechnics was introduced, and after 1892 estimate, agricultural economics, applied hygiene, and others.

Two-year courses of architectonic drawing for the graduated pupils of the Accademia di belle arti of Rome and Florence were kept, and those pupils could so get the title of architects. The initiative was abandoned, then taken over again in 1919, with the institution of a Higher School of Architecture.

The School of Application became Faculty of Engineering in 1935; in the same year, the year in which the University campus was inaugurated, the faculty of architecture was also opened.

[9] Our translation.

4.1.1.4 Curricula Studiorum

Casati's law originated from the necessity to give dignity and a right place to the technical and scientific culture, but, unfortunately, determined a strong differentiation between 'scientific' and 'human' knowledge. In the middle and high school, this differentiation turned into a true prevalence of human sciences. As a matter of fact, to enter university and, thus, make part of the future Italian ruling class, it was necessary to attend Middle and High School of a purely humanistic mold (ginnasio and liceo), where Latin and ancient Greek, believed to have a fundamental part in the educational process, were more important than mathematics and physics.

At the university, the differentiation was even stronger, but here 'scientific disciplines' had their own dignity. In the newborn faculty of mathematical, physical and natural sciences, very few classes of humanities were active, and were seen as secondary. This fact was also encouraged by the teachers of the scientific faculties themselves, who, following the positivistic thought of the time, believed that humanities, together with philosophy, were a bunch of 'useless sophistries'.

Professional courses of study—among them, engineering—considered their disciplines as important as humanities, encouraged by Casati's law, which gave great importance to basics, leaving to *post lauream* apprenticeship the acquisition of the specific knowledge relative to the various free professions.

In the beginning, each school had great freedom of choice in the teachings, but then the regulations for the Schools of Application for Engineers, approved by the Royal decree of October 8th, 1876 established the disciplines common to everybody. Pupils in engineering and architecture should attend a two-year preparatory period in the faculties of mathematics, plus some additional exams, among which we find drawing. Then the three-year period at the School of Application followed, where they should attend the following classes:

1. *First year*

The first year was common to both courses, and teachings were: Rational mechanics; Theoretical geodesy with exercises; Graphical statics and drawing; Applications of descriptive geometry; Docimastic chemistry[10] with manipulations. Some classes would be given by teachers of the faculty of sciences of the university.

2. *Second and third year*

From the second year, the courses for architects and engineers differentiated. Pupils in engineering should attend mandatorily classes in: Practical geometry; Mechanics applied to machines; Mechanics applied to buildings; Rural economics and estimate; Law matters; Technical physics; Mineralogy and geology applied to building materials; Practical hydraulics; Hydraulic machines; Agricultural machines; Thermal machines; Technical architecture; Civil and rural buildings; Foundations; Bridges in masonry, wood, and iron; Ordinary routes, railways, galleries; Hydraulic constructions and sea works; Agricultural hydraulics.

[10] This is a branch of applied chemistry, studying the nature and the composition of the substances and materials adopted in industry.

Table 4.3 Program of the teaching of graphical statics, 1878–1879 [83, p.33]

Principle of signs in geometry; Graphical sum; Products, power, root extraction; Change of plane figures
Graphical composition of forces in a plane; Moments of forces and couples; Equilibrium of non-free plane systems
Composition of given forces in space; Cremona's theory of reciprocal figures in graphical statics
Application of the theory of reciprocal figures to trusses
Centroids; Moments of parallel forces; Applications
Moments of 2nd order; Central ellipse and core of a plane figure; Inverse problem of the moments of inertia of a plane figure; Hints on central ellipsoid and core of some solids
Hints to the services that graphical statics may do the craftsmanship of the ship constructor
Antonio Fais

Table 4.4 Program of the teaching of mechanics applied to constructions, 1878–1879 [83, p.53–54]

Elastic bodies; Forces and elastic deformations
Deformations that may take place in a solid under the influence of external forces, admitting the principles of conservation of plane sections; Corresponding stress at various points
Results of experiments on various materials used in constructions; Experiments by Wöhler on the influence of repeated stresses on the strength of iron and steel—Methods by Winkler, Oerber, and so on
Resistance to extension or pressure; Influences of temperature on the stress in prisms; Work of elastic forces—Applications
Resistance to shearing; Calculation of joints
Theory of beams undergoing bending; Supported and clamped beams; Various cases; Beams of uniform resistance
Beams with straight axis resting on more than two supports; Determination of the reactions and of the moments at the supports
Inner forces and their distribution; Curves of the shearing forces and of the maximum tensions and pressures
Prisms subjected to forces parallel to the axis, central core; Case in which forces act along the axis
Beams subjected to forces acting slantwise with respect to the axis
Resistance to torsion
Resistance to bending and torsion
Theory of equilibrium of deck rafters
Theory of trusses; Load conditions determining maximum or minimum stresses in bars. Applications
Theory of metallic arches
Theory of domes
Theory of equilibrium of terrains
Stability of masonry; Conditions and equations of equilibrium, or of stability; Empirical formulas
Silvio Canevazzi

Other classes, those of the course in architecture of the Academies of fine arts, were added to these: Architectural styles, Manufacturing and clay modeling of architectural ornaments, Inner decorations, Drawing of perspective, Watercolor and aesthetics applied to architecture, and Measures from true. Each School of Application could, however, gather teachings and distribute them in the second and third year, according to the School councils.

In the following Tables 4.3 and 4.4 we show the programs of the teachings in structural subjects of the School of Application in Bologna for the school year 1878–1879, which, in any case, reflected that of other schools. Such programs would remain practically unaltered at least until 1900.

4.2 The Teaching

The history of the faculty of engineering and the organization of studies began to be a subject of historical investigation [62, 95] also from the point of view of understanding the relations among engineering, technology, and science [5].[11]

The problem of the relation between craftsmanship and science in building, and science properly said, is a part of the wider problem of the relation between science and technology, yet with some peculiarities. Strength of materials deals with phenomenologies that, however complex, are simpler than those met in other branches of engineering (mechanical, chemical, electric), and, thus, more easily subjected to the formalism of an axiomatic theory. In addition, an important subject dealt with in the strength of materials, that is the theory of elasticity, had been developed in depth at the beginning of the 19th century as a part of physics and mathematical physics, relative to the understanding of the constitution of matter, the transmission of light, electro-magnetism.

When, at the end of the 19th century, the development of industry and transportation asked to face structural problems, the craftsmanship and the science of building were well developed, and the problem arose of their integration. This, at least in Italy, was done quite quickly starting from the 1850s, also thanks to the School of Elasticity of Betti [12].

To have an idea of the evolution of the teaching of strength of materials we refer to the school in Turin, that is at the moment, one of the best known. Giovanni Curioni,[12]

---

[11] pp. 60–75.

[12] Giovanni Curioni (Invorio, 1831–Turin, 1887) graduated in engineering and architecture at the university of Turin and began his career as an assistant in Practical geometry in the Institute for technical teaching, that after Casati's law was turned into the School of Application for Engineers. In 1865, following the transfer to Florence of the regular professor Giulio Marchesi, he was appointed assistant professor in Constructions at the School of Application by Richelmy. In 1866 he became professor of Constructions (Costruzioni civili, idrauliche e stradali), and in 1868 he became full professor. At the same time, he taught Practical geometry, Constructions and estimate at the Professional junior high school. He was among the signers of the request for instituting the Society of engineers and architects of Turin, of which he became president some

professor in this school from 1865, was the first to rename *Scienza delle costruzioni*[13] his classes of Mechanics applied to constructions, in 1877.[14]

Curioni's reference textbook was *L'arte di fabbricare* [43], a monumental work in six volumes and five appendices, published from 1864, with various reissues. The part of the textbook relative to the strength of materials properly said is in the first volume, entitled *Resistenza dei materiali e stabilità delle costruzioni* (first edition 1867, third edition 1872), together with some appendices, written, however, after the first editions. The volume was directed at pupils of both technical junior high schools and engineering schools, thus the mathematical level was rather low. The textbook referred to the textbooks of the past about the *art of building* [1, 2, 51, 87], and to the classical *Résumé des leçons* by Navier [78, 79].

The textbook dealt briefly on strength of materials, without introducing the concept of stress precisely (which, however, also Navier did), and then passed to deal with the resistance of beams, considering traction first (Chapter II), then compression (Chapter III, without any reference to buckling), then torsion (Chapter IV), shear (Chapter V), bending (Chapter VI). It also dealt briefly with arches (Chapter X). A part devoted to the theory of structures is almost missing, a part from some hints on trusses (Chapter XIII).

At this time, a pupil of Curioni in the School of Application in Turin was Carlo Alberto Castigliano, who graduated in 1873. The role of Castigliano and his monograph *Théorie des systèmes élastiques et ses applications* [19], as a part of the drafting of textbooks on strength of materials, is still to be fully investigated. As a matter of fact, Castigliano wrote a modern textbook: he started with an extended discussion on continuum mechanics by a molecular approach following Saint Venant. Then, he provided a rather extended treatment of Saint Venant's cylinder; finally, there was an extensive chapter on structural mechanics, developed according to a discussion deserving to be considered as pioneering: for the first time, indeed, generalized procedures for the solution of redundant systems of beams were presented. The adopted technique was that of minimizing what we now call the complementary elastic energy, by an extensive use of the theorems now known as Menabrea's and Castigliano's first and second theorem. The *Théorie des systèmes élastiques et ses applications* was warmly welcome by the scholars of mechanics applied to constructions, both in Italy and abroad, but could not be accepted as a university textbook.

(Footnote 12 continued)
years later. From 1879 to 1893 he was the director of the Laboratory on Strength of materials and Theory of bridges. He was fellow of the Reale accademia delle scienze of Turin from 1873, and presented there some memoirs on the use and resistance of construction materials. As a part of these investigations, he had installed a great machine to test the strength of materials in the Laboratory of the School of Application. In 1879 he also taught at the Royal Industrial Museum; from 1881 to death, he was also director of the School of Application for Engineers. In 1881 he was elected to the parliament, representing the Council of Borgomanero, and the year after that of Biella [9].

13 Literally, 'Constructions science'. This denomination, which cannot be found elsewhere in Europe, has remained also nowadays.

14 Classes in structural engineering were held at the university until 1860, and the teacher was Menabrea; afterwards, the teaching assignment was given to Valentino Amò (1860–61) and Giulio Marchesi (1861–1865).

Castigliano's strangeness to the academic world only partly explains the partial fiasco of his theoretical text, as well as of the practical textbooks he had published before dying.

Curioni was followed in 1882 by Camillo Guidi (1853–1941), whose textbook *Lezioni di scienza delle costruzioni* [52] testified a synthesis now completed between the craftsmanship and the science of building. The setting of the textbook was deductive and axiomatic, even though its level of formalization was lower than that then adopted in books on mathematical physics. In the beginning, important notions of continuum mechanics were introduced: stress in a three-dimensional continuum, infinitesimal strain, constitutive relations. An extended discussion on prismatic solids followed, dealt with by a suitably simplified Saint Venant-like approach. A rather extended part dealing with structural mechanics followed this mechanical preambles. This part, however, was still an embryo if compared with the discussions of our time, and was limited to considering the approach according to Euler's *elastica*, which, however, made it possible solutions of practical application only for simple structures. Finally, a simplified discussion of buckling was found; use of graphical statics was remarkable.

After Guidi, Gustavo Colonnetti (1886–1968) took the chair in 1928, and his textbook *Principi di statica dei solidi elastici* [36] fixed a scheme for textbooks of the discipline, lasting until the 1970s. Colonnetti followed a deductive axiomatic approach, introducing also specific themes of his researches, in particular about the theory of distortions. Guidi's discussion was made more formal, and the part on structural mechanics assumed the form it has also nowadays, using energy theorems and the principle of virtual work for the solution techniques. The adopted method was that of forces, in that the method of displacements should seem, in the absence of quick and efficient means of automatic calculation, almost impracticable.

The following Table 4.5 shows some textbooks on strength of materials and theory of elasticity (Scienza delle costruzioni) diffused in Italy between the end of 19th and the beginning of the 20th century.

The older textbooks, among which those by Curioni (1867) and Pullino (1866), had a problem-oriented approach, referring partly to the French texts on the art of building. Starting from Guidi's book, the exposition turned into one, repeated by Castigliano, that would become standard, and saw the Scienza delle costruzioni divided into two parts. In the first one, we find a remarkably extended discussion on

Table 4.5 Italian textbooks on strength of materials and mechanics of structures

Giacinto Pullino	*Resistenza dei materiali e meccanica applicata*, Castellamare	1866
Giovanni Curioni	*L'arte di fabbricare*, Turin	1867
Alberto Castigliano	*Théorie des systèmes élastiques et ses applications*, Turin	1879
Emilio Almansi	*Introduzione alla scienza delle costruzioni*, Turin	1901
Cesare Ceradini	*Meccanica applicata alle costruzioni*, Milano	1910?
Camillo Guidi	*Lezioni sulla scienza delle costruzioni*, Turin	1891

continuum mechanics, defining strain (usually infinitesimal), stress and constitutive relations; then, a rather detailed and formalized discussion on Saint Venant's problem. In the second part, themes of structural mechanics are discussed, faced mainly by the method of forces. The resolution algorithms were based on the principle of virtual work, and/or on theorems of minimum energy (after Castigliano's book).

The maturity in the evolution of these textbooks was reached only at the mid 20th century, with the second edition (1941) of Colonnetti's textbook [37]. This book had the same setting as that of 1916 [36], but was more precise and rigorous, with more space dedicated to continuum mechanics and a deeper discussion on the mechanics of structures. There were important applications about influence lines and stress curves, developed by integral methods and energy reciprocity theorems. There were also: remarkable hints on distorsions, both thermal and mechanical (that is Volterra's distortions); some descriptions of elasto-plasticity; a rich bibliography; and several photographs related to experiments in photo-elasticity. However, there were not many examples on beam frames. Indeed, to catch the spirit of the textbook, is sufficient to read Colonnetti's preface:

> These pages—in which I have collected the lectures I have given this year to the pupils of the Polytechnic of Turin—faithfully reflect the didactic conception to which I inspire all of my teaching; which deliberately proposes high cultural purposes and, only secondarily, professional education. The choice of the subjects has been made with this only concern: to offer the scholar the occasion of knowing the fundamental principles, to deepen their meaning and their scope, to see how a rational body of doctrines could be built on them, and how this could then, time by time, be used to solve practical problems. The subjects better fitting this scope have been dealt with in full. Others, not less important in themselves, but less suggesting from this point of view, have been neglected in part or in full. The reader will not find here the usual collection of ready-made solutions, to be applied—properly, or inappropriately—to all the problems that the technical practice will let him face. On the other hand, he will be able to learn to analyze and solve each of those problems, realizing the value of the hypotheses on which the solution is based, and of the approximation degree it implies [37].[15] (A.4.5)

This way to intend Scienza delle costruzioni is usually accepted also nowadays; one deals not only with presenting methods of solution of structural problems, but also, most of all, with rationally discussing continuum and structural mechanics so that the professional engineer would be able to use techniques and methods with full awareness.

4.3 Luigi Federico Menabrea

Luigi Federico Menabrea (Chambéry 1809–1896) passed in 1828 the admission tests for the university of Turin. Among his teachers we quote in the first year abbot Bianchi for algebra, geometry, and trigonometry, in the second year Giovanni Plana for calculus, in the third and in the fourth year Plana again and Giorgio Bidone for hydraulics.

---

[15] Preface. Our translation.

He got the degrees in hydraulic engineering in 1832 and civil architecture in 1833, and was then appointed lieutenant in the corp of military engineers.[16]

In 1833 he replaced the lieutenant Camillo Benso count of Cavour, who later on became prime minister in the Kingdom of Sardinia and was a key person in the process of the unification of Italy, in the construction of Forte di Bard, a monumental fortress in the northwestern part of the kingdom, controlling the routes to and from France. In the spring of 1834 he was often sent to Genoa, the most fortified city of the kingdom, and began as a fortifier by compiling in 1837 a project for Alessandria, a fortified city considered as backing Genoa ("l'antemurale di Genova"). His stay in Genoa was very formative for Menabrea, not only as a designer of fortifications, but also as a cartographer, since he learned there the new system of topographic surveying conceived by major Ignazio Porro.

Menabrea's education in military architecture was influenced by the *editio princeps* of Francesco di Giorgio Martini (1841), edited by Cesare Saluzzo di Monesiglio and Carlo Promis, which, at the time, was a true compendium of defensive patterns adopted by the officers in artillery and military engineering. Between 1834 and 1842 Menabrea designed the extension of the new S. Antonio barracks and the buildings of the stable and the riding ground for the Accademia delle scienze of Turin, with a framing of the deck endowed with curved wooden trestles in rib, anchored to a wooden higher structure with iron stirrups. Curved wooden bows had already been used in bridges at the end of 18th century, but the structure above was inspired by a salon in the Louvre. In April, 1840, Menabrea designed a new kind of trestle, by reinforcing through underlying elements the Palladio truss.

He took part, as a lieutenant general of the corp of military engineers, in the campaigns of Lombardy (1859) and in the siege of the fortress of Gaeta (1860). On October 3rd 1860, he was given the honor of Great Officer of the Italian Military Order. From 1846 to 1860 he taught Constructions at the university of Turin. In 1848 he became member of the Parliament of the Kingdom of Sardinia, and was senator for thirty six consecutive years.

He was minister of the Navy in the cabinet Ricasoli (1861–62), and of Public works in the cabinet Farini-Minghetti (1862–64). From October 27th, 1867, to December 14th, 1869, he succeeded Urbano Rattazzi as a prime minister, heading three subsequent cabinets. In such a position, he found himself contrasting Giuseppe Garibaldi's attempts of conquering Rome, and, searching to achieve a balanced budget, he let the unpopular tax on flour be approved. He did not hesitate in inducing the Senate to confer extraordinary powers to general Raffaele Cadorna to suppress the revolts which agitated the whole country. Once he left government duties, he was appointed ambassador in London and then in Paris. He retired from public life only in 1892.

In spite of his political and military tasks, Menabrea had an intense scientific activity, and was a fellow of the Reale accademia delle scienze of Turin and of the Accademia dei Lincei. He was a forerunner of the introduction of energetic principles in continuum mechanics by his *Nouveau principe sur la distribution des tensions dans les systémes élastiques* of 1858 (hereinafter *Nouveau principe*), according to

[16] The news on Menabrea as an architect-engineer are taken in part from [48].

which the solution of the elastic problem is obtained by searching the minimum of the elastic work, when stress equilibrated with the external loads vary.

In 1840, Charles Babbage took part in the second congress of the Italian scientists in Turin, after an invitation by Plana, and presented his project of a calculating machine. Mossotti and Menabrea attended, with a lot of interest, his seminars, in which for the first time operations chain, that is programming, was discussed. Menabrea described Babbage's project in a paper that may be perhaps considered as the first scientific publication in the fields of informatics, that is *Notions sur la machine analytique de Charles Babbage*, published in 1842 in French [64], and translated into English some time later.

There is a strong connection between Menabrea's professional activities between 1834 and 1842 and his *Nouveau principe*: he also would say in his writings that such a principle had represented the core of his activities in designing and teaching. His approach to architecture, thus, contained *in nuce* a scientific conception, and his lectures on the principle of elasticity in constructions are in close relation with the design of a trestle for the reinforcing of the deck of the riding ground, by which he covered the remarkable length of 22 m [48].

In the following we quote Menabrea's works on mechanics:

- 1835. *Luigi Federico Menabrea da Ciamberì ingegnere idraulico e architetto civile luogotenente del genio militare per essere aggregato al Collegio amplissimo di filosofia e belle arti classe di matematica nella Regia università di Torino l'anno 1835 addì 10 dicembre alle ore 8 1/2 di mattina. Data a altri dopo il sesto la facoltà di argomentare*, Turin, Reale Tipografia.
- 1840. *Mouvement d'un pendule composé lorsqu'on tient compte du rayon du cylindre qui lui sert d'axe, de celui du coussinet sur lequel il repose ainsi que du frottement qui s'y développe*, Memorie della Reale Accademia delle scienze di Torino, s. 2, vol. 2, pp. 369–378 (read on March 3, 1839).
- 1855. *Études sur la théorie des vibrations*, Memorie della Reale accademia delle scienze di Torino, s. 2, t. 15, pp. 205–329 (read on June 12, 1853).
- 1858. *Nouveau principe sur la distribution des tensions dans les systémes élastiques* [65].
- 1864. *Note sur l'effet du choc de l'eau dans les conduites*, Memorie della Reale accademia delle scienze di Torino, s. 2, t. 21, pp. 1–10 (read on March 7, 1858).
- 1868. *Étude de statique physique. Principe général pour déterminer les pressions et les tensions dans un systéme élastique* [66].
- 1870. *Sul principio di elasticità. Dilucidazioni* (with remarks by A. Parodi, G. Barsotti, Bertrand, Y. Villarceau) [68].
- 1874. *Lettera all'Accademia delle scienze di Torino per una correzione da apportare al Principe général del 1868* [67].
- 1875. *Sulla determinazione delle tensioni e delle pressioni ne' sistemi elastici* [69].
- 1875. *Lettera al presidente dell'Accademia dei Lincei 27 marzo 1875* [70].
- 1884. *Concordances de quelques méthodes générales pour déterminer les tensions dans un systéme des points réunis par des liens élastiques et sollicités par des forces extérieures en équilibre* [71].
- 1971. *Memorie* [72].

We now report a commented synopsis of the most important papers by Menabrea on the *principle of elasticity*, that is the already quoted *Nouveau principe* of 1858, *Étude de statique physique* of 1868, and *Sulla determinazione delle tensioni e delle pressioni ne' sistemi elastici* of 1875.

4.3.1 1858. Nouveau principe sur la distribution des tensions

In this paper and in the following Menabrea considered a model made up of hinged elastic bars undergoing very small displacements as representative enough of an elastic body, thus reducing to a linear system. This procedure was not new: for instance, it was the one followed by Maxwell in his papers on elasticity. The motivations of the choice of this model were the simplicity from the points of view of both conceptual and mathematical treatment: algebraic equations, instead of the differential ones that would have arisen in continuous systems, were used. In addition, scholars were evidently convinced of the fact that the obtained results did not hold only for systems constituted by bars, like the steel trusses of the industrial constructions of the 19th century, but could also be extended,with minor changes, to more complex elastic systems.

Menabrea knew well, from the statics of rigid bodies, that equilibrium equations only are not sufficient to uniquely solve a system with n nodes and m bars, when $m > 3n-6$ (such a system is called redundant or hyperstatic nowadays). An indeterminacy remains:

> The number of equilibrium equations for the n points will be $3n$; if p is the number of the equations that must hold among the external forces, independent of stresses, so that one has equilibrium, the number of equations actually containing stresses will be reduced to $3n - p$. Thus, when m will be $> 3n - p$, the preceding equations will not be enough to determine all stresses.
>
> The same will happen when the system contains a certain number of fixed points. This indetermination means that there are infinite values of stresses that, combined with the given external forces, are suitable for keeping the system equilibrated. The actual values of stresses depend on the respective elasticity of the links, and, once this is determined, the same must be of the stresses [65].[17] (A.4.6)

The last sentence of this passage contains the key for the solution of the problem: it is necessary to account for the strain and the mechanical properties of the considered system, that provide the additional equations suitable for making the problem determinate. This result had been reached, among the first ones, by Navier in his paper of 1825 on two-span continuous beams [77].

Menabrea showed that the equations to be added to those of statics in order to solve the elastic problem are obtained using the *équation d'élasticité*, according to which:

---

[17] p. 1057. Our translation.

When an elastic system is equilibrated under the action of external forces, the work spent by the tensions, or compressions, of the links joining the various points of the system is a minimum [65].[18] (A.4.7)

Because of its shortness, we report Menabrea's proof in full.

Since in the case we are considering the tensions may vary and the equilibrium still holds, we must admit that these variations happen independent of all work of external forces; they are always accompanied by extensions or shortenings in the various corresponding links, which originates, in each of them, a work spent. The variations of length of the links shall be supposed very small, in order that the relative positions of the various points of the system are not sensibly altered. However, since during this small internal movement the equilibrium still holds, and the work of the external forces is nil, it follows that the total elementary work of the tensions so developed is also equally nil.

To express this conclusion, let T be the tension of any link, δl the elementary variation of the length of this link; the work spent because of the variation of the corresponding tension will be $T\delta l$, and, consequently, for the whole of the system, we will have

$$\sum T\delta l = 0. \tag{4.1}$$

Let l be the extension or the shortening that the link underwent formerly under the action of the tension T, we have, independent of sign,

$$T = \epsilon l, \tag{4.2}$$

where ϵ is a coefficient which I will call coefficient of elasticity, that is a function of the elasticity modulus, of the section, and of the length of the link.

The work spent to produce this variation of length l will be equal to $1/2\epsilon l^2$, and it follows that the total work of the system will be equal to $1/2\sum \epsilon l^2$.

However, by virtue of Eqs. (4.5) and (4.6) we have:

$$\sum T\delta l = \sum \epsilon l \delta l = \delta \frac{1}{2} \sum \epsilon l^2 = 0. \tag{4.3}$$

This is the proof of the stated principle, to which we may arrive also by other considerations. It is equally possible to express it in another way, because we have [65].[19] (A.4.8)

$$\sum T\delta l = \sum \frac{1}{\epsilon} T\delta T = \delta \frac{1}{2} \sum \frac{1}{\epsilon} T^2. \tag{4.4}$$

4.3.1.1 Analysis of the Proof

The proof is based on the observation that the internal forces T of the bars may vary in infinite ways without altering equilibrium; if the variation of T is infinitesimal,

[18] p. 1056. Our translation.

[19] pp. 1057–1058. Our translation.

such is also the variation δl of the length of the bars, and thus the variation of the position of the nodes is negligible. This reasoning was reiterated almost twenty years after, in the wake of proofs that were alternative to that referred to above:

> Given one of those equilibrium positions, if we suppose that the system gradually passes to another one, very near, the set of external forces (X, Y, Z) shall not stop being equilibrated for any of these dispositions, independent of the internal forces; and, since this equilibrium state does not depend only on the magnitude and direction of these forces, but also on the position of the points where they are applied, it follows that each node shall constantly stay in the same position, notwithstanding the variation that may happen in the tensions of the links corresponding to them [69].[20] (A.4.9)

Starting from this premise, Menabrea implicitly relied on the principle of virtual work: the sum of the work of internal ($L^i = \sum T\delta l$) and external ($L^e = \sum f\delta u$) forces shall vanish in a compatible variation of the configuration. If the displacements of the nodes are considered negligible, it is $L^e = 0$; since it must be $L^i + L^e = 0$ by the principle of virtual work, it is also $L^i = 0$ [Eq. (4.1); $\sum T\delta l = 0$]. Now the proof is substantially complete.

A modern reader contests the fact that the displacements of the nodes are negligible with respect to the variations of length of the bars, in that it is easy to verify, also on the basis of relations that Menabrea would write in following papers, that the displacements of the nodes and the stretches of the bars are of the same order of magnitude. There are other weak points in the proof: Eq. (4.1) of the above quotation, according to Menabrea, derives from the principle of virtual work, thus the displacements δl shall be compatible with the constraints. On the other hand, the minimum of the internal forces in Eqs. (4.3) and (4.4) is looked for by varying the internal forces in the range of those equilibrated with the external ones, hence the δl so obtained are, in general, not compatible. A minor difficulty, at least for the modern reader, is the statement according to which the work of internal forces is a minimum, and not a generic stationary point. Menabrea declared that "it is clear" that the work of inner forces is not a maximum, while he said nothing on the possibility that it could simply be a saddle point.[21] Another problem for a reader used to a high standard of rigor is whether or not the minimum of the work on deformations always provides the solution of the elastic problem; in other words, Menabrea claimed to prove only that, if a system is equilibrated and kinematically compatible, then the work on deformations is a minimum, but did not prove the opposite.

Such 'errors' were similar to others that occurred in the investigation of redundant systems, before and after Menabrea, due to a misuse in part of infinitesimals, in part of the principle of virtual work. For instance, Dorna committed errors of the first kind [47], while Cournot committed errors of the second kind [38]; these 'errors' shall, however, be framed in a context where the ideas used were not so well defined like nowadays. They were, however, errors of sequential thinking, and not of points

[20] p. 213. Our Translation.

[21] As a matter of fact, such a statement of Menabrea was not strange for his times: it was a proposition of a scholar who was not a perfect master of calculus, the notions of which had not yet been rigorously organized and diffused.

of view different from the modern ones; more careful scholars, like Bertrand and Castigliano, even though of the same time as Menabrea, did not commit them.

The various proofs, more or less inconsistent, of a principle believed as true are another demonstration of the fact that, when one thinks that a fact is true, the mathematical proof turns into a relatively secondary matter, and one may be happy with rhetorical artifices, more or less satisfactory; the certainty over the result reduces the critical capacities of the scholar: if a proof is wanted, it can in any case be found.

4.3.1.2 Immediate Criticisms to the Paper of 1858

The weakness of the proof in the paper of 1858 was immediately recorded; Menabrea himself, in a letter of 1870 to the president of the Reale accademia delle scienze of Turin reported some of the objection he received.

> It seems that my work was generally welcomed with favor by the scientists that had dealt more with that subject, and they did not doubt about the exactness of the method I proposed, except for Mr. Emilio SABBIA who, in a booklet entitled: Error of the principle of elasticity formulated by Mr. L. Federigo MENABREA, Critical hint by Emilio SABBIA, Turin 1869, contests, with a remarkable vividness, the truth of that principle [...].
>
> Going through the paper by Mr. SABBIA I believed to notice the misunderstanding in which he had incurred; and I would have not been late in answering his criticism, if other occupations, much more serious, had not held me back. Once given back to greater freedom, I was preparing to the task, when I was given notice of a writing by the talented lover of mathematical sciences, Mr. Comm. ADOLFO PARODI, General inspector of marine works, that has precisely the booklet by Mr. SABBIA as object. He talks back so neatly the points of Mr. SABBIA that I would not know better how to defend my theorem, than to make use of the same considerations brought by the illustrious author.
>
> [...] It will also not be disagreeable for the Academy to have an eye on two new proofs of the equation of elasticity, one given by Mr. BERTRAND and the other by Mr. YVON VILLARCEAU, both Fellows of the Institute of France, who, in the valuable letters of which I communicate the excerpts, present the question under points of view I would say new, and that lead to the same results [...].
>
> [...My proof] was judged, as it will be seen from the writings attached here, rigorous enough, and having, at least, the quality of simplicity and clarity [68].[22] (A.4.10)

Menabrea never admitted any doubt explicitly, and in his later works, that substantially perfected the proof, he was discreet, but firm, in defending his writing of 1858. For instance, here is what he wrote in 1875:

> Even though the coincidence of the results obtained by applying the principle of elasticity with those deduced by other special and uncontested methods was confirmed in my second memoir [that of 1865] by several examples, and should induce to admit that the principle and the method deriving from it were exact, in spite of this, both one and the other were object of harsh and strange denials by some, while many among the most eminent scientists of our time welcomed the principle with most benevolence. Despite the objections made, the applications of the principle of elasticity propagated and have increasingly confirmed the exactness, the simplicity, and the generality of the method deriving from it. Since this

[22] pp. 687–688. Our translation.

[principle] substantially contains in itself all the other ones, I believe it useful to try to eliminate, about the accuracy of the principle itself, any doubt that could still remain in the most scrupulous minds from the point of view of mathematical rigor [69].[23](A.4.11)

Menabrea behaved like the skilled politician he was: he avoided entering the scientific merit of the criticisms, and, on the contrary, he presented them as a support of his point of view, strengthened by the new proof of 1865, published in 1868. At least in part, this came from his desire to have a priority in the authorship of the proof, in particular toward Castigliano, with whom he had a quite vivacious dispute.

The strongest criticisms toward Menabrea, as he put into evidence in his letter of 1870 quoted above, came from Emilio Francesco Sabbia, who wrote many times on the principle of elasticity. We have not been able to search out a copy of his memoir of 1869, quoted by Menabrea [90]; we have, on the other hand, a copy of a memoir of 1870 [91], published after the indirect reply by Menabrea in the memoir *Sul principio di elasticità*, and a copy of a printed letter [89]. On the basis of these documents we can reconstruct Sabbia's critical arguments, that were well-founded according to us. The main objections were essentially two: the first was about the range of applicability of the principle, the second, about its formulation and its proof.

By adopting a modern terminology, in his first objection Sabbia claimed that the principle of minimum work is valid only in the absence of distortions[24] and of residual stresses; in his second objection, Sabbia claimed that Menabrea had confounded the elastic potential energy and the elastic complementary energy (modern meaning). According to Sabbia, the correct statement of the principle of minimum work is:

When an elastic system, *liable to a general neutral state* [emphasis added], finds itself equilibrated with external forces, among the various ways in which the tensions could be imagined distributed among the various links so as to equilibrate those forces, the way in which they are actually distributed satisfies the condition that the total work spent by the internal forces is a *minimum* [91].[25](A.4.12)

while Menabrea's definitions and proofs suggest the incorrect formulation

In any elastic system that came to equilibrium under the action of external forces, among the various positions that the mobile points could have taken, those that they actually took satisfy the condition that the total work spent by the internal forces in their relative displacements is minimum [91].[26](A.4.13)

Likely enough, this interpretation was somehow unnatural; Menabrea would not have recognized as his the principle of elasticity as reformulated by Sabbia (in modern terms a principle of minimum of the potential energy, unduly neglecting the potential energy of external forces). Much more likely, Menabrea did not clearly distinguish between variations of forces and of displacements. However, in his applications, as

---

[23] p. 203. Our translation.

[24] The term *distortion* would have been introduced only some years later, as we have seen, by Volterra.

[25] p. 3. Our translation.

[26] p. 6. Our translation.

Sabbia himself recognized, Menabrea used his principle without ambiguities, letting only forces vary.

Menabrea did not answer directly to Sabbia, who was simply a lieutenant, but entrusted his answer to Alfredo Parodi in an open letter [68].[27] In this letter, Parodi showed that he had not understood Sabbia's arguments and defended Menabrea's thesis, even though he admitted some ambiguities in the writings of the latter.

Yvon Villarceau and Joseph Louis François Bertrand in two letters to Menabrea, reported in [68],[28] advanced some veiled criticisms, suggesting refinements of the proof. Villarceau operated in a dynamical setting, by applying the equation of conservation of living forces:

$$L^a + L^i = \delta T \tag{4.5}$$

where L^a is the actual work of active forces, L^i the actual work of inner forces, δT the kinetic energy gained during the deformation. Since in this process the variation of kinetic energy is negligible, being an infinitesimal of higher order with respect to works, one is reduced to the equation of virtual work:

$$L^a + L^i = 0 \tag{4.6}$$

and Villarceau, thus, recovered Menabrea's arguments:

> Now, *if* [in italic in the original] we imagine that the work L^a remains constant [...], in spite of the possible variation of the work of the forces f, we will also have:
>
> $$L^a + L^i + \delta L^i = 0$$
>
> whence [68][29]
>
> $$\delta L^i = \sum f \delta \Delta \rho = 0.$$
>
> (A.4.14)

where f represents the inner forces and $\Delta\rho$ represents the compatible virtual displacements.

Villarceau provided correct arguments, but made a mistake when he thought that his result coincided with Menabrea's: indeed, he operated the variation of the work of inner forces ambiguously, without specifying if the variations referred to the inner forces or to the virtual displacements. To be correct, the variation should have operated on forces, hence the preceding equation in the quotation should have been re-written as:

$$\delta L^i = \sum \delta f \Delta \rho = 0 \tag{4.7}$$

[27] pp. 690–696.

[28] pp. 702–705.

[29] p. 705. Our translation.

which differs from that by Menabrea.

On the other hand, the content of the kind letter by Bertrand to Menabrea did not lead to misunderstandings. Because of its brevity, we report it in full:

> Following the proof and translating into ordinary language the consequences of the equation [...] [the principle of elasticity], we are led to the following statement, which does not yield any ambiguity.
>
> *The sum of the squares of the tensions, divided, respectively, by the coefficient of elasticity of the corresponding link is a minimum*; that is to say, his sum is less than for any other system of tensions capable of assuring equilibrium, once we neglect the conditions relative to the extensibility of the links.
>
> Let me, Sir, submit, in second place, a very simple proof of your equation [...].
>
> Let l be the length of one of the links, λ its elongation in the equilibrium configuration, T its tension, equal to $\epsilon\lambda$, $T + \delta T$ the tension of the same link for another solution of the equilibrium equations, if the links are supposed inextensible; the forces δT, if they were alone, would be self-equilibrated on the system, since the forces T and the forces $T + \delta T$ are, by hypothesis, equilibrated with the same external forces (the system is that in which the extensible links have disappeared). The sum of the virtual moments [that is works] of the forces δT is thus nil for all the displacements compatible with constraints other than the non-extensibility of the links. However, one of the displacements is that actually occurring, and for which the link l stretches by λ equal to T/ϵ, we have, consequently,
>
> $$\sum \frac{T\delta T}{\epsilon} = 0$$
>
> This is, precisely, equation [...], of which the principle of elasticity is the immediate translation [68].[30] (A.4.15)

Bertrand used the principle of virtual work (the virtual work of all forces vanishes at equilibrium), but, differently from Menabrea and Villarceau, made it clear that the variation of virtual work should be performed by considering fixed displacements and 'virtual' forces varying between T and $T + \delta T$. Since the virtual works of the forces T and $T + \delta T$ equal that of the external forces, which are implicitly assumed independent of the configuration, they have the same value, and thus, it follows that the work of the forces δT is nil. Bertrandas well, however, left the problem hanging, if the minimum of the work on strain always provides the solution of the elastic problem.

4.3.1.3 The Origins of Menabrea's Equation of Elasticity

Menabrea himself, in many occasions, tried to frame his 'equation' in the literature of his time. He reported what he considered the genesis of his principle in the clearest way perhaps in his paper of 1875:

> I did not, in the various occasions I recalled before, fail to expose the genesis of that theory, that originated, to my knowledge, in a memoir by Mr. *Vène*, higher official of the French corps of military engineers, who, since 1818, and then in 1836 (*Mémoire sur les lois que suivent les*

[30] pp. 702–703. Our translation.

> *pressions*) stated the following theorem for the special case of pressures exerted by weights on homogeneous footholds: *The sum of the squares of the weights shall be a minimum.* A hint was given on this new principle in the *Bulletin des Sciences Mathématiques de Ferussac*, tome ninth pag. 7 in a paper signed S. In another paper, following the preceding one in the same tome, pag. 10, signed A. C., the above said principle was extended to the case of non-homogeneous footholds, and to the case of pressures produced on the footholds by rigid bars. The Author of that paper, A. C., was supposed to be Augustin Cauchy; but, later on, it was more likely attributed to Mr. *A. Cournot.*—*Pagani* dealt with the special case of elastic cords, respectively fixed at one of their ends and joint at the other in a node where a force was applied. Mossotti in his *Meccanica* dealt with the aforementioned subjects [69].[31] (A.4.16)

The genesis suggested by Menabrea is carefully followed by Benvenuto [4],[32] [3, 76]. In what follows we investigate in some detail Cournot's argumentation, that seems to us fundamental for Menabrea's work, and was not thoroughly commented in [4]. We provide a hint also on Dorna's work.

Cournot considered a rigid body supported on various points by deformable supports [38]. If active forces are exerted upon the body, the contact points will exchange *pressures* (pressions). Cournot dwelled on this account, which evidently he did not consider as a standard one, even though more or less all the scholars of the problem of the body on many supports give the term pressure the meaning of 'concentrated contact force':

> These pressures [...] are quantities stranger to the forces by which they are generated [...].
>
> The determination of pressures must be considered as another branch of dynamics, that is of the science of the effects of forces; a branch that could assume the name of latent dynamics [...].
>
> If we deal with a system having various points constrained by fixed obstacles, each obstacle will undergo a pressure proportional to the infinitely small straight line that the corresponding point would describe during the elementary time [38].[33] (A.4.17)

Cournot's statement, that pressures are proportional to infinitesimal displacements, gives rise to some problems of interpretation. At a first sight, it would seem as it were the statement of a law of statics that does not depend on the linear elasticity of the support. Indeed, by adopting an ordinary linear elastic law, infinitesimal displacements would imply infinitesimal forces, but this is not what Cournot meant.

As a matter of fact, a careful reading of the paper makes it clear that actually Cournot postulated a law of elasticity, and that "the coefficient of elasticity" may vary "from a [point] to the other" [38].[34] Thus, the proportionality between force and displacement expressed a constitutive relation, rather than a new law of statics. In the end, Cournot admitted that pressures could be considered as forces, even though they are not forces in a strict sense. Thus, Cournot's attempt to define what we now call contact actions as heterogeneous to forces by means of a principle was indeed inconclusive, and truly reveals Cournot's embarrassment in accepting, at a metaphysical level, the idea of contact forces and constraint reactions, that at his time was accepted by the majority of the scholars in mechanics.

---

[31] p. 202. Our translation.

[33] pp. 11–12. Our translation.

[34] p. 18. Our translation.

These pressures, taken in the opposite sense, might be considered as forces applied to the system that keep it in equilibrium, once made abstraction of the obstacles [38].[35] (A.4.18)

Cournot applied the principle of virtual work to a rigid motion of the body subjected to the active forces F, F', ... and to the constraint reactions p, p', ..., opposite to the pressures. At the equilibrium, the sum of the virtual works of the active and reactive forces vanishes:

$$F\delta f + F'\delta f' + \cdots - (P\delta p + P'\delta p' + \cdots) = 0$$

formula that will give the equilibrium relations, after we will have reduced to the least possible number the independent variations, keeping into account the system own constraints, but not those resulting from the presence of the obstacles, now replaced by the forces P, P'[...] [38].[36] (A.4.19)

Cournot considered the virtual rigid displacements of a free body, because the constraints were replaced by the constraint reactions, and made an important assumtion, that let him reach a statement he considered satisfactory:

If we consider the presence of these obstacles in order to reduce the number of variations, it simply yields:

$$F\ \delta f\ + F'\ \delta f'\ + \cdots = 0,$$

hence also, in the same case:

$$P\ \delta p\ + P'\ \delta p'\ + \cdots = 0,$$

which immediately results from the fact that the two systems (F) and (P) are equivalent [38].[37] (A.4.20)

This last statement looks unintelligible in itself, and Cournot provided no further explanations. A justification of the passage could maybe be seen in what Mossotti wrote in a teaching book of his [74].[38] According to Mossotti, since the equality of works holds true for any virtual displacement, it is possible to choose one implying the vanishing of external forces, in such a way that the last two relations of the excerpt quoted above hold. In any case, this argument also seems inconsistent. On the other hand, it is interesting to compare Cournot's statement with that by Menabrea contained in his memoir of 1858, already quoted: "The variations of length of the links shall be supposed very small, in order that the relative positions of the various points of the system are not sensibly altered. However, since during this small internal movement the equilibrium still holds, and the work of the external forces is nil, it follows that the total elementary work of the tensions so developed is also equally nil". Menabrea provided an explanation, be it satisfactory or not, on why the virtual work

---

[35] p. 13.

[36] p. 18. Our translation.

[37] p. 18. Our translation.

[38] pp. 97–98.

of the external forces vanishes: that occurred as a consequence of the fact that the displacements of the points of their application are negligible.

Once admitted that the virtual work of external forces is nil, Cournot's following passages do not exhibit any difficulty. From the vanishing of the virtual work of all forces, and of the external forces, the vanishing of the virtual work of the internal forces follows, and, since pressures are proportional to the displacements, this yields:

> $$p\,\delta p + p'\,\delta p' + \cdots = 0,$$
>
> a relation by virtue of which [it results] that the sum of the quantities p^2, p'^2, etc., or, by hypothesis, that of the squares of the pressures P^2, P'^2, etc. is a *minimum*; since it is easy to be certain that the case of a *maximum* cannot take place here [38].[39] (A.4.21)

Cournot was so able to state the following *general theorem* (théoréme générale):

> As a consequence, the equations completing in all cases the number of those which are necessary for the full determination of the pressures result from the condition that the sum of the squares of these pressures be a minimum [38].[40] (A.4.22)

It is clear that Menabrea was deeply indebted with Cournot; the statement and the proof of the 'equation of elasticity' by Menabrea and of the 'general theorem' by Cournot are the same. There was, however, a step forward moved by Menabrea, who stated his equation of elasticity considering a general elastic system, and not only one or more bodies in rigid motion, connected among them and resting on a 'soil', disposing with the question as unreal, and thus without practical interest.

It is also worth considering the work of Dorna [47], a Menabrea's colleague, to show how an uncontrolled and not precise use of infinitesimals could induce errors. In his work, Dornaconsidered an elastic structure, just like Menabrea, thus a more general problem than that investigated by Cournot. Dorna wrote the equation of virtual work to characterize equilibrium, in the form $L^i + L^e + L^\nu = 0$, where L^i is the work of internal forces, L^e that of the external forces, and L^ν is that of the constraint reactions. Dorna stated that, since the virtual displacements of the constraints, made up of very rigid springs, are infinitesimals of higher order with respect to those determining L^i and L^e, L^ν may be neglected, thus one had $L^i + L^e = 0$; from $L^i + L^e + L^\nu = 0$ one then had $L^\nu = 0$. The argument contained two debatable statements: the first is that the work of constraint reactions be a higher order infinitesimal, the second is that, as a consequence of this and of the equality of works, that the work of constraint reactions be nil. The first statement is unconvincing because if the support is moving there is, in general, no reason for the displacements of the points to be infinitesimal. The second is a paralogism, since if in the sum $L^i + L^e + L^\nu = 0$, we insert the hypothesis that L^ν be a higher order infinitesimal, it correctly follows that $L^i + L^e \approx 0$, not that L^ν be nil.

---

[39] p. 18. Our translation.

[40] p. 18. Our translation.

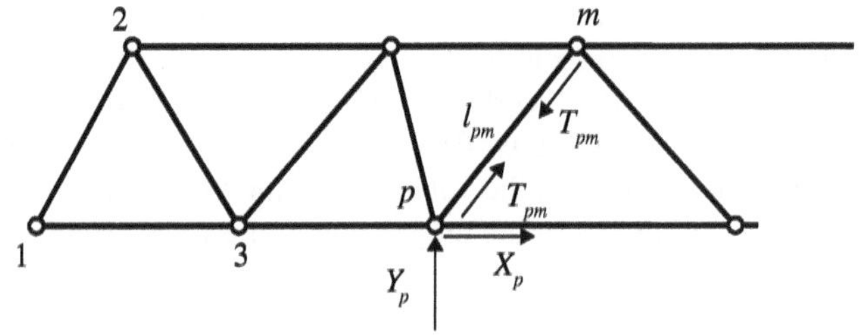

Fig. 4.1 The generic truss considered by Menabrea

4.3.2 1868. Étude de statique physique

Menabrea presented here a new proof of the equation of elasticity, starting with a rather tactful reference to his work of 1858:

> Since 1857 I let the Accademia delle scienze in Turin know the statement of this new principle; afterwards, in 1858 (meeting of May 31st) it was the object of a communication to the Institut de France (Académie des sciences). In the proof I gave I relied on considerations of the transmission of work in the body. Even though, according to me, that proof was rigorous enough, it looked too subtle to some geometers in order to be accepted without contestation. On the other hand, the implications of the equations deduced from this theorem were not sufficiently indicated. This is why I thought it better to resume this investigation, that was interrupted many times because of events which my position called me to take part in.[41] Nowadays I present these new investigations, that have had the result to lead me to a proof simple and rigorous at all [...] [66].[42] (A.4.23)

and then providing some hints to a proof based on physical arguments, adopting concepts of thermodynamics:

> To provide the question of the stress distribution all the generality it has from the physical point of view, we must account for the phenomena of *thermodynamics* that appear during the changes of shape of the body, or of the elastic system; however, I consider the body at the instant when equilibrium has established among *internal* and external forces, supposing that temperature did not vary. Then, we may admit that the work spent is collected in that concentrated at the *latent state* in the elastic system as an effect of external forces [66].[43] (A.4.24)

Menabrea's model was again a set of hinged elastic bars like that in Fig. 4.1, undergoing small displacements. To start with, he considered a system without external constraints. The first step of the new proof consisted in writing equilibrium equations at each hinge (*node*) *p* [66][44]:

---

[41] As we said above, from 1862 to 1869 Menabrea had important government responsibilities in a period of difficulties in Italy, and, in particular, from 1867 to 1869 he was prime minister.

[42] p. 8. Our translation.

[43] p. 9. Our translation.

[44] Equation (1), p. 29.

$$X_p = \sum T_{pm} \frac{x_m - x_p}{l_{pm}}; \quad Y_p = \sum T_{pm} \frac{y_m - y_p}{l_{pm}}; \quad Z_p = \sum T_{pm} \frac{z_m - z_p}{l_{pm}}. \quad (4.8)$$

Here X_p, Y_p, Z_p are the components of the external force applied at the node p, x, y, z are the coordinates of the nodes, l_{pm} and T_{pm} are the length and the elastic force, respectively, of the bar joining the nodes m and p; the sum ranges on the index m. The reference configuration is the present, deformed one, but for small displacements it coincides with the reference one, as it seems implicitly admitted by Menabrea.

If the nodes of the system of bars are n, the Eqs. (4.9) are $3n$. Since the system was supposed free, the 6 equilibrium equations of statics among the external forces shall hold for equilibrium to take place. This implies that only $n - 6$ of Eqs. (4.8) are independent. If the number N of the bars is such that $N > 3n - 6$, the system is redundant, and "we may conceive infinite ways of distribution of these tensions, that are all able to satisfy equilibrium conditions with external forces" [66].[45]

Menabrea considered infinitesimal variations δT_{pq} of the forces in the bars, such that the forces $T_{pq} + \delta T_{pq}$ be still equilibrated with external forces, and, thus, the forces δT_{pq} be self-equilibrated. If the δT_{pq} are infinitesimal, Menabrea said, the configuration of the system does not change. By variation of (4.8) one had the equilibrium equations for the δT_{pq}, self-equilibrated, for each node p [66][46]:

$$0 = \sum \delta T_{pq} \frac{x_q - x_p}{l_{pq}}; \quad 0 = \sum \delta T_{pq} \frac{y_q - y_p}{l_{pq}}; \quad 0 = \sum \delta T_{pq} \frac{z_q - z_p}{l_{pq}}, \quad (4.9)$$

where the sum ranges on the index q. In the (4.9) Menabrea inserted the linear elastic relation $T_{pq} = \lambda_{pq}\epsilon_{pq}$, where λ_{pq} and $\epsilon_{pq} = (E\omega/l)_{pq}$ are the absolute variation of the length and the stiffness, respectively, of the bar pq, with Menabrea's symbols (E is Young's modulus, and ω is the area of the cross section of the bar).

Menabrea linked the displacements of the nodes α, β, γ to the elongation λ_{pq} of the bars [66][47]:

$$\lambda_{pq} = \frac{(\alpha_q - \alpha_p)(x_q - x_p)}{l_{pq}} + \frac{(\beta_q - \beta_p)(y_q - y_p)}{l_{pq}} + \frac{(\gamma_q - \gamma_p)(z_q - z_p)}{l_{pq}}. \quad (4.10)$$

Such a relation would be written also nowadays, dealing with infinitesimal strain; at that time, we may assume that it was known.[48] Multiplying the expression of λ_{pq} by $\epsilon_{pq}\delta\lambda_{pq}$, and summing over both indices, Menabrea obtained [66][49]:

---

[45] p. 31.

[46] Equation (4), p. 32.

[47] Equation (7), p. 32.

[48] Indeed, Eq. (4.10) can be written in the form $\delta l = [(\mathrm{grad}\mathbf{u})\,\mathbf{n}]\cdot\mathbf{n}$, with $\mathbf{u}$ the displacement vector, and $\mathbf{n}$ the unit vector along the bar pq. A modern definition of strain was due to Saint Venant; Menabrea did not quote him, but he could not ignore him, for sure.

[49] Equation (8), p. 32.

$$\sum \epsilon_{pq}\lambda_{pq}\delta\lambda_{pq} = \sum \epsilon_{pq}\delta\lambda_{pq} \left\{ \begin{array}{l} \dfrac{(\alpha_q - \alpha_p)(x_q - x_p)}{l_{pq}} + \dfrac{(\beta_q - \beta_p)(y_q - y_p)}{l_{pq}} \\ + \dfrac{(\gamma_q - \gamma_p)(z_q - z_p)}{l_{pq}} \end{array} \right\}. \tag{4.11}$$

Afterwards, Menabrea multiplied all the equilibrium Eqs. (4.9) relative to the node p by $\alpha_p, \beta_p, \gamma_p$, and summed for all nodes. Since on the node p there is the term $(x_q - x_p)$, on the node q there will be the term $(x_q - x_p) = -(x_p - x_q)$, and Menabrea obtained [66][50]:

$$\sum \epsilon_{pq}\delta\lambda_{pq} \left\{ \frac{(\alpha_q - \alpha_p)(x_q - x_p)}{l_{pq}} + \frac{(\beta_q - \beta_p)(y_q - y_p)}{l_{pq}} + \frac{(\gamma_q - \gamma_p)(z_q - z_p)}{l_{pq}} \right\} = 0. \tag{4.12}$$

where the sum is over both p and q. This relation, keeping into account the preceding one, provided [66][51]:

> $$\sum \epsilon_{pq}\lambda_{pq}\delta\lambda_{pq} = \sum \frac{1}{\epsilon_{pq}} T_{pq}\delta T_{pq} = 0.$$
>
> that is the *equation of elasticity*, from which we deduce the theorem that we have stated at the beginning of this Memoir: *When an elastic system is in equilibrium under the action of external forces, the internal work spent on the change of shape deriving from it, is a minimum* [66].[52] (A.4.25)

Menabrea's new proof was satisfactory enough, also for modern standards; maybe the main dissonance was the presumption of having proved that the work is a minimum; indeed, it may only be said that it is stationary. Other drawbacks lie in the absence of some details: for instance, the assumption of small displacements is not well explicitly stated; in addition, the same can be said of the fact that the extremum is attained with respect to the variation of internal forces, as long as they are equilibrated with external forces. And, maybe the most important detail missing, the statement symmetric to the above quoted one was not proved, that is if the elastic work is a minimum, then the strains of the bars are compatible. In other words, only the necessary condition for equilibrium, and not the sufficient, was proved.

After having proved the principle of elasticity for a free system, Menabrea dealt with the case of a constrained one. If the constraints are smooth and fixed, he found easily the result holding for the free system.

---

[50] Equation (9), p. 33.

[51] Equation (10), p. 33.

[52] p. 33. Our translation.

4.3.2.1 The 'Inductive' Proof of the Principle

Menabrea's proof was preceded by the solution of a series of particular cases of trusses, in each of which he verified the validity of the principle of elasticity by comparing its result with that provided by *ad hoc* techniques, the application of which was not problematic. It looks like Menabrea was interested more in the truth than in the certainty of his statement. This derived, for sure, from his engineering education, giving scarce importance to formal aspects. Menabrea's procedure may also be read as a defense of his paper of 1858: what matters is to establish the truth of a principle, if the proof is not rigorous enough it does not matter, since it will not be a difficult task for people of good will to find the right proof.

Menabrea considered: a plane system of three bars; a plane system of six bars along the sides and the diagonals of a parallelogram; a three-dimensional system of 16 bars forming a regular dodecahedron; a bar loaded by n forces along its axis. In the following we give a hint on the first example only, which, however, is emblematic: three bars hinged with the "ground" and concurring in a single node, as shown in Fig. 4.2.

Menabrea found the solution by a method that we would call of displacements. The problem is not statically determinate, since we have two scalar equilibrium equations at the node D but three unknowns, the forces T_1, T_2 and T_3 in the bars. Menabrea, then, introduced the constitutive relation, linking the forces in the bars with their elongations; in such a way he could replace the forces in the equilibrium equations by their expressions in terms of the elongations λ_1, λ_2, λ_3, thus obtaining two equations in λ_1, λ_2, λ_3. Eventually, he wrote a compatibility equation among the elongations of the bars, imposing that they concur in the same node after strain. In such a way, he obtained three independent equations in three unknowns, that let him find the elongations of the bars. With a modern language, we would say that Menabrea wrote equilibrium, compatibility, and constitutive equations to solve the problem. After this, he applied the principle of elasticity, explaining the way in which it should be applied, and obtained the same expressions for λ_1, λ_2, λ_3.

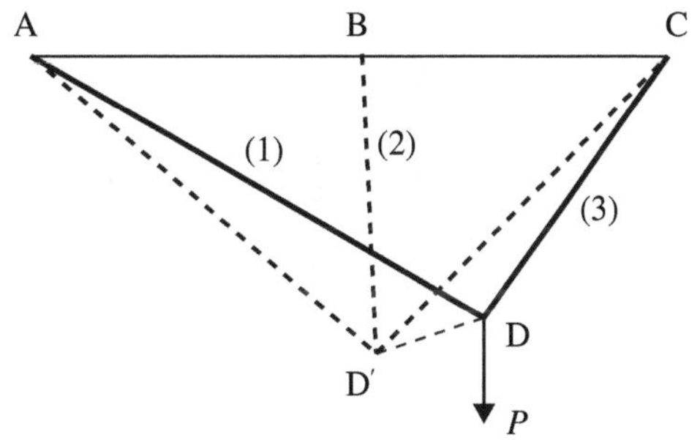

Fig. 4.2 Three hinged bars (Redrawn from [66, p.13])

4.3.3 *1875*. Sulla determinazione delle tensioni e delle pressioni ne' sistemi elastici

In this memoir the proof of the equation of elasticity, to which Menabrea referred also as the *principle of elasticity*, was developed inversely with respect to that of 1865. There Menabrea proved that the equations among the forces applied to nodes and the displacements of the latter, obtained by the method of displacements, implied the minimum of the strain work. Here he proved that such a minimum yields the same equations obtained by the method of displacements. Thus, the complete proof of his statement would be given by both arguments in the two papers. However, the difference between a necessary and a sufficient condition of equilibrium as defined by the minimum of the elastic work would not need to be highlighted, if a unique solution exists, which probably looked apparently natural to Menabrea as an engineer. Indeed, if a unique solution exists, proving that the minimum of the strain work is a necessary condition for equilibrium provides also a sufficient condition. Vice-versa, proving that a unique minimum of the strain work exists, that coincides with the solution of the elastic static problem (sufficient condition) also implies the necessity for equilibrium to exist; indeed, otherwise one could have equilibrium even in the case when the strain work is not a minimum, which is absurd because two different solutions of the elastic static problem would exist.

Menabrea's proof started again from the model of hinged bars: by kinematics, the elongations of the bars were determined as a function of the components $\alpha_m, \beta_m, \gamma_m$ of the (infinitesimal) displacement vector of the nodes [69][53]:

$$\lambda_{mn} = (\alpha_n - \alpha_m)\cos\phi_{mn} + (\beta_n - \beta_m)\cos\theta_{mn} + (\gamma_n - \gamma_m)\cos\psi_{mn}. \tag{4.13}$$

Here the angles $\phi_{mn}, \theta_{mn}, \psi_{mn}$ are those formed by the bar joining the nodes m, n with the coordinate axes, once the present and the reference configurations are assumed to coincide as a consequence of the hypothesis of infinitesimal displacements.

Menabrea then wrote the equilibrium equations for the nodes [69],[54] by expressing the forces in the bars as a function of their elongations, $T_{mn} = \epsilon_{mn}\lambda_{mn}$ [69][55]:

$$\begin{aligned} X_m &= \sum \epsilon_{mn}\lambda_{mn}\cos\phi_{mn} = \sum \epsilon_{mn}\lambda_{mn}\frac{x_n - x_m}{l_{mn}} \\ Y_m &= \sum \epsilon_{mn}\lambda_{mn}\cos\theta_{mn} = \sum \epsilon_{mn}\lambda_{mn}\frac{y_n - y_m}{l_{mn}} \\ X_m &= \sum \epsilon_{mn}\lambda_{mn}\cos\psi_{mn} = \sum \epsilon_{mn}\lambda_{mn}\frac{z_n - z_m}{l_{mn}}. \end{aligned} \tag{4.14}$$

[53] Equation (3), p. 205.
[54] Equation (6), p. 206.
[55] Equation (5), p. 206.

Keeping into account the Eq. (4.13), he came to the field equations in the displacement unknowns [69][56]:

$$
\begin{aligned}
X_m &= \sum \epsilon_{mn}[(\alpha_n - \alpha_m)\cos^2\phi_{nm} + (\beta_n - \beta_m)\cos\theta_{nm}\cos\phi_{nm} \\
&\qquad + (\gamma_n - \gamma_m)\cos\psi_{nm}\cos\phi_{nm}] \\
Y_m &= \sum \epsilon_{mn}[(\alpha_n - \alpha_m)\cos\phi_{nm}\cos\theta_{nm} + (\beta_n - \beta_m)\cos\theta^2_{nm} \\
&\qquad + (\gamma_n - \gamma_m)\cos\psi_{nm}\cos\theta_{nm}] \\
Z_m &= \sum \epsilon_{mn}[(\alpha_n - \alpha_m)\cos\phi_{nm}\cos\psi_{nm} + (\beta_n - \beta_m)\cos\theta_{nm}\cos\psi_{nm} \\
&\qquad + (\gamma_n - \gamma_m)\cos^2\psi_{nm}].
\end{aligned}
\tag{4.15}
$$

Menabrea then started a somehow tangled analysis of the number of equations and unknowns, which we will briefly summarize. Equations (4.15) constitute a system of $3p$ equations in which we find $3p$ unknowns, the components of the nodal displacement. These equations are not linearly independent, because the global equilibrium equations among the X_m, Y_m, Z_m hold. Thus, it would seem that there are more unknowns than independent equations in the considered system, but actually, since only the relative displacements are present in the Eq. (4.15), the solution is defined apart from a rigid motion; once defined the rigid motion, the solution becomes unique.

Obtained the solution of the elastic static problem by a standard, 'indubitable' technique, Menabrea considered the minimum of the work of inner forces, provided by [69][57]:

$$
\sum T_{mn}\delta\lambda_{mn} = \sum \epsilon_{mn}\lambda_{mn}\delta\lambda_{mn} = 0. \tag{4.16}
$$

where the meaning of the symbols is as usual, and the sum is extended over m and n.

The minimum of the work of inner forces is constrained, and should be searched by imposing that the tensions $T_{mn} = \epsilon_{mn}\lambda_{mn}$ be equilibrated, that is should satisfy the Eqs. (4.14), while the tensions $\delta T_{mn} = \epsilon_{mn}\delta\lambda_{mn}$ are self-equilibrated, that is should satisfy the variation of the Eqs. (4.14) [69][58]:

$$
\begin{aligned}
&\sum \epsilon_{mn}\cos\phi_{mn}\delta\lambda_{mn} = 0 \\
&\sum \epsilon_{mn}\cos\theta_{mn}\delta\lambda_{mn} = 0 \\
&\sum \epsilon_{mn}\cos\phi_{mn}\delta\lambda_{mn} = 0.
\end{aligned}
\tag{4.17}
$$

In these last, the sum ranges over n only.

[56] Equation (9), p. 207.

[57] Equation (25), p. 213.

[58] Equation (27), p. 213.

By multiplying each of the Eqs. (4.17) by the undetermined coefficients A_n, B_m, C_m, summing over m and then adding this to the Eq. (4.16), Menabrea obtained the free minimum problem [69][59]:

$$\sum \epsilon\delta\lambda_{mn}[\lambda_{mn} - (A_n - A_m)\cos\phi_{mn} - (B_n - B_m)\cos\theta_{mn} \\ - (C_n - C_m)\cos\psi_{mn}] = 0. \tag{4.18}$$

where the sum ranges over all the indexes, and account is taken of $\cos\phi_{nm} = -\cos\phi_{mn}$, and so on. Menabrea concluded:

> By equating to zero the coefficient of each variation, one has:
>
> $$\lambda_{mn} = (A_n - A_m)\cos\phi_{mn} + (B_n - B_m)\cos\theta_{mn} + (C_n - C_m)\cos\psi_{mn} = 0$$
>
> By comparing these expressions for λ with the (4.7) [our Eqs. (4.14)], one will see that they are identical, if we give the undetermined coefficients the values $A_m = \alpha_m$; $B_m = \beta_m$, $C_m = \gamma_m \ldots$, and so on. Thus, these expressions will lead to the same results obtained before. In this way, the exactness of the method deduced by the principle of elasticity is proved, and so the principle itself is confirmed [69].[60] (A.4.26)

4.3.4 Rombaux' Application of the Principle of Elasticity

In his paper of 1875 [69] Menabrea described applications of his principle of elasticity, referring to Giovanni Saccheri, and, most of all, to Giovanni Battista Rombaux, about whom he said in a note:

> Cavalier *Rombaux*, chief engineer of Roman railways, announced the publication of a memoir on the canopy of Arezzo railway station, whence he took inspiration to deal with the question of the distribution of tensions and pressures in elastic systems with the maximum amplitude. For sake of simplicity, he used the *principle of elasticity*, and, by several analytical and numerical examples, he showed the coincidence of the results obtained by it, with those obtained by other methods [69].[61] (A.4.27)

Indeed, Rombaux, engineer in the Roman railways, published the monograph *Condizioni di stabilità della tettoja della stazione di Arezzo* in 1876 [86], which is interesting and problematic at the same time. It is interesting because, in spite of the title, it is a treatise of good level on structural mechanics. It is problematic because its careful reading adds new elements in the controversy on the priority of the proof of the minimum of the elastic energy between Castigliano and Menabrea. This happened also because of some ambiguity in the date of the paper; the text was printed in 1876, but in the preface the date 1874 appears; in addition, the content of the paper appeared in serial form in the Italian journal *Giornale del Genio Civile* between 1875 and 1876.

---

59 Equation (28), p. 214.

60 p. 214. Our translation.

61 p. 203. Our translation.

4.3.4.1 *Condizioni di stabilità della tettoja della stazione di Arezzo*

Rombaux' monograph suggested two different methods to solve redundant trusses, which he called *method of flexibilities* (metodo delle flessibilità) and *principle of elasticity* (principio di elasticità); their difference was well described in the following passage:

> According to the method of flexibilities, one supposes that one of the supports be yielding, then, by the equations of the bending curves one calculates the analytical expression of the displacement appearing there, and by posing it equal to zero one has an equation of flexibility expressing the condition that the support reaction shall satisfy to reduce its point of application to the state of a fixed support. According to the principle of elasticity, when the prism is equilibrated under the action of external forces, the molecular work spent is a minimum, and, thus, its derivative with respect to the above said reaction shall be nil: whence an equation of elasticity results, to which the reaction itself shall satisfy to attain the minimum work. In the two ways of proceeding, the equations of flexibility and elasticity complete equilibrium equations and make indeterminacy come to an end [86].[62] (A.4.28)

Rombaux would then assert the equivalence of the two methods—at least in the case of smooth, fixed constraints—showing that the derivative of the molecular work (lavoro molecolare) with respect to the constraint reactions provides the displacements of the constrained points. By searching the minimum of the molecular work, imposing its derivatives with respect to the constraint reactions to vanish, equals to posing the displacements of the constrained points equal to zero, as it was done in the method of flexibilities.

The proof was not done in general, but only relative to continuous beams on several supports, showing the equivalence by inspection. For instance, for a beam with length b with an intermediate support, subjected to a concentrated force N at mid span, Rombaux found the following expression of the molecular work:

$$L'' = \frac{N^2}{6EI} b^3. \tag{4.19}$$

Here EI is the beam bending stiffness [86].[63]

The deflection at mid span, found by the method of the elastic curve, is given by [86][64]:

$$y = \frac{N}{3EI} b^3. \tag{4.20}$$

In any case, even if Rombaux' proof was not general, from certain points of view his approach is more advanced than Castigliano's proof, dated 1873 [13], where it was shown that the derivative of the molecular work with respect to a force F provides the displacement of the point of application of F. Castigliano obtained such a result

---

[62] p. 7. Our translation.

[63] p. 30.

[64] p. 35.

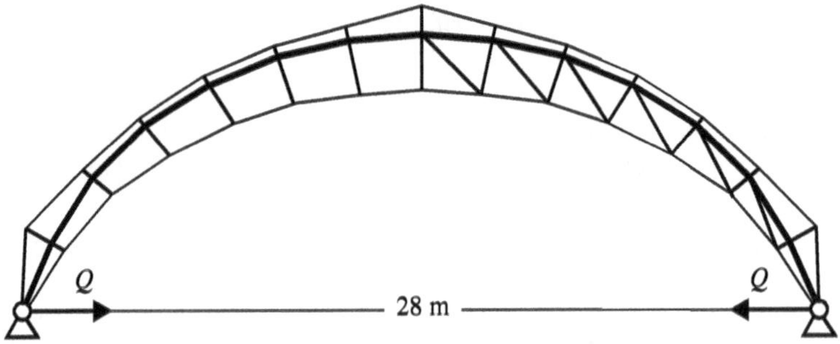

Fig. 4.3 The rib of the deck of the railway station of Arezzo (Redrawn from [86], Tav. II, Fig. 4.3)

as an intermediate step of his proof of the theorem of minimum molecular work and did not give it any particular relevance, as he would do in his later work of 1875 [14, 16].

We remark that Rombaux used the principle of elasticity to solve continuous beams on several supports, thus, differently from Menabrea, and like Castigliano, he applied the principle of elasticity to flexible elements as well. In the second part of the monograph, Rombaux applied the principle of elasticity for evaluating the forces in the wooden struts and iron rods in the rib of the deck of the railway station of Arezzo. This deck, sketched in Fig. 4.3, had shown an excessive subsidence, and a reinforcement was necessary. In the figure, on the left, the rib as it is was, on the right, the one with the proposal of reinforcement.

Rombaux considered the calculation of the rib before and after restoration, here, however, we describe the first case only, since it seems more interesting. In calculating the rib before restoration, Rombaux proposed to consider the rib as formed by two sub-structures in parallel. The first one was the arc of the struts, corresponding to the thick line in Fig. 4.3; the second one was the whole truss, as if the connections of the bars were effective, hence as if the whole structure were a reinforced beam (that is strengthened by a lower tie-beam). The first sub-structure bears a portion k_1 of the vertical loads, the second sub-structure bears the portion k, in such a way that $k + k_1 = 1$.

Each element of the arc of the struts was considered subjected to normal and bending stresses, while the elements of the reinforced beam were considered subjected to normal stress only. Rombaux evaluated stresses in all elements only by means of equilibrium equations, by a tricky procedure that eliminated redundancy.

The molecular work associated with the struts was evaluated by the relations [86][65]:

$$L = \frac{1}{6EI}\left\{\frac{3I}{\omega}\sum pP^2 + \sum p\left(m^2 + mm' + m'^2\right)\right\}. \tag{4.21}$$

65 p. 182.

Here l is the length of any strut, having moment of inertia I and cross-section area ω, under the axial force P and the bending moments m, m' at its ends; the work in the elements of the reinforced beam contains only the contribution of the normal force P. By summing over all the structural elements, Rombaux obtained the expression of the molecular work of the whole rib as a function of the coefficients k_1, k; by imposing the minimum of the molecular work with respect to k_1 and k, under the condition $k_1 + k = 1$, he determined k_1 and k and thus solved the structural problem.

4.3.4.2 The Question About the Priority

We know that Castigliano, after the publication of Menabrea's paper of 1875 [69], following his own paper [14] by a short while, asked the Accademia dei Lincei, the president of which was then Luigi Cremona, to be granted the priority of the proof of the principle/theorem of minimum molecular work, but obtaining only a partial positive answer. The debate between Castigliano and Menabrea is described and reconstructed with careful attention in [15, 32, 70, 75, 76]. Going back to this debate is beyond our purposes; we wish only to remark that maybe the conclusions drawn in [75, 76], according to which Castigliano would have been treated too badly, should be re-considered in the frame of what we have reported about Menabrea's work. This re-consideration should include Rombaux' contribution as well.

Indeed, Rombaux' monograph was written in the time of the debate, and, even though no explicit statement was made, the author clearly took Menabrea's parts. In his preface, one cannot but remark his extreme deference toward "His Excellency the Count Menabrea" ("Sua Eccellenza il Conte Menabrea"). Rombaux told about preceding applications of Menabrea's principle by Giovanni Saccheri, professor at the School of Application for Engineers in Turin; he did not make any mention, on the contrary, of Castigliano's work.

However, in the parts where Rombaux evaluated the molecular work of elements in bending, using the formula:

$$L'' = \frac{1}{2EI} \int_0^b M^2 dz, \tag{4.22}$$

a reference to the applications in Castigliano's master thesis [13] is apparent; even the symbols are the same. In addition, Rombaux used the term *molecular work* to denote elastic energy, and such a denomination was Castigliano's, not Menabrea's. Moreover, Rombaux was not alone at his time to ignore Castigliano: Lévy as well, in his treatise of 1874 [59], referred to Menabrea only talking about the new method of calculating redundant systems [32].

4.4 Carlo Alberto Castigliano

Carlo Alberto Castigliano (Asti 1847–Milano 1884)[66] received his diploma in mechanics at the Institute for industry and craftsmanship of Turin in 1866. In October of the same year he received the diploma of the Museo industriale, and at the end of 1866 he was appointed professor of Applied mechanics, Constructions, and Estimate at the Industrial institute of Terni (Central Italy), where he stayed four years in a very unsatisfactory economic situation. In spite of all, he managed to study assiduously and reached a strong mathematical education, to the point that in 1870 he passed with top grades the exam for the admission at the Faculty of mathematical, physical and natural sciences of the university of Turin, being also given the permission to take all the exams of the three-years course in Mathematics already at the end of the first year. In 1871 he got the degree and could subscribe the School of Application for Engineers in Turin.

Having Giovanni Curioni[67] as supervisor he graduated in 1873 defending the thesis *Intorno ai sistemi elastici*. The specific subject (the principle of minimum work) and the problem to be solved (the calculation of redundant structures) were for sure subjects dealt with in the School of Application, in the wake of Menabrea's works. We have already seen in Chap. 1 that Cerruti graduated in the same seat as Castigliano, with a thesis on the calculation of trusses [28]. Other theses of the School of Application about structural calculation were those by Annibale Gavazza (1874) and Moise Levi (1875), who applied Castigliano's procedures for calculating arch structures.

After his graduation, Castigliano was hired by the Railways of Northern Italy[68] as department head. He worked in Alba first, then in Turin; in 1875 he was moved to the central craftsman office in Milan and, after three years only, he was appointed section head. This promotion was also due to the reputation of his important publication on the applications of calculation of linear elastic structures [18].

Castigliano thus managed to put an end to his situation of poverty, and to be somehow successful, even at the European level, when he published in 1879 the monograph *Théorie de l'équilibre des systémes élastiques et ses applications* [20]. This monograph summarized and thoroughly completed his work on the theory of elasticity, pursued after his graduation thesis with the two papers *Intorno all'equilibrio dei sistemi elastici* [14] and *Nuova teoria intorno all'equilibrio dei sistemi elastici* [16], both in 1875. In 1882 he became corresponding fellow of the Accademia delle scienze in Turin.

---

[66] The following short biography is largely taken from [40, 49, 76].

[67] Curioni had an important role in Castigliano's life, showing him the way of the theory of structures and recognizing first his most important works, reading them at the Accademia delle science in Turin.

[68] The company called Railways of Northern Italy (Strade Ferrate dell'Alta Italia, SFAI) had been founded in 1865, when the unified Italian government sold the Piedmont railways to the company of the railways of Lombardy and central Italy. Later on, SFAI incorporated the Venetian railways as well: thus, a unique company, with seat in Milan, and managing all the railway net of Northern Italy, encompassing Emilia Romagna.

Unfortunately, destiny was not kind to him: he lost two of his four sons, Carlo after a few months of the birth (1883), and Emilia, aged 3 years (1884). Some months after, he also died of pneumonia, ironically after having been appointed chief of the craftsmanship office in Milan [76].

From this short biography, it is apparent that in some way Castigliano went again over the route of the great engineers of the *École polythecnique*: a solid theoretical background and a great attention to professional practice, providing motivations to theory and to design the necessary innovations for the solutions to the problems of a technological society. We have said 'in some way' because, even though the Italian Schools of Application had the École polythecnique as a model, yet they differed from it in the greater attention toward application, and in the less attention toward pure mathematics. In this, they resembled more the German schools of engineering, where the education of the engineer was completely held outside the universities, and inside the high technical schools.

Castigliano's main contribution to the theory of elastic structures was about the design of systems of bars and beams (*frames*). His results are exposed in a limited number of papers, due in part to his premature death, in part to his professional activity. The first paper was his graduation thesis of 1873 [13]: here Menabrea's principle of elasticity was extended from trusses to bent beams, and a proof different from that given by Menabrea in 1858 and 1868 was provided. We will later on discuss if Castigliano's proof was actually a progress with respect to Menabrea's.

In 1875 Castigliano published the two very important papers already referred to, *Intorno all'equilibrio dei sistemi elastici* and *Nuova teoria intorno all'equilibrio dei sistemi elastici* [14, 16]. The first went again over the ideas of his thesis of 1873, and represented its improvement and deepening; the second, on the other hand, proposed a rather different point of view. The two papers are separated by the known debate with Menabrea on the priority of the proof of the principle of elasticity [15], and by the publication of the monograph by Menabrea's 'pupil', Rombaux, recalled above. Rombaux, besides having applied the principle to an actual structure, had however realized the importance of the remark made by Castigliano in his thesis, according to which the derivative of the strain work with respect to a force equals the displacement in the direction of the force itself.

Castigliano gave a contribution to structural engineering also in other papers, among which we may quote the memoir *Formule razionali e esempi numerici per il calcolo pratico degli archi metallici e delle volte a botte murali* of 1876 [17]. Here he perfected a practical method for calculating arches without a direct use of integral calculus; the method would have been used to design the bridge on the Dora river, reported in the following.

A more theoretical paper was the memoir *Intorno a una proprietà dei sistemi elastici* of 1882 [21], where potential theory applied to elastic structures was dealt with in a simple way. Castigliano, by means of a more systematical approach, found here his results on the derivatives of the strain work again, as well as Betti's reciprocal theorem.

His last paper with theoretical basis was *Teoria delle molle* of 1884 on the theory of springs [23], containing some news with respect to the works at ease at his time;

its most important value was the use of the theorems of the derivatives of the strain work to evaluate the stiffness of springs. This made it possible to face rather complex problems, like those of the leaf springs with several leafs and various kinds of helical springs loaded in bending and torsion, in a simple way. In the paper Castigliano mentioned, even though marginally, the problem of impact in the ideal, yet realistic, situation when the distributed mass of the spring is negligible with respect to that of the bodies with which it interacts.

Among the works with an applicative aim we should quote the handbook *Manuale pratico per gli Ingegneri* [24], left unfinished by the death of the author. The textbook had a new outlook for his times, when the models were the 'Aide—mémoire' of the French: indeed, it was a 'rational handbook', and not a simple collection of technical formulas. Castigliano did not only collect already existing material, but also developed new formulas, where it was necessary; tables and figures as well had a rational presentation.

The issue of the handbook, undertook by the publisher Negro of Turin, begun in 1882, and in 1884 the third volume was given to press. The fourth volume was issued posthumously in 1888, edited by the *ingegner* Crugnola. As a matter of fact, it was also due to Castigliano's premature death that another professional handbook, the *Manuale dell'ingegnere civile e industriale* by Giuseppe Colombo, issued for the first time in 1877 by the publishing house Hoepli in Milan, could easily reach success among Italian professionals, lasting until nowadays.

In the following, we list the main works by Castigliano:

- 1873 *Intorno ai sistemi elastici*, Dissertazione presentata da Castigliano Carlo Alberto alla Commissione Esaminatrice della Reale Scuola d'applicazione degli Ingegneri in Torino, Turin, Bona.
- 1875 *Intorno all'equilibrio dei sistemi elastici*, Memorie della Reale Accademia delle scienze di Torino, v. 10, pp. 380–422.
- 1875 *Lettera al presidente dell'Accademia dei Lincei, 11 marzo 1875*. Memorie della Reale Accademia dei Lincei, s. 2, v. 2, pp. 59–62.
- 1875 *Nuova teoria intorno all'equilibrio dei sistemi elastici*, Memorie della Reale Accademia delle scienze di Torino, v. 11, pp. 127–286.
- 1876 *Formule razionali e esempi numerici per il calcolo pratico degli archi metallici e delle volte a botte murali*, L'ingegneria civile e le arti industriali, v. 9, pp. 120–135; v. 10, pp. 145–153.
- 1878 *Applicazioni pratiche della teoria sui sistemi elastici*, Strade ferrate dell'Alta Italia, Servizio della manutenzione e dei lavori, Milano, Crivelli.
- 1879–1880 *Théorie de l'équilibre des systémes élastiques et ses applications*, 2 v., Turin, Negro.
- 1881–1882 *Intorno a una proprietà dei sistemi elastici*, Memorie della Reale Accademia delle scienze di Torino, v. 17, pp. 705–713.
- 1882 *Esame di alcuni errori che si trovano in libri assai reputati*, Il Politecnico, nn. 1–2, pp. 66–82.
- 1884 *Teoria delle molle*, Turin, Negro.
- 1884–1889 *Manuale pratico per gli Ingegneri*, 4 vols, Turin, Negro.

4.4.1 *1873.* Intorno ai sistemi elastici

In the *Intorno ai sistemi elastici*, Castigliano considered at first a system composed by elastic beams, hinged with each other [13],[69] that is a truss.

4.4.1.1 The Method of Displacements

In Chap. 1 he used the method of solution of elastic structures that nowadays we would call of displacements, or of deformations, which was proposed by Navier and Poisson, and perfected by Clebsch in 1862 [34]; Castigliano, however, did not quote them. The paper is logically and clearly exposed, however, no hint on the track that the author wished to follow was provided. Equilibrium equations at the nodes were written first, then the compatibility equations among the elongations of the bars and the displacements of the nodes, and, in the end, the constitutive relations.

Castigliano remarked that, for a truss with n nodes, the equilibrium equations to be written are $3n - 6$, because 6 of the $3n$ total equilibrium equations for the nodes describe the global equilibrium with respect to rigid motions:

> [...] it follows that the necessary equations to determine these tensions are reduced to $3n-6$ and do not suffice in general to find all the unknowns but when the number of the bars equals $3n - 6$ [13].[70] (A.4.29)

The equilibrium equations for the node of a generic truss, like that in Fig. 4.4, have the form [13][71]:

$$X_p + \sum T_{pq} \cos \alpha_{pq} = 0, \quad Y_p + \sum T_{pq} \cos \beta_{pq} = 0, \quad Z_p + \sum T_{pq} \cos \gamma_{pq} = 0. \tag{4.23}$$

where X_p, Y_p, Z_p are the components of the external force at the node p with respect to a global Cartesian frame, T_{pq} is the elastic force[72] in the bar joining the nodes p, q. The following relations hold [13][73]:

$$\cos \alpha_{pq} = -\cos \alpha_{qp}, \quad \cos \beta_{pq} = -\cos \beta_{qp}, \quad \cos \gamma_{pq} = -\cos \gamma_{qp}. \tag{4.24}$$

Castigliano did not make it clear if the angles $\alpha_{pq}, \beta_{pq}, \gamma_{pq}$ should be taken in the reference or in the present configuration, however we infer from what follows that he considered the reference configuration and assumed infinitesimal displacements. Under such hypothesis, he obtained the following linear relation between the variation

---

[69] p. 8.

[70] p. 8. Our translation.

[71] Equation (1), p. 9.

[72] Castigliano, and we also in the following, adopted the term *tension* (tensione) to denote such a force.

[73] Unnumbered equation, p. 9.

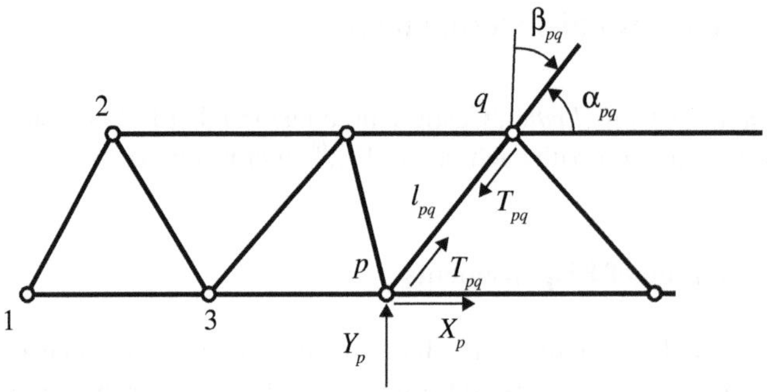

Fig. 4.4 The generic truss of Castigliano

in length of the bars and the displacements ξ, η, ζ [13][74]:

$$\lambda_{pq} = (\xi_q - \xi_p)\cos\alpha_{pq} + (\eta_q - \eta_p)\cos\beta_{pq} + (\zeta_q - \zeta_p)\cos\gamma_{pq}. \qquad (4.25)$$

where λ_{pq} is the elongation of the bar pq and ξ_p, η_p, ζ_p; ξ_q, η_q, ζ_q are the displacement components of the nodes p, q respectively.

Castigliano wrote the linear elastic constitutive relation as $T = \epsilon\lambda$, $\epsilon = E\omega/l$, with E Young's modulus, ω the area of the cross-section, and l the length of the bar, and obtained [13][75]:

> $$T_{pq} = \epsilon_{pq}\left\{(\xi_q - \xi_p)\cos\alpha_{pq} + (\eta_q - \eta_p)\cos\beta_{pq} + (\zeta_q - \zeta_p)\cos\gamma_{pq}\right\}$$
>
> By this formula we may express the tensions of all the bars as a function of the displacements of the vertexes parallel to the [coordinate] axes: these displacements would be $3n$, if all the vertexes could move, but, because of the conditions we have given to the three vertexes V_1, V_2, V_3, we have $\xi_1 = 0$, $\eta_1 = 0$, $\zeta_1 = 0$; $\eta_2 = 0$, $\zeta_2 = 0$; $\zeta_3 = 0$, whence the unknown displacements reduce to $3n - 6$ [13].[76] (A.4.30)

hence the number of unknowns coincides with the number of equations.

4.4.1.2 The Minimum of Molecular Work

In Chap. 2, Castigliano proved that

> If I determine the tensions T_{pq} in such a way that they make the expression $\sum T_{pq}^2/\epsilon_{pq}$ a minimum, by supposing that for those tensions Eq. (4.1) [that is the equilibrium equations] hold, in which, however, all the external forces X_p, Y_p, Z_p and all the angles $\alpha_{pq}, \beta_{pq}, \gamma_{pq}$

[74] Equation (3), p. 10.

[75] Equation (4), p. 11.

[76] p. 11. Our translation. We must remark that, in order to prevent rigid motions, Castigliano assumed node V_1 fixed, node V_2 fixed along the axes y, z and node V_3 fixed along the axis z. As a consequence, the pertaining equilibrium equations do not appear in (4.23).

are considered constant, the values of the tensions so obtained coincide with those obtained by the method of displacements [13].[77] (A.4.31)

There was no reference to the principle of elasticity by Menabrea, who was quoted only in the introduction. Castigliano affirmed that the above quoted statement was actually a theorem only incidentally, some lines after [13].[78] It seems as if Castigliano wanted to make the statement looks trivial and of minimal importance.

The proof followed linearly, even though it is not wholly satisfying, according to modern standards. By differentiating the expression $\sum T_{pq}^2/\epsilon_{pq}$, to which, in the following, he gave the name of *molecular work*. Castigliano obtained [13][79]:

$$\sum \frac{T_{pq}}{\epsilon_{pq}} dT_{pq} = 0. \tag{4.26}$$

This relation holds for tensions T_{pq} satisfying the equilibrium Eqs. (4.23) for the nodes, that is by differentiating, the relations (4.23) [13][80]:

$$\begin{aligned}
&\sum dT_{2q}\cos\alpha_{2q} = 0,\\
&\sum dT_{3q}\cos\alpha_{3q} = 0, \quad \sum dT_{3q}\cos\beta_{3q} = 0,\\
&\dots\\
&\sum dT_{pq}\cos\alpha_{pq} = 0, \quad \sum dT_{pq}\cos\beta_{pq} = 0, \quad \sum dT_{pq}\cos\gamma_{pq} = 0\\
&\dots
\end{aligned} \tag{4.27}$$

By summing (4.26) and (4.27), multiplied by suitable Lagrange multipliers A_p, B_p, C_p, Castigliano came to [13][81]:

$$\begin{aligned}
&\sum \frac{T_{pq}}{\epsilon_{pq}} dT_{pq} + A_2 \sum dT_{2q}\cos\alpha_{2q} + A_3 \sum T_{3q}\cos\alpha_{3q}\\
&\quad + B_3 \sum T_{3q}\cos\beta_{3q} + A_p \sum dT_{pq}\cos\alpha_{pq}\\
&\quad + B_p \sum dT_{pq}\cos\beta_{pq} + C_p \sum dT_{pq}\cos\gamma_{pq} = 0.
\end{aligned} \tag{4.28}$$

where he took into account only the independent equilibrium equations at the nodes.

Now, by equating to zero the coefficients of the differentials of all the tensions we obtain as many equations as tensions, and by adding the $3n-6$ equations (1) [our (4.23)] we will

---

[77] p. 14. Our translation.

[78] p. 14.

[79] Equation (5), p. 14.

[80] Equation (6), p. 15.

[81] Unnumbered equation, p. 15.

have as many equations as needed to determine all the tensions, and the $3n-6$ multipliers [13].[82] (A.4.32)

By equating to zero the coefficients of the differentials dT_{pq}, Castigliano obtained [13][83]:

$$\frac{T_{pq}}{\epsilon_{pq}} + A_p \cos\alpha_{pq} + B_p \cos\beta_{pq} + C_p \cos\gamma_{pq} + A_q \cos\alpha_{qp} + B_q \cos\beta_{qp} + C_q \cos\gamma_{qp} = 0. \tag{4.29}$$

This, keeping into account (4.24), and multiplying by ϵ_{pq}, yields [13][84]:

$$T_{pq} = \epsilon_{pq}\left[(A_q - A_p)\cos\alpha_{pq} + (B_q - B_p)\cos\beta_{pq} + (C_q - C_p)\cos\gamma_{pq}\right] \tag{4.30}$$

The equations like the (4.29), as many as the bars, are the equations to be added to the equilibrium Eq. (4.23). Castigliano remarked that the (4.29) are identical to the supplementary equations used in the methods of displacements, and that Lagrange multipliers are nothing but the displacements of the nodes. The solution that can be obtained by minimizing the molecular work thus coincides with that obtainable by the method of displacements, which is exact, in that it is given by 'undoubtable' mechanical methods.

Castigliano's procedure is elegant, and efficient in many points, yet it is not wholly rigorous, even though we are talking of some details. For example, there was no remark about the fact that the molecular work admits a minimum, and not simply an extremum; however this is irrelevant from an operative point of view. In addition, there was no explicit hint on the uniqueness of the solution of the elastic problem, which is the condition that could assure the coincidence of the results given by the two methods (displacements, and minimum molecular work). Castigliano proved only the implication:

Minimum molecular work → Solution of displacements method

and not the converse, which holds if the solution is unique. It would have been enough for Castigliano to remark that the variation of the molecular work with respect to the tensions of the bars, posed equal to zero, provides only, and always, the system of linear equations providing the solution according to the method of displacements.

4.4.1.3 Mixed Structures

Castigliano ended the Chap. 2 of his thesis by generalizing his 'theorem' to the case in which there are both bent and twisted elements, that is beams, in addition to elements

[82] p. 15. Our translation.

[83] First unnumbered equation, p. 16.

[84] Equation (7), p. 16.

subjected to axial stress only, that is bars: he referred to such structures as 'mixed'. We remark that Castigliano in this chapter called his statement a theorem right from the beginning. This may mean that Castigliano knew that considering elements in bending and torsion besides those in extension represented the real novelty of his work with respect to the result already obtained by Menabrea:

> 7. THEOREM.—Let us consider an elastic system formed by parts undergoing torsion, bending, or angular shearing, and by bars joined with those others, and among them: I say that if such a system is subjected to the action of external forces so that it deforms, the tensions of the bars after deformation are those, which make the expression of the molecular work of the system a minimum, keeping into account the [equilibrium] equations holding among these tensions, and supposing invariable the directions of the bars and of the external forces [13].[85] (A.4.33)

In this proof, Castigliano adopted a statement that did not belong to the 'classical' theory of elasticity, that is that the inner *molecular work* spent on the strain of the parts, which are not solely extensible, could be expressed uniquely as a function of the external forces $P, Q, R, \ldots$ and of the tensions $T_1, T_2, T_3, \ldots$ of the bars converging there [13][86,87]:

$$F(P, Q, R, \ldots, T_1, T_2, T_3, \ldots). \tag{4.31}$$

That is, he admitted that elastic forces are conservative: in modern terms, F is the elastic potential energy of the parts that are not purely extensible (Castigliano, for sake of brevity, choose to call them simply *flexible parts*, as a function of the active forces and of the tensions of the purely extensible parts, seen as external forces as well.

Castigliano proved this theorem analogously to the preceding one: he searched for the minimum (actually the stationarity) of the molecular work with respect to the tensions, and proved that one obtains equations, which are equivalent to those obtained by the method of displacements. The total molecular work of the system is the sum of the molecular works of the bars and of F. The equations provided by the differentiation of the total molecular work is [13][88]:

$$\left(\frac{T_1}{\epsilon_1} + \frac{dF}{dT_1}\right) dT_1 + \left(\frac{T_2}{\epsilon_2} + \frac{dF}{dT_2}\right) dT_2 + \cdots + \sum \frac{T_{pq}}{\epsilon_{pq}} dT_{pq} = 0. \tag{4.32}$$

Tensions with one index, $T_1, T_2, \ldots$, are those of bars which share a node with a bent, or twisted, part; tensions with two indexes, T_{pq}, are those of purely extensible bars. In this equation the tensions $T_1, T_2, \ldots, T_{pq}$ cannot vary arbitrarily, but they must be equilibrated among them and with the active forces P, Q, R.

---

[85] p. 17. Our translation.

[86] First unnumbered equation, p. 18.

[87] First unnumbered equation, p. 18.

[88] Equation (8), p. 18.

Starting from here, the text loses clarity and needs to be interpreted. Castigliano considered only the equilibrium equations of the nodes that can be modeled as hinges; in any case, in such equations the tensions of all elements are present. He implicitly assumed that each bar had at least a node-hinge; this is a necessary condition to obtain a simple form of the equilibrium equations of nodes as a function of the tensions T_j, T_{pq} in the bars only. Castigliano believed that equilibrium could always be verified for the nodes concurring to bent and twisted elements.

By differentiating the independent equilibrium equation at the nodes[89] that are less than $3n-6$ (they would be exactly $3n-6$ if there were no bent and twisted elements), multiplying the differentials by the Lagrange multipliers $A_1, B_1, C_1; A_2, B_2, C_2; \ldots;$ A_p, B_p, C_p; $A_q, B_q, C_q; \ldots,$ adding to (4.32) and equating the coefficients of the $dT_1, dT_2, \ldots, dT_{pq}, \ldots$ to zero, Castigliano obtained, without reporting all passages [13][90]

$$\begin{aligned}
&\frac{T_1}{\epsilon_1} + \frac{dF}{dT_1} - A_1 \cos\alpha_1 - B_1 \cos\beta_1 - C_1 \cos\gamma_1 = 0 \\
&\frac{T_2}{\epsilon_2} + \frac{dF}{dT_2} - A_2 \cos\alpha_2 - B_2 \cos\beta_2 - C_2 \cos\gamma_2 = 0 \\
&\ldots \\
&\frac{T_{pq}}{\epsilon_{pq}} - (A_q - A_p)\cos\alpha_{pq} - (B_q - B_p)\cos\beta_{pq} - (C_q - C_p)\cos\gamma_{pq} = 0.
\end{aligned} \tag{4.33}$$

> Meanwhile, if among the Eqs. (4.9) [our (4.33)] we consider those containing the tensions of the bars, which are not joined by any end with the flexible parts of the system, we recognize that they are exactly those that would be obtained by the method of displacements to express those tensions, only understanding that, in general, *A*, *B*, *C* represent the displacements of the vertex *V* parallel to the coordinate axes: I suppose that the three vertexes V_1, V_2, V_3 of which the first is posed at the origin of the coordinates, the second on the *x*-axis, and the third in the *xy*-plane, are among the ones to which only hinged bars concur.
>
> We are left to prove only that also those equations among the (4.1), which contain the tensions of the bars that are joined with one end with the flexible parts of the system, coincide with the equations provided by the method of displacements [13].[91] (A.4.34)

He remarked that, since he considered a quasi-static deformation, and he had admitted the external forces to be invariable, the work spent by them depends only on the initial and final configurations of the system, and its expression is [13][92]:

---

[89] The equilibrium equations at the nodes, once differentiated, assume the form:

$$\begin{aligned}
\delta T_1 \cos\alpha_1 + \cdots \sum \delta T_{1q} \cos\alpha_{1q} &= 0 \\
\sum \delta T_{pq} \cos\alpha_{pq} &= 0 \\
&\ldots
\end{aligned}$$

[90] Equation (9), p. 18.

[91] p. 19. Our translation.

[92] Equation (10), p. 21.

$$\frac{1}{2}\left(Pp + Qq + Rr + \cdots\right), \tag{4.34}$$

where P, Q, R are the final values of the external forces and p, q, r are the projections of the displacements of their points of application along their directions. For the equality of works, the expression (4.34) represents also the work spent by the inner forces:

> [...] but the work of the external forces shall be equal to the internal, or molecular, work, and this is independent of the law by which the external work have grown; thus, the formula (4.10) [our (4.34)] expresses the molecular work of deformation, whatever the law by which the forces producing it have varied [13].[93] (A.4.35)

By definition, the work of external forces in an infinitesimal deformation is [13][94]:

$$Pdp + Qdq + Rdr + \cdots \tag{4.35}$$

Also differentiating the expression (4.34) the work in an infinitesimal deformation is obtained [13][95]:

$$\frac{1}{2}\left(Pdp + Qdq + Rdr + \cdots\right) + \frac{1}{2}\left(pdP + qdQ + rdR + \cdots\right). \tag{4.36}$$

By equating (4.35) and (4.36), Castigliano deduced [13][96]:

$$Pdp + Qdq + Rdr + \cdots = pdP + qdQ + rdR + \cdots \tag{4.37}$$

Hence it follows that the work in an infinitesimal deformation of the bent and twisted elements is [13][97]:

$$pdP + qdQ + rdR + \cdots + t_1 dT_1 + t_2 dT_2 + \cdots \tag{4.38}$$

> [...] but we have seen that [the infinitesimal molecular work] is expressed also by the formula [13][98]
>
> $$\frac{dF}{dP} = dP + \frac{dF}{dQ}dQ + \frac{dF}{dR}dR + \cdots \frac{dF}{dT_1}dT_1 + \frac{dF}{dT_2}dT_2 + \cdots;$$
>
> then these two expressions, having to be equal whatever the values of the differentials $dP, dQ, dR, \ldots, dT_1, dT_2, \ldots$ it is necessary that [13][99] (A.4.36)

---

[93] p. 21. Our translation.

[94] Last but one unnumbered equation, p. 21.

[95] Last unnumbered equation, p. 21.

[96] First unnumbered equation, p. 22.

[97] Second unnumbered equation, p. 22. The t_i are the projections of the displacements of the flexible parts along the T_i.

[98] Third unnumbered equation, p. 22. Actually, Castigliano implicitly referred to the mechanical meaning of the F in the (4.29) and to its first variation.

[99] Equation (11), p. 22. Our translation.

$$\frac{dF}{dP} = p, \ \frac{dF}{dQ} = q, \ \frac{dF}{dR} = r, \ldots \frac{dF}{dT_1} = t_1, \ \frac{dF}{dT_2} = t_2, \ldots \tag{4.39}$$

Castigliano did not give any importance to this result, to which, on the other hand, he gave the title of 'theorem' in his following works.[100]

Once proved that the expressions dF/dT_i in the (4.33) coincide with the projections t_i of the displacements of the nodes of the flexible elements along the T_i, and keeping into account the geometrical relations that provide the t_i as a function of the components of the displacements ξ_i, η_i, ζ_i with respect to the coordinate axes [13][101]:

$$\begin{aligned} t_1 &= \xi_1 \cos\alpha_1 + \eta_1 \cos\beta_1 + \zeta_1 \cos\gamma_1, \\ t_2 &= \xi_2 \cos\alpha_2 + \eta_2 \cos\beta_2 + \zeta_2 \cos\gamma_2, \\ &\ldots \end{aligned} \tag{4.40}$$

One sees that, among the Eqs. (9) [our (4.33)], also those containing the tensions $T_1, T, \ldots$ fully coincide with those obtained by the method of displacements [13].[102] (A.4.37)

Indeed, take, for instance, the first of (4.33), rewritten taken into account (4.39–4.40):

$$\frac{T_1}{\epsilon_1} + (\xi_1 - A_1)\cos\alpha_1 + (\eta_1 - B_1)\cos\beta_1 + (\zeta_1 - C_1)\cos\gamma_1 = 0. \tag{4.41}$$

We obtain a compatibility equation, which states that the elongation of the bar 1 due to the tension T_1 equals the variation of the distance between the nodes of the bar. The proof of the theorem of minimum work is thus finished.

Castigliano extended the theorem also to purely flexible systems (in his sense, that is undergoing shear, torsion, and bending), without truss elements. Such an extension, however not general, was obtained by means of *ad hoc* procedures, and concerned only elements constrained to the 'ground', and not also among them, as usual in frames. His conclusion, that in these cases is about the determination of the redundant constraint reactions, was that, if the molecular work L can be expressed as a function of the undetermined reactions X_i (*redundant reactions*), the latter could be determined by searching the minimum of L with respect to the X_i.

4.4.1.4 Applications

Castigliano finished his thesis with a third part consisting of applications: a continuous beam, some trusses and some beams reinforced by a lower tie-beam. The interest is not in the applications *per se*, but rather in the introductory part, where Castigliano

[100] Nowadays (4.33) is known as Castigliano's (first) theorem.

[101] Last unnumbered equations, p. 22.

[102] p. 23. Our translation.

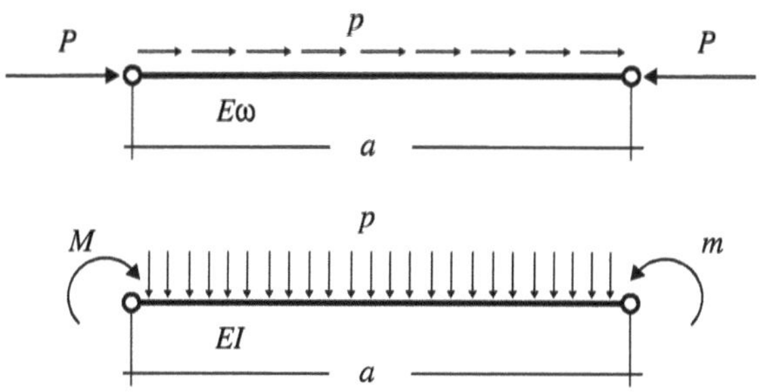

Fig. 4.5 One-span beam

determined the expression of the molecular work for the extended and bent elements as a function of the global inner actions, of the forces applied to the nodes, and of the external distributed loads.

The work associated with the extension of a bar of length a, Fig. 4.5, is [13][103]:

$$\frac{1}{2E\Omega}\int (P+px)^2\,dx = \frac{a}{2E\Omega}\left(P^2+Ppa+\frac{1}{3}p^2a^2\right), \tag{4.42}$$

where P is the end axial force, p the distributed axial load, E the Young modulus of the material constituting the element, Ω the area of the cross-section.

For an inflexed beam with length a, cross-section moment of inertia about the neutral axis I, end couples m, M, subjected to a uniformly distributed transverse load p, he provided the following expression of the molecular work [13][104]:

$$\frac{a}{2EI}\left[\frac{M^2+Mm-m^2}{3}-\frac{1}{12}pa^2\,(M+m)+\frac{1}{120}pa^4\right], \tag{4.43}$$

obtained by integrating $M(x)^2/2EI$ along the beam, $M(x)$ being the bending moment of the beam.

It is interesting to remark that Castigliano, shortly after, provided the expression of the work on shear—in any case, not fully equivalent to what we usually accept nowadays—but he did not provide that of the work on twist, because, he said, this case almost never happens in constructions [13].[105] In addition, we remark that Castigliano provided (4.42) and (4.43) without any comment, as he considered them as known results for the time. Probably, he knew the works of the European literature on the subject (Clebsch, Saint Venant, Lamé, Moseley—see also Chap. 1).

Castigliano used the expressions of the molecular work to solve topical issues: he re-obtained Clapeyron's equation for a continuous beam ('three moments equation'); he studied trusses formed by elements in bending and hinged beams (some Polonceau

[103] p. 33.
[104] p. 35.
[105] p. 35.

rafters first, then trusses of general shape, composed by some elements in bending supporting other hinged bars). For simplicity, we report in the following in full only the section about the analysis of continuous beams.

13. *Application to a beam supported by more than two points.*—I suppose the beam horizontal, rectilinear, homogeneous, with uniform stiffness, symmetrical with respect to the vertical plane through its axis, and loaded by a uniformly distributed weight on each part contained between two successive supports.

It is apparent that the values of the bending moments for the sections corresponding to the supports are functions of the weights distributed along the solid, and of the pressures, or reactions, of the supports; now, by keeping into account the two equations provided by statics involving the values of these reactions, we see that as many of these remain to be determined as the supports are, minus two, that is as many as the bending moments on the supports are, since the bending moments on the end supports vanish. Whence it follows that the support reactions can be expressed as a function of the bending moments relative to the same supports, and we can thus assume these moments as unknowns.

These unknowns shall be determined by the condition that the molecular work of the beam be a minimum; I neglect the work coming from the angular shear, whence the differential of the molecular work of the whole beam is equal to the sum of as many expressions analogous to (4.15),[106] as the parts in which the beam is divided by the supports, that is the spans, taking just care of the fact that for the right end span the expression [15] reduces to the first term only, and thus $dm = 0$, and for the left end span it reduces to the second term, because $dM = 0$.

In order that the molecular work be a minimum, we must determine the unknown bending moments, by equating the coefficients of the differentials of all these moments to zero. Now, the differential of the bending moment relative to the support B cannot enter but in one of the terms coming from the work of the span AB and in one of those coming from the work of the span BC; so that, calling a and a' the lengths of these two spans, p and p' the uniformly distributed weights on them, m, m', m'' the bending moments relative to the three supports A, B, C; E the coefficient of elasticity of the beam and I the cross-section moment of inertia, the two terms that contain the differential dm' in the differential expression of the molecular work are [see Fig. 4.6]:

$$\frac{a}{2EI}\left(\frac{m+2m'}{3}-\frac{1}{12}pa^2\right)dm'; \quad \frac{a'}{2EI}\left(\frac{2m'+m''}{3}-\frac{1}{12}p'a'^2\right)dm'.$$

Hence, equating the coefficient of dm' to zero, one obtains

$$am+2(a+a')m'+m''a'-\frac{1}{4}(pa^3+p'a'^3)=0.$$

This is exactly the equation due to Clapeyron [13].[107] (A.4.38)

[106] It is the variation of (4.43).

[107] pp. 35–36.

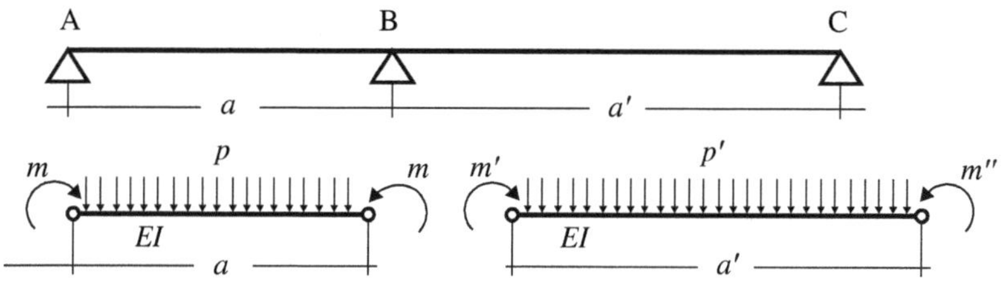

Fig. 4.6 A continuous beam

4.4.2 1875. Intorno all'equilibrio dei sistemi elastici

The *Intorno all'equilibrio dei sistemi elastici* referred to the graduation thesis, with the aim of improving its exposition and precise its contents. The introduction was much more complete, and provided a precise enough history of the principle of minimum work. Obviously, Castigliano quoted Menabrea also, even though his role appeared diminished; in particular when, commenting on Menabrea's paper of 1868 [66], with the new proof of the *équation d'élasticité*, he said:

> [The new proof], however, seems not to have been adjudged more rigorous than the first one, because, notwithstanding the great beauty, and the apparent usefulness of the theorem of the minimum work, nobody, as far as I know, believed to take advantage from it before the year 1872, when Eng. Giovanni SACHERI read, at the Society of the Engineers and Manufacturers in Turin a Memoir of his, in which he tried to apply that theorem [...]. In any case, I do not need to talk of this memoir, because, since it contains just a numerical example, it did not let the proof of the theorem proceed any further [14].[108] (A.4.39)

Castigliano did not criticize the merit in Menabrea's proof; he said that it was not rigorous, because the method had no applications, but he contradicted himself, since he quoted one. His proof of the theorem of minimum molecular work proceeded exactly as in his thesis; the main difference was perhaps in one section, *Spostamenti dei vertici in funzione delle forze esterne*, in which a result of the thesis, which would have been thoroughly reconsidered in a later paper [16] was put into better evidence.

In section 10 of the *Intorno all'equilibrio dei sistemi elastici* Castigliano made a statement that makes us think that to him (as well as to us) the superiority of his work with respect to Menabrea's mainly is in the extension to structural systems different from trusses:

> 10. *Usefulness of the theorem of minimum work.*—In practice, it almost never happens that we use elastic systems simply hinged, that is systems composed of elastic bars joined by pins: on the other hand, systems that I will call *mixed* are ever used, composed by beams reinforced by bolts or ties, that is by elastic rods joined by pins to the beams in various points of their lengths, and among them. For a theorem on elastic systems to be of practical usefulness, it is necessary that it be applicable to mixed systems. The theorem of minimum work has exactly this quality, and it is only because of this, that I worked as much as I could to prove its exactness and usefulness. Since, however, its properties about simple articulated

[108] pp. 3–4. Our translation.

systems are kept also for the mixed ones, as I will show in a while, I will as of now tell some of the advantages that it brings with respect to other methods of calculation of articulated systems [14].[109] (A.4.40)

4.4.2.1 Mixed Structures

The generalization of the theorem of minimum work to flexible elements followed a line completely different from that of 1873, toward which Castigliano himself expressed some doubts, when he wrote: "still it seems to me to have found some new proofs, simpler and more rigorous than those I had given before" [14].[110]

Castigliano started from the classical molecular hypothesis, which he considered undoubtable, in spite of the fact that it was questioned by many scholars of elasticity theory, and that he himself would doubt in his following paper. According to this assumption, matter is composed by particles, or molecules, exchanging opposite forces directed along the line joining them, proportional to the variation of their distance—at least for very small displacements. Everything goes as if molecules were linked by elastic bars: any structural element, bent, twisted, sheared, could be seen as a huge spatial truss.

It is thus reasonable to apply the theorem of minimum molecular work to this truss. Obviously, a practical difficulty arises: the minimum should be looked for by considering all intermolecular forces, and this is impossible in practice. Castigliano managed to overcome this difficulty by an argument containing some ambiguities; on the one hand, he maintained that:

> [...] if the state of the system after the deformation may be made to depend on a small number of quantities linked by some condition equations, and if the molecular work of the system in the deformation is expressed by means of those quantities only, we will obtain their values by considering them as variables linked by the condition equations, and seeking the arrangement of their values making the expression of the molecular work a minimum [14].[111] (A.4.41)

Thus, if for the flexible elements one is able to express the molecular work L as a function of the forces at the nodes X_l, such forces will be determined by the minimum of L with respect to X_l, considering the equilibrium equations for the X_l only.

On the other hand, it seems that Castigliano wanted to limit his focus on mixed systems, for which he could use a theorem on trusses, which he had proved before:

> [...] if we know, for an articulated system deformed by given forces, how to write the molecular work of a part contained inside a given surface S as a function of the tensions of the rods joining that part to the remaining one, we will obtain the tensions of these rods and of the ones outside the surface S by expressing that the molecular work of the whole system is a minimum, keeping into account the equilibrium equations of all the vertexes outside the surface S [only] [14].[112] (A.4.42)

[109] p. 29. Our translation.

[110] p. 6.

[111] p. 36. Our translation.

[112] pp. 24–25. Our translation.

Thus, if in a flexible element we take its outer boundary as the surface S, it is possible to look for the minimum of the molecular work of all the inner micro-rods by keeping into account the forces acting outside S only, obviously under the condition to be able to express the molecular work as a functions of such forces. If, then, in a mixed system of extensible and flexible elements, one is able to write the molecular work of the whole system as a function of the forces of the bars in extension only, then the structural solution may be found by imposing the minimum of the molecular work of the whole structure under the condition that the nodes to which only bars in extension concur only be equilibrated.

4.4.3 1875. Nuova teoria intorno all'equilibrio dei sistemi elastici

The *Nuova teoria intorno all'equilibrio dei sistemi elastici* was written after the debate with Menabrea on the priority of the proof of the theorem of minimum work, and the very short introduction was clearly influenced by this fact. Menabrea's role was diminished as much as possible, and, which is worse, no quote of Rombaux' monograph [86] appeared, even though in this monograph the author had put into evidence how the derivative of the molecular work with respect to the forces provides the displacements of their points of application.

Castigliano turned the setting of his preceding work of 1875 upside down, and moved the key point from the theorem of minimum work to that of the derivatives of work. The standard theory of articulated systems by the method of displacements was re-formulated with greater severity (maybe pedantry) with respect to what he had done in his previous works, for instance by making it clear which terms in the displacements, however small, may be neglected to arrive to a system of linear equations.

The theorems of the derivatives of work were formulated right at the beginning of the paper:

> That posed, the two new theorems are the following:
>
> 1st If, in any elastic system, the strain work expressed as a function of the external forces is differentiated with respect to one of these forces, the obtained derivative expresses the displacement of the point of application of the force along its direction. 2nd If the same expression of the strain work is differentiated with respect to the moment of a couple, the obtained derivative expresses the rotation of the line joining the two points of application of the couple.
>
> These theorems, the importance of which is apparent, hold only if the deformations are very small, in such a way that the powers of the displacements and of the rotations higher than the first are negligible with respect to it. They may be put together in a unique theorem, that I will call *theorem of the derivatives of the strain work*, or, in short, *theorem of the derivatives of the work*.
>
> 3. It will be seen in the following that it suffices to solve all the questions that appear in practice about the equilibrium of elastic systems. We will also see that it contains as an application, or better as a simple remark, the *theorem of minimum work of elastic strains*, or

principle of elasticity, which General MENABREA had stated first in all its generality in 1857 and 1858 at the Reale accademia delle scienze of Turin and Paris, and about which he has presented in 1868 another Memoir at the Accademia delle scienze in Turin [16].[113] (A.4.43)

The proof of the theorems was performed by limiting to trusses at first:

17. *Theorem of the derivatives of the strain work* – If the strain work of a truss is expressed as a function of the external forces, its derivative with respect to any one of these provides the displacement of its point of application along its direction [16].[114] (A.4.44)

[…] if the strain work of a [hinged] system is differentiated with respect to the moment M of the considered couple, the obtained derivative expresses the angle by which the straight line, joining the points of application of the two forces of the same couple, has rotated [16].[115] (A.4.45)

In the second theorem, the moment M is not the bending moment of an element in bending, but the moment of a couple of forces applied to distinct nodes of the truss. Castigliano proved the first theorem in a simpler way with respect to his graduation thesis, by referring to a truss. For brevity, we will not quote the proof, which is present in many monographs on Structural mechanics. The proof of the second theorem is more labored.

4.4.3.1 The Theorem of Minimum Work as a Corollary

Castigliano relegated the theorem of minimum work to the role of a simple corollary of that of the derivatives of work. The proof is simple[116] let p and q be two nodes of a truss; T_{pq} the tension of the bar joining them, supposed to be the only redundant one; F the strain work of the part of the truss without the bar, subjected not only to the external active forces, but also to the opposite forces T_{pq} applied at p, q. For the first theorem of the preceding article, the derivative:

$$-\frac{dF}{dT_{pq}}, \tag{4.44}$$

represents the relative displacement between p and q in the direction of the applied forces. The negative sign is necessary to provide a positive sign for the case of a stretched bar ($T_{pq} > 0$), when the nodes are subjected to forces tending to approach them. However, in the bar pq the quantity:

$$\frac{T_{pq}}{\epsilon_{pq}}, \tag{4.45}$$

113 p. 129. Our translation.

114 p. 146. Our translation.

115 p. 150. Our translation.

116 We summarize what Castigliano: wrote on pp. 150–152.

represents the positive or negative stretching of the nodes p, q thought belonging to it (positive stretching if $T_{pq} > 0$). By compatibility, it should be:

$$-\frac{dF}{dT_{pq}} = \frac{T_{pq}}{\epsilon_{pq}} \rightarrow \frac{dF}{dT_{pq}} + \frac{T_{pq}}{\epsilon_{pq}} = 0 \rightarrow \delta\left(F + \frac{1}{2}\frac{T_{pq}^2}{\epsilon_{pq}}\right) = 0. \tag{4.46}$$

In such a way, the theorem of the minimum strain work is proved when only one redundant bar is present, inasmuch the expression $F + 1/2T_{pq}^2/\epsilon_{pq}$ stands for the strain work of the complete structure. The generalization to several redundant bars is simple.

4.4.3.2 Generic Systems

To extend the theorems of the minimum work and of the derivatives of work to flexible elements, Castigliano went back to the positions of his graduation thesis, and to the principle of conservation of energy. He introduced also a meaningful change in terminology: the molecular work became strain work, or elastic work, two terms leaving aside any reference to the constitution of matter.

> I propose to show that also for these two classes of systems [containing flexible elements] both the *theorem of the derivatives of the strain work* and that *of the minimum work* hold.
>
> [...] To provide these proofs, I will invoke the *principle of conservation of energies*: I would not need to do it, if we would admit that, when an elastic body is deformed, the action developed between two near molecules is directed along the line joining their centers. This hypothesis has been accepted until now, and some distinguished authors like LAMÉ and BARRÉ DE SAINT VENANT continue to accept it, since, indeed, it is difficult to have a clear idea of another way of action.
>
> Since, however, the famous astronomer GREEN in his *Theory of light* has admitted that the action between two molecules could take place along a direction different from the straight line joining their centers, in such a way, though, that the principle of conservation of energies takes place, I will endeavor to show that the new theorems hold true independently of the direction where the action between the molecules of the bodies takes place [16].[117] (A.4.46)

By resorting to the principle of conservation of energy—or, better, to the first principle of thermodynamics—by an argument known at the time[118] Castigliano proved that the strain work is independent of the path of active forces.

> It is very important to convince oneself of the rigor of this argument [relative to central forces], since it is very likely that the case now considered is that occurring in nature.
>
> However, to introduce in our investigations no restrictions apart from the strictly necessary ones, we will report here the argument by GREEN, modified by Mr. BARRÉ DE SAINT VENANT, to prove the stated theorem, whatever the direction of the action between molecules be.

[117] p. 158. Our translation.

[118] See for instance [6], section 2.

> Let us suppose that the deformation of a body takes place in a bowl impermeable to heat, and that, after having let the forces from zero to their final values according to a given law, they are decreased again to zero according to another law, which is not exactly the inverse of the first one. Since the body is in a bowl impermeable to heat, it cannot have received heat, nor it can have given heat; and, on the other hand, since the body has taken its primitive state back, it will contain, at the end, the same amount of heat that it contained at the beginning.
>
> If, then, the work spent by the external forces in the period of their growth were not exactly equal to that gained in the period of their decrease, one would have a production or a consumption of work, which would not be compensated by an equivalent amount of heat consumed or produced. Which is against the *principle of conservation of energies* [16].[119] (A.4.47)

In this way, Castigliano could express the strain work as a function of the external forces only, in the same form obtained for trusses [16][120]:

$$\frac{1}{2}\left(Pp + Qq + Rr + \cdots\right). \tag{4.47}$$

By following the same passages of his thesis, Castigliano obtained the theorems of the derivative of work, and the theorem of minimum strain work. They can be formulated for bent structures of any kind, provided that one is able to express the strain work as a function of the external forces, of the tensions of the possible joint bars, and of the constraint reactions, linked among them by the equilibrium equations.

Castigliano was very careful, and preferred to provide an explicit formulation for a bent beam, presenting it as a theorem based on the conservation of plane cross-sections, modeled as rigid disks.

> Therefore, since, because of the perfectly rigid disk, the cross-section remains plane during the deformation of the system, [...] the following theorem holds, which is very important for the theory of the strength of solids.
>
> *The derivatives of the strain work with respect to the three forces X, Y, Z and the moments of the three couples, defined above, express the displacement of the cross-section center parallel to the directions of the forces, and the three rotations of the same cross-section about its principal axes of inertia and about its perpendicular through the center* [16].[121] (A.4.48)

It is worth remarking that in modern expositions of Castigliano's theorem one considers a general elastic system, for which the existence of the complementary elastic energy is proved. The theorem of minimum strain work, which is seen as a particular case of the theorem of minimum complementary elastic energy if the constraints are held fixed, refers to a general elastic system; applications to trusses are considered obvious.

---

[119] p. 162. Our translation.

[120] Equation (17), p. 164. Even though Castigliano did not quote him explicitly, this is the relation that Lamé attributed to Clapeyron in his Lectures, and that nowadays is known as Clapeyron's theorem. Castigliano, however, proved to know Lamé's textbook (for instance, quoting it on p. 158), thus we may suppose that he knew Clapeyron's formulation.

[121] p. 171. Our translation.

Castigliano found himself in a different condition: firstly, trusses were maybe the most important structures of his time, and deserved a special treatment; in addition, in the generalization to flexible elements some doubts on the rigor of the proof seemed to persist.

4.4.4 1879. Théorie de l'équilibre des systémes élastiques et ses Applications

The *Théorie de l'équilibre des systémes élastiques et ses applications* [19] (hereinafter *Théorie*) represented the top of all the preceding works on the subject. Even though it added almost nothing from the theoretical point of view, this publication was important because it permitted the diffusion, at an international level, of Castigliano's ideas. These would be particularly appreciated in Germany, where a strong School of Structural Engineering existed having as exponents Mohr, Grashof, Müller-Breslau. Besides the exposition of his theory, Castigliano reported also applications to real structures, which he had published in 1878 [18].

The *Théorie* is a true textbook of strength of materials and structural mechanics: Castigliano added elements of the theory of elasticity to the formulation of his theorems, which was presented in the first two chapters. The monograph, however, was not successful from the educational point of view: the language (French) made it easy for international diffusion, but limited it in the national setting; moreover, Castigliano was not an academic, and no professor of that time had the intellectual honesty to recognize its worth and to adopt it. Here is what Crotti wrote on the educational value of Castigliano's textbook:

> Having I asked him: Why in your book you did not think it better, as more general, Lamé's hypothesis? To this, he answered: And what would the second coefficient be useful for? do we have, for the generality of solid bodies, serious experiences that have established its value?
>
> The road kept by Castigliano in his main treatise is not always, rigorously speaking, what one would call the main road, and I believe that he has been induced in doing so by a very correct reason. Going down from the general to the particular is the main merit of the works, which are addressed to minds, in which the ideas on the subject one is dealing with are already mature; it is not the best road for a book that should be useful for the learned and, at the same time, for who desires to learn. And it is because of this that our Author puts beforehand the treatment of articulated systems, where solids are considered subject to tensile or compressive uniform forces for all of their cross-section. He starts, thus, from a very simple case to ascend to the reciprocal actions of a molecule with the others nearby, and each time he proves the principles of superposition of effects and of the theorem of the derivatives of work. After this preparation, which has let the reader get, little by little, familiar with certain ideas, he ascends to the general theory of the elementary parallelepiped and, once posed the general equations, he applies them to numerous cases of bending and flexure of solids of various form. Afterwards he skips to the part of approximated applications, justifying

the usual formulas for the beam, and preparing the arguments for a quick application of his theorem [40].[122] (A.4.49)

In the following, we report the important aspects of the *Théorie* according to us, that contain elements of novelty with respect to the design of one-dimensional elements, neglecting the aspects that are typical of a textbook on Structural mechanics. We will provide the new statement of the theorems of the derivatives of work, and, eventually, a part in which Castigliano presented the expression of the strain work as a sum of the works of single molecules. This part is particularly interesting because it uses an energetic, Green-like, approach, to yield the results obtained by Cauchy and Poisson by means of the equilibrium equations among molecules. In the end, we will report an application.

We begin by a synopsis of the introduction, which is particularly enlightening about Castigliano's conceptions about structural disciplines.

PREFACE

This work contains the *theory of equilibrium of elastic systems, presented according to a new method*, founded on some theorems that are brand new, or still little known.[123]

As a part of this theory, one will find here the *mathematical theory of the equilibrium of solid bodies*, considered, in particular, from the point of view of the strength of materials.

We believe that the moment has come, to introduce this rational way of presenting the strength of materials in teaching, thus abandoning the old methods that the illustrious LAMÉ has rightly defined as *semi-analitical and semi-empirical, useful only to hide the edge of the real science.*

We will now provide some historical information on the discovery of the theorems which we will almost continuously use through the entire course of this work.

These theorems are the following three

1st of the derivatives of the work, first part;

2nd id. id. second part;

3rd of minimum work.

The first one had already been used by the famous English astronomer GREEN, but only in a particular question, and had not at all been stated, nor proved, in a general way, as we do in the present work.

The second is the reciprocal of the first, and we believe that it been stated and proved for the first time in 1873, in our dissertation to obtain the diploma as Engineer in Turin: we have given it more extension in our memoir entitled *Nuova teoria intorno all'equilibrio dei sistemi elastici*, published in the Proceedings of the Reale accademia delle scienze of Turin in 1875. The third theorem may be regarded as a corollary of the second; but, just like in some other questions of *maxima and minima*, it has been, so to say, presented many times before the discovery of the main theorem.

[...] Here [are] now some information on the redaction of our work.

Since our goal is not only to present a theory, but also to let its advantages of brevity and simplicity be appreciated in practical applications, we have solved, following the new method, not only the majority of the general problems that one deals with in the teaching of

---

[122] p. 10. Our translation.

[123] We report in italics the parts that in the original were underlined.

strength of materials, but we have also added several numerical examples for the calculation of the most important elastic systems.

[...] As for the calculations, we will point that they are hardly longer than in the methods ordinarily adopted; and that, anyway, we will almost always be able to shorten them sensibly by neglecting some terms, which have small incidence on the result [19].[124] (A.4.50)

The theorems on the derivatives of work were so introduced:

11. *Theorem of the derivatives of the strain work.*

First Part—If we express the strain work of an articulated system as a function of the relative displacements of the external forces applied to its vertexes, we obtain a formula, the derivatives of which with respect to such displacements provide the values of the corresponding forces.

Second Part—If, on the other hand, we express the strain work of an articulated system as a function of the external forces, we obtain a formula, the derivatives of which with respect to such forces, provide the relative displacements of their points of application [19].[125] (A.4.51)

They nowadays are referred to as second, and first Castigliano's theorem, respectively (or Castigliano's complementary theorem and theorem, *tout court*), while the author presented them as parts of the single theorem of the derivatives of the strain work. The first part, as Castigliano rightly said in the introduction, may be attributed to Green, who, in 1828, put into relationship the derivatives of elastic potential with stress. Castigliano, on the other hand, attributed to himself the merit of extending it to the case of a structure, and also to have first discovered the unity of the two parts of the theorem.

Castigliano proved the first part in a very simple way: if a system of forces R_p acting on the nodes of an articulated system is incremented by dR_p, a compatible displacement increment dr_p will correspond to it. Following this increment of displacement, the external forces spend the work [19][126]:

$$\sum R_p dr_p, \tag{4.48}$$

where the sum ranges over all the loaded nodes.

If L represents the total strain work, expressed as a function of the displacements components of the nodes to which forces are applied (in modern terms, the elastic potential energy), it is clear that the increment of work due to the increment dr_p of the relative displacements of the vertexes will be expressed by the formula [19][127]:

$$\sum \frac{dL}{dr_p} dr_p. \tag{4.49}$$

The proof is obtained by equating the coefficients of dr_p in (4.48), (4.49).

[124] pp. 5–8. Our translation.

[125] p. 26. Our translation.

[126] First unnumbered equation, p. 26.

[127] p. 26; the equation is the second unnumbered one.

The second part of the theorem was proved as in the works of 1873 and 1875. For completeness, and since it was Castigliano's last word on the subject, we report it in full.

As for the second part, we observe that the strain work of the system due to the increments dR_p of the external forces shall also be represented by the differential of the formula (15),[128] which is

$$\frac{1}{2}\sum R_p dr_p + \frac{1}{2}\sum r_p dR_p:$$

we have then the equation

$$\sum R_p dr_p = \frac{1}{2}\sum R_p dr_p + \frac{1}{2}\sum r_p dR_p,$$

whence one has

$$\sum R_p dr_p = \sum r_p dR_p;$$

and, since the left hand side of this equation represents the strain work of the system by the increments dR_p of the external forces, it results that the right hand side represents also the same.

Now, if we call L the strain work of the system due to the forces R_p, it is apparent that the infinitely small work due to the increments dR_p will be represented by the formula

$$\sum \frac{dL}{dR_p} dR_p$$

Since this formula shall be identical to the other $\sum r_p dR_p$, it follows that one shall have for each force

$$\frac{dL}{dR_p} = r_p$$

that proves the second part of the theorem [19].[129] (A.4.52)

4.4.4.1 Flexible Systems

The results obtained for trusses were all contained in Chap. 1, and are reported above. In Chap. 2, they were extended to mixed, or even simply flexible, systems. Here Castigliano went back to the first proof of 1875 and considered a generic structural element, or even a whole frame, as a huge truss, the vertexes of which are the molecules, and the bars of which model intermolecular forces. For such kind of systems, the theorems of minimum strain work 'undoubtedly' hold. To transfer the result in a global way to a beam, eliminating the considerations on the molecules constituting it, Castigliano adopted a trick he had already used in his second paper of 1875: he considered the beam cross-sections as a unique body, a bunch of molecules

[128] Which expresses the elastic work in the form $L = \frac{1}{2}\sum R_p r_p$.

[129] p. 27. Our translation.

moving rigidly. The forces and the displacements entering the discussion, then, are those reduced to the center of gravity of the cross-section. By expressing the strain work with respect to these quantities, Castigliano came to formulate the theorem:

> 1. The resultants $\mathscr{X}$, $\mathscr{Y}$, $\mathscr{Z}$ and the resultant moments M_x, M_y, M_z are the derivatives of the strain work of the system with respect to the displacements ξ_0, η_0, ζ_0 and to the rotations $\theta_x, \theta_y, \theta_z$.
>
> 2. The three displacements ξ_0, η_0, ζ_0, and the three rotations $\theta_x, \theta_y, \theta_z$, are the derivatives of the strain work of the system with respect to the resultants $\mathscr{X}$, $\mathscr{Y}$, $\mathscr{Z}$ and the resultant moments M_x, M_y, M_z [19].[130] (A.4.53)

4.4.4.2 The Costitutive Relationship

Castigliano obtained the constitutive equation starting from the molecular work, arriving to a relation with 15 coefficients:

> 4. Strain work of a very small parallelepiped.
>
> In the elementary parallelepiped, the ridges of which are $\delta x, \delta y, \delta z$, let us consider the small straight line r joining two molecules closely spaced. In the deformation of the body, this straight line grows, starting from the initial length to the value $r(1 + \partial_r)$, and the tension between the two molecules grows proportionally to the stretch, in such a way that, when the straight line will have the length $r + \rho$, ρ being a quantity smaller than $r\partial_r$, the tension between the two molecules will be $\epsilon\rho$, calling ϵ a coefficient that is constant for any pair of molecules, but different for the various pairs.
>
> The strain work of the straight line r will be[131]
>
> $$\int_0^{r\partial_r} \epsilon\rho d\rho = \frac{1}{2}\epsilon r^2 \partial_r^2$$
>
> that is to say, replacing $\partial^2 r$ by its value given by formula (4.8),[132]
>
> $$\frac{1}{2}\epsilon r^2 (\partial_x \cos^2\alpha + \partial_y \cos^2\beta + \partial_z \cos^2\gamma + g_{yz}\cos\beta\cos\gamma + g_{xz}\cos\gamma\cos\alpha + g_{xy}\cos\alpha\cos\beta)^2$$
>
> where one shall remark that developing the square, and gathering the terms containing the same products of the cosines $\cos\alpha, \cos\beta, \cos\gamma$ *the distinct terms reduce to fifteen.*[133] To have the strain work of the whole parallelepiped, we must add the expressions analogous to that, for all the molecular pairs that it contains [19].[134] (A.4.54)

After having summed and put into evidence the distinct terms, Castigliano came to the expression:

---

[130] p. 52. Our translation.

[131] $\partial_x, \partial_y, \partial_z$ represent the axial infinitesimal strains, and g_{xy}, g_{xz}, g_{yz} the angular distortions.

[132] Not reported here.

[133] Italics is ours.

[134] p. 59. Our translation.

$$\frac{\overline{\omega}}{2}\begin{bmatrix} a_1\partial_x^2 + a_2\partial_y^2 + a_3\partial_x^3 + b_1(2\partial_y\partial_z + g_{yz}^2) + b_2(2\partial_z\partial_x + g_{zx}^2) \\ +b_3(2\partial_x\partial_y + g_{xy}^2) + 2c_1(\partial_x g_{yz} + g_{zx}g_{xy}) \\ +2c_2(\partial_y g_{zx} + g_{yz}g_{xy}) + 2c_3(\partial_z g_{xy} + g_{yz}g_{zx}) \\ +2e_1\partial_z g_{yz} + 2e_2\partial_x g_{zx} + 2e_3\partial_y g_{xy} + 2f_1\partial_y g_{yz} \\ +2f_2\partial_z g_{zx} + 2f_3\partial_x g_{xy} \end{bmatrix}. \tag{4.50}$$

Here $\overline{\omega} = \delta x \delta y \delta z$ represents the volume of the parallelepiped, and the 15 coefficients a_i, b_i, c_i, etc. contain sines and cosines to various powers. One has, for example:

$$a_1\overline{\omega} = \sum \epsilon r^2 \cos^4 \alpha. \tag{4.51}$$

By deriving the strain work with respect to the strain measures, Castigliano obtained stresses as a function of 15 distinct coefficients.

In his analysis Castigliano, as it was usual in the French tradition, assumed a discrete model (molecules) to balance energies, and a continuous model to write displacements, that are dealt with as a regular vector field. To him, the potential energy of a given pair of molecules depended on the mutual distance only, and did not depend on the placement of the other molecules. The variation of the distance of two molecules, supposed small, was then expressed as a function of the components of the symmetric part of the gradient of the displacement field.

Some years later, Poincaré [81][135] would make it clear that the peculiarity of the molecular model, leading to a constitutive relation with 15 coefficients, lays right in the assumption of the independence of the action between two molecules on the position of the others and thus, by using an energetic language, in the assumption that the potential energy of the whole body be the sum of the potential energy of pairs of molecules, $U = U_1(r_1) + U_2(r_2) + \cdots + U_n(r_n)$. To obtain a constitutive relation with 21 coefficients one must admit, always keeping a point-like molecular scheme, that two molecules exert on each other a force depending on the position of the other molecules as well, and, thus, the potential energy of the whole body be not uncoupled, but expressed as a function of the mutual distance of all the molecules, $U = U(r_1, r_2, \cdots, r_n)$.

4.4.4.3 Applications: The Dora Bridge

Castigliano ended the textbook with a series of applications to real cases, which he had already published in Italian in 1878 [18]. Their function is not only an example (to let us understand how to apply the theory), but it also has an apologetic value: he wanted to persuade professional engineers that his approach to the theory of structures was not only theoretical, but it was also the most suitable one to analyze structures.

In the following, for reasons of space, we report an application only. It is the calculation report of the street bridge on the Dora river in Turin, a granite arch bridge

[135] Before Poincaré, other scholars had made the same remark; on the subject, see [55], pp. 179–217.

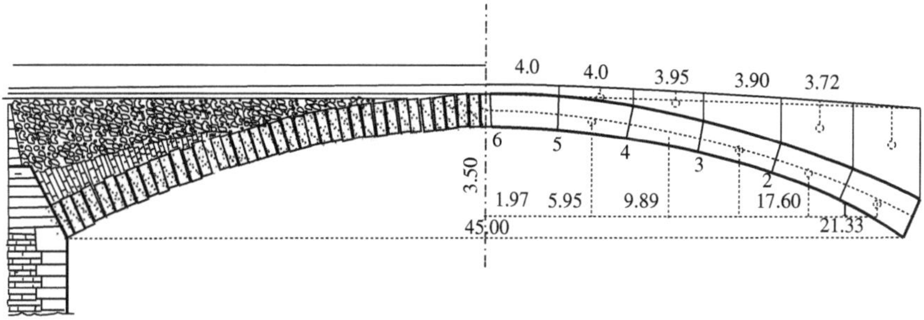

Fig. 4.7 The Dora bridge (Redrawn from [19], Planche XIV, Fig. 4.2)

slender enough (span 45 m, height 5.5 m). The bridge was built by the engineer Carlo Bernardo Mosca in 1828, with an accuracy well appreciated by Castigliano; in addition, the notoriety of the building influenced for sure Castigliano's choice to analyze it by his methods.

The archway of the bridge, shown in Fig. 4.7, was made up of granite blocks from Malanaggio, near Pinerolo, the weight of which was 27.5 KN/m^3, arranged in 93 orders. The abutments were made of bricks, the weight of which was 23 KN/m^3, the filling was in dirt, with weight 16 KN/m^3. On this dirt, there was a roadbed, the weight of which was 18 KN/m^3. The blocks at keystone and at the abutments were made of mortar instead of granite.

Castigliano made an accurate analysis of the loads and of the mechanical characteristics, of the cross-section areas, and of the moments of inertia, by dividing the bridge axis into twelve equal segments of length 4m. The result of the analysis of the loads led to the values of the bending moment and normal force in the seven cross-sections of the various segments of half the bridge, reported in Table 4.6, numbered from 0 to 6 starting from the abutment, with reference to Fig. 4.8. We avoided reporting the values of the shearing force, since the bridge was supposed not deformable in shear. The values of the geometrical characteristics, cross-section area and moments of inertia, of the seven sections are reported in Table 4.7.

Table 4.6 Inner actions in the arch; M, Q are the moment and the normal force at the keystone (thrust of the arch) [18, p.128]

Section	Bending moment	Axial force
0	$M_0 = M - 5.36Q + 2,021,937$	$P_0 = 0.895Q + 102,217$
1	$M_1 = M - 3.76Q + 318,295$	$P_1 = 0.925Q + 63,085$
2	$M_2 = M - 2.40Q + 800,470$	$P_2 = 0.950Q + 35,915$
3	$M_3 = M - 1.35Q + 429,935$	$P_3 = 0.973Q + 18,177$
4	$M_4 = M - 0.60Q + 185,955$	$P_4 = 0.988Q + 7,480$
5	$M_5 = M - 0.15Q + 46,392$	$P_5 = 0.966Q + 1,815$
6	$M_6 = M$	$P_6 = Q$

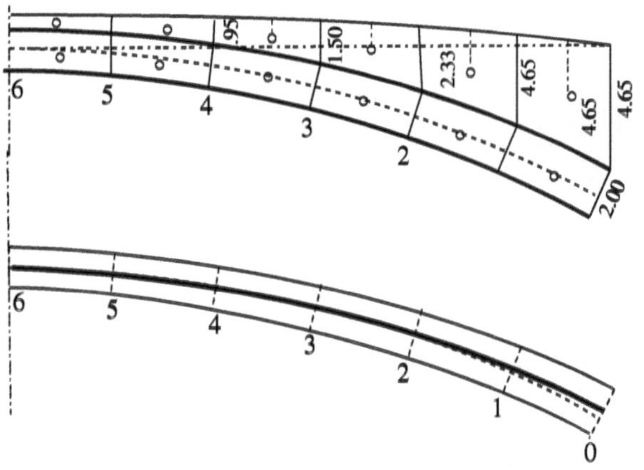

Fig. 4.8 Model for calculations (Redrawn from [20], Planche XIV, Figs. 4.2 and 4.3)

Table 4.7 Geometrical characteristics of the arch cross-sections (m) [18, p. 131]

Section	Area	Moment of inertia
0	2.01	0.67672
1	1.85	0.52764
2	1.72	0.42404
3	1.62	0.35429
4	1.55	0.31032
5	1.51	0.28691
6	1.50	0.28125

Castigliano calculated the strain work of the arch with the formula, though not made explicit:

$$L = 2\left(\int \frac{M}{2EI}\,dx + \int \frac{P^2}{2E\Omega}\,dx\right) \tag{4.52}$$

where the symbols are usual and the integral is extended over one half of the arch. If the bridge had been all in granite, blocks at the abutments and at keystones included, the strain energy would have been given by (4.37), with $E =$ constant, equal to the modulus of elasticity of granite. By applying formulas of numerical integration,[136] Castigliano provided this expression of the strain work:

[136] He assumed a quadratic function for M^2/EI, linear for $P^2/E\Omega$, and used Simpson formulas, introduced on p. 202, and the formulas of trapeze, respectively. The intervals in which the integration domain is divided are equal to the length of the segments, that is 4 m.

$$2 \times \frac{4,00}{2E} \left\{ \frac{1}{3} \left(\frac{M_0^2}{I_0} + 4\frac{M_1^2}{I_1} + 2\frac{M_2^2}{I_2} + \cdots \frac{M_6^2}{I_6} \right) \right.$$
$$\left. + \left(\frac{P_0^2}{2\Omega_0} + \frac{P_1^2}{\Omega_1} + \frac{P_2^2}{\Omega_2} + \cdots \frac{P_6^2}{2\Omega_6} \right) \right\}.$$

The values of M_i, P_i are reported in Table 4.6, as a function of M, Q.

The bridge was not entirely in granite: there were parts in mortar, hence the strain work of the joints, thought made of granite, had to be subtracted from the preceding expression, and the corresponding strain work of the mortar joints had to be added. The strain work of granite joints was[137]:

$$\frac{1}{2E} \frac{a' + a''}{2} \left(\frac{P^2}{\Omega} + \frac{M^2}{I} + \frac{a'' - a'}{a'' + a'} \frac{4MP}{h\Omega} \right).$$

Here a', a'' are the thicknesses of the blocks at at the extrados and the intrados, respectively; P, M, I, Ω are referred to the middle section, and h is the height of the middle section. For a block of mortar equal to one of granite, the strain work was obtained by replacing the modulus of elasticity E of the granite with that of the mortar, E'. The difference between the two works is given by:

$$\frac{1}{2} \left(\frac{1}{E} - \frac{1}{E'} \right) \frac{a' + a''}{2} \left(\frac{P^2}{\Omega} + \frac{M^2}{I} + \frac{a'' - a'}{a'' + a'} \frac{4MP}{h\Omega} \right).$$

For each of the two joints at the abutments, one has:

$$P = P_0, \quad M = M_0, \quad \Omega = \Omega_0, \quad I = I_0, \quad h = h_0 \quad a' = 0.0283, \quad a'' = 0.064$$

while at the keystone one has:

$$P = P_6, \quad M = M_6, \quad \Omega = \Omega_6, \quad I = I_6, \quad h = h_6 \quad a' = 0.09, \quad a'' = 0$$

The following quantity, corresponding to the blocks of the abutments and key, must be added to the strain work of the granite bridge:

$$\frac{1}{2} \left(\frac{1}{E} - \frac{1}{E'} \right) \times$$
$$\left\{ 0.0025 \left(\frac{P_0^2}{\Omega_0} + \frac{M_0^2}{I_0} + 1.547 \frac{M_0 P_0}{h_0 \Omega_0} \right) + 0.045 \left(\frac{P_6^2}{\Omega_6} + \frac{M_6^2}{I_6} - 4 \frac{M_6 P_6}{h_6 \Omega_0} \right) \right\}.$$

Replacing $M_0, M_1, \ldots, P_0, P_1, \ldots$ by their values in function of M, Q given by Table 4.6, and inserting the values of the geometrical characteristics, Castigliano

137 Castigliano integrated the product of axial stress and strain over the sections of the joints.

came to the following expressions of the strain work of the arch, supposed of granite, and of the additional terms due to the contribution of the blocks:

$$2 \times \frac{4,00}{EI}\left(16.34M^2 - 22.98MQ + 70.06Q^2 + 7\,818\,893 \cdot 2M - 23\,249\,885 \cdot 2Q\right) +$$
$$\frac{1}{2}\left(\frac{1}{E} - \frac{1}{E'}\right)\left(0.2984M^2 - 0.766 \cdot 2MQ + 3.876Q^2 + 281\,765 \cdot 2M - 1\,466\,630 \cdot Q\right).$$

To sum the two expressions, one should have known the values of the elasticity moduli E, E'. Castigliano knew that it was difficult to provide reliable values for the two moduli. He limited himself to provide an estimate of their ratio, by supposing $E = 100E'$, so that the total strain work became:

$$2 \times \frac{4.00}{2E}\left(20.03\,M^2 - 32.46 \cdot 2MQ + 118.00\,Q^2 + 11\,305\,728 \cdot 2M - 41\,398\,510 \cdot 2Q\right).$$

The geometrical conditions that the archway shall satisfy during deformation are:

(a) The section corresponding to the key shall stay vertical, that is with zero rotation.
(b) The same section shall undergo no horizontal displacement.

The values of M, Q satisfying these two conditions make the derivatives of the strain work with respect to M, Q vanish, respectively:

$$\begin{aligned} 20.03\,M - 32.46\,Q + 11\,305\,728 &= 0 \\ -\,32.46\,M + 118.00\,Q + 42\,398\,510 &= 0. \end{aligned} \tag{4.53}$$

4.4.5 A Missing Concept: The Complementary Elastic Energy

In Castigliano and Menabrea's works a modern scholar of structural mechanics remarks the absence of the concept, common in the present teaching and research, of complementary elastic energy. This leads to true errors (like in the first proof by Menabrea of the principle of minimum work), or to inaccuracies, or, at least, to ambiguities in the terminology (like in Castigliano). We have already commented Menabrea's theoretical drawbacks, while for Castigliano the situation is more complex.

For instance, in his proof of the theorem of least work, in his thesis of 1873, a modern reader could meet difficulties, for both the use of principles exterior to the theory of elasticity (the conservation of elastic forces, the equality of inner and outer works), and an ambiguous use of the idea of strain work.

Castigliano called strain work what we now call the elastic potential energy of each bar. The sum of the strain work of all the bars, referred by Castigliano still as strain work, is not however the total elastic potential energy of the system (modern meaning), because in principle no-one assures equilibrium, nor compatibility.[138]

The same drawbacks, less evident because the language is less precise than in the proof, lay in the statement of the theorem of least work. The molecular work should be minimized by varying the tensions in the bars, keeping them equilibrated among them and with the external forces. However, in this way, the strain work could not have the meaning of potential energy that Castigliano seemed to provide it with, because, in general, compatibility is not verified. It is not clear whether Castigliano realized his ambiguity in terms; he managed to obtain a correct proof without the concept of complementary energy because he would consider only linear elastic structures.

Castigliano showed the same ambiguity in the proof, contained in his second memoir of 1875, of Menabrea's principle of elasticity starting from the theorem of the derivatives of work. The latter was proved in an unexceptionable way; but, when Castigliano separated the structure from the redundant bars, he kept on calling *strain work* the sum of the strain work of the system without the redundant bars and the strain work of redundant bars:

$$F + \frac{1}{2}\frac{T_{pq}^2}{\epsilon_{pq}}. \tag{4.54}$$

This would be erroneous if one wished to identify such a sum with the elastic potential energy of the system. Indeed, if T_{pq} would vary arbitrarily, compatibility, that is the equivalence of the displacements of p, q thought as belonging to the bar and to the system without the bar, could not always be assured: the sum of strain works, thus, would not have any physical meaning. If equilibrium—that is the equivalence of the tension T_{pq} applied to the bar and the same applied to the truss without the bar—is imposed, then the sum of the strain works is what we nowadays call complementary elastic energy.

These 'errors', 'inaccuracies', or 'ambiguities' were, at least in part, reported to Castigliano by his friend Crotti, who so commemorated him:

> Let us stop for a while to consider what is, from the scientific point of view, the novelty, the scope, and the usefulness of this theorem of the derivatives of work, and of the other, we may say its twin, of minimum work.
>
> Well then, these theorems, if considered from the point of view of the general theory, do not constitute substantially new statements. Legendre already had proved that, given a function ϕ of n variables x, one can form by its partial derivatives a function ψ the partial derivatives of which are equal to the variables x, respectively. It had also been recognized that, if ϕ is quadratic, it turns out that $\phi = \psi$. Later on, the famous English mathematician George Green was lead, by considerations on the impossibility of perpetual motion, to establish that the work of an elastic system was represented by a potential of the displacements, and this in the two illustrious memoirs on light of 1839.

138 Using a modern language, if compatibility is added, the sum of strain works coincides with the elastic potential energy, if equilibrium is added, the sum of strain works coincides with the elastic complementary energy.

> The analytical background expressing the properties of the two theorems of which I talk was, then, completely known; I do not believe, however, that they have been formally stated, maybe perhaps they did not concur to the progress of the general theory, which, by the considerations on the displacements, comes to use the same formulas to which those two theorems lead [40].[139] (A.4.55)

Actually, Crotti's clarification, even if it had been put into practice by Castigliano, would not have eliminated ambiguities. Crotti's concept of complementary energy is more restricted than that necessary to simplify Castigliano's argument. In his monograph of 1888 [41] Crotti introduced the elastic potential energy of a linear and of a non-linear system, using the term *work function*, expressed in terms of displacements $f_1, f_2, \ldots, f_n$ [41][140]:

$$L = \phi(f_1, f_2, \ldots, f_n) \tag{4.55}$$

or in terms of the forces $F_1, F_2, \ldots, F_n$ that produced these displacements:

$$L = \psi(F_1, F_2, \ldots, F_n). \tag{4.56}$$

Thus, Crotti considered a structural system without decomposing it into elements, and the work function he introduced corresponded to a situation with internal compatibility.

Crotti introduced the elastic complementary energy, without attributing any name to it, but only the symbol λ[141]:

> Let us now make
>
> $$\lambda = F_1 f_1 + F_2 f_2 + \cdots - L.$$
>
> By differentiating, we have
>
> $$d\lambda = f_1 dF_1 + f_2 dF_2 + \cdots + (F_1 df_1 + F_2 df_2 + \cdots - dL)$$
>
> but, being nil the quantity between parenthesis, it remains:
>
> $$d\lambda = f_1 dF_1 + f_2 dF_2 + \cdots$$
>
> Considering then λ as a function of the forces F, we have [41][142]:
>
> $$\frac{d\lambda}{dF_1} = f_1, \quad \frac{d\lambda}{dF_2} = f_2 \quad \text{etc.}$$
>
> (A.4.56)

[139] pp. 5–6. Our translation.

[140] p. 60.

[141] In this way, Crotti would so assign an instrumental value to λ, and not the meaning of a physical quantity.

[142] pp. 61–62. Our translation.

Crotti remarked that in linear elasticity $\lambda = L$ ond one re-obtains Castigliano's result; he, then, went on presenting dual laws, following a fashion of the time borrowed from projective geometry.[143]

By the function λ Crotti threw new light on Castigliano's results, and extended them to the non-linear case, but he was not able to eliminate the linguistic drawbacks. To this purpose, the modern concept of complementary energy would have been needed, which, to understand Crotti's position better, we show in the following. For simplicity, we consider the discrete linear elastic case.

Given a linear elastic system made up of elements connected by n nodes, let $u_1, u_2, \ldots$ be their displacements, and $f_1, f_2, \ldots$ the active forces acting on them. For some nodes, the constraints displacements $\underline{u}_1, \underline{u}_2, \ldots$ are assigned, corresponding to the constraint reactions $\underline{f}_1, \underline{f}_2, \ldots$. In addition, let U_E be the elastic energies of the elements as a function of the $u_1, u_2, \ldots$, and V_E the same quantities as a function of the $f_1, f_2, \ldots$. That posed, the total potential energy of the system is defined as the function of the congruent displacements:

$$U = \sum_E U_E - \sum_i f_i u_i \tag{4.57}$$

where f_i are assumed as known.

On the other hand, we call total complementary energy the function of equilibrated forces:

$$V = \sum_E V_E - \sum_i \underline{f}_i \underline{u}_i, \tag{4.58}$$

where $\underline{u}_i$ are assumed as known.

The expressions of U and V in general do not coincide even in the case of linear elasticity, because the external forces and the distortions appear in a 'complementary' way. The expressions $\sum_E V_E$, $\sum_E U_E$ instead coincide if variables are are related by means of the constitutive (linear) relationship, but they are not meaningful at the same time. Indeed forces constitutively linked to compatible displacements are not, in general, equilibrated with external forces and *vice versa*.

The following two theorems of minimum hold for the functions U, V: the set of admissible displacements minimizing U is associated to inner tensions equilibrated with external forces; the set of inner forces equilibrated with external forces minimizing V is associated with a field of compatible displacements. The solution of the elastic problem can thus be obtained by minimizing one function or the other. By searching the minimum of U and V we obtain, respectively:

[143] In the first part of the 19th century, the existence of dual laws had been underlined by Poncelet and Plücker. Later on, Chasles [33], Culmann [42], and Cremona [39] had given great importance to duality laws. Nowadays the duality considered by Crotti leads to the two approaches, called of displacements (or deformations) and of forces (or tensions).

$$\frac{\partial \sum_E U_E}{\partial u_i} = f_i \qquad \frac{\partial \sum_E V_E}{\partial \underline{f}_i} = \underline{u}_i \tag{4.59}$$

that is Castigliano's two theorems.[144]

By the concept of complementary energy, Castigliano's proofs may be reformulated with a language without ambiguities. The sum

$$F + \frac{1}{2}\frac{T_{pq}^2}{\epsilon_{pq}} \tag{4.60}$$

may be considered sum of complementary energies, and called total complementary energy.

Obviously, it would be unhistorical to accuse Castigliano and Crotti not to have introduced such a concept; apart from the fact that a pioneer never perfects his theory, we must consider that, in the field of problems in which Castigliano was interested in, that is the calculation of linear elastic structures without distortions, the introduction of the concept of complementary energy appears immaterial at all, or even a baroquism.

4.5 Valentino Cerruti

Valentino Cerruti (Crocemosso di Biella (near Turin) 1850–1909) was still a student, when he published works on analytical geometry in the Giornale di matematiche by Battaglini.[145] After his graduation at the School of Application for Engineers in Turin in 1873, Cerruti moved to Rome, where he became private teacher of the children of Quintino Sella and became close friend with him. In 1873 he also had an appointment as assistant professor in Hydraulics at the School of Application for Engineers in Rome; he became professor of Rational mechanics in 1877, full professor in 1881. In 1888 he became rector of the university of Rome, then dean of the faculty of sciences in 1892 and rector of the university again from 1900 to 1903. In 1901 he was elected at the Senate of the Kingdom of Italy and promoted the bill of law which transformed the two different technical schools in Turin into a polytechnic school, which still operates under the name of Politecnico di Torino. In 1903 he became director of the School of Application for Engineers in Roma.[146]

[144] Remark that in (4.59) $\frac{\partial \sum_E U_E}{\partial u_i} = f_i$ is calculated at equilibrium, while $\frac{\partial \sum_E V_E}{\partial \underline{f}_i} = \underline{u}_i$ is calculated in a compatible configuration.

[145] This journal, founded in 1863 by the Italian mathematician Giuseppe Battaglini, was intended to spread mathematical knowledge and up-to-date research in new branches of mathematics in Italian universities, as its complete title says: *Giornale di matematiche ad uso degli studenti delle Università italiane, pubblicato per cura del professore G. Battaglini.*

[146] A more detailed biography on Cerruti may be found in [56, 58, 92].

Even if his education was in the field of engineering, the scientific work of Cerruti was mainly devoted to rational mechanics. This of course emphasizes his strong mathematical background, which should have been a main feature of the school in Turin. Some of Cerruti's contributions were referred to in Chap. 3. Here we will concentrate on his thesis *Sistemi elastici articolati*, that as already mentioned, was presented in 1873 in the same session and with the same supervisor (Curioni) as Castigliano.

4.5.1 Sistemi elastici articolati. *A Summary*

Cerruti's treatment appears elegant and written with intelligence and mastery of mathematical instruments. On the other hand, the presentation of the subject is not always precise and uniform—sometimes one feels like a conceptual jump from one of his main subjects to the other, and conversely—and a series of typographical errors are present and quite apparent. It looks like Cerruti did not spend too much time on a very precise preparation of this work, which seems reasonable since from his biography it turns out that at that time he was deeply busy with other research and activities. Here we will summarize the thesis, expounding the main points of interest. Some of them will be however highlighted in the subsequent Sects. 4.5.2 and 4.5.3.

After a short introduction in section 2, Cerruti introduced articulated systems as structures composed of bodies (which are understood to be rectilinear, even if this statement is never openly made) connected by nodes that are frictionless spherical hinges. Cerruti said that his results hold also for cylindrical hinges (that is plane trusses) and admitted that the absence of friction was an idealization:

> I will suppose that at the contact surface of different elements friction does not exist, or is at least negligible: if this happens, the elements will bear only longitudinal stress. I must at once add, that frictionless articulated systems do not actually exist and are merely abstractions. In spite of this, their study leads to applications, at least in those cases in which the effects of bending may be neglected [28].[147] (A.4.57)

Cerruti distinguished among 'simple' and 'complex' articulated systems: the former are those in which nodes connect only two elements and, he says, reduce to chains and may easily be treated by means of graphical methods like that of the funicular polygon. The latter are those in which nodes connect more than two elements and are those dealt with by Cerruti.

4.5.1.1 Counting of Equations and Constraints

In his section 3, Cerruti defined a truss as a system of points in space connected by rectilinear bars. This view, focusing on the nodes seen as body-points rather than on bars, is the same as that of Menabrea [65, 66, 68, 70], who for sure left his mark

[147] p. 6.

in Turin. Cerruti specified the minimum condition number on the total $n(n-1)/2$ 'distances' between the couples of points of the system to have a unique well-defined shape, resulting as $3n-6$. These conditions, as well as those on some terminal nodes (*vertici*, vertexes) to fix the configuration may easily be interpreted, in contemporary language, as a definition of a statically determined truss.

External constraints, that is the conditions imposed on the vertexes, must be at least in number of six to have a well constrained body. Their expressions f_i were given by Cerruti in terms of the coordinates $x, y, z;\ \xi, \eta, \zeta;\ \ldots$ of the points on which they act [28][148]:

$$f_1 = 0,\ f_2 = 0,\ f_3 = 0, \ldots f_6 = 0. \tag{4.61}$$

Constraint reactions are the Lagrange multipliers of the first variation of condition equations [28][149]:

$$\begin{array}{ccc} \Pi_i \dfrac{df_i}{dx}, & \Pi_i \dfrac{df_i}{dy}, & \Pi_i \dfrac{df_i}{dz}, \\ \Pi_i \dfrac{df_i}{d\xi}, & \Pi_i \dfrac{df_i}{d\eta}, & \Pi_i \dfrac{df_i}{d\zeta}, \\ \ldots & \ldots & \ldots \end{array} \tag{4.62}$$

Internal constraints are given by the presence of bars which impose a constraint on the distance between couples of points i, j. They have the form [28].[150]

$$F_{ij} = (x_i - x_j)^2 + (y_i - y_j)^2 + (z_i - z_j)^2 - l_{ij}^2 = 0 \tag{4.63}$$

and forces in bars are again seen as constraint reactions:

> The conditions compelling the system because of bars connecting nodes are also expressed by equations: [...] in the same way as Eq. (4.5) imply the forces represented by the expressions (4.6), so Eq. (4.7) will imply other forces, that will be nothing else than tensions in bars: among these forces there is [...] the fundamental difference between external and internal forces, in that if a force A is given by Eq. (4.7), then also $-A$ exists [28].[151] (A.4.58)

Though the static problem, that is the problem to find internal and external constraint reactions, was not openly stated by Cerruti, he presented the correct number of equilibrium equations and of unknown inner forces and displacement of nodes. However, he did not use a proper terminology that could have avoided circumlocutions.

---

[148] Equation (1), p. 7. Cerruti's equation numbering is actually written in square brackets, but here in order to avoid confusion with references we will use parentheses.

[149] Equation (2), p. 8.

[150] Equation (3), p. 8.

[151] p. 8. Our translation.

4.5.1.2 Evaluation of External Constraint Reactions. Statically Determinate Systems

Cerruti examined the case when one or more nodes are fixed and unknown constraint reactions are present, and stated the condition for which statics of rigid bodies is sufficient to determine them. He however considered only the legacy and not their actual evaluation, consequently no equilibrium equations were written.

When there is one fixed node only, he said, the solution is always possible and unique.[152] When the nodes are two or three, he stated that the system of applied forces must reduce to a resultant only, that is either a force or a couple.[153]

Cerruti stated that this requirement coincides with the vanishing of the trinomial invariant characteristic of the system of active forces [28][154]:

$$\sum X \cdot \sum M_x + \sum Y \cdot \sum M_y + \sum Z \cdot \sum M_z = 0 \qquad (4.64)$$

where the symbols are usual; the condition is only necessary. Even if this is not stated by Cerruti, it may be inferred from the fact that he immediately provided sufficient conditions in the case of two and three fixed nodes. The resultant must be orthogonal either to the line joining the nodes, for two fixed nodes, or to the plane defined by the nodes, for three different nodes. Notice that Cerruti's reasoning was consistent only if the fixed node are interpreted as simply supports and not as spherical hinges; he was however ambiguous in this point.

In section 4 Cerruti examined an example, shown in Fig. 4.9 with some slight changes with respect to the original. The truss is statically determined and presents a recursive sequence of both geometry and load: this made it possible for Cerruti to find, by means of ordinary equilibrium equations, recursive formulas[155] for constraint reactions and inner forces in all the elements of the truss. These formulas are for sure compact and elegant from a mathematical point of view, but also have practical utility, because they would be a basic tool for Cerruti in order to solve redundant trusses.

Cerruti was able to present a complete and exhaustive description of the distribution of inner forces in bars: where they attain extrema, where they vanish, where they change sign so that some bars are in tension and other in compression and which are the geometrical parameters influencing this change.[156]

[152] From statics, any system of self-equilibrated forces may be reduced to two opposite forces passing through a given point, of which one is the resultant active force and the other is the constraint reaction.

[153] In general a system of forces in space can be reduced to both a force and a couple lying on a plane orthogonal to it.

[154] Equation (4), p. 10.

[155] Equations (8–15), pp. 14–15. There are actually two Eqs. (4.12) in [28], one on p. 12, the other on p. 14. This is an example of the apparent misprints in Cerruti's thesis, which we talked of in Sect. 4.5.1.

[156] His formulas (8–15).

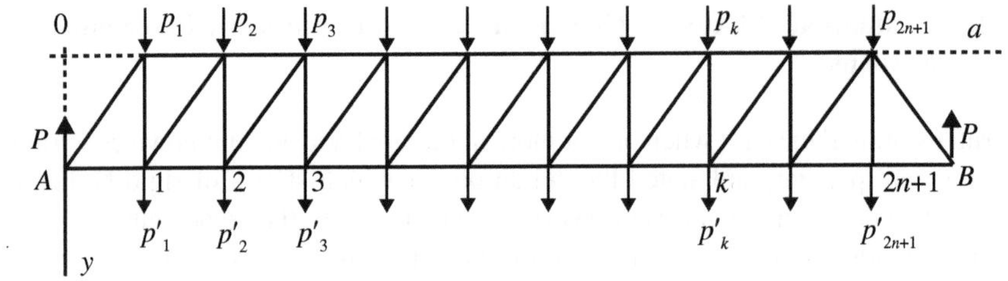

Fig. 4.9 A statically determined truss (Redrawn from [28, p.13])

In the last part of section 4, Cerruti showed how, for a truss of the kind depicted in Fig. 4.9 (designed in order to have uniform resistance), it is easy to find the displacement of all nodes. The results[157] are obtained by elementary geometry and the linear elastic law of extension for the bars connecting each pair of nodes. Calculations are simplified since the design of uniform resistance lets the strain of the bars be the same, and Cerruti obtained new recursive and elegant formulas, still showing his skills in what we nowadays would call 'automatic implementation'.

4.5.1.3 Redundant and Uniform Resistance Trusses

At the end of his section 4, Cerruti began dealing with the core of his treatment, that is redundant trusses and the possibility of designing such trusses in order to have uniform resistance. These topics continue until section 11.

> The formulas found here would provide [...] the shape of the deformed truss, [...] the variation of angles and the work spent by external forces during the deformation. But I will quit this subject and will discuss [...] the distribution of tensions and pressures in those cases where statics of rigid bodies throws us into indeterminacy and it is necessary to adopt the laws of elasticity [28][158] (A.4.59)

Since there is not a clear thread in his work, we prefer to distinguish between the treatment of redundant trusses and the uniform resistance trusses, and to deal with them in two following separate sections. In these cases we will not follow the order in which the subjects were given by Cerruti. We will rather follow a logical order.

4.5.1.4 Final Sections

The last two sections of Cerruti's thesis do not have a direct link with articulated systems and trusses, but rather represent an outlook of Cerruti's interests in the open questions of rational mechanics, which would constitute the main field of his future research.

[157] Equations (16–20), pp. 19–22.

[158] p. 22. Our translation.

Cerruti so began section 12:

One could ask which is the reason why the problem of the distribution of tensions and pressures in an articulated system can be solved by the preceding methods in a quite quick way and with all the rigor of the mathematical theory of elasticity, while many other problems remain unsolved because of the amount of difficulties they present. This is due to the fact, that in the considered case the laws of the displacement of the points of the system are known, and indeed this is the general problem of the theory of elasticity: "Provided the forces acting on a body, find the displacements parallel to three axes, which they make any of its molecules undergo". When these displacements are known, as it was said in section 8, it will be very easy to find the expression of the elastic forces originated at each of its points. Yet all difficulties actually lie in finding the law of these displacements. The nature of systems sometimes indicates a priori this law: an example is in elastic articulated systems: in these cases nothing else remains, than to find their magnitude, once external forces are given. However, hypotheses will never lead us to the true knowledge of displacements; and indeed, when a certain kind of deformation appears likely, it is easy to verify if it be possible or not; it is enough to prove, if by such an hypothesis the equations for internal equilibrium are verified. I state this, because the theories on the strength of materials, as usually exposed, rely on a particular hypothesis about the law of displacements; and this hypothesis is almost never verified, as I will now try to prove [28].[159] (A.4.60)

Cerruti considered a linear elastic homogeneous cylindrical body and wrote the equilibrium equations in terms of displacement components [28][160]:

$$\begin{aligned}(\lambda+\mu)\frac{d\theta}{dx}+\mu\delta^2 u=0,\\ (\lambda+\mu)\frac{d\theta}{dy}+\mu\delta^2 v=0,\\ (\lambda+\mu)\frac{d\theta}{dz}+\mu\delta^2 w=0,\end{aligned} \tag{4.65}$$

with λ, μ the Lamé elastic constants for the material, θ the volumetric strain and δ the Laplace operator.

Then Cerruti remarked that in the standard theories of elasticity the small displacements are supposed to consist of a translation and a rigid rotation, generally varying from place to place in the body[161]:

$$u=a+qz-ry,\ \ v=b+rx-pz,\ \ w=c+py-qx \tag{4.66}$$

where the components a, b, c of the infinitesimal translation and p, q, r of the infinitesimal rotation are actually fields defined on the cylinder.

By inserting Eqs. (4.66) into (4.65), Cerruti obtained a set of equations[162] which

[...] cannot hold in other cases than when a, b, c are linear functions of z and p, q, r are uniform; a very special case, that for sure does not embrace all those occurring in practice.

---

[159] pp. 50–51. Our translation.

[160] Equation (66), p. 52.

[161] Reference [28], Equation (67), p. 52.

[162] Reference [28], Equation (68), p. 53.

> But [...] I will not consider further this question. It might perhaps be useful, at least that of showing once more, that no remarkable perfections and advantages to the theories of the strength of materials will be brought without getting free from many (unjustified) hypotheses from which they start [28][163] (A.4.61)

In section 13, Cerruti showed that the methods of solution of redundant problems he presented in section 5 are not limited to the mechanics of structures but are of more general application. He decided to show one of these applications, that is the well-known problem of the 'pressures' (that is constraint reactions) of the support points of a massy body over a plane. This problem is standard and had among its solvers Euler, Cournot and Menabrea.[164] Cerruti replicated Euler's treatment with elegance and precision. However he forgot citing Cayley [25] who made a treatment very similar to Cerruti's.[165] Cerruti put into evidence some remarkable features of such a relationship and, by the ordinary equilibrium of force and moment, easily obtained the same results as Euler [28].[166]

4.5.2 Trusses with Uniform Resistance

Cerruti at the end of Section 3 stated that a truss will have uniform resistance if [28][167]:

$$\frac{T_{ij}}{E_{ij}\sigma_{ij}} = \text{uniform} = T\,benot \tag{4.67}$$

where T_{ij} is the stress in the bar joining the nodes i, j, E_{ij} is its Young's modulus and σ_{ij} is the area of its cross-section. Remark that Eq. (4.67) defines a limit *strain*, hence it seems that Cerruti adopted a maximum strain (Navier) criterion. In Section 3 Cerruti limited to state that, if the truss is statically determined, there is a unique distribution of the T_{ij} and hence of the required cross-sections of the bars.

The subject was considered again in section 6, after Cerruti had somehow dealt with redundant trusses, and let the author state some interesting theorems. Indeed, if there are m redundant external constraints and k redundant inner constraints, they must be described by constraint equations similar to (4.61), (4.63). By differentiating constraint equations with respect to the coordinates of the nodes and the length of the bars, respectively, Cerruti was able to insert the condition of uniform resistance (4.67) in both sets of equations. He then obtained [28][168]:

---

[163] p. 53. Our translation.

[164] See, among others [10].

[165] Reference [28], Equation (72), p. 55.

[166] pp. 55–57.

[167] Equation (7), p. 12.

[168] Equation (28), p. 28.

$$\sum \frac{dF_1}{dl_{ij}} l_{ij} = 0, \sum \frac{dF_2}{dl_{ij}} l_{ij} = 0, \ldots \sum \frac{dF_k}{dl_{ij}} l_{ij} = 0, \tag{4.68}$$

which must be verified so that the problem be compatible: this depends on the functions F and we will also see later that depends on the stresses underwent by the bars. This is not enough: in equation (25)[169] let us replace the variations of the coordinates by their expressions via the elongations of the bars, then by the relevant stresses. Let us introduce the condition of uniform resistance: after having eliminated the six variations still present, m conditions remain, independent of constraint reactions, to be satisfied in order that the problem be compatible: but it will in general not be so, the functions f_i being at will. We conclude, then, that *a linear elastic truss cannot have uniform resistance, if the number of condition equations [simple external constraints] exceeds six* [28].[170] (A.4.62)

This result is very interesting, since, without solving the linear elastic problem for a redundant truss, Cerruti could provide a design suggestion: if the system has redundant external constraints, no structure with uniform resistance can be obtained. Still, the requirement that the simple external constraints do not exceed six is only a necessary condition:

If the number of these conditions does not exceed six, it is necessary to check if equations (4.32) [our Eqs. (4.68)] hold or not. In the latter case we can say that it is impossible to fix the cross-sections of the bars in order to compose a system with uniform resistance: on the contrary, in the former case this will be [...possible] in k infinite ways; indeed, by replacing T_{ij} in equilibrium equations by his value $TE_{ij}\sigma_{ij}$, $3n-6$ equations in the cross-sections of the $3n-6+k$ bars of the system will result: yet choosing arbitrarily k of those cross-sections the above quoted equations will uniquely provide the remaining $3n-6$, and each of these cross-section may be arbitrarily attributed infinitely different values [28].[171] (A.4.63)

Thus, Cerruti concluded, a truss with n nodes may be designed to be with uniform resistance in a unique way only if it has not more than six simple external constraints and the number of bars does not exceed $3n-6$, that is when it is statically determined (modern classification).

When there are k redundant bars and Eqs. (4.68) hold, there are k different solutions for the design of a truss with uniform resistance, Cerruti easily proved that [28][172]:

$$\delta l_{ij} = T l_{ij}, \tag{4.69}$$

which says that *the variations in length of the elements are independent of the way in which the arbitrary k cross-sections are chosen* [28].[173] (A.4.64)

From this result it also easily follows a theorem on the work spent by the stress, clearly influenced by Menabrea's school:

---

[169] The differentiated equations of external constraints, p. 27.

[170] p. 28. Our translation.

[171] pp. 28–29. Our translation.

[172] Equation (29), p. 29.

[173] p. 29. Our translation.

[...] the work of external forces, and thus also that of molecular forces [inner work] during the deformation do not depend at all on the way the choice of those k cross-sections was made [28].[174] (A.4.65)

A very interesting theorem from the point of view of applications follows from the last result; from the expression of inner work L and the condition (4.67) of uniform resistance, Cerruti obtained [28][175]:

$$L = \frac{1}{2}\sum \frac{T_{ij}^2}{E_{ij}\sigma_{ij}} l_{ij} = \frac{T}{2}\sum E_{ij}\sigma_{ij}l_{ij}$$
$$\frac{2L}{T} = \sum E_{ij}\sigma_{ij}l_{ij}. \tag{4.70}$$

Now, L and T do not change with the choice of the k cross-sections, the right hand side [of the last equation of our (4.70)] shall then remain constant for any of those choices. Then we can say, that in this case *the sum of the products of the volumes of each bar times the relevant coefficient of elasticity is independent of the choice of the k arbitrary cross-sections*. If the coefficient E_{ij} is the same for each bar we may also add that, in any way we make this choice, *the weight of the employed material will always be the same* [28].[176] (A.4.66)

The technological consequences of this result are apparent and any comment is straightforward.

Cerruti in section 7 was interested in the conditions for which Eqs. (4.68) are satisfied. He then examined a particular case, the consequences of which are general enough: a spatial system with five nodes is geometrically determined by means of nine bars, and the length of a tenth is necessarily dependent on the other nine. Cerruti gave credit to Cayley,[177] for providing the condition [28][178]:

$$C = \det\left(l_{ij}^2\right) = 0, \quad i,j = 0,1,2,\ldots,5 \tag{4.71}$$

where $l_{00} = 0$, $l_{i0} = l_{0j} = 1$, $l_{ii} = 0$, $l_{ij} = -l_{ji}$. The first variation of Eq. (4.71) is the compatibility condition for the solution of the considered redundant system. After lengthy passages which show his mastery, Cerruti could state:

The system may be reduced to have uniform resistance in a simply infinite way, when all bars undergo stresses of the same kind [28].[179] (A.4.67)

and, in general:

The number of nodes is n, that of bars is $n(n-1)/2$, it is necessary to make some distinctions: either it is possible to select $(n^2 - 7n + 12)/2$ groups of five nodes, through which ten bars

---

[174] p. 29. Our translation.

[175] Equations (30) and the following unnumbered, p. 30.

[176] p. 30. Our translation.

[177] Cerruti did not make a precise quotation. We may infer that he referred to [27].

[178] Equation (31), p. 31.

[179] p. 34. Our translation.

> linked with each other undergo stresses of the same kind (which in any case may vary from a group to the other), then the system may have uniform resistance in $(n^2 - 7n + 12)/2$ infinite different ways; or this is impossible, and then, unless very special cases, it is impossible to design the system in order to have uniform resistance. In any case, if it is possible to select some groups of five nodes whiche satisfy the above mentioned conditions, the bars composing them may be designed to have uniform resistance in as many infinite ways as these groups are [28].[180] (A.4.68)

Cerruti appeared to be satisfied with these conclusions: he had indicated some design prescriptions and that seemed enough, so that he skipped to the other main subject of his thesis.

4.5.3 Statically Indeterminate Trusses

4.5.3.1 Poisson's and Lévy's Approaches

Cerruti began to examine redundant trusses in section 5, where he declared his intentions in order to find the solution, in terms of inner forces, of the linear elastic problem.

The first approach presented by Cerruti looks for as many auxiliary unknowns as the equilibrium equations, like in the problem of linear elastic continua, where the auxiliary unknowns are the displacement components:

> [...] the trick for the solution consists in letting the search for the unknown pressures and tensions depend on the search for $3n - 6$ other quantities, as many as the independent equilibrium equations, which is [...] always possible. This trick holds not only for the problem I consider, but for many other general questions [...]: it is known indeed that the knowledge of molecular forces in a body depends on that of six functions [the stress components] related by three partial differential equations, that are not sufficient to determine them, if the mentioned functions could not be expressed by means of other three only. The nature of these three functions is determined by the model we make of molecular forces: for elastic forces, these three functions are the orthogonal components of the displacement of any molecule of the body [28].[181] (A.4.69)

As an example of this approach for an articulated system, Cerruti assumed the node displacement components as auxiliary unknowns. First he wrote the length l_{ij} of a bar in terms of the differences among the coordinates of its terminal nodes i, j and differentiated it [28][182]:

$$\begin{aligned}&(x_j - x_i)(\delta x_j - \delta x_i) + (y_j - y_i)(\delta y_j - \delta y_i)\\&+ (z_j - z_i)(\delta z_j - \delta z_i) - l_{ij}\delta l_{ij} = 0\end{aligned} \tag{4.72}$$

[180] pp. 35–36. Our translation.

[181] pp. 22–23. Our translation.

[182] Equation (21), p. 24.

then, by using the linear elastic constitutive relation for the force originated in the bar to evaluate l_{ij}, he obtained [28][183]:

$$T_{ij} = \frac{E_{ij}\sigma_{ij}}{l_{ij}^2}[(x_j - x_i)(\delta x_j - \delta x_i) + (y_j - y_i) \\ \times (\delta y_j - \delta y_i) + (z_j - z_i)(\delta z_j - \delta z_i)] \tag{4.73}$$

> Inserting the expressions for stresses given by (4.26) [our (4.73)] into the equilibrium equations, these will contain the variation of the coordinates only, that can so be determined: once known their values, by Eq. (4.26) the forces can be calculated [28].[184] (A.4.70)

Notice that in the whole of his thesis Cerruti did not write down explicitly the equations of equilibrium of nodes, as in Menabrea [65] and differently from Castigliano [13].

The second approach pursued by Cerruti avoids the use of auxiliary unknowns. It is based on the choice of k independent relations among the $3n - 6 + k$ distances. If the displacements, as supposed, are small, all variations may be written in the reference configuration. Thus, the k relations among distances may be differentiated and the expressions of forces in terms of displacement variations may be inserted in them. These conditions, in addition to the $3n - 6$ independent equilibrium equations for the first, will determine the forces in the $n - 6 + k$ bars.

Cerruti gave credit to Poisson [82][185] for the first approach and to Lévy for the second [57]. Poisson, as referred to in Chap. 1, actually studied the motion of a body-point P subjected to a given active force and constrained to fixed points A_i by means of elastic threads. The strain ζ_i of each thread i with initial length l_i is given by

$$\zeta_i = \frac{1}{l_i}\left[(\alpha - a_i)u + (\beta - b_i)v + (\gamma - c_i)w\right], \tag{4.74}$$

where u, v, w are the displacement components of P, and the equilibrium equations are

$$\sum_i \frac{(\alpha - a_i)\zeta_i}{l_i\epsilon_i} = X, \quad \sum_i \frac{(\beta - b_i)\zeta_i}{l_i\epsilon_i} = Y, \quad \sum_i \frac{(\gamma - c_i)\zeta_i}{l_i\epsilon_i} = Z, \tag{4.75}$$

where $\alpha, \beta, \gamma; a_i, b_i, c_i$ are the cartesian coordinates of P and A_i, respectively, and ϵ_i are the extensibilities of the threads. By inserting Eqs. (4.74) into (4.75) Poisson obtained three independent equations of motion in the unknown coordinates α, β, γ, and then provided the conditions for equilibrium. In modern language, such an approach is a version of the method of displacements, in which the unknowns are the components of the displacement of the nodes and the equations to be solved are those of equilibrium. For an engineer of the 1800s it was straightforward to interpret

[183] Equation (22), p. 24.

[184] pp. 24–25. Our translation.

[185] vol. 2, pp. 402–404.

P as a node of a truss and the linear elastic threads as the bars connecting the node to the others of the truss.

Lévy presented and perfected a method to solve redundant trusses which is a version of what we now call the method of forces (see Chap. 1) [59]. The method is based on the possibility, in a truss with k redundant bars, to write k compatibility equations linking the lengths of the redundant bars to those of the remaining m:

$$F_j(l_1, l_2, \ldots, l_n) = 0,\ j = 1, 2, \ldots, k, \quad n = m + k. \tag{4.76}$$

Equation (4.76) hold in the reference configuration and for small deformations in its neighborhood, so the first variation of Eq. (4.76) provides:

$$\frac{\partial F_j}{\partial l_1} dl_1 + \frac{\partial F_j}{\partial l_2} dl_2 + \cdots + \frac{\partial F_j}{\partial l_n} dl_n = 0. \tag{4.77}$$

By inserting the forces f_j in the bars in terms of the variation of length dl_j into Eq. (4.77), one obtains k independent compatibility equations which, together with the $3n - 6$ independent equilibrium equations for the nodes, completely determine the forces in all the $3n - 6 + k$ bars.

The reference to Lévy's paper, of the same year as Cerruti's thesis, puts into evidence how the School of Engineering in Turin was up-to-date and well documented on the most important researches on the subject in France. On the other hand, it is strange and worth remarking that no credits were given neither to Clebsch [35], who for sure perfected the method of displacements, nor to Navier [78], who quite likely introduced it. The first omission, which may at first glance seem the most serious, is in part justified by the fact that Clebsch's' treatise was translated into French, a kind of second mother language for educated people in Piedmont, only in 1883. As a confirmation of this hypothesis, reading the early works of Castigliano [13, 19] puts into evidence the same omission, which lets us think that Clebsch was not at all known in the School of Engineering in Turin. The omission of the reference to Navier has no easy interpretation, though: even if Poisson's treatise was for sure well known, Navier's should have been better known, since it had successive editions until the most famous one with notes by Saint-Venant. Maybe an explanation for the quotation of Poisson only is due to the fact that Cerruti in his thesis often referred to the links between nodes seen as body-points connected by elastic forces and the general view of continua seen as molecules interacting by mean of central forces. This view, present also in Navier's treatise, was for sure perfected and better explained by Poisson, one of the fathers of the molecular theory of elasticity.

4.5.3.2 Cerruti's Contribution to Solution of Redundant Trusses

After having presented examples in literature, Cerruti advanced a method of his own, that appears to be a variant of Lévy's method:

> Let us consider the case in which the system shall satisfy certain geometrical conditions, that is the case in which a certain number of *surface equations* exist, to which the coordinates of the vertexes of the system shall obey (we will suppose, however, that no fixed points exist, or, if they exist, the conditions indicated in § 3 are also verified). Let these conditions be $m+6$: if $m = 0$ no difficulty exists and this subject was already dealt with in the quoted § 3; if $m > 0$ the rules expressed there are no more sufficient. But on this purpose we will note that the surface equations will hold for any value that the coordinates attain during deformation, hence if differentiated will also be satisfied when the variations of the coordinates will be replaced by the actual values they have attained under the action of external forces. This posed, let us find by one of the above quoted methods the forces in function of the external forces and of the $m+6$ constraint reactions: let us express the variations of the coordinates by means of these forces and let us insert these expressions in the differentiated $m+6$ equations of condition: we will thus have $m+6$ equations among the constraint reactions and six variations of the coordinates, in fact by means of forces we can express but the values of $3n-6$ variations, and in our case all the variations are determined and no one remains arbitrary. Yet by combining equilibrium equations one obtains six of them relating external forces and constraint reactions that, in conjunction with the first $m+6$ makes $m+12$ among $m+6$ reactions and six variations of coordinates, that is as many as the unknowns of the problem. In any case, one can have $m+6$ equations among the constraint reactions only by eliminating among the first $m+6$ the six variations of the coordinates [28][186] (A.4.71)

Cerruti applied his approach to the truss of Fig. 4.10, composed by six bars along the sides and the diagonals of a plane four-sided polygon. There are four external forces applied to the nodes, fulfilling the conditions of global equilibrium. The eight independent scalar components of the equilibrium equations for the four nodes in terms of the six unknown bar forces reduce to five [28],[187] since three equations are needed for global equilibrium. Another independent equation is needed to close the problem and Cerruti indicated that such equation has the form given by Cayley's condition (4.71) on the distances among a given number of points in space:

$$C = \det\left(l_{ij}^2\right) = 0, \quad i,j = 0,1,2,\ldots,4 \tag{4.78}$$

where this time the indexes i, j range from 0 to 4 [28].[188]

Indeed, by differentiating this condition and replacing the variations δl_{ij} by their expressions in terms of the elastic forces in the bars, Cerruti obtained a compatibility equation in the form [28][189]:

---

[186] pp. 26–27. Our translation.

[187] Equations (40), p. 37.

[188] Equation (41), p. 37.

[189] Equation (44), p. 38.

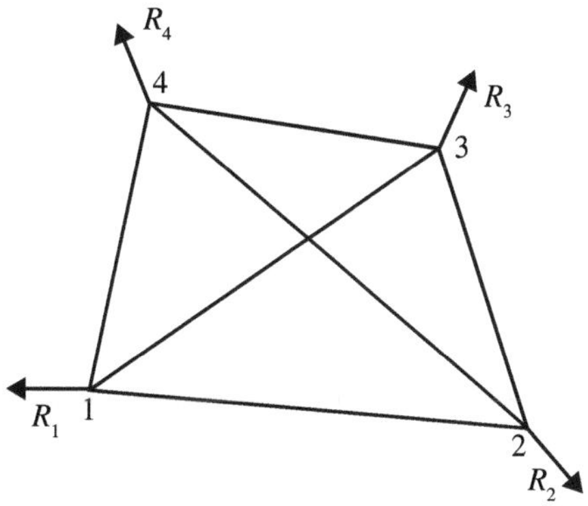

Fig. 4.10 A redundant truss with no fixed points (Redrawn from [28, p.36])

$$\begin{aligned}&\frac{\sin(3\hat{1}4)\sin(3\hat{2}4)}{l_{34}}\frac{T_{12}}{\epsilon_{12}}-\frac{\sin(2\hat{1}4)\sin(2\hat{3}4)}{l_{24}}\frac{T_{13}}{\epsilon_{13}}\\&+\frac{\sin(2\hat{1}3)\sin(2\hat{4}3)}{l_{23}}\frac{T_{14}}{\epsilon_{14}}+\frac{\sin(1\hat{2}4)\sin(1\hat{3}4)}{l_{14}}\frac{T_{23}}{\epsilon_{23}}\\&-\frac{\sin(1\hat{2}3)\sin(1\hat{4}3)}{l_{13}}\frac{T_{24}}{\epsilon_{24}}+\frac{\sin(1\hat{3}2)\sin(1\hat{4}2)}{l_{12}}\frac{T_{34}}{\epsilon_{34}}=0.\end{aligned}\tag{4.79}$$

The use of Eq. (4.78), in our opinion, represents the major contribution by Cerruti to the solution of redundant trusses. The equation represents indeed an easy and well established algorithm to obtain Lévy's Eqs. (4.76). The reference of (4.78) to a truss with only six bars, as depicted in Fig. 4.10, is not a serious limitation because most trusses of civil and industrial architecture can be decomposed into meshes similar to that of Fig. 4.10.

When the system reduces to three bars joined by the node 4 and fixed at their other extremities 1, 2, 3, that is a standard problem in the study of redundant structures, Cerruti remarked that he might obtain [28][190]:

$$\sin(2\hat{4}3)\frac{T_{14}}{\epsilon_{14}}-\sin(1\hat{4}3)\frac{T_{24}}{\epsilon_{24}}+\sin(1\hat{4}2)\frac{T_{34}}{\epsilon_{34}}=0,\tag{4.80}$$

which, he observed, by taking into consideration the geometrical properties of the triangles, is the solution for the same problem obtained by Menabrea in [66].[191] Another tribute to Menabrea in a work, where almost no hint on work and energy is given, is a meaningful sign of how Menabrea's teaching was deeply rooted in the School of Engineering in Turin.

Always with the aim of providing an iterative procedure, Cerruti reduced the linear algebraic system composed by the five equilibrium equations for the nodes and the

[190] Equation (45), p. 39.

[191] pp. 16–20. Here Menabrea studied a six bar system similar to that considered by Cerruti. See also [10].

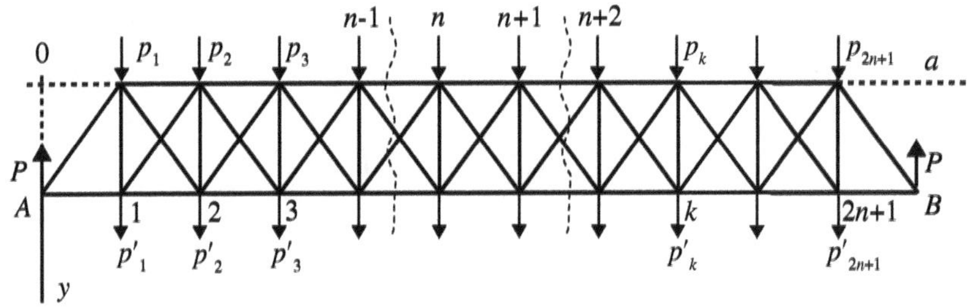

Fig. 4.11 A redundant truss obtained starting from a simple one (Redrawn from [28, p.13])

compatibility condition (4.79) to a linear algebraic system of two equations in the two unknowns T_{23}, T_{14} [28],[192] so that he could examine a series of particular cases of interest in the applications.[193] He began by examining the case where the external forces are directed along the diagonals, then went on to study the case when the polygon reduces to a trapeze and to a parallelogram. In this last case, he obtained a universal formula for the forces in the bars in terms of the external ones [28].[194] This formula is simplified again when the parallelogram becomes a rectangle or a square [28].[195] Cerruti remarked that this result is now independent of possible symmetries both in geometry and external load which could otherwise provide an answer for the problem of redundant trusses. Indeed, he remarked, these symmetry considerations fail as soon as deformation begins and may lead to errors, while the formulation of additional compatibility conditions based only on the geometry of distances among points is not affected by such errors.

In section 9, Cerruti applied his recursive formulas obtained for the truss of Fig. 4.9 for the redundant truss in Fig. 4.11, which is obtained from the simple one in Fig. 4.9 by adding bars along the other diagonal of each rectangular element composing the truss. He examined the rectangle between the nodes n and $n + 1$, which is but a particular case of the polygon he studied in his section 8. By means of ordinary equilibrium equations on the portion of the truss with respect of transverse imaginary sections between the nodes $n + 1$, $n + 2$ and $n - 1$, n, he obtained the external forces applied to the vertexes of the considered rectangle [28].[196] He then wrote the equilibrium equations for the considered rectangle as another portion of the truss [28],[197] as well as the equilibrium equations for the nodes of the rectangle, even if, he remarked, not all of these equations are linearly dependent [28].[198] He then

192 Equation (46), p. 39 and (47), p. 40.

193 pp. 41–44.

194 Equation (51), p. 43.

195 Equation (52) and (53), p. 44.

196 Equations (54–57), p. 45.

197 Equations (58), (59), p. 46.

198 Equation (60), p. 46.

provided the compatibility equation corresponding to (4.79) [28][199] and remarked that one has nine equations in ten unknowns, which

> will let us determine nine of the forces in terms of the tenth. After that, by using the already obtained results, one finds the forces of the different elements of the preceding rectangle, and so on until one reaches the [external] support. All forces will be expressed in terms of that tenth, which had been left undetermined in the calculation of the first rectangle, the value of which will be found at the end of the procedure. By replacing its value found in this way into the preceding expressions, all the forces in the elements will be known [28][200] (A.4.72)

In section 10 Cerruti left the applications aside for a while and considered again Poisson's study of body-points motion, to which he referred as inspiring one of the methods of solution for redundant systems. He framed the method directly into a structural environment, by considering the points as nodes-hinges and the threads as elastic bars, and studying the standard case of three bars hinged at fixed points and joined at a common node to which an external force is applied. Cerruti put into evidence the meaning of the constraint equations in terms of length of the bars and applies the rules of determinants to solve the linear system of equilibrium equations in terms of displacements. As a matter of fact, he actually did not add anything original to Poisson, but suggested that a matrix approach to a method of displacements is fruitful:

> This example is useful to show how simple is to solve the problem of the stress distribution by letting it depend on the search of as many quantities as equilibrium equations [28].[201] (A.4.73)

In section 11 Cerruti spent some more words on the possible cases in which the k redundant bars cannot be described by the additional compatibility condition (4.78). Indeed, this is effective when considering five nodes (only one redundant distance which can be expressed by means of the other nine), but even in the case of six nodes the situation is more complicated. Indeed, twelve bars are sufficient for a statically determined truss, but when considering a single redundant bar, it is not possible to express a single compatibility condition for it in terms of other distances. Cerruti remarked that this possibility depends much on how the truss is actually built.

References

1. Belidor BF (1782) Architecture hydraulique, ou l'art de conduire, d'élever et de ménager les eaux pour les différens besoins de la vie (4 vols). Cellot L, Paris, pp 1737–1739
2. Belidor BF (1819) Architecture hydraulique, ou l'art de conduire, d'élever, et de ménager les eaux pour les différents besoins de la vie (4 vols). Didot F, Paris (Navier CLMH (ed))
3. Benvenuto E (1984) A brief outline of the scientific debate that preceded the works of Castigliano. Meccanica 19(1 Suppl):19–32

[199] Equation (61), p. 46.

[200] p. 47. Our translation.

[201] p. 49. Our translation.

4. Benvenuto E (1991) An introduction to the history of structural mechanics (2 vols). Springer, New York
5. Benvenuto E (1995) Dall'arte del fabbricare alla scienza delle costruzioni. In: Zorgno AM (ed) Materiali tecniche progetto. Franco Angeli, Milan
6. Betti E (1874) Teoria della elasticità (1872–1873). Soldaini, Pisa
7. Bongiovanni M (2009) Formazione tecnica e ingegneri a Torino. Cultura industriale e formazione tecnica: la nascita degli studi d'Ingegneria. Hevelius' webzine
8. Bongiovanni M, Florio Plà N (2010) Emma Strada, ingegnere dal 1908. La vita della prima donna ingegnere attraverso le fonti archivistiche istituzionali e private. In: Proceedings of the 3^0 Convegno Nazionale di Storia dell'Ingegneria. Cuzzolin, Naples, pp 1037–1046
9. Bongiovanni M, Stella F (2014) Giovanni Curioni e la didattica della Scienza delle costruzioni alla Scuola d'applicazione per gl'Ingegneri di Torino. In: Proceedings of the 50 Convegno Nazionale di Storia dell'Ingegneria. Cuzzolin, Naples, pp 729–740
10. Capecchi D, Ruta G (2010) La scienza delle costruzioni in Italia nell'Ottocento. Springer, Milan
11. Capecchi D, Ruta G (2011) Polytechnic schools in the 19th century Europe. The polytechnic school in Karlsruhe. Meccanica 49:13–21
12. Capecchi D, Ruta G, Tazzioli R (2006) Enrico Betti. Teoria della elasticità, Hevelius, Benevento
13. Castigliano CA (1873) Intorno ai sistemi elastici. Dissertazione presentata da Castigliano Carlo Alberto alla Commissione Esaminatrice della R. Scuola d'applicazione degli Ingegneri in Torino. Bona, Turin
14. Castigliano CA (1875) Intorno all'equilibrio dei sistemi elastici. Memorie della Reale Accademia delle Scienze di Torino 10:380–422
15. Castigliano CA (1875) Lettera al presidente dell'Accademia dei Lincei, 11 marzo 1875. Memorie della Reale Accademia dei Lincei, s 2, 2:59–62
16. Castigliano CA (1875) Nuova teoria intorno all'equilibrio dei sistemi elastici. Memorie della Reale Accademia delle Scienze di Torino 11:127–286
17. Castigliano CA (1876) Formule razionali e esempi numerici per il calcolo pratico degli archi metallici e delle volte a botte murali. L'ingegneria civile e le arti industriali 9:120–135; 10: 145–153
18. Castigliano CA (1878) Applicazioni pratiche della teoria sui sistemi elastici. Strade ferrate dell'Alta Italia. Servizio della manutenzione e dei lavori, Crivelli, Milan
19. Castigliano CA (1879) Théorie des systèmes élastiques et ses applications. Negro, Turin
20. Castigliano CA (1879) Théorie des systèmes élastiques et ses applications. Planches, Negro, Turin
21. Castigliano CA (1881–1882) Intorno a una proprietà dei sistemi elastici. Memorie della Reale Accademia delle Scienze di Torino 17:705–713
22. Castigliano CA (1882) Esame di alcuni errori che si trovano in libri assai reputati. Il Politecnico 1–2:66–82
23. Castigliano CA (1884) Teoria delle molle. Negro, Turin
24. Castigliano CA (1884–1889) Manuale pratico per gli Ingegneri (4 vols). Negro, Turin
25. Cayley A (1857) Solution of a mechanical problem. Q Math J 1:405–406
26. Cayley A (1858) A memoir on the theory of matrices. Philosophical Transactions of the Royal Society of London 148:17–38
27. Cayley A (1892) On a theorem relating to five points in a plane. In: Cayley A (1892) Cayley's collected papers, vol 5. Cambridge University Press, Cambridge, pp 480–483
28. Cerruti V (1873) Intorno ai sistemi elastici articolati. Dissertazione presentata alla Commissione Esaminatrice della Real Scuola d'Applicazione per gli Ingegneri in Torino. Bona, Turin
29. Cesaro E (1964–1968) Opere scelte (2 vols in 3 tomes). Cremonese, Rome
30. Cesaro E (1889) Sulle variazioni di volume nei corpi elastici. In: Cesaro E (ed) (1964–1968) Opere scelte (2 vols in 3 tomes). Cremonese, Rome
31. Charlton TM (1971) Maxwell, Jenkin and Cotterill and the theory of statically-indeterminate structures. Notes Rec Roy Soc Lond 26(2):233–246

32. Charlton TM (1995) Menabrea and Lèvy and the principle of least work. Eng Struct 17(7):536–538
33. Chasles M (1875) Aperçu historique sur l'origine et le développement des méthodes en géométrie (2 édition). Gauthier Villars, Paris
34. Clebsch RFA (1862) Theorie der Elasticität fester Körper. Teubner BG, Leipzig
35. Clebsch RFA (1883) Théorie de l'élasticité des corps solides, Traduite par MM. Barré de Saint Venant et Flamant, avec des Notes étendues de M. Barré de Saint Venant. Dunod, Paris
36. Colonnetti G (1916) Principi di statica dei solidi elastici. Spoerri, Pisa
37. Colonnetti G (1941) Principi di statica dei solidi elastici. Einaudi, Turin
38. Cournot A (1828) Sur la théorie des pressions. Bulletin de Férussac 9:10–22
39. Cremona L (1872) Le figure reciproche nella statica grafica. Hoepli, Milan
40. Crotti F (1884) Commemorazione di Alberto Castigliano. Atti del Collegio degli ingegneri ed architetti in Milano, pp 5–6
41. Crotti F (1888) La teoria dell'elasticità ne' suoi principi fondamentali e nelle sue applicazioni pratiche alle costruzioni. Hoepli, Milan
42. Culmann C (1866) Die graphische Statik. Meyer und Zeller, Zürich
43. Curioni G (1864–84) L'arte di fabbricare. Ossia corso completo di istituzioni teoriche pratiche per ingegneri (6 vols and 5 appendixes). Negro, Turin
44. Curioni G (1884) Cenni storici e statistici della Scuola d'applicazione per gl'ingegneri. Candeletti, Turin
45. Della Valle S (2008) La Facoltà di ingegneria di Napoli. Website of the Faculty of Engineering, Naples
46. Discussione sul regolamento ministeriale 6 settembre 1913 per le Scuole di applicazione degli ingegneri (1914). Atti della società degli ingegneri e architetti in Torino, anno XLVIII
47. Dorna A (1857) Memoria sulle pressioni sopportate dai punti d'appoggio di un sistema equilibrato e in istato prossimo al moto. Memorie dell'Accademia delle Scienze di Torino s 2(17):281–318
48. Fara A (2011) Luigi Federico Menabrea (1809–1896). Scienza, ingegneria e architettura militare dal Regno di Sardegna al Regno d'Italia. Olschki, Florence
49. Ghisalberti AM (1960) Dizionario biografico degli italiani. Istituto enciclopedia italiana, Rome
50. Giuntini A (2001) II paese che si muove. Le ferrovie in Italia tra '800 e '900. Franco Angeli, Milan
51. Girard PS (1798) Traité analytique de la resistance des solides et des solides d'égale résistance. Didot, Paris
52. Guidi C (1891) Lezioni sulla scienza delle costruzioni (2 vols). Salussolia, Turin
53. Guidi C (1910) Lezioni sulla scienza delle costruzioni dall'ing. prof. Camillo Guidi nel Real Politecnico di Torino. Bona, Turin
54. Grioli G (1960) Lezioni di meccanica razionale. Borghero, Padua
55. Jellett JH (1852) On the equilibrium and motion of an elastic solid. Trans Roy Soc Acad s 3(22):179–217
56. Lauricella G (1912) Commemorazione di Valentino Cerruti. Giornale di Matematiche di Battaglini 50:329è336
57. Lévy M (1873) Mémoire sur l'application de la théorie mathématique de l'élasticité à l'étude des systèmes articulés formés de verges élastiques. Comptes Rendus Hebdomadaires des Sèances de l'Académie des Sciences 76:1059–1063
58. Levi-Civita T (1909) Commemorazione del socio Valentino Cerruti. Rendiconti della Reale Accademia dei Lincei, s 5, 18:565–575
59. Lèvy M (1874) (1886–1888) La statique graphique et ses applications aux constructions (4 vols). Gauthier-Villars, Paris
60. Lorgna AM (1794) Dell'azione di un corpo retto da un piano immobile esercitata ne' punti di appoggio che lo sostengono. Memorie di matematica e fisica della Socità italiana delle scienze 7:178–192
61. Mangone F, Telese R (2001) Dall'accademia alla facoltà. L'insegnamento dell'architettura a Napoli. Hevelius, Benevento

62. Marchis V (2007) Le radici di una scuola politecnica. Progettare 309:27–30
63. Mele C (2006) Origini e formazione del Politecnico di Torino. In: Proceeding of the 1° Convegno Nazionale di Storia dell'ingegneria, Napoli, pp 369–378
64. Menabrea LF (1842) Notions sur la machine analytique de M. Charles Babbage. Bibliothéque universelle de Gèneve. Nouvelle Série 41:352–376
65. Menabrea LF (1858) Nouveau principe sur la distribution des tensions dans les systèmes élastiques. Comptes Rendus Hebdomadaires des Séances de l'Académie des Sciences 46: 1056–1060
66. Menabrea LF (1868) Étude de statique physique. Principe général pour déterminer les pressions et les tensions dans un système élastique (1865). Imprimerie Royale, Turin
67. Menabrea LF (1874–75) Lettera all'Accademia delle Scienze di Torino per una correzione da apportare al Principe général del 1868. Memorie della Reale Accademia delle Scienze di Torino, s 2, 10:45–46
68. Menabrea LF (1869–70) Sul principio di elasticità. Dilucidazioni. Con dichiarazioni di A. Parodi, G. Barsotti, J. Bertrand, Y. Villarceau. Memorie della Reale Accademia delle Scienze di Torino, s 2, 5:686–710
69. Menabrea LF (1875) Sulla determinazione delle tensioni e delle pressioni ne' sistemi elastici. Atti Reale Accademia dei Lincei, s 2, 2:201–220
70. Menabrea LF (1875) Lettera al presidente dell'Accademia dei Lincei, 27 marzo 1875. Memorie della Reale Accademia dei Lincei, s 2(11):62–66
71. Menabrea LF (1884) Concordances de quelques méthodes générales pour déterminer les tensions dans un système des points réunis par des liens élastiques et Sollicitiés par des forces extérieures en équilibre. Comptes Rendus Hebdomadaires des Séances de l'Académie des Sciences 98:714–717
72. Menabrea LF (1971) Memorie, a cura di Briguglio L e Bulferetti L. Centro per la storia della tecnica in Italia del CNR, Giunti-Bré, Genova
73. Meriggi M (1996) Breve storia dell'Italia settentrionale dall'Ottocento a oggi. Donzelli, Milan
74. Mossotti OF (1858) Lezioni di Meccanica razionale. D'Arcais [s.n.]
75. Nascè V (1984) Alberto Castigliano, Railway engineer: his life and times. Meccanica Suppl 19:5–14
76. Nascè V, Benvenuto E (eds) (1984) Alberto Castigliano. Selecta 1984. Levrotto e Bella, Turin
77. Navier CLMH (1825) Sur des questions de statique dans lesquelles on considére un corps supportè par un nombre de points d'appui surpassant trois. Bulletin des Sciences par la Société Philomatique 11:35–37
78. Navier CLMH (1826) Résumé des leçons donnés à l'École royale des ponts et chaussées sur l'application de la mécanique à l'établissement des constructions et des machines (2 vols). Didot F, Paris
79. Navier CLMH (1864) Résumé des leçons données à l'École de ponts et chaussées sur l'application de la mécanique à l'établissement des constructions et des machines, avec des notes et des appendices par M. Barré de Saint Venant. Dunod, Paris
80. Novello P, Marchis E (2010) Emma Strada. Temi forme e maestri della formazione politecnica, progetti, disegni e opere della professione di progettista. In: Proceedings of the 3° Convegno Nazionale di Storia dell'Ingegneria. Cuzzolin, Naples, pp 1047–1056
81. Poincaré H (1892) Leçons sur la théorie de l'élasticité. Carré, Paris
82. Poisson SD (1833) Traité de mécanique (1831). Bachelier, Paris
83. Programma della Real Scuola di applicazione per gli ingegneri in Bologna, (1879) Anno scolastico 1878–1879. Società tipografica dei compositori, Bologna
84. Pugno GM (1959) Storia del Politecnico di Torino. Stamperia Artistica Nazionale, Turin
85. Richelmy P (1872) Intorno alla Scuola di applicazione per gl'ingegneri fondata in Torino nel 1860. Fodratti, Turin
86. Rombaux GB (1876) Condizioni di stabilità della tettoja della stazione di Arezzo. Tipografia e litografia del Giornale del Genio Civile, Rome
87. Rondelet J (1802–1817) Traité théorique et pratique de l'art de batir (5 vols). Rondelet, Paris

88. Ruta G (1995) Il problema di Saint Venant. Studi e Ricerche n. 7/95 del Dipartimento di ingegneria strutturale e geotecnica, Università La Sapienza, ESA, Rome
89. Sabbia E (place and date of publication unknown) Al Professore Sig. S. L.
90. Sabbia E (1869) Errore del principio di elasticità formolato dal sig. Menabrea. Cenno critico di Emilio Sabbia, Tipografia Baglione, Turin
91. Sabbia E (1870) Nuove delucidazioni sul principio di elasticità. In: Risposta ad un opuscolo contenente delucidazioni di Luigi Federico Menabrea sullo stesso principio. Tipografia Baglione, Turin
92. Silla L (1910) Commemorazione di Valentino Cerruti. Il Nuovo Cimento s 5, 19:5–9
93. Silvestri A (2006) La nascita delle facoltà di ingegneria e di architettura in Italia. In: Proceedings of the 1° Convegno Nazionale di Storia dell'Ingegneria. Cuzzolin, Naples, pp 223–234
94. Website of the Faculty of Engineering, Sapienza, University of Rome, Rome
95. Zorgno AM, (a cura di) (1995) Materiali tecniche progetto. Franco Angeli, Milan

Chapter 5
Computations by Means of Drawings

Abstract The second half of the 19th century saw a very quick diffusion of graphical statics. Lectures on graphical statics were given in Switzerland (Zurich); in Germany (Berlin, Munich, Darmstadt, Dresden); in the Baltic regions (Riga); in the Austrian-Hungarian empire (Vienna, Prague, Gratz, Brunn); in the United States; in Denmark. The author that mainly developed its techniques was the German scholar Carl Culmann, who placed graphical statics besides the newborn projective geometry. Culmann's approach was enthusiastically followed in Italy, where, first in Milan at the Higher technical institute, then, after 1870, in many Schools of application for engineers, among which those of Padua, Naples, Turin, Bologna, Palermo, Rome, and, eventually, also in the universities of Pisa and Pavia, courses of graphical statics were activated. The Italian scholar who collected Culmann's inheritance, and extended it, was Luigi Cremona.

5.1 Graphical Statics

In the second half of the 19th century there was a very quick diffusion of the techniques of graphical calculation to solve engineering problems, above all the determination of the forces in the trusses frequently used for industrial buildings and for bridge construction. The term used to denote these techniques was *graphical statics*.

Classes on graphical statics were held in all Europe (Zurich, Berlin, Darmstadt, Munich, Dresden, Riga, Vienna, Prague, Gratz, Brunn) and in the United States. In Italy there were classes in Milan (at the Higher technical institute) and, after 1870, in many Schools of application for engineers, among which those of Padua, Naples, Turin, Bologna, Palermo, Rome, and also in the universities of Pisa and Pavia.

The meaning of the locution 'graphical statics' is rather nuanced, and has undergone changes in time. At the beginning of the 1800s, and in our time, the term simply indicated a part of geometrical statics, that is of statics developed by geometrical means. We may say that geometrical statics, properly stated, deals with the geometrical deduction of the laws of statics, while graphical statics deals with the geometrical procedures that let the problems of statics of engineering practice be

D. Capecchi and G. Ruta, *Strength of Materials and Theory of Elasticity in 19th Century Italy*, Advanced Structured Materials 52,
DOI 10.1007/978-3-319-05524-4_5

solved by graphical means. Among these practical problems, besides the structural verification of trusses, there were those of geometry of masses (centers of gravity, moments of inertia, principal axes, and so on), of verification of beam cross-sections undergoing eccentric tension and of retaining walls, of metrics in ground movements, and so on. In the second half of the 19th century, after Culmann's fundamental monograph in 1866 [28],[1] the term 'graphical statics' was used in a restricted sense, to indicate a discipline unifying graphical calculus and projective geometry, or geometry of position, as it was then called.

Geometrical statics may go back to Stevin [76], but a fundamental role was played by Pierre Varignon, who, in the *Nouvelle mécanique ou statique* [81], besides using in an extensive way the rule of parallelogram, taught us how to construct both the polygon of forces and the funicular polygon, basic ingredients of graphical statics. Barthelemy Edouard Cousinery developed the *calcul par le trait* (that is graphical calculus) [18, 19].[2] Jean Victor Poncelet established the bases of projective geometry, in order to generalize the results of descriptive geometry obtained by Gaspard Monge; he was among the first to use graphical methods for the verification of retaining walls [69].[3] The studies by Lamé and Clapeyron on the funicular polygon were important [54]. The reciprocity of the polygons of forces and funicular according to a well determined relation was analyzed in a paper by Maxwell in 1864 [13], a result in some way anticipated by Rankine who, in 1858 [71], proved a reciprocity theorem for trusses.

Culmann undoubtedly theorized graphical statics at most, and for many years influenced its methods with the monograph *Die graphische Statik* [28], a textbook difficult to read nowadays because of the extensive use of projective geometry, not cultivated much anymore. It was very successful, and inspired many textbooks for engineers at once.[4] Culmann was not happy about these textbooks; indeed, in the French translation of the second edition of his textbook (1875) he wrote:

---

[1] The issue was anticipated by the print of handouts of his lectures on graphical statics at the Polytechnic of Zurich, in 1864 and 1865. In 1875 the first volume of the second edition, designed in two volumes, was issued, but Culmann died in 1881, before completing the second volume. A French translation of the first volume of the second edition was issued in 1880 [29].

[2] The expression *graphical calculus* was introduced for the first time by Louis Ézéchiel Pouchet [70].

[3] Projective geometry after Poncelet would be developed in France by Chasles, in Germany by Karl Georg Christian von Staudt, Plücker, Möbius, Steiner, and Clebsch.

[4] We report, as examples, the references quoted by Cremona:

> K. Von Ott, *Die Grundzüge des graphischen Rechnens und der graphischen Statik*, Prag, 1871; J. Bauschinger, *Elemente der Graphischen Statik*, München, 1871; F. Reauleaux, *Der Constructeur* (3rd edition) (2 Abschnitt), Braunschweig, 1869; L. Klasen, *Graphische Ermittelung der Spannungen in den Hochbau-und Brückenbau-Construction*, Leipzig, 1878; G. Hermann, *Zur graphischen Statik der Maschinengetriebe*, Braunschweig, 1879; S. Sidenam Clarke, *The principles of graphic statics*, London, 1880; J. B. Chalmers, *Graphical determination of forces in engineering structures*, London, 1881; K. Stelzel, *Grundzüge der graphischen Static und deren Anwendung auf den continuirlichen Träger*, Graz, 1882; M. Maurer, *Statique graphique appliquée aux constructions*, Paris, 1882 [22, pp. 341–342].

> Our diagrams were more successful than our methods. Our publication was followed by a great number of Elementary statics, in which, while reproducing our simplest diagrams (most of the time without changing them), the authors tried to provide their analytical proofs [29].[5] (A.5.1)

Indeed, only some scholars in structural mechanics followed Culmann's conceptions directly, while others—although taking a strong inspiration from his work—'betrayed' its spirit, mainly because the part of graphical statics that is important for applications may be exposed with rather superficial notions of projective geometry.

Some important exponents of the Italian school followed Culmann, among them Luigi Cremona, Carlo Saviotti, Antonio Favaro. The latter admonished us not to mistake *graphical statics* with the consolidated *geometrical statics* in the French translation [33] of his treatise [32], even though the borders between the two, he maintained, are not rigid:

> We agreed to keep the name of *Graphical statics* to a whole category of recent researches, which constitute a body of doctrines now well coordinated, and that, taken as a whole, are characterized by the double condition of implementing the construction processes of Linear, or graphical, calculus, and to rest on the fundamental relation existing between the polygon of forces and the funicular polygon.
>
> The domain of *Graphical statics* being so not rigorously defined, but [only] indicated, we agree to call by the name *Geometrical statics* the set of other applications of Geometry, and, more in particular, of Ancient geometry, to statics [33].[6] (A.5.2)

Saviotti's considerations are similar:

> And it is well obvious that, because of this, *Graphical statics* shall increasingly detach from the analytical one, not because of a different scope, but because the former abandons, little by little, those subjects that are dealt with by the latter in a simpler, and at the same time more general, way.
>
> It is so, in fact, that the subject of strictly non-deformable [that is statically determined] trusses has almost been suppressed in Analytical statics, and reserved to the graphical one, while the latter still keeps in its domain the investigation of trusses with redundant elements.
>
> Graphical statics shall be cultivated in the right address, conforming its method to Modern geometry. And it is to be deplored that it, appeared for the first time in the masterly work by CULMANN, already based on Projective geometry, from which one derived the most beautiful and elegant of its proofs, has then had authors that wanted to rest it on the resources of Elementary geometry only.
>
> These authors may be excused only if they wished to address themselves to those who are not able to know Projective geometry. Their treatises, while not lacking usefulness, are, however, at the same level of the treatises on *Analytical mechanics* developed by elementary calculus only.
>
> Not few works on Statics that saw the light in these last years, with titles alluding to the application of the graphical method, are actually developed in part analytically and in part graphically, that is to say, by a method that could be said mixed. We do not believe in a future for this procedure, which does not presume any character of generality, and which

[5] p. XI. Our translation.

[6] p. XXIV. Our translation.

> may be thought acceptable only on a transitional, or in the additional applications of Statics to constructions [72].[7] (A.5.3)

Not everybody agreed; for example, Johann Bauschinger stated that:

> I believe that the low diffusion that the application of Graphical statics has had until now among engineers, mainly derived from the lack of an ordered book, styled to the teaching of this new science. And it would be my great satisfaction if, were my opinion correct, my book could repair this defect. Since Graphical statics is for sure of such importance for the study of the science of professional engineers, that its greater diffusion shall be desired, which will for sure happen.
>
> Maybe to such a diffusion will also contribute a quality of my book, which is, that it is not necessary the knowledge of the so-called new geometry to understand it. I have not done this on purpose, but it happened by itself, and I am pleased by this, since I believe that in this way a useful service is given to engineers, who had not the occasion before to get familiar with the new methods of geometry, being thus able to embrace the study of Graphical statics at once, without the necessity to deal with an auxiliary science [1].[8] (A.5.4)

A substantially definite perfecting of graphical statics based on the theory of reciprocal figures was due to Cremona in 1872 [22, 26], in a paper in which a sufficiently accurate report of its history was given. Toward the mid 1900s graphical statics went into disuse and, after the arrival of computers, substantially has disappeared from teaching and professional practice. There are attempts to propose it again nowadays, at least in the educational field, placed side by side with CAD techniques [35].

The reasons of the decline of graphical statics are partly 'ideological' and partly 'objective'. The former are associated with the diffusion of mathematical analysis and to the teaching in depth of continuum mechanics in engineering schools. In such a cultural climate, the engineer progressively left drawing out in favor of calculus, which permits a better control of the results and greater precision; graphical statics is substantially useless for designing purposes, but it satisfies those who consider the designing phase in itself and not only to simplify calculations. An objective aspect is represented by the low usefulness of graphical statics in dimensioning the framed structures with rigid nodes of the end of the 19th century. The problems of geometry of masses and of cross-sections optimal design lost importance because the fundamental results had been acquired and encoded in textbooks. Another objective aspect has been the introduction of vector calculus, developed in the second half of 19th century; graphical statics, indeed, overlaps vector calculus, and may in part be supplanted by it. The decline of graphical statics was in any way very slow, and at least until the 1950s in many Italian faculties of engineering graphical statics would have some space.

The contemporary decline in the faculties of mathematics of the teaching of descriptive geometry, which in the second half of the 19th century was placed side by side with projective geometry, shall be attributed to analogous reasons. Here as well the theoretical aspect was preferred to the applicative one.

---

[7] p. X. Our translation.

[8] Introduction. Our translation.

5.2 Graphical Statics and Vector Calculus

The concept of a vector as an element constituting an algebraic structure goes back to the first half of the 19th century; the term vector was introduced by Hamilton in 1844 in a paper [47] about the theory of quaternions:

> On account of the facility with which this so called imaginary expression, or square root of a negative quantity, is constructed by a right line having direction in space, and having x, y, z for its three rectangular axes, he[9] has been induced to call the trinomial expression itself, as well as the line which it represents, a VECTOR. A quaternion may thus be said to consist generally of a real part and a vector [47].[10]

In the same paper, shortly after, the term scalar was introduced:

> It is, however, a peculiarity of the calculus of quaternions, at least as lately modified by the author, and one which seems to him important, that it selects no one direction in space as eminent above another, but treats them as all equally related to that extra-spacial, or simply SCALAR direction, which has been recently called 'Forward' [47].[11]

Actually, since ancient times there had been a distinction among scalar and vector quantities, the former characterized by a numerical value only, the latter also by a direction and an orientation. Some operations, like those of sum and vector product, were already used in geometry, kinematics, and statics: indeed, the parallelogram rule goes back at least to Aristotle, and the idea of moment of a force with respect to a point (which is a particular case of vector product) goes back at least to Leonardo da Vinci and Giovanni Battista Benedetti [7].

One of the first works of the 19th century which fitted into the genre of the 'algebrized' vector calculus was due to Möbius, who in 1827 [61] introduced the concepts of oriented segments and of sum of collinear segments. The formalization operated by Möbius represented a first detachment from a purely graphic representation of segments, a differentiation which was, on the other hand, almost absent in the substantially coeval book by Barthelemy Edouard Cousinery [19] where the concept of sum of segments was introduced, keeping, however, the description to a graphical level only:

> The linear, or graphical, sum and subtraction are two elementary operations so simple that it will be enough for us just mention them here for memory: to sum two linear quantities a and b, is to put them endwise on an infinite [straight] line cd; the total is the quantity which is comprised between the two limits that are not in contact. The rule is the same whatever be the number of terms to sum among them [19].[12] (A.5.5)

The history of modern vector calculus embraces the period ranging from the work by Möbius until the first years of the 20th century, and is described in detail in the monographs by Crowe [27] and Caparrini [5]. Here we summarize only the principal

---

[9] Hamilton was talking about himself in third person.

[10] p. 3.

[11] p. 3.

[12] p. 12. Our translation.

dates that let us monitor the influence of the development and of the diffusion of vector calculus on graphical statics and on structural mechanics.

Crowe identified two main traditions, one derived from William Rowan Hamilton's theory of quaternions, linked to the representation of complex numbers, and a more geometrical one associated to Hermann Günter Grassmann. Together with these two traditions there are many others, testifying the need, especially by the physicists, of creating some new form of vector calculus; the contributions by Saint Venant, Giusto Bellavitis e Michel Chasles are meaningful [27].[13]

The works of Hamilton [47–49] on quaternions, of which he had announced the 'discovery' in 1843 at a seat of the Royal Irish academy, though containing a great part of modern vector calculus, still kept it somehow hidden. Quaternions are *hyper-complex numbers* with the form $w + ix + jy + kz$, where x, y, z, w are real numbers, while i, j, k are unit vectors directed along x, y, z obeying the rules:

$$\begin{aligned} &ij = k; \quad jk = i, \quad ki = j \\ &ji = -k; \quad kj = -i, \quad ik = -j \\ &ii = jj = kk = -1. \end{aligned} \tag{5.1}$$

Such relations let the operations of sum and product between quaternions be defined, which substantially coincide with those of modern vector calculus. Remark that the product of a unit vector by itself, somehow providing its norm, is negative, contrarily to what is assumed in the modern theory.

The results obtained by Grassmann [43–45] are in themselves closer to modern vector calculus than Hamilton's. The main differences with respect to modern calculus probably lie in terminology, in symbols, and, most of all, in the definition of vector product, the result of which is not a vector, but the oriented area of a surface.

The fundamental notion of Grassmann's theory is that of hyper-number with n components. A hyper-number with three components is written as:

$$\alpha = \alpha_1 e_1 + \alpha_2 e_2 + \alpha_3 e_3, \tag{5.2}$$

where $\alpha_1, \alpha_2, \alpha_3$ are real numbers, and e_1, e_2, e_3 are primary units geometrically represented by unit oriented segments forming a right-handed system. Sum, inner (scalar) and outer (vector) products are defined on hyper-numbers.

Despite the greater proximity of Grassmann's approach to the modern one, Crowe thought that modern vector calculus derived in large part from the calculus of quaternions, thanks to Josiah Willard Gibbs' [37][14] and Oliver Heaviside's [50] contribution at the end of 19th century.[15] They, and many physicists with them, considered

[13] Chapter 3.

[14] This volume was privately issued in New Haven in two parts, the first in 1881, the second in 1884.

[15] Besides the volumes [50], Heaviside published several papers on the subject in the journal "The Electrician" during the 1880s.

calculus by quaternions uselessly complex, and simplified it by introducing the vector parts of quaternions as fundamental quantities, together with operations on them.

Vector calculus found application at first in many papers on electro-magnetism, especially by Heaviside, while its applications to geometry and mechanics, which had stimulated its birth, came later on. In Italy, after the pioneer work of Bellavitis [3], the first organic paper on vector calculus was that by Cesare Burali-Forti and Marcolongo in 1909 [2], after Grassmann's tradition.

The affirmation of vector calculus began to be felt in graphical statics as well. In spite of the fact that in *La statica grafica* of Saviotti [72] the term vector was not used to denote forces (the term equipollence, introduced by Bellavitis, was used, however), and the author talked about composition instead of sum, the section devoted to graphical calculus departed from the purely geometrical formulation adopted by Cousinery and repeated by Culmann to adopt a more algebraic symbology, at least in the definition of the resultant of segments:

> **21.** If two concurring segments are identified in magnitude and direction, we intend by their resultant segment the one closing the line formed by the two segments with their own direction. Such a resultant segment AC is so identified in magnitude, orientation, and direction. If the lines of the component segments are also given, the segment AC is said to be *equipollent*, in the same way as AB and BC are [equipollent] of the two components [72].[16] (A.5.6)

The introduction of vectors remarkably simplified the presentations and the justifications of the constructions of graphical statics, to such a point that some parts lost their graphic design, because the algebraic form is perspicuous enough.

5.3 The Contributions of Maxwell and Culmann

Among the authors who, directly or not, provided remarkable contributions to the development of graphical statics and to its applications in the field of structural mechanics, we must remember Maxwell, with his investigations on reciprocal figures in the mid-1860s. However, the most influential scholar on all the future developments was for sure Culmann, professor at the Federal technical high school in Zurich (founded in 1855), who influenced all graphical statics after his monograph *Die graphische Statik* of 1866. In the following we will provide some details on the contributions of these two authors.

5.3.1 Reciprocal Figures According to Maxwell

In 1864 Maxwell published two fundamental papers on structural mechanics in the Philosophical magazine: the first, which we have already discussed, was about

[16] v. 1, p. 16. Our translation.

the solution of the statical problem for trusses with redundant constraints [12]; the second, the subject of this section, was about reciprocal figures, and may be used for the graphical investigation of trusses [13].

We said in Chap. 1 that the first paper was almost ignored by engineers, in particular by those in the Continent, and was re-discovered only after Mohr published his method, based on the principle of virtual work and leading to the same results, in 1874 [65, 66]. The second paper had a better fortune and was followed by another paper in 1872 [14], in spite of the fact that it had been essentially ignored at first, for reasons analogous to those of the first; the Philosophical magazine was usually not read by engineers, and the theory developed by Maxwell could be applied to situations of modest relevance in structural mechanics, mainly trusses with one redundancy, subjected to a coaction state. In addition, Maxwell did not provide a definite procedure for drawing reciprocal figures. One of the few exceptions to this carelessness toward the works by Maxwell was represented by Fleeming Jenkin, who even introduced improvements in the theory, and used it in some applications [34]. Maxwell's paper was, on the other hand, read and appreciated by Cremona, who was inspired by it for his *Le figure reciproche nella statica grafica* [22] and came to a graphical construction much used by engineers. For this, we report in the following a large excerpt of Maxwell's paper.

It took inspiration, or, at least, so it is written in the introduction, from a memoir published shortly before by Rankine, native of Edinburgh like Maxwell:

> The properties of the 'triangle' and 'polygon' of forces have been long known, and the 'diagram' of forces has been used in the case of the funicular polygon; but I am not aware of any more general statement of the method of drawing diagrams of forces before Professor Rankine applied it to frames, roofs, &c. in his 'Applied Mechanics' p. 137, &c. The 'polyhedron of forces', or the equilibrium of forces perpendicular and proportional to the areas of the faces of a polyhedron, has, I believe, been enunciated independently at various times; but the application to a 'frame' is given by Professor Rankine in the Philosophical Magazine, February 1864 [13].[17]

Maxwell turned the static problem suggested by Rankine at once in a purely geometrical one, proceeding in an axiomatic way by definitions and proofs. Only after having proved the properties of reciprocal geometrical figures did he go back to statics, using the physical model of the truss, defined as a system of bars connecting a set of nodes.

The general definition of reciprocal figures opened Maxwell's paper:

> RECIPROCAL figures are such that the properties of the first relative to the second are the same as those of the second relative to the first. Thus inverse figures and polar reciprocals are instances of two different kinds of reciprocity.
>
> The kind of reciprocity which we have here to do with has reference to figures consisting of straight lines joining a system of points, and forming closed rectilinear figures; and it consists in the directions of all lines in the one figure having a constant relation to those of the lines in the other figure which correspond to them [13].[18]

---

[17] p. 251.

[18] p. 250.

Such a definition was restricted shortly after, before entering the details. Maxwell limited himself, for the time being, to consider plane figures and chose a particular kind of reciprocity, leading to the desired mechanical interpretation, even though he did not anticipate anything on the subject. The first part is about the relations among points and lines for the possibility of drawing reciprocal figures:

On Reciprocal Plane Figures.

> *Definition.* Two plane figures are reciprocal when they consist of an equal number of lines, so that corresponding lines in the two figures are parallel, and corresponding lines which converge to a point in one figure form a closed polygon in the other.
>
> *Note.* If corresponding lines in the two figures, instead of being parallel are at right angles or any other angle, they may be made parallel by turning one of the figures round in its own plane.
>
> *Since every polygon in one figure has three or more sides, every point in the other figure must have three or more lines converging to it*[19]; and since every line in the one figure has two and only two extremities to which lines converge, every line in the other figure must belong to two, and only two closed polygons [13].[20]

Denoting by e the number of lines, s the number of points, and f the number of distinct figures contained in the figure, Maxwell provided the relations, which always hold:

$$e = s + f - 2, \quad e = 2s - 3, \tag{5.3}$$

for strictly determined figures.[21]

Now, it is not difficult to arrive at the conclusion that an assigned figure admits, in general, a unique reciprocal figure, or infinite, or none, depending on the number of its points being equal, greater than, or less than the number of the closed polygons, respectively, and that in these cases we have $e = 2s - 2$, $e < 2s - 2$, $e > 2s - 2$, respectively.

The simplest case of reciprocal figures, according to Maxwell, is the one shown in Fig. 5.1, in which we have four points ($s = 4$), six lines ($e = 6$), and four polygons ($f = 4$), corresponding to the three inner triangles and to the outer triangle. In order to read Fig. 5.1 it is not important to consider as the basic figure that marked by capital letters, or that marked by small-case letters, since the reciprocity relation is reflexive. As one may see, the reciprocal figure should be rotated by $\pi/2$ so that the corresponding lines are parallel. It should also be clear that reciprocal figures are defined modulo a similarity, since metric properties do not enter the definition of reciprocity.

After the geometrical analysis, Maxwell provided a static interpretation:

---

[19] Italics is ours.

[20] p. 251.

[21] Remark that the first one corresponds to Euler's theorem for polyhedra in space; the second, mechanically interpreted, provides the necessary condition for a plane truss to be kinematically and statically uniquely determined.

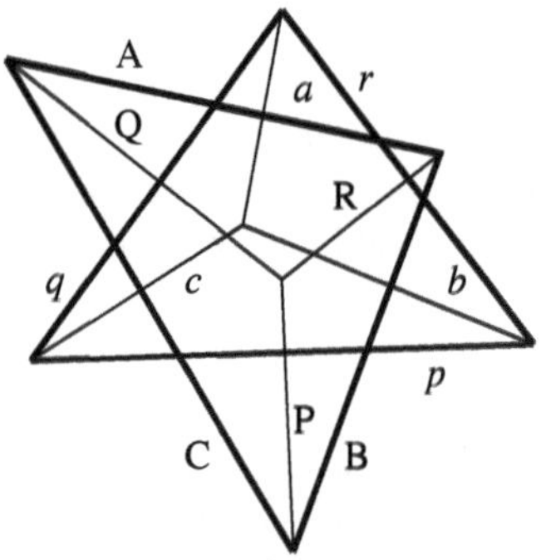

Fig. 5.1 Example of reciprocal figures (redrawn from [13, p. 165])

> The doctrine of reciprocal figures may be treated in a purely geometrical manner, but it may be much more clearly understood by considering it as a method of calculating the forces among a system of points in equilibrium [...] [13].[22]

The theorem—Maxwell did not call it that, but it actually is—holds:

> If forces represented in magnitude by the lines of a figure be made to act between the extremities of the corresponding lines of the reciprocal figure, then the points of the reciprocal figures will all be in equilibrium under the action of these forces [13].[23]

the proof of which is laconic:

> For the forces which meet any point are parallel and proportional to the sides of a polygon in the other figure [13].[24]

Indeed, the proof of the theorem is very simple if we think that a point of the reference figure, to which n lines concur, corresponds to a closed polygon with n lines in the reciprocal figure, with lines parallel to the lines concurring to the node in the reference figure. If the lines of this polygon are interpreted as forces acting on the nodes (points) of the first along the lines concurring to it, it is apparent that such forces, suitably oriented, are equilibrated, because the polygon of the forces acting on each node is closed if the node is equilibrated.

In order for the mechanical interpretation to be complete, or, at least, satisfactory for an engineer, reference should be made to an actual structure. Maxwell did it by considering a very particular case: a truss with one redundancy, free in space and not subjected to external forces, which finds itself in a state of coaction. The polygon of forces for each of its nodes, closed since the node is equilibrated, has lines parallel to the bars concurring to it because there are no external forces, and the tensions in the bars are directed along the bars. This makes the mechanical interpretation simple, while for a uniquely determined truss, or for an articulated mechanism, in which, in the absence of external forces the inner tensions of the bars vanish, the mechanical interpretation is not possible.

---

[22] p. 258.

[23] p. 258.

[24] p. 258.

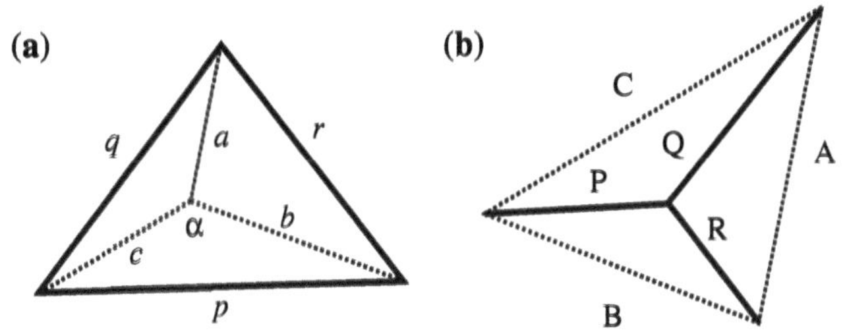

Fig. 5.2 Correspondence among summits and polygons in reciprocal figures [13, p. 207]

Consider, for example, Fig. 5.2, where the reciprocal figures represented in Fig. 5.1 have been split, and one has been rotated by $\pi/2$. Let us interpret Fig. 5.2a as representing a truss on which only coaction states are present. The reciprocal Fig. 5.2b may then be seen as an 'aggregate' of polygons of forces, one for each node of the truss in Fig. 5.2a. Consider, for instance, the point α of Fig. 5.2a, to which the lines a, b, c concur. They find correspondence in Fig. 5.2b with the lines A, B, C that are parallel to them, and form a closed polygon. If we interpret the lines a, b, c as the lines of action of the forces acting on α, the polygon A, B, C provides values of equilibrated forces. It is apparent from the construction that the magnitude and the orientation of the forces are defined only if we fix the magnitude and the orientation of one of them.

Maxwell's mechanical interpretation ended here: he never spoke about polygons of forces, or funicular polygons, like Culmann and Cremona would have done. Things would go in a different way when he faced the problem for the second time in 1872 in a memoir with a title similar to that of 1864 [14]. Here the mechanical interpretation became dominant with respect to the geometrical one; figures became frames, and the names polygon of forces and funicular polygon were present. The treatment as well was more general, and, at least for plane figures, the cases of statically determined frames loaded at the nodes were considered. Reciprocal figures were set within the scopes of projective geometry, even though we cannot find here an explicit reference to such a discipline. In particular, the closed polygons of the paper of 1864 became projections of faces of polyhedra:

> The diagram, therefore, may be considered as a plane projection of a closed polyhedron, the faces of the polyhedron being surfaces bounded by rectilinear polygons, which may or may not, as far as we yet know, lie each in one plane [14].[25]

In addition, a systematic solution for the construction of reciprocal figures was suggested, letting a point of the reciprocal figure correspond to each line of the reference figure, and vice versa. In some points, Maxwell's contribution to the following work by Cremona is apparent, in particular, the above evidenced idea of considering polygons as projections of polyhedra.

[25] p. 7.

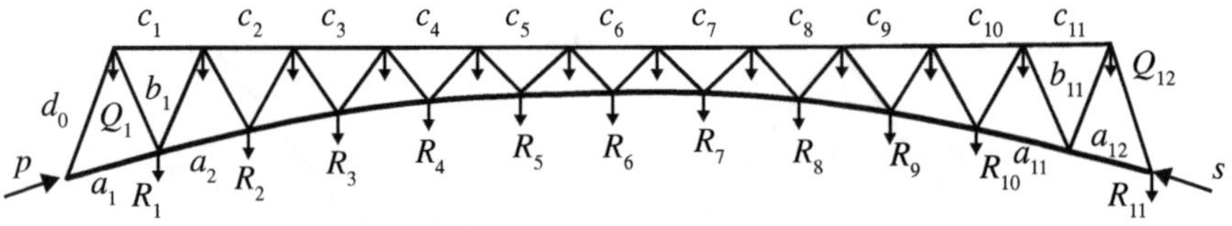

Fig. 5.3 Bridge designed by Fleeming Jenkin; truss model (redrawn from [14], plate XII)

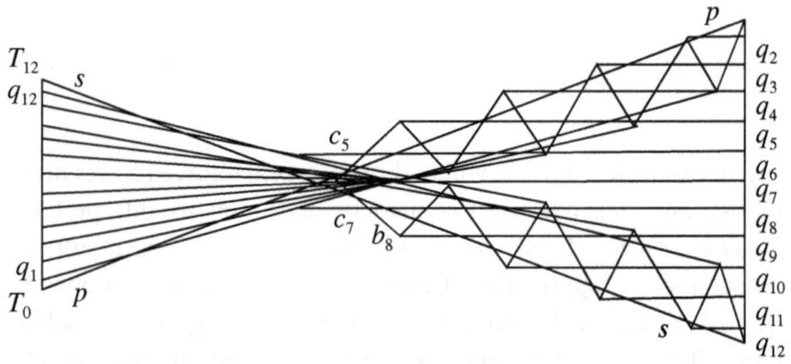

Fig. 5.4 Bridge designed by Fleeming Jenkin; forces on bars (redrawn from [14], plate XIII)

We report a truss in Fig. 5.3, which would have been later on considered also by Cremona; on this purpose, we also report a graphical construction anticipating the construction of Cremona's diagram in Fig. 5.4.

Besides investigating plane trusses, Maxwell faced the spatial case, and tried to extend the use of reciprocal figures to the analysis of stresses in three-dimensional continua. In addition, he posed the bases for an analytical foundation of the theory of reciprocal figures, thus detaching in some way from graphical statics properly said.

Even though we think it was very important, Maxwell's paper of 1872 was given modest attention by engineers, and we can say it is little known also nowadays. Exceptions were represented, as already said, by Jenkin and by Cotterill, who published a textbook of applied mechanics in which Maxwell's results were applied [16], and Cremona.

5.3.2 *Culmann's* Graphische Statik

Carl Culmann (Bad Bergzabern 1821–Zurich 1881) completed his studies in Karlsruhe at the local High technical school in 1841,[26] then joined the Bavarian civil service, where he worked to design and realize railway constructions in mountain regions; he was afterwards (1847) moved to office work in Munich. In summer 1849 he was sent to a study tour of the United Kingdom and the United States,

[26] We have examined some features of this school in [9].

where he visited railways. Indeed, the period of his tour coincided with that of the construction of the Britannia Bridge by Robert Stephenson in the United Kingdom, and with the beginning of the construction of iron bridges in the United States. A résumé of Culmann's tour of the United States appeared in 1851: it was so well received, and Culmann's reputation grew so much, besides of his fame as a designer, that in 1855 he was offered a chair at the newly founded Federal technical high school in Zurich, where he would have had the possibility to suitably combine theory and practice. Culmann was very dedicated to teaching and spent all his energies to develop graphical methods of calculus [57, 79].[27]

In the second half of the 19th century Culmann published the masterpiece *Graphische Statik* [28],[28] where an essentially thorough theory of graphical calculus for a large part of engineering problems was presented: it ranged from structural analysis to geometry of masses, from the analysis of stress in solids to the thrust of terrains.

In the following, we report a long excerpt of the preface that Culmann wrote for the French translation [29] of the second edition of *Graphische Statik*, which is particularly enlightening about Culmann's ideas on graphical statics and its educational motivations, besides providing a short summary of the contents.

Author's preface

The first systematic applications of graphic methods for the determination of the dimensions of the various parts of constructions are due to Poncelet. In fact, these methods, of which the nice works by Monge had in some way posed the bases, were professed for the first time by Poncelet at the school of application of civil and military engineers in Metz, in front of an audience formed by former pupils of the École polytechnique de Paris, the only one where graphical sciences had been taught at that time.

Poncelet had first realized that these methods, though being much more expeditious than the analytical methods, offered, however, an approximation more than sufficient in the practice, since, whatever we do, it will never be possible to obtain, in a design reported on paper, an accuracy higher than that provided by a graphical working drawing.

These methods, applied to the theory of vaults and of retaining walls, have been published in the *Mémorial de l'officier du génie* (tomes XII and XIII, years 1835 and 1840).

However, to determine the resultants Poncelet has not utilized the funicular polygon, the use of which provides so precious resources to graphical statics,[29] and it was left to his successor at the school of Metz, Mr. Michon, to first operate its application to the determination of the centers of gravity of portions, in his *Théorie des voûtes*.[30]

---

[27] In this book the uncertainty on the correct writing of Culmann's Christian name (with a capital K or C) is also told to derive from the non-uniqueness of German language of the first half of the 19th century.

[28] The second edition of the work is dated 1875.

[29] (*) Varignon mentioned it in its Nouvelle mécanique published in 1687 (Note by Culmann).

[30] (**) It is by chance that in 1845 an autographed course without the name of the author, having title: Instruction on the stability of constructions, has fallen into our hands. He who gave it to us attributed it to Mr. Michon. That course contains six lessons on the stability of vaults and four on that of coating walls (Note by Culmann).

The geometry of position, to which Poncelet let make a lot of progress, was not advanced enough at that time though, to make it possible to replace ordinary geometry (*Geometry of Masses*) by it in the development and proof of working drawings. Also Poncelet resorted, as often as possible, to ordinary geometry, and when elementary methods were not sufficient any more for him for his proofs, he merely translated algebraic formulas into drawings.

We must point out, however, that the first Treatise on geometry of position, in which complete abstraction was made of the idea of measure, has not been published until 1847, by G. de Staudt, professor of mathematics in Erlangen (*Die Geometrie der Lage*, Nürenberg, 1847).

When we were called, in 1855, at the time of the creation of the Polytechnic school of Zurich, to profess the course of constructions (including earthworks, and construction of bridges, routes, and railways), we were compelled to introduce Poncelet's graphical methods in our teaching to fill the gaps of the course of applied mechanics. That course did not include other than analytical methods in Zurich then; it was the same, at that time, at the École des ponts et chaussées in Paris, and it would be in vain if we looked in the *Cours de résistance des matériaux* by Mr. Bresse for the working drawings by Poncelet and by Mr. Michon.

This introduction of the theories of Graphical statics in the course of constructions did not fail to present some drawbacks, unduly delaying the course of studies; we obtained, in 1860, the creation of a winter course (two lessons per week) mandatory for engineers, in which we dealt those among the problems of statics applied to constructions, which were susceptible of graphic resolutions, and the teaching of which did not find any place, for lack of time, in the course of technical mechanics (then professed by Mr. Zeuner).

Such was the origin of Graphical statics. Since the courses of constructions (bridges and railways) which were more particularly comprised in our specialties, and that of statics, thus found themselves united in the same teaching, we were often led to give pupils supplementary explications on the parts that they had not perfectly understood. Under these circumstances we have always found that it had been simpler to recall theorems of geometry of position, the proof of which may be made by means of the lines of the drawing themselves, than to resort to analytical calculations, the passages of which would have required the use of a different sheet of paper.

This is how we were brought, irresistibly so to say, to replace algebra by geometry of position as much as possible. During the first years, the pupils' knowledge of this subject left much, to be true, to be desired; but, after a special course in geometry of position professed by Mr. Fiedler (to which the *Géométrie descriptive* of this author had already prepared the pupils), had been introduced in the study schedule, we have not found any difficulty more in our teaching.

It was when this teaching had taken some development, that we have published the first edition of our *Graphical statics*. (The first half was issued in 1864 and the second in 1865.)

Our diagrams were more successful than our methods. Our publication was followed by a great number of Elementary statics, in which, while reproducing our simplest diagrams (most of the time without changing them), the authors tried to provide their analytical proofs.

We believe that the truth is not there at all; since we will never come to draw the lines of a scheme and, at the same time, perform the algebraic operations that the explication of this scheme implies, nor to well penetrate the meaning of each line, and to represent the static relationships, if we limit to translate a formula, the passages of which are not present any more in our memory.

We must, however, except the Italian authors from the reproach we believe in right to address to our successors, and in particular Cremona, who has introduced Graphical statics in the teachings of the Polytechnic school in Milan. This scientist, to whom graphical sciences are indebted for beautiful works, of which we took advantage, did not disdain to teach he himself geometry of position to his students. Although Cremona has nowadays left Milan for Rome,

> the teaching of Graphical statics has continued with the same spirit at the Polytechnic school in Milan.
>
> The preceding explanations seemed to us necessary for an historic of Graphical statics; it remains to indicate, in some words, the order we followed in our work.
>
> The first chapter of the first part deals with the *calculus by segments*. Although it is alien to Statics properly said, it is necessary that students know it, and, since it is not taught in preparatory courses, we have thought that it be necessary to make these methods known, which are borrowed from French authors, most of all Cousinéry. To the calculus by segments we added the calculation of earthworks volumes, the movements of terrains, the theory of the slide rule, the so ingenious methods by Mr. Lalanne (nowadays general inspector of bridges and structures, and director of the École des ponts et chaussées in Paris) on graphical representations and on logarithmic squares.
>
> The second part deals with the composition and the decomposition of forces in general.
>
> The third part is devoted to parallel forces and to their moments of first and second order, the applications of which to the theory of elasticity, which form the fourth part of the work, are so numerous [29].[31] (A.5.7)

It is apparent from this preface that, even though he had set targets of rigor, and geometrical and algebraic precision, Culmann did not lose sight of the ultimate goal of the discipline, that of providing the technicians of his time with a powerful means of calculation, design and verification of a large number of engineering realizations in many fields of application.

Culmann's textbook was issued in two editions. The second, expected in two volumes, should have been, in the author's intentions, a more extended and more organic version, containing more applications.[32] Unfortunately, Culmann died before completing the second volume, and the second edition is indeed less extensive than the first, even though the published parts are more complete, and also account for the theoretical developments that graphical statics underwent after 1866. In the quotation of the preface to the French edition reported above, the cross-reference to the diffusion work operated by Cremona and his pupils in Milan and Rome should be noticed.

In the following we first summarize the tables of contents of the two editions of the *Graphische Statik*, then we briefly comment the contents, with reference mainly to the second edition, also in its French translation, very popular at the time [29].[33] We will consider only those parts directly related to mechanics of structures, in particular calculation of trusses and beams in flexure, largely neglecting the problems of geometry of masses.

The first edition, in German, was divided into 8 sections (*Abschnitt*). The first, *Das graphische Rechnen* (graphical calculus), of about 70 pages, concerned the graphical techniques for performing the operations of sum, difference, multiplication, raising to power, root extraction, integration, and derivation. Everything was done by refined

[31] pp. IX–XII. Our translation.

[32] Culmann's intentions, as we read in the preface in French to the second edition, reported above, were: "The second volume will contain a series of applications to beams, to frameworks, to arches, and to retaining walls".

[33] The copy of the German edition of *Graphische Statik* that we have consulted was property of Saviotti, as written in pen in one of the first pages, and contains some annotations, quite likely by Saviotti himself.

and very precise, at least for the standards of the time, constructions based on the properties of plane lines such as straight lines and parabolas.[34] It is apparent that Culmann, wishing to join rigor and applications, put at the base of his following applicative arguments a foundation of precise calculus, based on undebatable properties.

The second section, *Die graphische Statik* (graphical statics), of about 130 pages, began with the definition of forces and their composition rules. The *Kräftepolygon* (polygon of forces) and the *Seilpolygon* (funicular polygon) were introduced, and in a dedicated chapter their relations of projectivity and reciprocity were discussed. Parallel forces were then considered, since they are very important for engineering applications, in that they represent a very good scheme for distributed weights, thrusts of terrains, pressures of fluids on dams and containing walls, and so on. The last part of the section dealt with geometry of masses, which was also aimed at applications, in that the geometrical properties of the cross-sections of beams are essential for the processes of calculation, design, and verification of beams. We found particularly interesting the illustrations of the graphic calculation of the properties of area of the cross-section of a rail; this example may be found also in many other textbooks on Graphical statics following Culmann's, thus witnessing the importance of the investigation of railways in the second half of the 19th century.

The third section, *Der Balken* (the beam), of about 60 pages, concerned inflected beams. In it we find the principles of analytical and graphical calculation to determine the inner actions in one-dimensional structural elements, with numerical examples as well as examples of application of calculus and graphical statics, thus confirming the twofold aim of the textbook, theoretical and practical. In particular, we wish to remark that there was a chapter devoted to applications to cranes.

The fourth section, *Der continuirliche Balken* (the continuous beam), of about 90 pages, concerned the investigation of continuous beams on several supports, a very fashionable subject in structural mechanics, since it represented one of the archetypes of redundant structures, of very large use in applications. In particular, we find the applicative example of a continuous beam with four spans, suitable for railway transportation.

The fifth section, *Das Fachwerk* (the truss), of about 90 pages, investigated trusses; in this case also we deal with a very fashionable subject in the second half of 19th century, since a large set of application in buildings (decks, railway bridges, towers, cranes) was realized in with trusses. And, indeed, in the section the author presented, besides the general arguments, the structural realizations that were called *Pauli's bridge* (*Pauli'sche Brücken*), *English deck* (*englische Dachstuhl*), and *Belgian deck* (*belgische Dachstuhl*).

The sixth section, *Der Bogen* (the arch), of about 80 pages, dealt with masonry (hence, discrete) and continuous arches, and of bolts. It dealt with the pressure lines (*Drucklinie*) that may be defined on these structural elements to determine inner

[34] Still until twenty or thirty years ago, almost identical techniques of graphic integration and derivation were taught in some engineering schools in Italy.

stresses. We also find a chapter on the stability of the arch ends, and chapters where curved beams were dealt with.

The seventh section, *Der Werth der Constructionen* (the value of constructions), of about 20 pages, dealt with a subject that nowadays we would call project engineering, that is the cost and, consequently, the economic impact of a construction. Hints were also given on questions of structural optimization from the economic point of view, that is how to choose the beam cross-sections that make the cost a minimum, without diminishing the loading capacity.

The eighth, and last, section, *Theorie der Stütz und Futtermauern* (theory of retaining and containing walls), of about 80 pages, concerned subjects that nowadays are part of geotechnical engineering: the calculation of the thrust of terrains and of incoherent material, the introduction of cohesion, the first-attempt design of retaining and containing walls. This whole branch of applications, which was part of the higher education of any engineer at Culmann's time, is nowadays typical of civil engineers, and is taught in a group of academic teachings that are detached from traditional structural mechanics.

The textbook was completed by almost two hundred pages of tables and figures, examples of graphical calculus, abacuses, and geometrical constructions that illustrated all the procedures exposed in the eight sections of the work, which thus presents itself as a tome of more than 800 pages, a true compendium of theoretical and applied engineering in the field of constructions.

The only published volume of the second edition of the *Graphische Statik* is divided into 4 sections. The first, *Das Graphische Rechnen* (graphical calculus), of about 150 pages, concerned again, as in the first edition, the procedures and the constructions of graphical calculus. The second, *Die Zusammensetzung der Kräfte* (the composition of forces), of about 150 pages, concerned the definition of forces and the graphic operations on them. Besides introducing the concepts of polygon of forces (*Kräftepolygon*) and funicular polygon (*Seilpolygon*), in this section forces in space were studied, and the idea of reciprocity between forces was introduced; in the end, the reciprocity relations between *Kräftepolygon* and *Seilpolygon* in projective geometry were discussed. The third section, *Momente paralleler Kräfte* (moments of parallel forces), of about 180 pages, concerned the geometry of masses, considered as the study of the geometrical distribution of the weights of the areas and, thus, as moments of the weights of successive order, that is regular densities of parallel forces with respect to the measures of area of regions of space. The fourth, and last, section, *Elemente der Elasticitätstheorie* (elements of the theory of elasticity), of about 120 pages, was relative to the application of graphical statics to continuum mechanics and theory of elasticity.

Before entering the core of the subjects dealt with in this section, we wish to remark a meaningful change in the titles of the sections. In the first edition, the second section had the title *Die graphische Statik*; in the second edition, this title disappeared. In this way, Culmann for sure wanted to specify what he meant by the locution Graphical statics (*graphische Statik*): since he left it only as the title of the book, he wanted to make it clear that it meant not only the graphical procedures useful for solving the problems relative to the equilibrium of forces, but all the graphical procedures in engineering.

Another aspect to underline is that in the second edition we find parts of analytical statics:

> We have tried in the second edition [of the *Graphische Statik*] to link the analytical solutions to the purely geometrical solutions as briefly as possible. The new analytical methods have the great advantage to lead directly to the goal, and, in addition, to be in accord with the geometrical methods. In the majority of cases, we have been able to deduce the formulas from the geometrical passages preceding them. This way of proceeding has the advantage to give the theorems a form that, in a lot of cases, takes immediately off the geometrical constructions, and, in addition, to leave the choice, each time we provide the two solutions, between the graphic construction and the calculation; in the practice, it is as many times one of the methods than the other that leads to goal more quickly. [...]
>
> Thanks to the method that we have followed, we have shown to those who search to explain a working drawing analytically, how it is necessary to apply analysis to make the identity of formulas and drawings emerge [29].[35] (A.5.8)

To investigate the equilibrium of forces, already in the first edition of his book Culmann made large use of the *Kräftepolygon* and of the *Seilpolygon*, for which he recognized a relation of reciprocity in projective geometry, which would be made clear in the second edition. We must remark that, differently from the first edition, in the second edition Culmann did quote Maxwell's paper of 1864 and also presented Cremona's theory of the reciprocal figures, that he made his own:

> The reciprocal properties between the polygon of forces and the funicular polygon, which we have made known up to now, and which have been indicated for the first time by professor *Clerk Maxwell* in the *Philosophical Magazine*, 1864, p. 250, are relative to plane systems only. If we consider these polygons as the projections of skew polygons, the latter may be considered on their turn as reciprocal forms of a focal system. This theory has been developed by Cremona in his remarkable memoir titled: *Le figure reciproche nella Statica grafica, Milano, Bernardoni*, 1872. We will follow here mainly this last work [29].[36] (A.5.9)

In the second edition the analysis of reciprocity relations was preceded by the presentation of the projective properties of systems of forces in space, finally known at the time. It came out that any system of forces in space could be made equivalent to two non-coplanar forces, called reciprocal. The two forces, or better their lines of action, define a relation of polarity with respect to a second-order hyperboloid, which makes a point correspond to a plane. Since Culmann followed Cremona, we will postpone to a following section, dedicated to Cremona, an investigation of these aspects.

Culmann presented here an interesting use of the funicular polygon (see next section) in drawing the diagram of the bending moments (*Biegungsmomente*) of a simply supported beam, or of a span of a continuous beam, of which, besides the applied forces, the moments of continuity are known, determined for instance by the equation of the three moments. In the following we briefly report the way that

[35] p. XIII. Our translation.

[36] p. 291. Our translation.

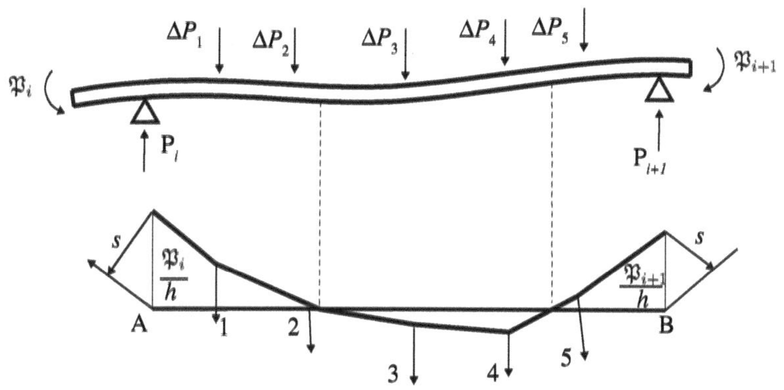

Fig. 5.5 Diagram of the bending moments in a beam in flexure (redrawn from [29, p. 310])

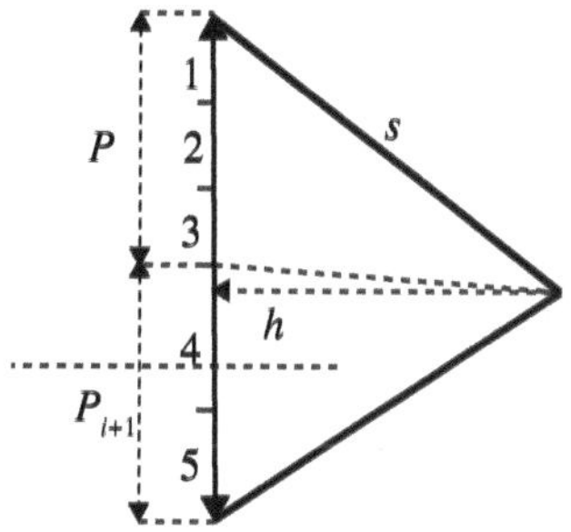

Fig. 5.6 Polygon of forces in a beam in flexure (redrawn from [29, p. 310])

Culmann went, making reference to the simply supported beam of Fig. 5.5, subjected to the two end moments $\mathfrak{P}_i$, $\mathfrak{P}_{i+1}$.

We first construct the polygon of forces, as in Fig. 5.6. Then we begin to draw the funicular polygon starting from the left, according to the construction reported in the lower part of Fig. 5.5, starting from a distance equal to $\mathfrak{P}_i / h$ from the reference point A, h being the distance from the pole O to the line of action of the forces of Fig. 5.6. Once arrived to the intersection of the last edge of the funicular polygon with the vertical line led from the right support, we obtain the point B, by drawing a vertical segment equal to $\mathfrak{P}_{i+1}/h$ downwards. The straight line AB divides the funicular polygon, which represents the diagram of the moments modulo the constant h, in the parts in which the top or the low fibers are in tension, while its parallel through the pole O in Fig. 5.6 provides the values of the constraint reactions P_i, P_{i+1}.

Besides the aspects of structural mechanics, Culmann's work is important because it anticipated some themes which were taken over again and diffused later on by his compatriot Christian Otto Mohr, whom we have extensively talked of in Chap. 1. Culmann introduced the elasticity ellipsis for the calculation of strains in beams in flexure and in arches already in the first edition of *Graphische Statik* (1866), thus anticipating the work by Mohr on arch trusses in 1881 [68]. In the second edition (1875), in the section dedicated to continuum mechanics, Culmann introduced a

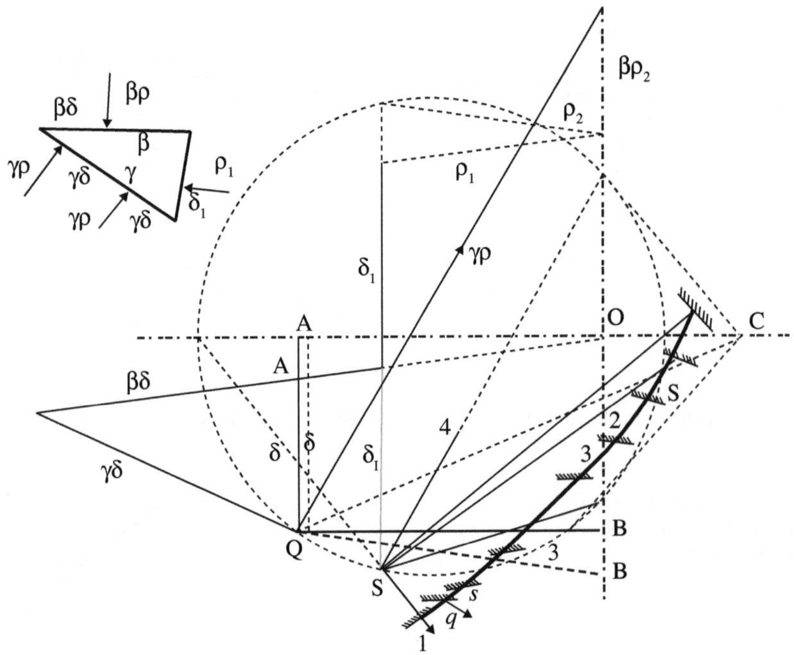

Fig. 5.7 Culmann's circle of stresses (redrawn from [29, p. 437])

graphical construction for representing the state of stress at an internal point of an elastic body. This is based on the *coniugio* property among points and straight lines induced by an ellipsis, which Culmann called ellipsis of forces.[37] Culmann's graphical construction is similar to the well known representation usually attributed to Mohr, who reported it in a paper of 1872 [64].

Figure 5.7 shows the construction of the circle of stresses at a point of a two-dimensional continuum.

Figure 5.8 shows the use of the elasticity ellipse, which Culmann called *central ellipsis*, for which the following theorem holds:

> Any force at will loading an arch, makes the extremity of the arch rotate around its anti-pole with respect to the central ellipsis,[38] of $\Delta s/\epsilon\mathfrak{I}$; and the magnitude of the rotation equals the force, multiplied by the static moment of these $\Delta s/\epsilon\mathfrak{I}$ with respect to its direction [29].[39] (A.5.10)

[37] Remember that a one-to-one correspondence exists among the matrices that, with respect to a basis, are the image of symmetric tensors of order two, and plane conics. As a consequence, this representation is a natural graphic representation of Cauchy's theorem on the state of stress at a point.

[38] The standard font is ours. It is a central ellipsis similar to that of inertia, the coefficients of proportionality being different; ϵ is the elastic modulus, $\mathfrak{I}$ is a moment of inertia, Δs is an element of arc.

[39] pp. 530–531. Our translation.

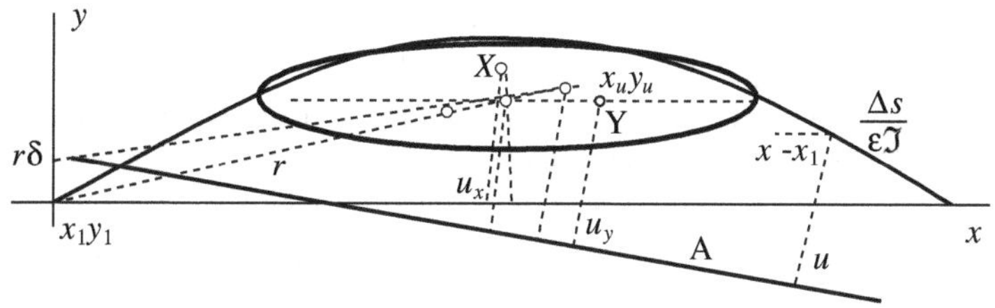

Fig. 5.8 Use of the ellipse of elasticity (redrawn from [29, p. 530])

5.4 The Contribution of Luigi Cremona

Luigi Cremona (Pavia 1830–Roma 1903) was the first of the four children of Gaudenzio Cremona and Teresa Andreoli.[40] His brother Tranquillo was a famous painter belonging to the *Scapigliatura* art movement, established for his originality.

In 1849 Cremona completed middle classic studies and enrolled in the course of civil engineering at the university of Pavia, where Bordoni and Brioschi were among his teachers. In 1853 he obtained, *cum laude*, the title of "dottore negli studi di ingegnere civile e architetto" (doctor in civil engineering and architecture), and soon after he got busy in the same university as lecturer in applied mathematics until 1856, when he took the necessary exams of mathematics and physics to be appointed as substitute teacher at the Gymnasium in Pavia. Two years later he was moved to Cremona and appointed full professor at the Gymnasium, lecturing from arithmetics to algebra, from plane and spatial geometry to trigonometry. In 1858 he moved to the Lyceum S. Alessandro (nowadays Lyceum Beccaria) in Milan, and he began to weave a net of international connections from there.

Under suggestion by Brioschi and Genocchi, in 1860 Cremona was called by the Ministry of public education to hold the chair of Geometria superiore in Bologna, established on purpose for him, first in Italy with that name. Cremona's stay in Bologna lasted until 1867, when Brioschi, aware of how Cremona was the most suitable person to teach the new techniques of graphical statics that spread through Europe, called him to Milan to teach Graphical statics at the Royal higher technical institute, which would later on become the Polytechnic of Milan.

When in Milan, Cremona had to take up the teaching of graphical statics only, thus abandoning, at least in part, his purely geometrical subjects of investigation. In 1873 he moved to Rome to be the director of the School of engineering, taking up also the chair of Graphical statics, which was transformed in a chair of Higher mathematics in 1877.

On March 16th, 1879, Cremona was appointed Senator of the Kingdom of Italy. He so began his political career, which definitely took him off his studies. He directed a lot of sessions of the Higher council of public education for many years, and he

[40] Gaudenzio Cremona had already had three children from his preceding marriage with Caterina Carnevali.

was vice-president of the Senate between 1897 and 1898. In 1898 he accepted the position of Minister of public education, offered to him by Rudinì, but he stayed in charge for thirty days only because of the troubled political situation. He managed, however, to propose a project of law, made up of a few articles, to modify those parts of Casati's law about disciplinary sanctions toward teachers.

He was fellow of the most famous Academies, in Italy and abroad, doctor *honoris causa* in Dublin and Edinburgh. He was appointed Cavaliere dell'Ordine dei Savoia, and in 1903 he was awarded the Ordre pour le mérite by the Emperor of Germany, an honour bestowed on few in Italy [36, 82].

Cremona published his first paper, *Sulle tangenti sfero-coniugate*, in 1855 in Barnaba Tortolini's[41] Annali di scienze matematiche e fisiche. Two more papers appeared in 1858, entitled *Sulle linee a doppia curvatura* and *Teoremi sulle linee a doppia curvatura* (he would later on call these lines 'skew cubics', and devote many other papers on their subject, the last of which in 1879), where he managed to derive some theorems that Chasles had only stated in his *Aperçu historique* [10], completing them with the proof and showing some more meaningful properties. Cremona let then these papers be followed by original investigations on conics, and published some papers on homofocal quadrics, and on conjugated conics and quadrics.

Even though the analytical method prevailed in his first works, of which we have given some hints until now, slowly a lively interest toward 'pure' geometry began to emerge. This geometry was that learned by Chasles, which would lead him to write *Considerazioni di storia della Geometria* published on *Il Politecnico* in 1860, where he exposed a rich historical picture of geometrical researches, both ancient and modern, with remarkable ability.

His evolution toward purely geometrical methods began, indeed, after having learned the teachings of the German school, with which he got in touch after his call at the university of Bologna. Bologna's time was for him the most fruitful and productive: it was, indeed, in these years that Cremona published his most innovative and important papers: *Introduzione a una teorica delle curve piane* in 1861, and *Preliminari di una teoria generale delle superfici* in 1866.

Other fully original papers are two notes with the same title, *Sulle trasformazioni geometriche di figure piane*, published in 1863 and 1865, respectively, where Cremona presented the transformations that have become his most important contribution, and that bring his name also nowadays. For our scopes, we must quote the memoir of 1872 entitled *Le figure reciproche nella statica grafica*, now considered a classic of graphical statics [22].

According to Cremona, the education of engineers should aim to build a class of highly qualified technicians that could also be culturally part of the new Italian ruling class; thus, he defended the basic education role of scientific culture, absolutely inseparable from the purely practical one, and underlined the importance of learning geometry as a basis on how to learn to proceed rationally. In this order of ideas, it is

[41] Barnaba Tortolini (1808–1874), priest and mathematician, professor of mathematics at the University of Rome, founded the first Italian scientific journal with international diffusion.

well understandable how he and Brioschi fought a battle to go back to the study of Euclid's *Elements*.

Another interesting collaboration between Brioschi and Cremona was the direction of the Annali di matematica pura ed applicata, which had seen a decadence in both style and interest in the 1860's. The direction was based on the common will to create, together with the country's political unity, a scientific, in particular mathematical, culture that would represent Italy, placing it at the same level of the other European countries. Their expectations were fully realized, and the *Annali* became again an important scientific journal at the European level.

5.4.1 The Funicular Polygon and the Polygon of Forces as Reciprocal Figures

Maxwell had already considered the polarity between the funicular polygon and the polygon of forces in the particular case in which forces were self-equilibrated and all converged to a single point. Cremona would consider, on the other hand, reciprocity in the more general case of non-concurring forces, also modifying the construction rules in such a way that a rotation by $\pi/2$ between the reciprocal figures was not necessary any more for the mechanical interpretation. Reaching this result, however, needed extensive use of projective geometry.[42]

5.4.1.1 The Funicular Polygon and the Polygon of Forces

A system of plane forces, however assigned, may be reduced to another one with simpler representation, by means of a suitable reduction. With a modern language, such a reduction is based on the equality of the virtual work spent on any rigid body motion, or, equivalently, on the equality of the resultant and of the resultant moment

[42] In the following we will consider the two notions of polarity and reciprocity as equivalent, even though in projective geometry they are distinct in general. Let us give some definitions:

> A reciprocity in a plane, where any two homologous elements correspond to each other in a double way (by an involution), that is a reciprocity equivalent to its inverse, is called a polar system, or a polarity; a point and a straight line corresponding to each other in a plane polarity are called pole and polar one of the other.
>
> Polarity in a plane may also be defined as a one-to-one correspondence among points and straight lines, such that: if the straight line corresponding to a point A (its polar) passes through a point B, the corresponding (polar) of B passes through A.
>
> Remark. Analogously (in space) we may define polarity in a star [31], p. 186. Our translation. (A.5.11)

To each polarity we may associate a conic, ellipsis, hyperbole, or parabola. "A set of the points and of the straight lines conjugated with themselves is said fundamental conic of the polarity" [31], p. 204.

of the two systems. On the basis of such a criterion, it may be proved that any non-vanishing system of forces may be reduced to a force (*resultant* of the system) applied to a well determined straight line in space (*central axis* of the system).[43]

The search of the resultant and of the central axis of a system of forces may be easily made by the modern techniques of linear algebra, considering the vector representation of force in the Euclidean ambient space. In times when the linear algebraic and vector calculus was not yet well developed, purely geometrical reduction techniques were developed.

The funicular polygon and the polygon of forces are graphic constructions, very suggestive also nowadays, which are used to determine the resultant of a system of forces and a point of its line of action (the central axis). The resultant is obtained by the polygon of forces, which is an extension of the idea of the composition of two concurring forces; a point of the central axis is obtained by means of the funicular polygon. In what follows we will present some historical hints relative to these two constructions, and a short explanation of them.

Historical background

The first notions of the funicular polygon and the polygon of forces are due to Simon Stevin, who used and proved, although incompletely, the rule of the parallelogram for the composition of forces [77].[44] Stevin, however, summed only two forces, and did not introduce the composition of more forces, hence the polygon of forces, explicitly; on the other hand, he introduced the idea of funicular polygon very clearly, although not as a tool of graphical statics, but as a law (theorem) of mechanics. In the following, we quote a passage by Stevin in which the introduction to the funicular polygon was presented.

> However, if we had several weights hanging on the same line, as the line ABCDEF here, [being] fixed its end points A, F, to which 4 known weights G, H, I, K are suspended; it is apparent that we may say what is the effort they act on the rope, to each of its parts AB, BC, CD, DE, EF: Since for example, moving GB up toward L, & MN parallel to BC: I say BN gives BM, like the weight G will give the effort that is done on AB (Fig. 5.9).
>
> Once again BN gives MN, like the weight G that will come will be the effort that is done on BC [78].[45] (A.5.12)

Thus, Stevin remarked, the broken line formed by the rope, fixed at its ends and to which a given number of weights are suspended, forms an open polygonal with edges parallel to the forces acting on the rope. The determination of the forces in the single portions of the rope is simple, and is based on the decomposition of each weight in the directions of the two adjacent portions of rope, as explained by Stevin when he moved upwards the segment representing the weight G and he decomposed it ideally along the directions of the two adjacent portions AB, BC. More precisely, according to the geometrical construction of decomposition of two concurring forces, the segment BN is to BM as the weight G is to the tension to which AB is subjected, and is to MN as the weight G is to the tension to which BC is subjected.

---

43 If the resultant vanishes, the central axis coincides with the line at infinity of the plane.

44 See also [5, 6, 29, 60].

45 p. 505. Our translation.

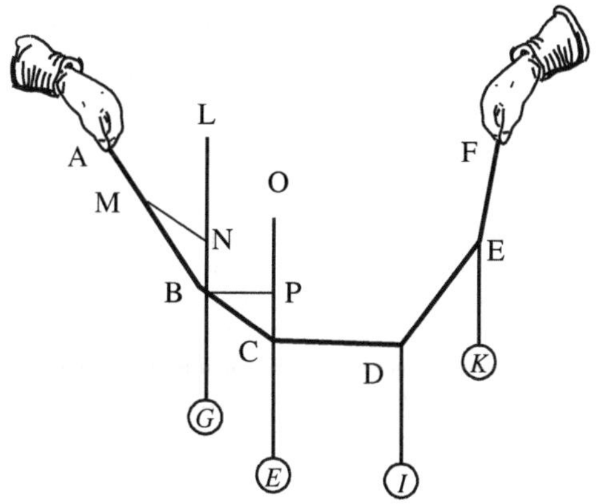

Fig. 5.9 The funicular polygon according to Stevin (redrawn from [78, p. 505])

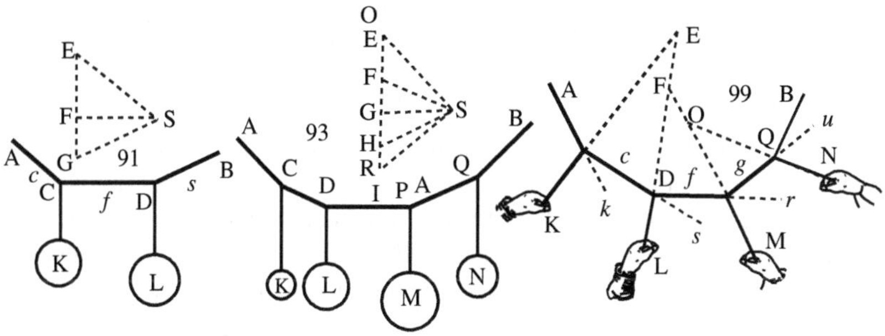

Fig. 5.10 The funicular polygon according to Varignon (redrawn form [81], vol. 1, between pp. 220–221)

Stevin's ideas were taken up and completed by Varignon [81],[46] who used the polygon of forces and the *polygone funiculaire* [81].[47] In this second case he also used the modern terminology, although not in the technical sense, alternating it with the more neutral term *polygon formed by the rope*.

All the considerations and the graphical representations on forces of Varignon's textbook are about the actions on ropes in tension. Forces were seen, in a way typical of 18th century statics, as weights that put in tension the various parts of a rope to which they are applied, deforming it according to a polygonal broken line (the "polygon formed by the rope", precisely). Sometimes, in Varignon's figures the forces were represented also as hands pulling a rope, in order to avoid introducing constraints and constraint reactions, that could create some problems for Varignon's restricted idea of force [8].

[46] Explicit figures, practically coinciding with those drawn also nowadays, are on pp. 190–191. Varignon had already expressed these ideas in [80].

[47] p. 202.

In the following, we quote Varignon's passages that describe the idea of the polygon formed by the rope, in the sense described above.

THEOREM X.

> I. [Since] Any two powers K, L, directed at will, & applied at any two points C, D, of a loose and perfectly flexible rope ACDB, attached by its ends at two nails or hooks A, B, remain in equilibrium between them as in Th. 8. 9., let us draw from any point S [the segments] SE, SF, SG parallel to the three edges AC, CD, DB, of the polygon ACDB that these powers let this rope turn into; & let us draw from a point F, taken at will along SF, [the segments] FE, FG, parallel to the directions CK, DL, of the powers K, L, until these two lines meet SE, SG, in E, G. This done, I say that in this case of equilibrium the powers K, L, are to each other as EF, FG, that is to say, K.L::EF.FG.
>
> II. Reciprocally, the configuration ACDB of the rope being given, that is to say, the polygon that it assumes being given, if from a point S taken at will we draw SE, SF, SG, parallel to the three edges AC, CD, DB, of this polygon; & from a point F taken also at will, along SF, we draw any two straight lines FE, FG, that meet SE, SG in E, G: two powers K, L, that are to each other as these two lines FE, FG, & that will have their directions CK, DL, parallel to each of these lines themselves, will keep the rope ACDB in that given configuration, there remaining equilibrated between them [81].[48] (A.5.13)

The construction of the funicular polygon is still nowadays taught according to Varignon's original procedures, taken up, among others, by Saviotti in his textbook (see *infra*). The key concept is that of using 'invariant operations' only, ideal transformations of primarily graphic nature, usually attributed to Varignon. These do not modify the resultant and the resultant moment of a system of forces, and make possible a reduction to systems that are more simple to represent. These elementary invariant operations, the validity of which is justified in an intuitive graphic manner, are the following:

1. Two forces applied at the same point may be replaced by their resultant applied at that point, according to the parallelogram rule.[49]
2. A system of forces may be equivalently replaced by another one, if we add or subtract couples with null arm.[50]
3. A force may be shifted along its line of action.[51]

In this way, the construction of the funicular polygon is reduced to a given succession of invariant operations, ending when we arrive to a unique force applied to a well determined direction (the central axis of the system).

---

[48] vol. 1, pp. 190–191. Our translation.

[49] Conversely, a force applied at a point may be decomposed in the sum of two forces applied at the same point, along two assigned directions.

[50] A couple is a system composed by two parallel forces of equal magnitude and opposite sign, the distance of their lines of actions being called *arm* of the couple. Just like a force causes the variation of a translational motion (that is it is the prototype of interactions spending power on translations), a couple causes the variation of a rotary motion (that is it is the prototype of interactions spending power on rotations). A couple with null arm is thus a system composed by two collinear opposite forces, trivially equivalent to a null system.

[51] This operation is, however, implied by the operation 2.

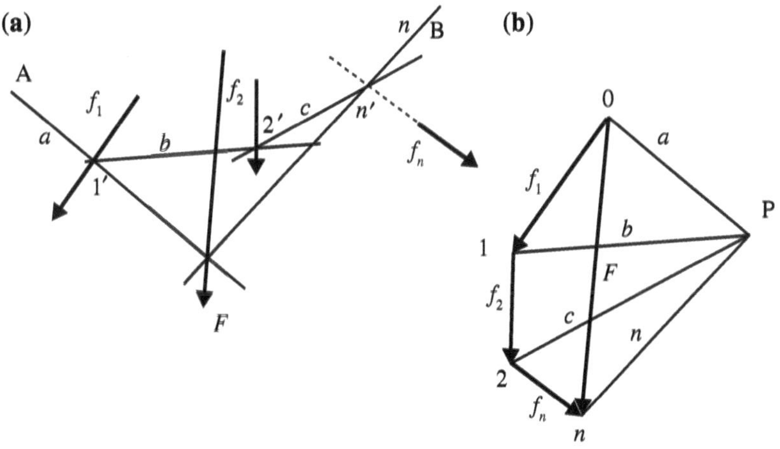

Fig. 5.11 The funicular polygon (**a**) and the polygon of forces (**b**)

Consider a generic number of forces $f_1, f_2, \ldots, f_n$, which, for sake of simplicity, are contained in a plane π (the extension to three-dimensional space may obviously be operated by linear superposition of what happens on three independent planes). We obtain the resultant of the system by operating the graphic construction called polygon of forces, or of the successive resultants, or of composition according to Saviotti, which is but an extension of the parallelogram of forces. As shown in Fig. 5.11b, starting from the force f_1, which is represented in a given scale by the oriented segment 01, we successively draw the force f_2, represented in the same graphic scale by the oriented segment 12, and so on, until the last force f_n; the resultant F of the system is, in the same graphic scale, represented by the oriented segment 0n in magnitude, direction, and orientation.

To determine the central axis, that is the straight line defining the direction of application of the resultant F to reduce the given system to a unique force, we choose a pole P to the right of the polygon of forces, called *projection pole*, and we join P with the summits $0, 1, 2, \ldots, n$ of the polygon of forces. We then choose an arbitrary point A of the plane (see Fig. 5.11a) and we draw through A the parallel to the projector radius P0, which intersects the line of action of f_1 at the point 1′. We then draw through 1′ the parallel to the projector radius P1, which intersects the line of action of f_2 at the point 2′, and so on until the force f_n, the line of action of which is intersected by the projector radius $Pn - 1$ at the point n'. The intersection of the parallels to the first and the last projector radii P0 and Pn passing through A and n', determines a point of the central axis.

The line A1′2′ … n' is the funicular polygon and its construction is justified in terms of elementary invariant operations. What we actually do is to decompose all the forces along the directions of two successive projector radii. In this way, we obtain a series of couples of forces with null arm, one for each inner projector radius, which may be eliminated. The system is so reduced to the sum of the components along the outer projector radii only, the composition of which provides the resultant and its line of application, which is what we do when we intersect the parallels to the first and last projector radii.

It is not difficult to prove that, if the system of forces is self-equilibrated, that is, if it has vanishing resultant and resultant moment, then both the polygon of forces and the funicular polygon are closed. Conversely, if both polygons are closed, the system of forces is self-equilibrated.

5.4.1.2 The Null Polarity

Cremona's contribution to the theory of reciprocal figures may be seen as aimed to perfect a procedure to evaluate the tensions in the bars of a statically determined truss, which at that time were the key components of many industrial buildings. This procedure was made simple by Cremona by means of a particular form of reciprocity, or polarity, that exists between the polygon of forces and the funicular polygon. Indeed, Cremona introduced this relation in space, called null polarity, to let the polygon of forces and the funicular polygon in a plane correspond. In this way, the two figures may be drawn by either method, that is Varignon's procedure and their polarity relation.

The polarity used by Cremona may be introduced either formally, by using analytical geometry, or by direct considerations on the entities exhibiting the property. Cremona, taking inspiration from preceding works [61, 75], chose the second approach.

In the following, we briefly examine Cremona's results, and also present the analytical definition of polarity. As already suggested it may be easily proved that a system S of forces $f_1, f_2, \ldots, f_n$ in space may be reduced, by repeated applications of the parallelogram rule and by shifting forces along their line of application, to a force F directed along a straight line a and to a couple in a plane π normal to F, that is having a moment M parallel to F.

Alternatively, the same system S may be reduced to two forces f, f', usually not in the same plane, and one of the two forces may be chosen arbitrarily.[52] The forces f, f' are called conjugated, or reciprocal, and the rules for their construction define a polarity. The straight lines r, r' of action of f, f', respectively, are equally called conjugated, or reciprocal.

The characteristic element of this polarity is the straight line a of action of the resultant F that comes from the reduction of S to a force and a couple, called the *axes of the polarity*; the plane $\pi \perp a$ is called the *orthographic plane*, and the parallels to a are called *principal directions*.

The polarity induced by S enjoys the following properties:

---

[52] The idea is that the sum of the two forces provides in any case the resultant of the system, while the arbitrary choice of the line of action of one of the two let it be posed at a distance such as to warrant the equivalence of the resultant moment M.

1. If the straight line r describes a star with center P, the reciprocal straight line r' describes a plane π_P, called polar of P, which is said to be the pole of π_P.
2. The pole P is contained in its polar plane π_P.
3. The projections of the reciprocal straight lines r, r' on the orthographic plane according to the principal directions are parallel.
4. If n straight lines concur to a point P of the orthographic plane, they correspond by reciprocity to a closed polygon with n edges parallel to the n lines concurring to P.

The polarity may be expressed in analytical form [74]. Let us assume the general case in which the system of forces is not reducible to a force, or to a couple, only; that is the relation:

$$\mathfrak{T} = F \cdot M \neq 0 \tag{5.4}$$

holds, where $\mathfrak{T} = F_x M_x + F_y M_y + F_z M_z$, the factors of which are the components of the resultant force and moment of the system S, is called *invariant trinomial*, and is the algebraic characteristic of the system of forces, not depending on the chosen frame of coordinates.

In the projective space and with homogeneous coordinates

$$x_0 : x_1 : x_2 : x_3 = 1 : x : y : z \tag{5.5}$$

assuming the axis of polarity a coinciding with the z-axis, the polarity is expressed by the relation:

$$-cx_3^* x_0 - x_2^* x_1 + x_1^* x_2 + cx_0^* x_3 = 0 \tag{5.6}$$

where $c = \mathfrak{T}/||F||$.

A polar plane π_P with the above defined equation corresponds to a given pole $P \equiv (x_o^*, x_1^*, x_2^*, x_3^*)$. Such correspondence defines a polarity, since it is turned into itself by replacing x_i with x_i^*. It is not, however, a standard polarity, because the pole $P \equiv (x_o^*, x_1^*, x_2^*, x_3^*)$ belongs to its polar plane π_P. Such a polarity is often called *null* or *focal* polarity $\mathfrak{N}$ (equivalently, null or focal system). It is not difficult to prove that the properties listed above hold for it.

The null polarity was used by Cremona without specifying the characteristic parameters (the axis a and the coefficient c), which are all independent on the plane system of forces he considered (for which we would in any case have $\mathfrak{T} = 0$), to obtain projective relations between the polygon of forces and the funicular polygon. This was made by considering these two figures as projections of two reciprocal polyhedra on the orthographic plane.

Two reciprocal polyhedra are such that the summits of the one are poles for the faces of the other, and vice versa. The projections of the reciprocal polyhedra onto the orthographic plane are reciprocal figures, and they are characterized as follows:

> *Reciprocal figures*. Once projected these reciprocal polyhedra onto an orthographic plane, to each edge of the first figure a parallel edge in the second will correspond. Then, since to ρ edges forming the contour of a face of the one, the edges concurring to the summit

corresponding [to this face] in the other correspond, then in a projection ρ parallel edges forming a closed polygon will correspond to ρ edges concurring in a summit. Each edge of both polyhedra is common to two faces, and joins two summits; each face has three edges at least, and to each summit at least three edges concur. It follows that in their projections each edge will be common to two polygons and will join two summits, and, since each polygon will have three edges at least, three edges will concur to each summit. The elements of a polyhedron are connected by Euler's relation: $v + f = s + 2$ (1), where v indicates the number of summits, f of the faces, s of the edges. Since to the v summits of one polyhedron v faces in the other correspond; to the f faces in the one, f summits in the other, and to the s edges in the one the s edges in the other, so the relation (1) holds also for the reciprocal polyhedron. As to the two orthographic figures, if one consists of v summits, f closed polygons, and s edges, the other will consist of f summits, v closed polygons, and s edges. If a polyhedron has a summit at infinity, the other has a face orthogonal to the orthographic plane; whence, if one of the orthographic figures has a summit at infinity, the corresponding polygon in the other reduces to a straight segment on which are marked the points corresponding to the summits of the face of which it is a projection. The orthographic projections of two reciprocal polyhedra are said *reciprocal figures* [72].[53] (A.5.14)

5.4.1.3 Reciprocity

Cremona dealt with self-equilibrated forces, that is with nil resultant and resultant moment, for which the funicular polygon and the polygon of forces are both closed, and considered two cases. The first, illustrated in Fig. 5.12, is about the case when the self-equilibrated force meet at a unique point. Fig. 5.12a reports a polygon of 6 forces of this kind, together with the pole O and the necessary construction lines for drawing the funicular polygon; Fig. 5.12b reports the funicular polygon, together with the lines of action of the forces, concurring to a point by hypothesis. We see at once by the construction that the two figures are reciprocal, according to the null polarity, and that they may be seen as the projections of two pyramids, the first with summit O, the second with summit the common point of intersection of the forces, which are reciprocal polyhedra.

If the lines of actions of the forces do not meet at a unique point, showing the reciprocity is more complex, since the funicular polygon does not appear as the reciprocal figure of the polygon of forces, plus the projection radii, as it was in the preceding case. We may, in any case, still obtain reciprocal funicular polygons by considering two different projections of the polygon of forces, as shown in Fig. 5.13. Six forces form a polygon of forces still closed (Fig. 5.13a), but are not converging to a unique point. The reciprocal figure (Fig. 5.13b) is obtained by considering on the one hand two poles O, O′ and the polygons of forces derived from these poles; on the other hand, the corresponding funicular polygons, together with the lines of actions of the forces.

The two figures may be seen as projections of two reciprocal polyhedra onto the orthographic plane. The first polyhedron (Fig. 5.13a), associated to the polygon of forces and to the two poles, is defined by n non-concurring straight lines that meet two by two and form a closed skew polygon, and by the two poles O, O′, in such a

[53] vol. 2, p. 72. Our translation.

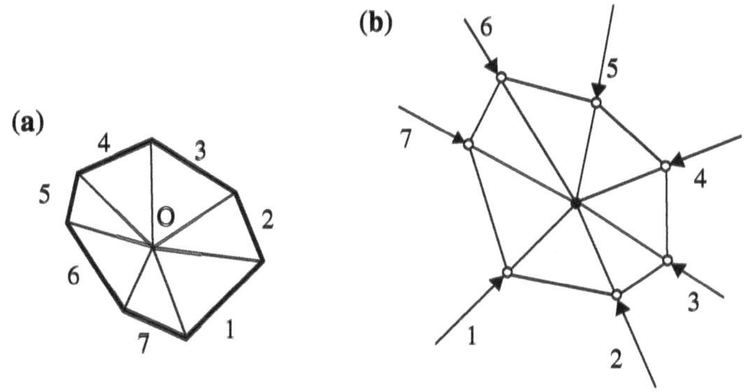

Fig. 5.12 The first case of reciprocal figures (redrawn from [22, p. 343])

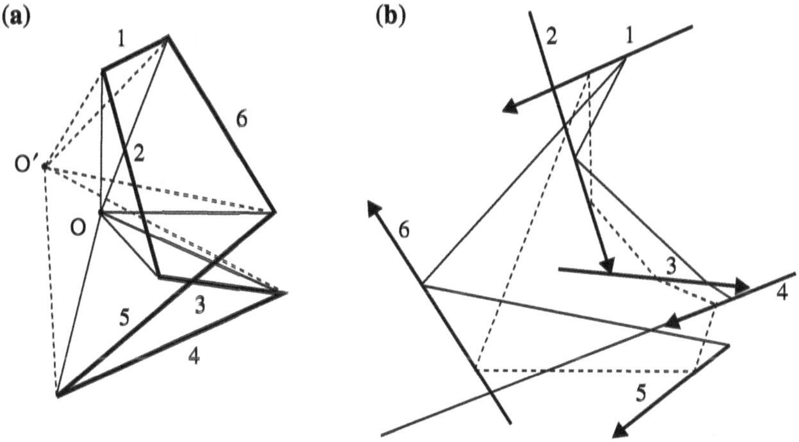

Fig. 5.13 The second case of reciprocal figures (redrawn from [22, p. 344])

way to obtain a polyhedron formed by two pyramids, the faces of which intersect on the skew polygon. The reciprocal polyhedron (Fig. 5.13b) is a prismoid defined by two plane faces as bases, the polar planes of O, O′, which are the funicular polygons associated to O, O′, and by n lateral faces, the polar planes of the summits of the skew polygon, the edges of which are parallel to the six forces.

We may simplify the drawing of the reciprocal figures by imagining the pole O′ to infinity, orthogonally to the orthographic plane. The first polygon then reduces to a pyramid with pole O and to a prismoid, the second polyhedron is the infinite portion of space contained by a plane polygon and as many planes that cross the edges of the polygon. The reciprocal diagrams, projection of the polyhedra, are then formed, one by the polygon of forces, completed by the radii projecting its summits from O, the other by the lines of action of the forces, the funicular polygon, and the straight line at infinity.

The possibility to consider the funicular polygon and the polygon of forces as projection of reciprocal polyhedra lets us clearly prove their reciprocity, and lets us establish in a rather simple way the rules for the construction of Cremona's diagram, of which we will talk in the next section. In addition, it lets us prove the theorems about the funicular polygon with relative simplicity, which would be difficult operating with the rules of the traditional geometrical statics. Among these theorems we quote:

> Two corresponding edges (r s), (r s)′ of the two polygons cut each other over a fixed line, that is parallel to the line joining the two poles O, O′. This theorem is fundamental in Culmann's methods [22].[54] (A.5.15)

> Let a plane polygon with n edges $1, 2, 3, \ldots, n-1, n$ be given; and, moreover, in the same plane let $n-1$ straight lines $1, 2, 3, \ldots, n-1$ be given, parallel to the first $n-1$ edges of the polygon, respectively. From a point, or pole, mobile in the plane (with no restriction), let the summits of the given polygon be given. Now let us imagine a variable polygon with s edges, the first $n-1$ summits of which $1, 2, 3, \ldots, n-1$ should be found in their order in the given straight lines with the same label, while the n edges $(n \,.\, 1), (1 \,.\, 2), (2 \,.\, 3), (n-1 \,.\, n)$ should be parallel to the radii that project the summits with the same label of the given polygon from the pole. The point to which any two edges $(r \,.\, r+1), (s \,.\, s+1)$ of the variable polygon concur in a determined straight line, parallel to the diagonal between the summits $(r \,.\, r+1), (s \,.\, s+1)$ of the given polygon.
>
> This theorem, the proof of which by means of plane geometry only does not seem obvious, results apparent instead, if we consider plane figures as orthographic projections of reciprocal polyhedra [22].[55] (A.5.16)

5.4.1.4 Cremona's Diagram

The reciprocity relation for the polygon of forces and the funicular polygon may be extended to polygons related to the mechanical behavior of real structures. The natural application is for statically determined trusses. Cremona, indeed, showed the existence of a reciprocity according to the null polarity for two figures relative to these structures. The first is formed by the bars of the trusses and the lines of action of the external active and reactive forces; the second consists of the set of the forces in the bars. This second figure, which is reciprocal of the first, is called Cremona's diagram, or *Cremonian*.

Both figures may be obtained as projections onto the orthographic plane of reciprocal polyhedra, more complex than those of the preceding article. A polyhedron P is composed by (a) a polyhedral surface having a skew contour S formed by as many rectilinear segments as the external forces are, and by a lateral surface presenting as many edges as the bars of the truss we wish to study are; (b) a pyramid having pole O and the skew contour of the surface S as base. The other polyhedron P', reciprocal of P, is obtained by applying the rules of the null polarity. The polyhedron P' represents the funicular polygons of the external and internal forces, while the polyhedron P represents their polygons of forces.

---

[54] p. 345. Our translation.

[55] p. 348. Our translation.

Cremona assumed then that the pole O goes to infinity in the direction orthogonal to the orthographic plane. In this way the projection of P' consists of the truss and the lines of action of external forces. The projection of P, since the pole goes to infinity, does not contain the funicular polygon, but reduces to segments that provide the values of external and internal forces.

Actually, the theory of reciprocal figures is not indispensable for drawing the Cremonian. As for the polygon of forces and the funicular polygon, indeed, Cremona's diagram may be constructed by elementary considerations on the equilibrium of forces. And, actually, in the Italian teachings of structural mechanics the reciprocity relations are not present anymore. For instance, Odone Belluzzi wrote:

> The reciprocal diagram of a truss, or Cremona's diagram, re-unites in a single figure the polygons of equilibrium of all its nodes. In it, any segment that measures the stress S of a bar, run once in a direction, and then in the other, is the common edge to the two polygons of equilibrium of the end nodes of the bar; then, it appears only once. Omitting the general theory, let us restrict ourselves to indicate the relations that link the reciprocal diagram to the scheme of the truss, its main properties, and the practical rules to build it [4].[56] (A.5.17)

It is apparent, however, that Cremona reached his results by using reciprocity: recognizing that the Cremonian is reciprocal of the truss, it may be drawn by applying the rules of the null polarity, neglecting in part those of statics.[57]

Construction rules and an example of a Cremonian

Cremona's diagram is a plane figure consisting of closed polygonal circuits, each with edges that represent the forces acting on each node P of a statically determined truss. To each node of the truss, to which n straight lines-representing bars and external forces-concur, a closed polygonal circuit corresponds in the Cremonian, with n edges parallel to the n forces concurring to P.

Let us now provide an example showing how to draw such a diagram for the very simple statically determined truss in Fig. 5.14a composed of two equal triangular circuits which collaborate to form a simply supported system.

Suppose that external forces f_1, f_2, f_3 be assigned; by means of simple graphical techniques it is immediate to find the reactions of the constraints with the 'ground'. First of all, we may shift the active force f_1, applied at B, along its line of action passing through BD (elementary invariant operation) until it is applied at D, where also the forces f_2, f_3 act. By another elementary operation, the three active forces may be replaced by their resultant f_R applied at D. The reaction r_C of the simple support at C has a known line of action, through C and orthogonal to the segment AC. Since the other constraint reaction r_A, exerted by the hinge at A, shall form a self-equilibrated system with f_R and the reaction of the support r_C, it shall go through A and the point E of intersection of the lines of action of f_R and r_C. Once known the lines of action of the constraint reactions and the active forces, the polygon of forces in Fig. 5.14b lets us fully determine the constraint reactions.

---

56 p. 535.

57 Cremona's reciprocal figures were investigated until relatively recent years, See, for instance, [5, 58, 74, 83].

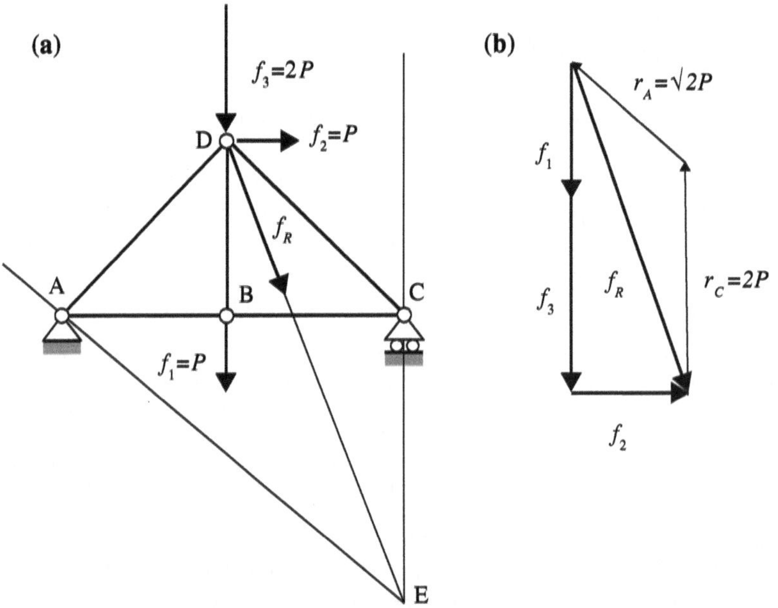

Fig. 5.14 The truss (**a**) and the polygon of the external forces (**b**)

To get a reciprocal figure it is necessary to follow a cyclic order for the external and internal forces, according to the numeration, or, in any case, the ordered labels of the nodes of the truss. This stems from the necessity to run the circuits of the truss always in the same orientation to respect the projections onto the orthographic plane. From the point of view of elementary invariant operations, this stems from the need to have a uniform, iterative procedure.

> Supposed that all the nodes to which the external forces are applied find themselves at the boundary of the scheme of the truss, these forces shall be taken in the order they are met by those who follow the above said contour. When we do not follow these rules, and the others exposed below, we may still solve the problem of the graphic determination of inner stresses, and we have no longer reciprocal figures, but rather more complicated and disconnected figures, where the same segment, not finding itself at its convenient place, shall be repeated or re-drawn to give place to the following constructions, as it happened in the old method of constructing the polygon of forces and the funicular polygon separately for each node of the truss [22].[58] (A.5.18)

Cremona's diagram relative to the truss of Fig. 5.14 is shown in Fig. 5.15. To draw it, we start from a node where only two bars concur, for instance, A (the mechanical motivation of such a choice is obvious: graphical equilibrium in the plane is possible when we have to equilibrate a given force according to two assigned directions). The reaction r_A and the tensions N_{AB}, N_{AD} of the bars concurring to A, which, by the hypothesis of dealing with a truss, are directed along the segments joining

[58] p. 352. Our translation.

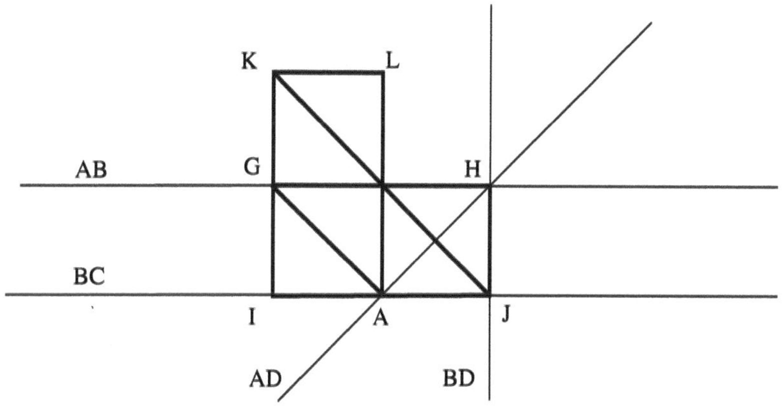

Fig. 5.15 The Cremonian of the considered truss

the nodes, cyclically concur to A. Let us draw AG $\equiv r_A$; let us draw from G the parallel to AB, from A the parallel to AD; we so close the circuit AGH, in which GH $\equiv N_{AB}$ (traction) and HA $\equiv N_{AD}$ (compression). We then skip to the node B, on which only two unknown forces act, that is the tensions N_{BC}, N_{BD}. Starting from HG $\equiv N_{BA}$ let us draw, following the usual cyclic order, GI $\equiv f_1$, from I the parallel to BC and from H the parallel to BD. We so close the circuit HGIJ, in which IJ $\equiv N_{BC}$ (traction) and JH $\equiv N_{BD}$ (compression). We proceed similarly for the node C; the circuit relative to it in the Cremonian is JIK, where IK $\equiv r_C$. The Cremonian is eventually closed by examining the node D, which is represented by the circuit JKLAH, where KL $\equiv f_2$ and LA $\equiv f_3$.

We easily see that all the edges of the circuits composing the Cremonian are walked twice in opposite directions, thus are equivalent to couples of null arm, which may be added or subtracted without altering the considered system of forces—and, indeed, they represent inner forces—with the exception of KL $\equiv f_2$, LA $\equiv f_3$, AG $\equiv r_A$, GI $\equiv f_1$, IK $\equiv r_C$, which represent the external forces.

The example we brought clearly shows apparently how the Cremonian diagram may be interpreted by means of elementary invariant operations on forces, and it is for this reason that, in spite of the undoubtable elegance and the numerous possibilities of application, the theory of reciprocal figures has lost attraction in the teaching of graphical statics and, more in general, of geometry applied to mechanics.

Cremona would not have agreed on this conclusion; here is his comment on the two different ways of drawing a Cremonian:

> This method, that could be said static, is sufficient in itself for the graphic determination of the inner tensions, like the geometric method exposed previously, which is deduced from the theory of reciprocal figures, and consists in the successive construction of the polygons corresponding to the various nodes of the truss. The static method, however, looks less simple to me, and might rather be helpful in combination with the other, most of all to verify the accuracy of the graphic operations already performed [22].[59] (A.5.19)

[59] p. 356. Our translation.

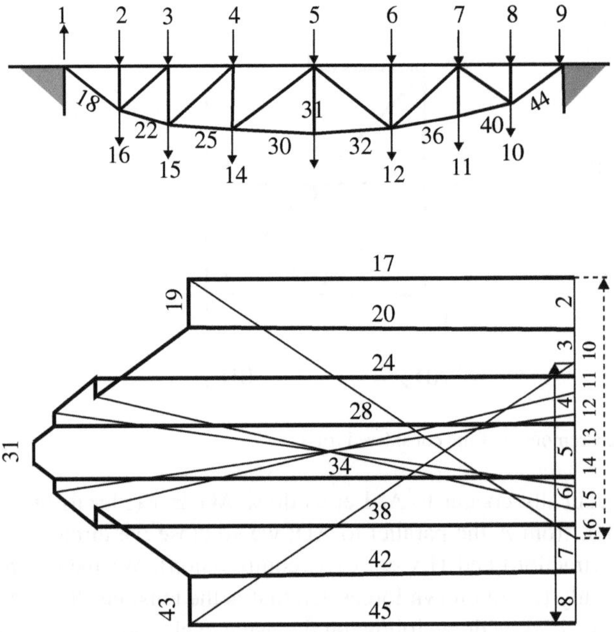

Fig. 5.16 The Cremonian of a bridge beam (redrawn from [22, p. 359])

Figures 5.16 and 5.17 report some more complex cases reported by Cremona [22].

Remark that we may not draw the Cremonian for all statically determined trusses. For instance, it may happen that there is no node to which only two bars belong and whence the construction can begin. Or it may happen that we meet nodes with more than three unknown tensions, like in the composed Polonceau truss. In such cases we either abandon the drawing of the Cremonian or integrate it by other methods, among which that of Ritter's sections [4].[60]

5.4.2 The Lectures on Graphical Statics

We saw that Brioschi, director of the Royal higher technical institute in Milan, appointed Cremona in 1867 to the chair of *Graphical statics*, esteeming him as the right person to start this teaching, which was gaining importance throughout Europe. Cremona did not live up to expectations, and after a short time he published his fundamental textbook, *Lezioni di statica grafica* [21], which collected his lectures of the academic year 1868–1869. Such a textbook, even though not as complete as Culmann's, and not fully original, was however very important, being the first work in Italian on the subject. In 1873 he published the *Elementi di geometria*

---

[60] v. 1, pp. 540–541.

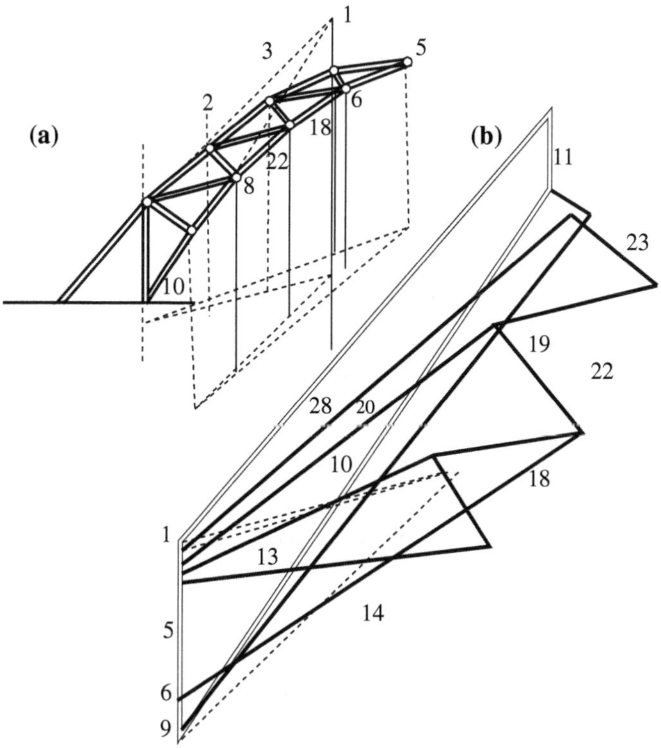

Fig. 5.17 The Cremonian of the truss for a crane (redrawn from [22], p. 363)

proiettiva [23], with the aim of providing the elements of projective geometry he believed
necessary for good mastership of the methods of graphical statics in a simple way. The book was fairly successful and was translated into French and English [24, 25].

It seemed useful to us to speak of this textbook after having presented the most original part of Cremona's investigations on graphical statics, to better understand his contribution to the teaching of the discipline. The course was divided into three parts and two volumes; we limit ourselves to presenting the title and the table of contents of each part.

Part I. Projective geometry

§ 1. Fundamental geometric forms.

§ 2. Harmonic systems.

§ 3. Projective forms.

§ 4. Involutions.

§ 5. Generation of conics.

§ 6. Poles and polars.

§ 7. Diameters of conics: ellipses, hyperbolae, parabolae.

§ 8. Exercises and constructions.

§ 9. Theorem by Desargues. Projective forms in conics.

§ 10. Exercises and constructions.

§ 11. Problems of 2nd order.

§ 12. Foci of conics.

§ 13. Other problems and constructions.

§ 14. Cones and skew surfaces of 2nd order.

§ 15. Exercises.

§ 16. Projectivity of the fundamental geometrical forms of second kind.

§ 17. Affinity and similitude of plane figures.

§ 18. Exercises.

§ 19. Generation of surfaces of 2nd order.

§ 20. Poles and polar planes with respect to a surface of 2nd order. Diameters, centers, axes.

§ 21. Projectivity of the fundamental geometrical forms of 3rd kind.

§ Exercises.

Part II. Graphic calculus

§ 1. Addition and subtraction of straight lines. Properties of vector sum, polygon of vectors. Subtraction. Parallel vectors.

§ 2. Multiplication by a scalar.

§ 3. Raising to power, root extraction.

§ 4. Multiplication of straight lines by straight lines.

§ 5. Transformation of areas with rectilinear boundary.

§ 6. Graphic tables.

§ 7. Transformation of circular figures.

§ 8. Transformation of curvilinear figures in general.

§ 9. Theory of the planimeter.

§ 10. Cubage of regular masses of cut and carry.

§ 11. Cubage of irregular masses.

§ 12. Graphic calculation of terrain movings.

Part III. Graphical statics

§ 1. Composition of the forces applied at a point.

§ 2. Composition of several forces, placed at will in a plane.

§ 3. Projective correspondence between the polygon of forces and the funicular polygon.

§ 4. Examples and particular cases.

§ 5. Moments of forces in a plane.

§ 6. Infinitely small and infinitely far forces.

§ 7. Equilibrium of plane forces.

§ 8. Equilibrium of spatial forces.

§ 9. Parallel forces in a plane.

§ 10. Centers of gravity.

§ 11. Moments of inertia.

§ 12. Central ellipsoid.

§ 13. Ellipsoid of inertia.

§ 14. System of parallel forces the intensity of which be proportional to the distances of their points of application from a plane.

§ 15. Ellipses of inertia.

§ 16. System of parallel forces acting on a plane section.

§ 17. Construction of the central ellipsis and the core of a plane figure.

§ 18. Central ellipsis and the core of a rail profile.

§ 19. Central ellipsis and the core of an angular section.

§ 20. Distribution of the inter forces in the sections of a frame.

§ 21. Construction of the inner forces [21]. (A.5.20)

5.4.3 Cremona's Inheritance

Cremona's experience and teaching spread throughout all the Scuole d'applicazione per gl'ingegneri. He directly taught in Rome, where he moved in 1873 with the role of director of the local school for engineers, taking also care of the chair of Graphical statics. We may say that he did not give origin to a proper school of thought, like Betti did, but many scholars were influenced, either directly or indirectly, by him. His lectures and his scientific production made a mark, among others, on GiuseppeVeronese, Eugenio Bertini, Guido Castelnuovo, Federigo Enriques, Francesco Severi, active in the investigations on algebraic geometry.

As of graphical statics, Cremona had an immediate and important successor in Carlo Saviotti who wrote a fundamental treatise, following Cremona's spirit, but with greater attention to engineering applications; the following section is devoted to it.

5.4.3.1 Carlo Saviotti

We have little information on the life of Carlo Saviotti (1845–1928), apart from his place of birth, Calvignano near Pavia, where he also died. We also know that he taught *Statica grafica* in the Scuola d'applicazione per gl'ingegneri in Rome, and, thus, that he was a direct pupil of Cremona after the move of the latter from Milan. His textbook on graphical statics of 1888 [72], in our opinion the fundamental textbook of the discipline written in Italian, was divided, like many other textbooks of the time, in several volumes; the second tome of the second volume, and a good half of the third, present extremely detailed and precise drawings that illustrate the graphical concepts and constructions. Saviotti himself, in the introduction to his work, was proud of this richness, that he thought indispensable for the aim of learning the discipline:

> While issuing this course now for printing, with didactic purpose, we hope to make some service to any student of this discipline. It contains several examples and problems and many many figures, more than 1050. This may seem superfluous to persons who are already in possession of the matter, but not for the beginners, to whom the book is especially dedicated [72].[61] (A.5.21)

It is remarkable that, in spite of the textbook dating back to the end of the 19th century, the tables illustrating the realization of the polygon of forces and the funicular polygon were still suggestively represented by means of hands pulling ideal ropes taken up, as Saviotti said, from the *Nouvelle mécanique ou statique* of Varignon, printed more than a century and a half before [72].[62]

The first volume of the textbook was, like in the works by Culmann and Cremona, about graphical calculus: operations on oriented segments, measures of angles, surfaces, volumes, centers of area and of volume; this is a testament to how, in spite of the fact that more than twenty years had passed since the first edition of Culmann's book, the procedures of graphic calculus were thought reliable as much, and maybe more, than the algebraic ones to the aim of applications in calculations for design. Anyway, Saviotti was not a purist of the discipline, since he admitted that:

> We do not mean that in a book on *Graphical statics* we should refuse all that be not conformed to pure Geometry. There are cases, in which the geometrical method has not, and maybe never will, come to replace the analytical one, and there are then other [cases] in which the geometrical method would be non-expeditious, presenting itself either less general, or less simple, than the analytical one.
>
> We wished, for instance, to report as an example in the first part the geometrical method by *Archimedes* to determine the center of weight of a parabolic segment [...]. However, anybody sees how in this case the analytical method is simpler [72].[63] (A.5.22)

On the other hand, as previously mentioned, it was only when the automatic calculus, made possible by the electro-mechanical calculating machines of the first half of the 20th century, that the precision, the reliability, and the velocity of graphical calculus started to be overpassed and replaced by numerical methods on analytical bases.

The second volume of Saviotti's textbook was divided into two tomes, the first of which has a descriptive aim, we could say it deals with theory, while the second, as we hinted, contained all the figures and table necessary for understanding the text and the procedures exposed in the first. Omitting this second tome, of which we will present some images only, in the following we report the table of contents of the first tome, for a comparison with the corresponding parts of the works by Culmann and Cremona.

---

[61] vol. I, p. XI. Our translation.

[62] vol. II, footnote on p. 21.

[63] vol. I, p. X.

Graphical statics. – External forces

Introduction to the second part

CHAPTER ONE.

CONCENTRATED FORCES.

§ 1. Preliminary notions.

§ 2. Composition of concurring forces.

§ 3. Composition of non-concurring forces in a plane.

§ 4. Properties of funicular polygons.

§ 5. Composition of non-concurring forces in a plane by the method of the funicular bundle; its properties.

§ 6. Composition and center of parallel forces in space.

§ 7. Forces acting by rotation.

§ 8. Graphical determination of the resultant moment of a system of forces in a plane.

§ 9. Decomposition of a force in others, coplanar.

§ 10. Composition of couples in space.

§ 11. Composition of non-concurring forces in space.

§ 12. Other two methods of composing forces in space.

§ 13. Central axis; its determination; its properties.

§ 14. Polar system singled out in space by a system of forces.

§ 15. Mechanical interpretation of reciprocal figures.

§ 16. Application of reciprocal figures to the design of decks.

§ 17. Decomposition of forces in space.

CHAPTER TWO.

DISTRIBUTED FORCES AND EQUILIBRIUM OF BODIES WITHOUT FRICTION.

§ 1. Distributed forces.

§ 2. Conditions for the equilibrium of a constrained body and constraint reactions.

§ 3. Systems of bodies in equilibrium.

§ 4. Problems on the equilibrium of systems of bodies.

§ 5. Systems in indifferent equilibrium.

CHAPTER THREE.

EQUILIBRIUM OF SUPPORTED BODIES WITH FRICTION.

§ 1. Friction – Stability.

§ 2. Equilibrium of minimum stability for a body.

§ 3. Systems of supported bodies equilibrium of minimum stability.

§ 4. Friction in bolts.

CHAPTER FOUR.

STABILITY OF SUPPORTED BODIES.

§ 1. Stability ratios. – Thrust of water. – Stability of a dam with respect to sliding, rotation, and compression. – Dam with triangular profiles. – Chimneys.

§ 2. Thrust of semi-fluid matters and terrains with no cohesion.

§ 3. Systems of bodies resting on extended surfaces forming closed circuits. – Vaults. – Pressure line tangent to a given line.

§ 4. Funicular curves.

CHAPTER FIVE.

TRUSSES.

§ 1. Generation of strictly non-deformable trusses.

§ 2. Calculation of non-deformable trusses loaded at the nodes.

§ 3. Second problem. Method of the diagram for the calculation of non-deformable trusses.

§ 4. Applied trusses.

§ 5. Trusses with loaded elements.

§ 6. Strictly non-deformable trusses with elements with more than two nodes.

CHAPTER SIX.

EFFECTS OF THE EXTERNAL FORCES IN THE SECTIONS OF SOLIDS.

§ 1. Fixed external forces.

§ 2. Diagrams of the component actions of the resultants relative to all sections of the solid.

§ 3. Solids with curvilinear axis.

§ 4. Horizontal beams supported at the ends and subjected to movable loads.

§ 5. Diagram of the shearing forces in a section of a beam crossed directly by a uniformly distributed load.

§ 6. Diagrams of the bending moments.

§ 7. Application of the funicular polygon for the search of the maximum moments in the sections of a beam crossed by a system of loads.

§ 8. Trusses indirectly subjected to moving loads.

§ 9. Three-hinged trusses subjected to moving loads [72].[64] (A.5.23)

Even from the table of contents only it is apparent how Saviotti, though adopting Cremona's teaching on reciprocal figures, did not focus the discussion of graphical statics on them, but rather on their mechanical interpretation in terms of forces. In addition, as apparent in the titles of the sections of Chaps. 2, 3, and 4, Saviotti's treatise was not primarily directed to structural mechanics—as it usually happened when dealing with statical graphics—but also to the applications of those disciplines that nowadays would be called Mechanics applied to machines, and Geotechnics.

[64] vol. II, pp. V–IX.

That is, we may remark how, in spite of the fact that he did not abandon rigor in the setting, Saviotti saw, with a rather strong technical spirit, theory aimed to applications in as many fields of engineering as possible. Indeed, Saviotti wrote in his introduction:

> The direction that CULMANN has given to *Graphical statics* makes us think that its field be limited to mechanics applied to civil constructions. The graphical method, however, does not lend itself less usefully in the investigation of the minimum stability of supported bodies. In this way *Graphical statics* is useful also as a preparatory course for that of mechanics applied to machines, into which it infiltrates with ever growing advantage, and to which it serves, in addition, for everything is related to the fundamental notions on the strength of materials [72].[65] (A.5.24)

Indeed, Saviotti had declared the key points of his treatise already in the introduction to the second tome. All the rest was nothing but a consequence of a valid and undoubtable tool of calculus:

> The object of the second part consists of the investigation of the external forces, that is of the actions, not excluding gravity, that a body, considered separately, receives by others.
>
> In Graphical statics we seldom take the angles between forces into consideration, since they do not enter, like in analytical mechanics, as necessary elements to single forces in space out.
>
> We also make a limited use of moments, since it is more comfortable and quick to operate with forces (segments) rather than couples (surfaces).
>
> The appealing simplicity of the methods that characterize graphical statics let us tackle, after a few notions, several problems that in the past could be dealt with only after many other applications.
>
> We give space to *distributed forces*, usually neglected in the standard treatises on Mechanics. *All the forces of nature are composed of elements, the only forces that really exist*, says *Belanger* in his *Course of Mechanics* on pag. 37; *the others are conceptions of our mind, that enter science under the names of sum, or of resultant.*
>
> The investigation of distributed forces directly enters the statics of bodies.
>
> We consider systems of bodies *deformable, non-deformable* and *soluble*. The first ones are met especially in the machines, and we consider their *equilibrium of minimum stability*. The non-deformable and soluble systems are met especially in static constructions. Of these latter, ordinarily constituted by chains of bodies resting on plane surfaces, we examine the *stability* and among the non-deformable systems we study those strictly non-deformable, like the *trusses*, having particularly aim at the determination of the *mutual reactions* among the bodies that compose them.
>
> We produce exempts and applications on the equilibrium of the chains of resting bodies, by which one acquires easiness in noticing where and how the resting bodies of various configurations and in various conditions transmit pressures.
>
> The investigation of the actions that equilibrated forces exert on the various sections of a solid to which they are applied forms the last subject of this second part. It soon leads to the investigation of inner forces, that are dealt with in the third part [72].[66] (A.5.25)

Saviotti's graphical statics was then, most of all, an investigation of forces and systems, considered as aggregations of free bodies subjected to external forces, active

---

[65] vol. I, p. XI. Our translation.

[66] vol. II, pp. 3–4. Our translation.

and reactive, that maintain their equilibrium. The investigation of the stresses due to the actions of extension, bending, and torsion in the sections of the elements of machines or of structures was dealt with by Saviotti separately in the third volume, as it were other than the theory of equilibrium of free bodies subjected to forces. Moreover, we must be aware that, in Saviotti's jargon, *deformable* systems were actually mechanisms, that is systems of bodies each of which undergoes rigid motion, *non-deformable* systems were non-movable structures and, in particular, *strictly non-deformable* systems were statically determined structures; to end with, *soluble* systems were incoherent ("a system of bodies [...that] may be separated") [72].[67]

Saviotti presented the composition of two forces at first, relying on the notions of equilibrium and remarking that he who wanted to deal with statics independently of dynamics should have done like him.[68] The discussion was extended to several forces, non-coplanar as well, and Saviotti obtained the composition by means of a polygon he called of connection, or of composition, even though the latter is commonly called funicular polygon in the contemporary treatises on graphical statics. Actually, their graphic constructions are practically the same, and the different denomination was due only to the fact that the imaginary ropes might be loaded both in tension and in compression, while for actual ropes this is obviously impossible.

Saviotti then went on to illustrate the various properties of the polygon of composition of forces and of the funicular polygon, and the 'forces acting by rotation'. Indeed, since force was the only cause of motion for him, then the study of moments followed that of the forces.[69] Consequently, after having defined moment as the entity causing rotation, Saviotti studied its properties of composition, recalling a theorem by Varignon. Then, he went on to investigate the decomposition of forces and the composition of forces and couples in three-dimensional space.

Saviotti presented in the end some projective properties of reciprocal figures, and provided a mechanical interpretation for them, but his discussion was extremely

---

[67] vol. II, p. 99.

[68] Recalling many of his predecessors, among whom Varignon, Venturoli, Clebsch, Mossotti, Belanger, Ritter, he also remarked that the law of composition of forces could have been deduced from that of the motions if one had put dynamics before statics, and also provided an interesting historical resume on the subject ([72, vol. II, pp. 12–14]). Similar remarks had also been provided by Gabrio Piola, who had started precisely from this point of view.

[69] In the first article of the second volume Saviotti declared how force was a primitive element for him:

> A body cannot shift itself by its own if it is at rest, nor can modify the movement it has without the intervention of a *cause* exterior to it. [...] We do not investigate its origin; we only *evaluate its effect* [...] [72, vol. II, p. 5.] Our translation (A.5.26)

This vision was substantiated by the model of matter:

> In Statics we consider ideal bodies, that have all their dimensions infinitesimal, without having a determined shape, and that are called *elements* or *material points*. In addition, we consider ideal forces applied to them, with *finite magnitude*, which, being concentrated onto a point, are called *concentrated forces*. [...]

concise (from page 63 to page 82 of the second volume): this gives a clear idea of how, already a few years after the issues by Culmann and Cremona, the purely geometrical theory of reciprocal figures was given less attention by those teaching Graphical statics. Saviotti, in particular, was a direct pupil of Cremona, and in the introduction of his textbook he exalted his investigations and his brilliancy, but in reality he preferred an approach more linked to the traditional approaches by Varignon than to the pure and perfect projective geometry of his maestro.

In the rest of the volume, Saviotti dealt shortly with distributed forces, and then passed to usual subjects of mechanics of solids, structures, and machines: constraint devices, constraint reactions, articulated systems, drawbridges, friction, equilibrium in the presence of friction (which he called *of minimum stability*, meaning that the system is infinitely next to motion), belts, brakes, ropes, chains, thrusts of fluids, dams, incoherent terrains. The last part of the volume is devoted to trusses, which were the principal components of all the civil and industrial buildings of the time, and to the introduction to problems of mechanics of beams, seen as solids generated by the movement of a plane figure along a line passing through its center of area.

The investigation of trusses was dealt with by the Cremonian diagram for statically determined structures ("strictly non-deformable"); he reduced to symmetric structures, symmetrically loaded, in the case of redundant frames. The drawing of the diagrams of bending moments for beams was realized uniquely by graphical methods.

The third volume, entitled *Inner forces*, dealt with the application of graphic calculus to the investigation of inner stresses in Saint-Venant solids. The first chapter, indeed, dealt with the geometrical theory of the moments of inertia: conics of inertia, circles, ellipses, and cores of inertia, centers of pressure. The second chapter dealt with stresses of extension, compression, simple and compounded bending, torsion, flexure, with examples of calculations and verification of resistance. Chapter three was about the investigation of infinitesimal elastic deformation, the linear elastic constitutive laws, the formulation of the equation of the 'elastica' and the graphic calculation of some transverse deflections, coming to the solution, by means of compatibility, of the beam on several supports, as originally done by Navier.

(Footnote 69 continued)

Several points of application are called *rigidly connected* when they be linked in such a way that their relative distances always remain unchanged, or when they are part of a non-deformable body.

Even though we consider bodies as material in Statics, still at first we will make abstraction of their weight, that is, we will consider them as *geometrical* bodies, or *rigid joints, infinitely resistant*, of the points of application [of outer and inner forces] that concur to form a system with invariable shape [72, vol. II, pp. 6–7]. Our translation. (A.5.27)

The mechanical model we already saw in Maxwell and Menabrea is apparent.

5.4.3.2 The Overcoming of the Maestro

Cremona's teaching was continued at the Technical institute in Milan by Giuseppe Jung,[70] whose handbook *Appunti al corso di statica grafica* [53] had some success.

At the Scuola d'applicazione per gl'ingegneri in Turin, maybe the most important in Italy at the time, Cremona's influence was less remarkable, and graphical statics took an address that was partially independent of projective geometry. Ferdinando Zucchetti, in charge of the chair of Graphical statics probably starting from 1876,[71] in 1878 published a successful textbook, *Statica grafica. Sua teoria e applicazioni* [84]. It was a rather complete text on the discipline, even though it was shorter than Saviotti's (only 250 pages).

Zucchetti's approach was still further from projective geometry than Saviotti's, and that is an indication of an ongoing trend in the teaching of statics. He stated this fact already in the introduction to his book:

> Culmann wants to establish the investigation of Graphical statics on the Geometry of position, which believes to be necessary for the *perfect*[72] discussion of the theory of Graphical statics. Others, like Bauschinger, Lévy, believe it sufficient to follow simpler and more elementary methods in the exposition of Graphical statics.
>
> In these two years, during which I have had the honour of teaching Graphical statics at the Royal School of Application for Engineers in Turin, because of the conditions of the studies performed by my audience I believed it better to track Bauschinger and Lévy. And I have drawn up this writing, that I have decided to publish, hoping it can be useful to those who wish to proceed by the easiest ways in the study of graphical statics [84].[73] (A.5.28)

In the following we report the content of Zucchetti's textbook, as he himself summarized in the introduction.

> I now expose briefly the program of the subject that I tried to develop in this writing. In the first chapter some elementary notions of graphical calculus are given, that form an introduction useful for the study of graphical statics. The second chapter deals with the geometrical properties of the funicular polygons. In the third chapter we define the reciprocal figures of Graphical statics, and we examine some of its examples, deduced from the geometrical theory of the funicular polygons. In this chapter we state a theorem relative to the figures, which can be seen as plane projections of polyhedra.[74] Such figures always admit reciprocal figures. The proof of the same theorem is given later in the eleventh chapter by means of the theory of the systems equivalent to two forces in space. *Such a way of considering reciprocal figures as plane projections of polyhedra in space is due to the most famous Professor Cremona* [Italics is ours]. In the fourth chapter we consider the systems of forces concurring to a point. Chapter

[70] Giuseppe Jung was born in Milan in 1845 and died there in 1926. In 1867 he graduated in Naples and soon after, he went back to Milan, where he became Cremona's assistant. In 1876, when at the Higher technical institute in Milan also the two-year period preparatory for the studies in engineering was introduced, he was appointed professor of Projective geometry and Graphical statics, but he became full professor only in 1890. He was fellow of the *Regio istituto Lombardo.*

[71] Zucchetti would have later on be an assistant to the chair of Steam engines and railways in Turin.

[72] Italics is ours.

[73] p. 6. Our translation.

[74] Here is the theorem to which Zucchetti refers: "The figures that may be seen as plane projections of polyhedra always admit reciprocal figures"; [84, p. 46].

five deals with the composition of forces on a plane by means of the use of the polygon of forces and the funicular polygon, of which the usefulness and the importance in Graphical statics become apparent. In the sixth chapter we expose the theory of the moments of the forces and of the couples lying on a plane, and we explain the constructions to reduce the moments to a basis. In chapter seven we solve different problems on the decomposition of the forces in a plane. In the eighth chapter we study various problems relative to the equilibrium of a body subjected to particular conditions and under the action of forces all contained in a plane, as well as various problems relative to the equilibrium of the articulated polygons. Chapter nine is about the diagrams of the stresses of extension and compression in systems of rods. And we made it apparent the usefulness of reciprocal figures for the description of the diagrams in the case of trusses, of which we provide many examples drawn from buildings. We also hint on the description of the diagrams of the stresses of extension and compression for other systems of rods different from trusses. The tenth chapter deals with the diagrams of the shearing stresses and of the bending moments for a horizontal beam placed on two supports and subjected to either fixed or movable loads. Chapter eleven is devoted to the investigation of the systems of forces in space. It deals with the reduction of a system of forces whatsoever to a force and a couple, of the moments of forces with respect to an axis, of the reduction of a system of forces whatsoever to an equivalent system of two forces, of the property of the equivalent systems of two forces in space, and of reciprocal polyhedra. In this chapter we give the proof of the theorem stated in the third chapter, relative to the figures that may be regarded as plane projections of polyhedra. The twelfth chapter deals with the determination of the center of gravity of a system of parallel forces. - Chapter thirteen has the subject of the determination of the centers of gravity of lines, areas, and volumes. - The fourteenth chapter is about the moments of second order, and especially about the moments of inertia, and the ellipsis of inertia of a plane area. In this chapter we determine the center of a system of parallel forces again [84].[75] (A.5.29)

Remark the quote to Cremona about reciprocal figures. Their discussion, however, was reduced to the bare necessity, aiming at the construction of the Cremonian.

In 1882 the chair of *Statica grafica* in Turin was given to Guidi,[76] who kept it for a few years, before moving to the more prestigious chair of *Scienza delle costruzioni* in 1887. Guidi introduced graphical statics in his lectures on *Scienza delle costruzioni* [52], turning it from a secondary discipline to the main core of its teaching, that is structural mechanics. Guidi's example would have been gradually followed by all the Schools of application for engineers of the Kingdom of Italy.[77]

[75] pp. 6–7. Our translation.

[76] Camillo Guidi was born in Rome on July 24th, 1853. He graduated at the School of Application for Engineers in Rome in 1873. Guidi was professor of Graphical statics from 1882 at the School of Application for Engineers in Turin. He was appointed with the chair of *Scienza delle costruzioni* (Structural mechanics) from 1887, and in 1893 he became director of the Cabinet of Structural mechanics and theory of bridges. He died in Rome on October 30th, 1941. His investigations on reinforced concrete have remained famous.

[77] We remark that, besides the textbooks by Saviotti and Zucchetti, two other books were particularly considered in the Schools of application for engineers, the already quoted one by Bauschinger [1] and that by Maurice Lévy [59].

References

1. Bauschinger J (1871) Element der graphischen statik. Roldenbourg, Munich
2. Burali-Forti C, Marcolongo R (1909) Elementi di calcolo vettoriale: con numerose applicazioni alla geometria alla meccanica e alla fisica-matematica. Zanichelli, Bologna
3. Bellavitis G (1835) Saggio di applicazioni di un nuovo metodo di geometria analitica (Calcolo delle equipollenze). Ann delle Sci del Regno Lombardo-Veneto 5:244–250
4. Belluzzi O (1941) Scienza delle costruzioni (4 vols). Zanichelli, Bologna
5. Caparrini S (2009) Early theory of vectors. In: Becchi A et al (eds) Essay in the history of mechanics. Birkhäuser, Basel, pp 179–198
6. Capecchi D (2012) Historical roots of the rule of composition of forces. Meccanica 47(8):1887–1901
7. Capecchi D (2012) History of virtual work laws. Birkhäuser, Milan
8. Capecchi D, Drago A (2005) On Lagrange's history of mechanics. Meccanica 40:19–33
9. Capecchi D, Ruta G (2014) Polytechnic schools in the 19th century Europe. The polytechnic school in Karlsruhe. Meccanica 49:13–21
10. Chasles M (1875) Aperçu historique sur l'origine et le développement des méthodes en géométrie (2 édition). Gauthier Villars, Paris
11. Drago A (2011) The birth of the non-Euclidean geometries as the more significant crisis in the foundations of modern mathematics. Logic Philos Sci 9(1):103–110
12. Clerk Maxwell J (1864) On the calculation of the equilibrium and stiffness of frames. Philos Mag 27:294–299
13. Clerk Maxwell J (1864) On reciprocal figures and diagrams of forces. Philos Mag 27:250–261
14. Clerk Maxwell J (1872) On reciprocal figures, frames, and diagrams of forces (1870). Proc Roy Soc Edinb 7:53–56
15. Clerk Maxwell J (1873) A treatise on electricity and magnetism (2 vols). Clarendon, Oxford
16. Cotterill JH (1884) Applied mechanics. Macmillan, London
17. Cournot A (1828) Sur la théorie des pressions. Bull de Férussac 9:10–22
18. Cousinery BE (1828) Géométrie perspective ou principes de projection polaire appliqués à description des corps. Carilian-Goeury, Paris
19. Cousinery BE (1839) Le calcul par le trait—ses éléments et ses applications. Carilian-Goeury et Dalmont, Paris
20. Cremona L (1914–1917) Opere Matematiche (3 vols). Hoepli, Milan
21. Cremona L (1868–1869) Lezioni di statica grafica. Regio stabilimento Roncih, Milan
22. Cremona L (1872) Le figure reciproche nella statica grafica. In: Cremona L (1914–1917) Opere matematiche (3 vols). Hoepli, Milan, pp 336–366
23. Cremona L (1873) Elementi di geometria proiettiva. Paravia, Rome
24. Cremona L (1875) Élements géométrie projective. Gauthier Villars, Paris
25. Cremona L (1885) Elements of projective geometry. Clarendon, Oxford
26. Cremona L (1885a) Les figures réciproques en statique graphique. Gauthier-Villars, Paris
27. Crowe MJ (1967) A history of vector analysis. Dover, New York
28. Culmann C (1866) Die graphische Statik. Meyer und Zeller, Zürich
29. Culmann C (1880) Traité de statique graphique. Dunod, Paris
30. Curioni G (1864–1884) L'arte di Fabbricare. Ossia corso completo di istituzioni teoriche pratiche per ingegneri (6 vols and 5 appendixes). Negro, Turin
31. Enriques F (1898) Lezioni di geometria proiettiva. Zanichelli, Bologna
32. Favaro A (1877) Lezioni di statica grafica (1873). Sacchetto, Padua
33. Favaro A (1879) Leçons de statique graphique. Gauthier-Villars, Paris
34. Fleming Jenkin HC (1869) On practical applications of reciprocal figures to the calculation of strains on frameworks. Trans Roy Soc Edinb 25:441–447
35. Gerhart R, Kurrer KE, Pichler G (2003) The methods of graphical statics and their relation to the structural form. In: Proceedings of first international congress on construction history, Madrid, pp 997–1006

36. Ghisalberti AM (1960–) Dizionario biografico degli italiani. Istituto enciclopedia italiana, Rome
37. Gibbs JW (1881) Elements of vector analysis. Published in private form at New Haven into two parts; the former in 1881, the latter in 1884
38. Girard PS (1798) Traité analytique de la resistance des solides et des solides d'égale résistance. Didot, Paris
39. Giulio I (1841) Expérience sur la résistence des fers forgés dont on fait le plus d'usage en Piémont. Mem della Reale Accad delle Sci di Torino, s 2, 3:175–223
40. Giulio I (1841) Expérience sur la force et sur l'élasticité des fils de fer. Mem della Reale Accad delle Sci di Torino, s 2, 3:275–434
41. Giulio I (1842) Sur la torsion des fils métallique et sur l'élasticité des ressort en hélice. Mem della Reale Accad delle Sci di Torino, s 2, 4(1):329–383
42. Giuntini A (2001) Il paese che si muove. Le ferrovie in Italia tra '800 e '900. Franco Angeli, Milan
43. Grassmann HG (1844) Die Wissenschaft der extensiven Grösse oder die Ausdehnungslehre, eine neue Disziplin dargestellt und durch Anwendungen erläutert. In: Erster Theil: die lineale Ausdehnungslehre enthaltend. von Otto Wigand, Leipzig
44. Grassmann HG (1855) Sur les différents genres de multiplication. Crelles J 49:123–141
45. Grassmann HG (1862) Die Ausdehnungslehre. Vollständig und in strenger Form bearbeitet. Verlag von Th. Chr. Fr. Enslin, Berlin
46. Green G (1827) An essay of the application of the mathematical analysis to the theories of electricity and Magnetism. Green G (1871) Mathematical papers. McMillan, London, pp 1–82
47. Hamilton WR (1847) On quaternions (1844). Proc Roy Ir Acad 3(1–16):273–292
48. Hamilton WR (1853) Lecture on quaternions. University Press, Dublin
49. Hamilton WR (1854–1855) On some extensions of quaternions. Philos Mag Ser 4 7:492–499; 8:125–137, 261–269; 9:46–51, 280–290
50. Heaviside O (1893–1912) Electromagnetic theory (3 vols). The Electrician Publishing Company, London
51. Hellinger E (1914) Die allgemeinen Ansätze der Mechanik der Kontinua. Enzyklopädie der mathematischen Wissenschaften. Teubner BG, Leipzig, Bd. IV/4, pp 602–694
52. Hooke R (1678) Lectures de potentia restitutiva or Of spring explaining the power of springing bodies. John Martyn, London
53. Jung C (1890) Appunti al corso di statica grafica presi dagli allievi Tullio Imbrico e Enrico Bas. Luigi Ferrari, Milan
54. Lamé G, Clapeyron EBP (1827) Mémoire sur les polygones funiculaires. J des Voies de Commun Saint-Petersbourg 1:43–56
55. Lamé G (1852) Leçons sur la théorie mathématique de l'élasticité des corps solides. Bachelier, Paris 1852
56. Lamé G (1859) Leçons sur les coordonnées curvilignes et leurs diverses applications. Bachelier, Paris
57. Lehmann C, Maurer B (2005) Karl Culmann und die graphische Statik–Zeichnen, die Sprache des Ingenieurs. Ernst & Sohn, Berlin
58. Levi Civita T, Amaldi U (1930) Lezioni di meccanica razionale. Zanichelli, Bologna
59. Lévy M (1886–1888) La statique graphique et ses applications aux constructions (1874) (4 vols). Gauthier-Villars, Paris
60. Mach E (1883) Die Mechanik in ihrer Entwickelung. Brockhaus FA, Leipzig
61. Möbius AF (1827) Der barycentrische calculus, ein neues Hülfsmittel zur analytischen Behandlung der Geometrie. Barth JA, Leipzig
62. Möbius AF (1837) Lehrbuch der Statik. Göschen, Leipzig
63. Mohr O (1860–1868) Beitrag zur Theorie der Holz und Eisen-Konstruktionen. Zeitschrift des Architekten-und Ingenieur Vereins zu Hannover, vol 6 (1860), col 323 et seq, 407 et seq; vol 8 (1862), col 245 et seq; vol 14 (1868) col 19 et seq
64. Mohr O (1872) Über die Darstellung des Spannungszustandes und des Deformationszustandes eines Körperelementes und über die Anwendung derselben in der Festigkeitslehre. Der Civilingenieur 28:113–156

65. Mohr O (1874) Beitrag zur Theorie der Bogenfachwerksträger. Zeitschrift des Architekten-und Ingenieur Vereins zu Hannover, 20:223 et seq
66. Mohr O (1874) Beitrag zur Theorie des Fachwerks. Zeitschrift des Architekten-und Ingenieur Vereins zu Hannover, 20:509 et seq
67. Mohr O (1875) Beitrag zur Theorie des Fachwerks. Zeitschrift des Architekten-und Ingenieur Vereins zu Hannover 21:17–38
68. Mohr O (1881) Beitrag zur Theorie des Bogenfachwerks. Zeitschrift des Architekten-und Ingenieur Vereins zu Hannover 27:243 et seq
69. Poncelet JV (1840) Mémoire sur la stabilité des revêtements et leurs fundations. Mémorial de l'Officier du Génie 13:7–270
70. Pouchet LE (1797) Métereologie terrestre ou table des nouveaux poids, mesures et monnaies de France etc. Guilbert et Herment, Rouen
71. Rankine WJM (1858) A manual of applied mechanics. Griffin, London
72. Saviotti C (1888) La statica grafica (3 vols). Hoepli, Milan
73. Signorini A (1930a) Sulle deformazioni finite dei sistemi continui. Rend della Reale Accad dei Lincei 6(12):312–316
74. Signorini A (1952) Meccanica razionale con elementi di statica grafica. Società tipografica Leonardo da Vinci, Città di Castello
75. Staudt KGF von (1856) Beiträge zur Geometrie der Lage. Verlag von Bauer und Raspe, Nürnberg
76. Stevin S (1586) De Beghinselen der Weeghconst. Plantijn, Leyden
77. Stevin S (1605) Tomus quartus mathematicorum hypomnematum de statica. Patii I, Leiden
78. Stevin S (1634) Les oeuvres mathematiques de Simon Stevin, par Albert Girard. Elsevier, Leiden
79. Timoshenko SP (1953) History of strength of materials. McGraw-Hill, New York
80. Varignon P (1687) Projet d'une nouvelle mechanique avec un examen de l'opinion de M. Borelli, sur les propriétes des poids suspendus par des cordes. Veuve d'Edme Martin, Boudot J, Martin E, Paris
81. Varignon P (1725) Nouvelle mécanique ou statique (2 vols). Claude Jombert, Paris
82. Veronese G (1903) Commemorazione del Socio Luigi Cremona. Rend della Reale Accad dei Lincei 5(12):664–678
83. Viola T (1932) Sui diagrammi reciproci del Cremona. Ann di Mat Pura ed Appl 10:167–172
84. Zucchetti F (1878) Statica grafica: sua teoria ed applicazioni con tavole illustrative per l'ingegnere. Negro, Turin

Appendix A
Quotations

A.1 Quotations of Chap. 1

1.1 *Tant qu'il ne s'agit que de petits déplacements, et quelle que soit la loi des forces que les molécules du milieu exercent les unes sur les autres, le déplacement d'une molécule dans une direction quelconque produit une force répulsive égale en grandeur et en direction à la résultante des trois forces répulsive produites par trois déplacements rectangulaires de cette molécule égaux aux composants statiques du premier déplacement.*
Ce principe, presque évident par son énoncé même, peut se démontrer de la manière suivante.

1.2 On regarde un corps solide élastique comme un assemblage de molécules matérielles placées à des distances extrêmement petites. Ces molecules exercent les unes sur les autres deux actions opposées, savoir, une force propre d'attraction, et une force de répulsion due au principe de la chaleur. Entre une molécule M, et l'une quelconque M′ des molecules voisines, il existe une action P, qui est la difference de ces deux forces. Dans l'état naturel du corps, toutes les actions P sont nulles, ou se détruisent réciproquement, puisque la molécule M est en repos. Quand la figure du corps a été changée, l'action P a pris une valeur differente Π, et il y a équilibre entre toutes les forces Π et les forces appliqué es au corps, par lesquelles le changement de figure a été produit.

1.3 Les molécules de tous les corps sont soumises à leur attraction mutuelle et à la répulsion due à la chaleur. Selon que la première de ces deux forces est plus grande ou moindre que la seconde, il en résulte entre deux molécules une force attractive ou répulsive; mais dans les deux cas, cette résultante est une fonction de la distance d'une molécule à l'autre dont la loi nous est inconnue; on sait seulement que cette fonction décroît d'une manière très rapide, et devient insensible dés que la distance a acquis une grandeur sensible. Toutefois nous supposerons que le rayon d'activité des molécules est très-grand par rapport aux intervalles qui les séparent, et nous admettrons, en outre, que le décroissement

D. Capecchi and G. Ruta, *Strength of Materials and Theory of Elasticity in 19th Century Italy*, Advanced Structured Materials 52,
DOI 10.1007/978-3-319-05524-4

rapide de cette action n'a lieu que quand la distance est devenue la somme d'un très-grande nombre de ces intervalles.

1.4 Soit M un point situé dans l'intérieur du corps, à une distance sensible de la surface.
ce point menons un plan qui partage le corps en deux parties, et que nous supposerons horizontal [...]. Appelons A la partie supérieure et A′ la partie inférieure, dans laquelle nous comprendrons les points matériels appartenant au plan même. Du point M comme centre, décrivons une sphère qui comprenne un très-grand nombre de molécules, mais dont le rayon soit ce-pendant insensible par rapport au rayon d'activité des forces moléculaires. Soit ω l'aire de sa section horizontale; sur cette section élevons dans A′ un cylindre vertical, dont la hauteur soit au moins égale au rayon d'activité des molécules; appelons B ce cylindre: l'action des molécules de A′ sur celles de B, divisée par ω, sera la *pression* exercée par A′ sur A, rapportée à l'unité de surface et relative au point M.

1.5 On suppose que dans les corps de cette nature, les molécules sont régulièrement distribuées, et quelles s'attirent ou se repoussent inégalement par leurs différens cotès. Par cette raison il n'est plus permis, en calculant l'action exercée par une partie du corps sur une autre, de regarder l'action mutuelle de deux molécules comme une simple fontion de la distance qui les sépare [...]. S'il s'agit d'un corps homogène qui soit dans son état naturel, où il n'est soumis á aucune force étrangère, on pourrá le considérer comme un assemblage de molécules de même nature et de même *forme*, dont les sections homologues seront parallèles entre elles.

1.6 Les composants P, Q, &c., étant ainsi réduites á six forces différentes, et la valeur de chaque force pouvant contenir six coefficiens particulaires, il en résulte que les équations générales de l'équilibre, et par suite, celles du mouvement, renferment trente-six coefficients, q'on ne pourra pas réduire a un moindre nombre sans restreindre la généralité de la question.

1.7 Il suit de là que si l'on considère deux parties A et B d'un corps non cristallisé qui soient d'une étendue insensible, mais dont chacune comprenne cependant un très-grand nombre de molécules, et qu'on veuille déterminer l'action totale de A sur B, on pourra supposer dans ce calcul que l'action mutuelle de deux molécules m et m' se réduise, comme dans le cas des fluides, à une force R dirigée suivant la droite qui joint leurs centres de gravité M et M′, et dont l'intensité ne sera fonction que de la distance MM′. En effet, quelle que soit cette action, on peut la remplacer par une semblable force, qui sera la moyenne des actions de tous les points de m' sur tous ceux de m, et que l'on combinera avec une autre force R', ou, s'il est nécessaire avec deux autres forces R' et R'', dépendantes de la disposition respective des deux molecules. Or, cette disposition n'ayant par hypothèse aucune sorte de régularité dans A et B, et les nombres de molécules de A et B étant extrêmement grands et comme infinis, on conçoit que toutes les forces R' et R'' se compenseront sans altérer l'action totale de A sur B, qui ne dépendra par conséquent que des forces R. Il faut d' ailleurs ajouter que pour un même accroissement dans la distance, l'intensité des forces R' et R'' diminue plus rapidement en général que celle des forces R ce qui contribuera encore à faire disparaître l'influence des premières forces sur l'action mutuelle de A sur B.

1.8 Par cette méthode de réduction, on obtient définitivement, pour les N_i, T_i, dans le cas des corps solides homogènes et d'élasticité constante, les valeurs [...] contenant deux coefficients, λ, μ. Quand on emploie la méthode indiquée à la fin de la troisième Leçon, on trouve $\lambda = \mu$, il ne reste plus qu'un seul coefficient. Nous ne saurons admettre cette relation, qui s'appuie nécessairement sur l'hypothèse de la continuité de la matière dans les milieux solides. Les résultats des experiences de Wertheim font bien voir que le rapport de λ à μ n'est pas l'unité, mais ne semblent pas assigner à ce rapport une autre valeur fixe et bien certaine. Nous conserverons donc les deux coefficients λ et μ, en laissant leur rapport indéterminé.

1.9 Telle est la méthode suivie par Navier et autres géomètres, pour obtenir les équations générales de l'élasticité dans les milieux solides. Mais cette méthode suppose évidemment la continuité de la matière, hypothèse inadmissible. Poisson croit lever cette difficulté, [...] mais [...] il ne fait, en réalité, que substituer le signe Σ au signe $\int$ [...]. La méthode que nous avons suivie [...] dont on trouve l'origine dans les travaux de Cauchy, nous paraît à l'abri de toute objection [...].

1.10 L'élasticité des corps solides et même des fluides, [...], toutes leurs propriétés mécaniques prouvent que les molécules ou les dernières particles qui les composent exercent les unes sur les autres des actions répulsives indéfiniment croissentes pour les distances mutuelles les moindres, et devenant attractives pour des distances considérables, mais relativement insensibles quand ces distances, dont elles sont ainsi fonctions, acquièrent une grandeur perceptible.

1.11 Je me ne refuse pas pourtant à reconnaître que les molécules intègrantes dont les arrangements divers composent la texture des solides, et dont les petits changement de distance produisente les déformations perceptibles appelées ∂, g ne sont pas les *atoms* constituants del la matière, mais en sont des groupes inconnous. Je reconnais en conséquence, tout en pensant que les actions entre atoms sont régie par la loi des intensités fonction des seules distances ou elles s'exercent, qu'il n'est pas bien certain que les actions *résultantes* ou entre *molécules*, doivent suivre tout à fait la même loi vis-à-vis des distances de leurs centres de gravité. On peur considérer aussi que les groupes, en changeant de distances, peuvent changer d'orientation [...].

1.12 Les 36 coefficients [...] ne sont pas indépendants les uns des autres, et il est facile de voir qu'il y a entre eux vingt et une égalités.

1.13 Les trente-six coefficients [...] se réduisent à deux [...] et on peut dire même à un seul [...] en vertu de ce que les trente-six coefficients sont réductibles à quinze.

1.14 Mais les experiences [...] et la simple considération de la manière dont s'opèrent le refroidissement et la solidification des corps, prouvent que l'isotropie est fort rare [...]. Aussi, plutôt que de prendre, au lieu des formules [...] à un seul coefficient [...], les formules [...] à deux coefficients [...], qui ne sont composées comme celles-ci que pour des corps parfaitement isotropes, il conviendra de se servir le plus qu'on pourra des formules [...] relatives au cas plus général d'une élasticité inégale dans deux ou trois sens.

1.15 Si dans un corps solide élastique ou non élastique on vient à rendre rigide et invariable un petit élément du volume terminé par des faces quelconques, ce petit élément éprouvera sur ses differentes faces, et en chaque point de chacune d'elles, une pression ou tension déterminé. Cette pression ou tension sera semblable à la pression qu'un fluide exerce contre un élément de l'envelope d'un corps solide, avec cette seule différence, que la pression exercée par un fluide en repos contre la surface d'un corps solide, est dirigée perpendiculairement á cette surface de dehors en dedans, et indépendante en chaque point de l'inclinaison de la surface par rapport aux plans coordonnés, tandis que la pression ou tension exercée en un point donné d'un corps solide contre un très petit élément de surface passant par ce point, peut être dirigée perpendiculairement ou obliquement à cette surface, tantôt de dehors en dedans, s'il y a condensation, tantôt de dedans en dehors, s'il y a dilatation, et peut prendre de l'inclination de la surface par rapport aux plans dont il s'agit.

1.16 Dilatation, en un point M d'un corps, dans le sens d'une droite Mx qui y passe, la proportion de l'allongement (positif ou négatif) qu'éprouve une portion quelconque trés-petite de cette droite, en vertu des déplacements moyens du corps, tels qu'on les a définis à l'article precedent; Glissement, suivant deux petites droites primitivement rectangulaires Mx, My, ou suivant l'une d'elles et dans le plan qu'elle fait avec l'autre, la projection actuelle, sur chacune, de l'unité de longueur portée dans la direction de l'autre. Nous désignerons cette quantité, qui en grandeur n'est autre chose que le cosinus de l'angle actuel des deus droites, par

$$g_{xy} \quad \text{ou} \quad g_{yx}$$

selon qu'on la regardera comme désignant le glissement relatif des diverses lignes parallèles à Mx situées dans le plan xMy, ou comme le glissement relatif des lignes parallèles a My, située dans ce même plan. *Elle est positive quand l'angle primitivement droit yMx est devenu aigu.*

1.17 Mais Green, en 1837–1839, et, d'après lui, divers savants d'Angleterre et de l'Allemagne ont cru pouvoir lui [la legge di azione molecolare funzione della sola distanza fra ogni coppia di particelle-punti materiali] en substituer f une autre plus générale, ou qualifiée plus générale parce qu'elle est *moins déterminée* [...], loi dont la conséquence analytique immédiate est la possibilité que l'intensité de l'action entre deux molécules dépende non seulement de leur distance mutuelle propre, mais encore de leurs distances aux autres molécules, et même des distances de celles-ci entre elles; en un mot, de tout l'ensemble actuel de leurs situations relatives ou de l'état présent complet du système dont *f* font partie les deux molécules dont on s'occupe, dût-on l'étendre à l'univers entier.

1.18 Cette vue de Green constitue une *troisième* origine [...] de l'opinion qui domine aujourd'hui et que nous combattons.

1.19 Si la prudence scientifique prescrit de ne pas se fier à toute hypothèse elle n'ordonne pas pas moins de tenir pour fortement suspect ce qui est manifestement contraire à une grande synthèse reliant admirablement la généralité des faits [...]. Aussi repoussons-nous toute formule théorique en contradiction formelle avec la loi des actions fonctions continues des distances des points matériels et dirigées suivant leurs lignes de jonction deux à deux. Si, en recourant à une telle formule, on explique plus facilement certains faits, nous la regarderons toujours comme un *expédient* trop commode [...].

1.20 Les cristaux sont des assemblages de molécules identiques entre elles et semblablement orientées, qui, réduites par la pensée à un point unique, leur centre de gravité, sont disposées en rangées rectilignes et parallèles, dans chacune desquelles la distance de deux points est constante.

1.21 Cessant de regarder les molécules comme des points et le considérant comme des petits corps.

1.22 Les molécules des corps cristallisés seront donc pour nous dorénavant des polyèdres dont les sommets, distribués d'une manière quelconque autour du centre de gravité, seront les centres, ou pôles, des forces émanées de la molécule.

1.23 Il Voigt (1887) supponendo il corpo come formato da un aggregato di corpuscoli (e quindi discontinua la materia costituente il corpo): supponendo che ogni corpuscolo risenta dagli altri delle azioni riducibili a una forza e una coppia, decrescenti indefinitamente col crescere della distanza; ha ritrovate le equazioni generali della elasticità sotto la stessa forma ottenuta dalla teoria del potenziale, senza che siano necessariamente verificate le relazioni di Cauchy-Poisson.

1.24 La théorie moléculaire ou des actions à distance, fondée par Navier, Cauchy et Poisson [...] fait dépendre, en effet, les propriétés élastiques des corps isotropes d'un seul paramètre, alors que de nombreuses observations semblant être en désaccord avec ce résultat.

1.25 C'est alors que fut généralement adoptée pendant quelque temps une nouvelle théorie [...] en supposant la matière continue et les actions mutuelles entre les portions de matière voisines localisées dans leur surface de séparation [...] [qui] donne, contrairement à la précédente, deux constantes caractéristiques des milieux isotropes, et tous ses résultats se sont trouvés d'accord avec l'observation.

1.26 La théorie moléculaire ancienne de l'élasticité part d'une conception fondamentale inutilement spécialisée, à savoir l'hypothèse d'actions moléculaires centrales et ne dépendant que de la distance [...] la formation régulière d'un cristal [...] n'est compréhensible que si un *moment directeur* agit sur la particule [...].

1.27 Nous avons considéré exclusivement les pressions des forces agissant entre les molécules, mais il est que les moments, ou les couples, qui agissent entre elles peuvent être traités de la même manière.

1.28 On a les neuf pressions de moments particulières

$$L_x,\ L_y,\ L_z, \qquad M_x,\ M_y,\ M_z, \qquad N_x,\ N_y,\ N_z,$$

qui correspondent exactement aux $X_x, \ \ldots, \ Z_z$.

1.29 Eine molekulare Theorie der elastischen Vorgänge, die Aussicht bietet, alle Beobachtungen zu erklären, wird eine so allgemeine Grundlage verlangen, wie sie etwa der *Bravais*sche Strukturtheorie [...] an die Hand gibt. Der Kristall ist nach ihr aus unter einander identischen und parallel orientierten Bausteinen oder Elementarmassen aufgeführt zu denken, die so angeordnet sind, daß jeder von diesen innerhalb der Wirkungssphäre in gleicher Weise von anderen umgeben ist.

1.30 Die Wechselwirkungen ein Potential haben sollen; [...] wird es [...] erlaubt sein, die Elementarmassen wie starre Körperchen zu behandeln.

1.31 Die zwischen zwei starren Körper (h) und (k) stattfindenden Wechselwirkungen geben Veranlassung sowohl zu Gesamtkomponenten X_{hk}, X_{kh}, ..., als auch zu Drehungsmomenten L_{hk}, L_{kh}, ..., die wir je um den Schwerpunkt des betreffenden Körpers rechnen wollen.

1.32 L'intention primitive de l'auteur était de ne mettre dans ce livre que ce dont il avait besoin pour se guider dans les leçons qu'il professe à l'école polytechnique de Carlsruhe. Mais bientôt il sentit tellement la nécessité de fonder sur une base solide les recherches dont les résultats servent aux applications techniques, qu'il se détermina à entrependre la rédaction d'un traité de la théorie de l'élasticité qui, autant que cela était possible dans une étendue modérée, présentât un système complet de principes et de usages de cette théorie: travaille devenu possible aujourd'hui grâce aux belles recherches de MM. Kirchoff et de Saint-Venant. Il fallait assurément, pour cela, traiter brievement bien des points, mais il convenait, avant tout, d'exposer en détail ce qui est désirable pour une connoissance suffisante de cette branche nouvelle de la science. Ainsi, pour tout ce qui regarde les transformations analytiques que M. Lamé a enseigné à opérer avec une si grande élégance sur les équations fondamentales de l'élasticité, il fallait renvoyer à l'ouvrage si connu et si répandu de cet illustre savant.

1.33 Verum si pondus quattuor pedibus plano insistat, determinatio singularum pressionum non solum multo magis ardua deprehenditur, sed etiam prorsus incerta et lubrica videtur.

1.34 Ne autem perfectissima illa pedum aequalitas, qualem vix admittere licet, negotium facessat concipiamus planum sive solum cui pondus incumbit, non adeo esse durum, ut nullam plane impressione recipere posset, sed quasi panno esse obductum, cui pedes illi aliquantum se immergere queant.

1.35 Sive pondus pluribus pedibus innitatur sive basi incumbat plana cuiscunque figurae, sit punctum M sive extremitas cuiuspiam pedis, sive elementum quodpiam basis pro quo pressio quaeritur. Concipiatur ibi perpendiculariter erecta linea Mμ ipsis pressioni proportionalis, atque necesse est omnia ista puncta μ in quopiam plano terminari, hoc igitur principio stabilito, quaedmodum pro omnibus casibus pressionem in singulis basis punctu definiri oporteat, hic sum expositurus.

1.36 Quand une verge rigide chargée de poids est soutenue sur un nombre de points d'appui plus grand que 2, les efforts que chacun de ces points d'appui doit supporter sont indéterminés entre certaines limites. Ces limites peuvent toujours être fixées par les principes de la statique. Mais, si l'on suppose la verge élastique, l'indétermination cesse entièrement. On considérera seulement ici une des questions de ce genre le plus simples qui puissent être proposées.

1.37 Cette méthode consiste à chercher les déplacements des points des pièces en laissant sous forme indéterminée les grandeurs, les bras de levier et les directions des forces dont nous parlons. Une fois les déplacements exprimés en fonctions de ces quantités cherchées, on pose les conditions définies qu'ils doivent remplir aux points d'appui on d'encastrement, ou aux jonctions des diverses pièces, ou aux points de raccordement des diverses parties dans lesquelles il faut diviser une même pièce parce que les déplacements y sont exprimés par des équations différentes. De cette maniere, on arrive à avoir autant d'équations que d'inconnues, car il n'y a, dans les questions de mécanique physique, évidemment aucune indétermination.

1.38 J'ai eu à m'occuper de cette question pour la première fois comme ingénieur à l'occasion de la reconstruction du pont d'Asnières, près Paris, détruit lors des événements de 1848. Les formules aux quelles je fus conduit furent appliquées plus tard aux grands ponts construits pour le chemin de fer du Midi, sur la Garonne, le Lot et le Tarn, dont le succès a parfaitement répondu à nos prévisions. C'est le résultat de ces recherches que j'ai l'honneur de soumettre au jugement de l'Académie.
Dans ce premier Mémoire, dont voici le résumé, j'examine d'abord le cas d'une poutre droite posée sur deux appuis à ses extrémités, sa section est constante, elle supporte une charge répartie uniformément; on se donne en outre le moment des forces agissant aux deux extrémités au droit des appuis. On en conclut l'équation de la courbe élastique qu'affecte l'axe de la poutre, les conditions

mécaniques auxquelles tous ses points sont soumis, et la partie du poids total supportée par chaque appui.

La solution du problème général se trouve ainsi ramenée à la détermination des moments des forces tendant à produire la rupture de la poutre au droit de chacun des appuis sur lesquels elle repose. On y parvient en exprimant que les deux courbes élastiques correspondant à deux travées contigues sont tangentes l'une à l'autre sur l'appui intermédiaire, et que les moments y sont égaux.

1.39 Si l'on ajoute au quadruple d'un moment quelconque celui qui le précède ou celui qui le suit sur les deux appuis adjacents, on obtient une somme égale au produit du poids total des deux travées correspondantes par le quart de l'ouverture commune. Si les ouvertures sont inégales, la même relation subsiste, sauf de légères modifications dans les coefficients.

1.40 Pour en donner un example, on supposera le poids Π supporté par les trois pièces inclinées AC, A′C, A″C contenues dans le même plan vertical, et l'on nommera

$\alpha, \alpha', \alpha''$ les angles formés par la direction des trois pièces avec la corde vertical CΠ;

p, p', p'' les efforts exercés, par suite de l'action du poids Π dans la direction de chacune des pièces;

F, F', F'' les forces d'élasticité des trois pièces;

a la hauteur du point C au-dessus de la ligne horizontale AA″;

h, f les quantités dont le point C se déplace horizontalement et verticalement, par l'effet de la compression simultanée des trois pièces.

[...] Cela posé, les conditions de l'équilibre entre le poids Π et les trois pressions execées suivant les pièces donneront d'abord

$$p\cos\alpha + p'\cos\alpha' + p''\cos\alpha'' = \Pi p\sin\alpha + p'\sin\alpha' + p''\sin\alpha'' = 0.$$

1.41 On en conclut les trois équations

$$p = F\frac{f\cos^2\alpha - h'\sin\alpha\cos\alpha}{a};$$
$$p' = F'\frac{f\cos^2\alpha' - h'\sin\alpha'\cos\alpha'}{a};$$
$$p'' = F''\frac{f\cos^2\alpha'' - h'\sin\alpha''\cos\alpha''}{a}$$

qui, réunies avec les deux précedentes, donneront les valeurs des déplacements h et f, et les efforts p, p' et p''.

1.42 Aus dieser Verlängerung entsteht eine elastische Kraft, mit welcher der Stab sich wieder zusammenzuziehen bestrebt ist; dieselbe ist im Punkte i gegen den

Punkt k gerichtet und umgekehrt, ihre Grösse ist [...]

$$\frac{E_{ik}q_{ik}\rho_{ik}}{r_{ik}}.$$

wenn E_{ik} den Elasticitäts modulus, q_{ik} den Querschnitt des betreffenden Stabes bezeichnet.

1.43 Setzen wir nun die Gleichgewichtsbedingungen an, d. h. lassen wir die Summen entsprechender Componenten verschwinden, so ergeben sich die drei Gleichungen:

$$(56) \quad \ldots \quad \begin{cases} X_i + \sum_k \dfrac{E_{ik}q_{ik}\rho_{ik}(x_k - x_i)}{r_{ik}^2} = 0 \\ Y_i + \sum_k \dfrac{E_{ik}q_{ik}\rho_{ik}(y_k - y_i)}{r_{ik}^2} = 0 \\ Z_i + \sum_k \dfrac{E_{ik}q_{ik}\rho_{ik}(z_k - z_i)}{r_{ik}^2} = 0. \end{cases}$$

In diesen Gleichungen ist nichts unbekannt als die in den ρ vorkommenden Grossen u, v, w.

1.44 Pour le faire voir, supposons, pour fixer les idées, que la force qui agit sur le point m soit la pesanteur, que nous représenterons par g. En prenant l'axe des z vertical et dirigé dans le sens de cette force, ses trois composantes seront $X = 0$, $Y = 0$, $Z = g$. Appelons $\epsilon, \epsilon', \epsilon'', \epsilon'''$, les extensions que les quatre fils l, l', l'', l''', éprouveraient si le poids mg était suspendu verticalement à leur extrémité inférieure; soient $\zeta, \zeta', \zeta'', \zeta'''$, les extensions de ces mêmes fils au bout du temps t, pendant le mouvement; leurs tensions au même instant auront pour valeurs (n 288)

$$\frac{gm\zeta}{\epsilon}, \quad \frac{gm\zeta'}{\epsilon'}, \quad \frac{gm\zeta''}{\epsilon''}, \quad \frac{gm\zeta'''}{\epsilon'''}.$$

Le mobile m n'étant plus assujetti à demeurer à des distances constantes de A, A′, A", A"', on devra supprimer les termes des équations (4), qui ont $\lambda, \lambda', \lambda''$, λ''', pour facteurs, et qui provenaient de ces conditions; mais, d'un autre côté, il faudra joindre au poids de ce point materiel les quatre forces précédentes, dirigés de m vers A, de m vers A′, de m vers A″, de m vers A‴; ce qui revient à substituer, dans les équations (4), les valeurs précédentes de L, L', L'', L''', en y faisant, en même temps,

$$\lambda = -\frac{gm\zeta}{\epsilon}, \quad \lambda' = -\frac{gm\zeta'}{\epsilon'}, \quad \lambda'' = -\frac{gm\zeta''}{\epsilon''}, \quad \lambda''' = -\frac{gm\zeta'''}{\epsilon'''}$$

Au bout du temps t, soient aussi

$$x = \alpha + u, \qquad y = \beta + v, \qquad z = \gamma + w$$

α, β, γ, étant les mêmes constantes que précédentement, et u, v, w, des variables tr ès petites, dont nous négligerons les carré et les produits; il en résultera

$$\zeta = \frac{1}{l}\left[(\alpha - a)u + (\beta - b)v + (\gamma - c)w\right],$$

$$\zeta' = \frac{1}{l'}\left[(\alpha - a')u + (\beta - b')v + (\gamma - c')w\right],$$

$$\zeta'' = \frac{1}{l''}\left[(\alpha - a'')u + (\beta - b)v + (\gamma - c'')w\right],$$

$$\zeta'' = \frac{1}{l'''}\left[(\alpha - a''')u + (\beta - b)v + (\gamma - c''')w\right];$$

et, relativement à ces inconnues u, v, w, les équations (4) seront linéaires, et se réduiront à

$$\frac{d^2u}{dt^2} + g\left[\frac{(\alpha - a)\zeta}{l\epsilon} + \frac{(\alpha - a')\zeta'}{l'\epsilon'} + \frac{(\alpha - a'')\zeta''}{l''\epsilon''} + \frac{(\alpha - a''')\zeta'''}{l'''\epsilon'''}\right] = 0,$$

$$\frac{d^2v}{dt^2} + g\left[\frac{(\beta - b)\zeta}{l\epsilon} + \frac{(\beta - b')\zeta'}{l'\epsilon'} + \frac{(\beta - b'')\zeta''}{l''\epsilon''} + \frac{(\beta - b''')\zeta'''}{l'''\epsilon'''}\right] = 0,$$

$$\frac{d^2w}{dt^2} + g\left[\frac{(\gamma - c)\zeta}{l\epsilon} + \frac{(\gamma - c')\zeta'}{l'\epsilon'} + \frac{(\gamma - c'')\zeta''}{l''\epsilon''} + \frac{(\gamma - c''')\zeta'''}{l'''\epsilon'''}\right] = 0;$$

[...] Si l'on suppose nulles les quantités u, v, w, et q'on supprime en conséquence, les premiers termes des trois dernières des sept équations précédentes, les valeurs de u, v, w, ζ, ζ', ζ'', ζ''', qu'on déduira de ces sept équations, répondront àl létat d'équilibre du poids de m et des quatre fils de suspension.

1.45 Lorsqu'une force tire ou presse un corps solide, dont au moins trois points sont fixes, le produit de cette force par la projection, sur sa direction, du déplacement total qu'elle a fait subir à son point d'application, représente le double du travail effectué, depuis l'instant où le déplacement et la force étaient nuls, jusqu'à celui où le déplacement et la force ont atteint leurs valeurs finals. [...] M. Clapeyron a trouvé une autre expression du même travail, dans laquelle interviennent toutes les forces élastiques développées dans l'intérieur du corps solide. L'égalité de ces deux expressions constitue un théorème, ou plutôt un principe, analogue à celui des forces vives, et qui parait avoir une importance égale pour les applications.
[...] on arrive facilement à l'équation

$$\sum(Xu+Yv+Zw)\varpi$$

$$(2)\qquad =\iiint dx\,dy\,dz\left\{\begin{aligned} &N_1\frac{du}{dx}+T_1\left(\frac{dv}{dz}+\frac{dw}{dy}\right)\\ &+N_2\frac{dv}{dy}+T_2\left(\frac{dw}{dx}+\frac{du}{dz}\right)\\ &+N_3\frac{dw}{dz}+T_3\left(\frac{du}{dy}+\frac{dv}{dx}\right)\end{aligned}\right\}.$$

Le premier membre est la somme des produits des composantes des forces agissant sur la surface du solide, par les projections des déplacements subis par leur points d'application; c'est la première expression connue [...] du double du travail de la déformation; le second membre en est donc une autre expression. Lorsque le corps est homogène et d'élasticité constante, [...] au second membre de l'équation (2), cette parenthèse [...] peut se mettre sous la forme

$$(4)\qquad \left\{\begin{aligned} &\frac{1+\dfrac{\lambda}{\mu}}{3\lambda+2\mu}\left(N_1^2+N_2^2+N_3^2\right)\\ &-\frac{1}{\mu}\left(N_1N_2+N_2N_3+N_1N_3-T_1^2-T_2^2-T_3^2\right)\end{aligned}\right\}.$$

Posons, pour simplifier,

$$(5)\qquad \left\{\begin{aligned} N_1+N_2+N_3&=F,\\ N_1N_2+N_2N_3+N_1N_3-T_1^2-T_2^2-T_3^2&=G,\end{aligned}\right.$$

et rappelons la valeur du coefficient d'élasticité E, [...] l'équation (2) prend la forme

$$(6)\qquad \sum(Xu+Yv+Zw)\varpi=\iiint\left(EF^2-\frac{G}{\mu}\right)dx\,dy\,dz.$$

C'est cette équation qui constitue le théorème de M. Clapeyron. Il faut remarquer que [...] F, G, et, par suite, la parenthèse $\left(EF^2-\frac{G}{\mu}\right)$ conservent les mêmes valeurs numériques quand on change d'axes coordonnés. C'est-à-dire que cette parenthèse [...] représente le double du travail intérieur [...]; et la moitié du second membre de l'équation (6) est la somme des travaux de tous les éléments, ou le travail du volume total du corps. C'est ainsi que toutes les forces élastiques développées concourent à former la seconde expression du travail de la déformation.

1.46 La théorème de M. Clapeyron, à proprement parler, en ce que le travail en question est esprimé, avec nos notation, par

$$\frac{1}{2}\left(p_{xx}\partial_x + p_{xx}\partial_y + p_{yy}\partial_x + p_{zz}\partial_z + p_{yz}g_{yz} + p_{zx}g_{zx}p_{xy}\cdot g_{xy}\right)$$

Nous mettons 1/2 parce que ce travail est produit par des forces dont les intensités commencent par zero et croissent uniformément.

1.47 Voici d'abord la règle générale à laquelle je suis arrivé:
étant donné une figure (plane on non) formée par des barres articulées en leurs extrémités et aux points d'articulation desquelles est appliqué un système quelconque de forces les maintenant en équilibre, pour trouver les tensions développées dans les diverses barres on commence par écrire que chaque point d'articulation est séparément en équilibre sous l'action des forces extérieures qui y sont appliquées et des tensions des barres en nombre quelconque qui y aboutissent. Si l'on obtient ainsi autant d'équations distinctes qu'il y a de tensions inconnues, le problème est résolu par la Statique pure (1). Si l'on obtient k équations de trop peu, on peut être certain que la figure géométrique formée par les axes des barres contient k lignes surabondantes, c'est-à-dire k lignes de plus que le nombre strictement nécessaire pour la définir; que, par suite, entre les longueurs des lignes qui la composent, c'est-à-dire entre les longueurs des barres, il existe nécessairement k relations géométriques (c'est un problème de Géométrie élémentaire). écrivez ces relations, différentiez-les en regardant toutes les longueurs qui y entrent comme variables; remplacez les différentielles par des lettres représentant les allongements élastiques des barres; remplacez à leur tour ces allongements élastiques par leurs expressions en fonction des tensions et des coefficients d'élasticité des barres (2); vous aurez ainsi k nouvelles équations auxquelles devront satisfaire ces tensions et qui, avec les équations déjà fournies par la Statique, formeront un total égal à celui des tensions à déterminer.

1.48 Soient

$$a_1,\ a_2,\ a_3,\ldots,a_m$$

les longueurs des m barres à l'état naturel, c'est-à-dire lorsque aucune force n'agit sur elles. Sous l'influence des forces appliquées aux divers points d'articulation, ces barres prendont des allongements

$$\alpha_1,\quad \alpha_2,\quad \alpha_3,\ldots,\alpha_m$$

ens sorte que leurs nouvelles loguers seront

$$a_1+\alpha_1,\quad a_2+\alpha_2,\quad a_3+\alpha_3,\ldots,a_n+\alpha_m$$

puisque entre ces longueurs il existe k relations algébriques, soit

$$F(a_1 + \alpha_1, a_2 + \alpha_2, a_3 + \alpha_3, \ldots, a_n + \alpha_m).$$

une de ces relation.

1.49 Telles sont les k relations à joindre à celles fournies par la Statique pour définir les tensions t_i.

1.50 Wir bezeichnen nun die von der Lage der Last unabhängigen Zahlen:

$$\begin{aligned} &\pm\frac{c}{b} \quad \text{mit} \quad u \\ &\pm\frac{d}{b} \quad \text{mit} \quad v \\ \text{und} \quad &\pm\frac{1}{b} \quad \text{mit} \quad w. \end{aligned}$$

so daß für jeden Konstruktionstheil des Trägers zwischen $x = 0$ und $x = a$

$$S = (2zu + v) \quad \text{Tonnen}$$

und für jeden Konstruktionstheil zwischen $x = a$ und $x = 1/2$

$$S = (2zu + aw) \quad \text{Tonnen}$$

wird.

1.51 Man denke sich nun, der Träger sei so aufgestellt, daß die Auflager in horizontaler Richtung frei ausweichen können und daß die Längenänderungen der einzelnen Konstruktionstheile nicht gleichzeitig sondern nach einander eintreten. Jede Längenänderung Δl eines Konstruktionstheils wird alsdann eine bestimmte von der geometrischen Form des Trägers abhängige Veränderung Δs der Spannweite s zur Folge haben. Die Summe der vor allen Konstruktionstheilen herrührenden Werthe von Δs muß, da die Spannweite in Wirklichkeit ihre Größe nicht verändert, gleich Null sein. Da ferner die hier betrachtete Formveränderung des Trägers in Bezug auf die Trägermitte symmetrisch ist, so muß die Summe der Werthe von Δs auch für die Trägerhälfte gleich Null sein:

$$\sum_{x=0}^{x=1/2} \Delta s = 0.$$

1.52 Man kann diese Bewegung auch hervorrufen durch einen Horizontalschub H gegen die Auflager, welcher nach dem Obigen in den elastischen Stange CD die Spannung $u \cdot H$ erzeugt. Während die Kraft H den Weg Δs zurücklegt und sonach die mechanische Arbeit $-H \cdot \Delta s$ leistet,[1] wird die wiederstehende Spannung $u \cdot H$ der Stange CD auf dem Wege Δl überwunden und dadurch die mechanische Arbeit $u \cdot H \cdot \Delta l$ absorbirt.[2] Nach dem Princip der virtuellen Geschwindigkeit sind diese Arbeiten gleich groß und demnach:

$$-H \cdot \Delta s = u \cdot H \cdot \Delta l$$

oder

$$-\Delta s = u \cdot \Delta l.$$

Durch Einsetzen der Werthe von Δl [...] ergibt sich

$$0 = 2z \sum_{0}^{s/2} ru^2 + \sum_{0}^{a} rvu + \sum_{a}^{s/2} rawu$$

oder

$$-z = \frac{\sum_0^a rvu + \sum_a^{1/2} rawu}{2 \sum_0^{s/2} ru^2}.$$

1.53 Die Bestimmung der genannten Auflagerreaktionen und Spannungen geschieht durch Rechnung oder auf graphischem Wege unter Anwendung sehr einfacher Methoden, die wir als bekannt voraussetzen dürfen.

1.54 Indem man diese Werthe von Δl in die durch die Gleichungen (4) ausgedrückten Beziehungen zwischen den Längenänderungen der überzähligen und denjenigen der notwendigen Konstruktionstheile einführt, ergeben sich die Bedingungen:

$$\begin{cases} \sum u_1 \cdot S \cdot r = 0 \\ \sum u_2 \cdot S \cdot r = 0 \\ \sum u_3 \cdot S \cdot r = 0 \\ \cdots \end{cases}$$

und wenn man den Werth von S nach Gleichung (6) einsetzt:

[1] Wir haben im Obigen den Horizontalschub H das positive Vorzeichen beigelegt; die Verkürzung Δs der Spannweite hat das negative Vorzeichen; demnach ist $-H \cdot \Delta s$ eine positive Größe (Original note by Mohr).

[2] Die Größe $u \cdot H \cdot \Delta l$ ist immer positiv, weil Δl eine Verlängerung oder Verkürzung bezeichnet, je nachdem $u \cdot H$ eine Zug-oder eine Druckspannung ist. Die Größen Δl und $u \cdot H$ haben demnach in der hier vorliegenden Betrachtung dasselbe Vorzeichen (Original note by Mohr).

$$\begin{cases} 0 = \sum u_1 \cdot \mathfrak{S} \cdot r + S_1 \sum u_1^2 \cdot r + S_2 \sum u_1 \cdot u_2 \cdot r + S_3 \sum u_1 \cdot u_3 \cdot r + \cdots \\ 0 = \sum u_2 \cdot \mathfrak{S} \cdot r + S_1 \sum u_1 \cdot u_2 \cdot r + S_2 \sum u_2^2 \cdot r + S_3 \sum u_2 \cdot u_3 \cdot r + \cdots \\ 0 = \sum u_3 \cdot \mathfrak{S} \cdot r + S_1 \sum u_1 \cdot u_3 \cdot r + S_2 \sum u_2 \cdot u_3 \cdot r + S_3 \sum u_3^2 \cdot r + \cdots \\ \cdots \end{cases}$$

Die Gleichungen (9) dienen zu Bestimmung der Spannungen der überzähligen Konstruktionstheile [...].

1.55 La prima deduzione delle equazioni e leggi di *Maxwell* seguendo la via piùbreve dell'utilizzazione del principio degli spostamenti virtuali si deve a *Mohr*. I suoi *Contributi alla teoria del traliccio* nel Zeitschrift des Architekten-und Ingenieur Vereins zu Hannover del 1874 e 1875 contengono le prime importanti applicazioni della teoria di *Maxwell*. *Mohr* rappresentò per primo anche la linea elastica dell'asta dritta e la linea di inflessione del traliccio con l'aiuto del poligono funicolare.

1.56 Forse un unico nome, quello di Gabrio Piola, si può citare, come autore di ricerche che si connettono colla teoria generale fondata dal Navier.

1.57 Ogni mutazione di forma in ogni parte infinitesima di un corpo solido dà origine a forze che tendono a restituire a ciascuna parte infinitesima la sua forma primitiva. Qual è l'origine di queste forze? Qual è la legge con cui queste forze agiscono? è noto il concetto che domina nella Fisica relativamente alla costituzione dei corpi. Si riguardano composti di un numero infinitamente grande di punti materiali separati che si attraggono o si respingono secondo la retta che li unisce con una intensità che è funzione della loro distanza. Quando questa distanza ha un certo valore piccolissimo l'azione è nulla, è ripulsiva a distanze minori, attrattiva a distanze maggiori, nulla a distanze sensibili. Questo concetto non è in accordo con un altro che ha avuto origine dalla teorica del calorico, cioè che le parti infinitesime dei corpi non siano mai in quiete, ma siano animate da movimenti rapidissimi. Quindi tutte le teoriche fondate supponendo che le particelle dei corpi siano in quiete non possono più ammettersi, anche se rendessero conto degli altri fenomeni, il che non è. Il concetto dovrebbe modificarsi e riguardare invece un corpo come costituito di un numero infinito di sistemi di punti materiali in ciascuno dei quali esistono rapidissimi moti intorno a un centro [...]. Ma per sottoporre a calcolo i fenomeni che presenta un corpo solido quando è stato deformato, non è necessario fondarsi sopra questa ipotesi. Una legge generale della Natura dà il modo di fondare una teoria generale che permette il calcolo di tutti i fenomeni della elasticità. Questa legge generale è la seguente: Il lavoro meccanico che si fa per passare un corpo da uno a un altro stato senza perdita né acquisto finale di calore è indipendente dagli stati intermedi per i quali si fa passare il corpo stesso. Questo principio non è altro che quello della conservazione della forza.

1.58 Mi è grato il pensare che il dotto ingegnere, il quale aveva riconosciuta tutta l'importanza del concetto di potenziale elastico, avrebbe probabilmente approvata la mia proposta di fondare sovr'esso anche la deduzione delle condizioni anzidette.

1.59 Lorsqu'un système élastique se met en équilibre sous l'action de forces extérieures, le travail développé par l'effet des tensions ou des compressions des liens qui unissent les divers points du système est un minimum.

A.2 Quotations of Chap. 2

2.1 Si dimanda un'applicazione de' principi contenuti nella Meccanica analitica dell'immortale Lagrange ai principali problemi meccanici e idraulici, dalla quale apparisca la mirabile utilità e speditezza dei metodi lagrangiani.

2.2 Gli elementi della prima non possono essere che una particolare determinazione degli elementi della seconda, e le formole di questa non si potrebbero aver per buone e generali se il caso non comprendessero dell'equilibrio con tutti gli accidenti che a esso appartengono. La pratica stessa dei ragionamenti che impiegasi nel premettere la statica alla dinamica ci fa sentire questa verità coll'irregolarità e con la contraddizione [...]. Perciocché vedesi costretta a mettere in campo il ripiego di certo meccanico movimento infinitesimale.

2.3 [...] il faut convenir qu'il n'est pas assez evident par lui-même pour être érigé en principle primitif [...].

2.4 È dunque necessario abbandonare alquanto le nostre pretese, e, seguendo il gran precetto di Newton, cercare nella natura que' principi con che spiegare gli altri fenomeni naturali [...]. Queste riflessioni persuadono che sarebbe un cattivo filosofo chi si ostinasse a volere conoscere la verità del principio fondamentale della meccanica in quella maniera che gli riesce manifesta l'evidenza degli assiomi. [...] Ma se il principio fondamentale della meccanica non può essere evidente, dovrà essere non di meno una verità facile a intendersi e a persuadersi.

2.5 Io, educato da Brunacci alla scuola di Lagrange, ho sempre impugnato l'infinitesimo metafisico, ritenendo che per l'analisi e la geometria (se si vogliono conseguire idee chiare) vi si deve sempre sostituire l'indeterminato piccolo quanto fa bisogno: ma ammetto ciò che potrebbe chiamarsi l'infinitesimo fisico, di cui è chiarissima l'idea. Non è uno zero assoluto, è anzi tal grandezza che per altri esseri potrebbe riuscire apprezzabile, ma è uno zero relativamente alla portata dei nostri sensi.

2.6 Ecco il maggiore vantaggio del sistema della Meccanica Analitica. Esso ci fa mettere in equazione i fatti di cui abbiamo le idee chiare senza obbligarci a considerare le cagioni di cui abbiamo idee oscure [...]. L'azione delle forze attive o passive (secondo una nota distinzione di Lagrange) è qualche volta tale che possiamo farcene un concetto, ma il più sovente rimane [...] tutto il dubbio che il magistero della natura sia ben diverso [...]. Ma nella M. A. si contemplano gli effetti delle forze interne e non le forze stesse, vale a dire le equazioni di condizione che devono essere soddisfatte [...] e in tal modo, saltate tutte le difficoltà intorno alle azioni delle forze, si hanno le stesse equazioni sicure ed esatte che si avrebbero da una perspicua cognizione di dette azioni.

2.7 INTRODUZIONE
La meccanica de' corpi estesi secondo le tre dimensioni, solidi e fluidi di ogni sorta è stata recentemente promossa mediante le ricerche dicanica de' corpi estesi secondo le tre dimensioni, solidi e fluidi di ogni sorta è due insigni geometri francesi, Poisson e Cauchy, i quali trattarono problemi assai difficili per l'addietro non toccati. Il secondo di essi ne' suoi Esercizi di Matematica diede alcune soluzioni in doppio, cioè nell'ipotesi della materia continua, e nell'ipotesi della materia considerata come l'aggregato di molecole distinte a piccolissime distanze: il primo invece, credendo che la supposizione della materia continua non basti a rendere ragione di tutti i fenomeni della natura, si attenne di preferenza all'altra supposizione, bramando rifare con essa da capo tutta la Meccanica. Prima dei sullodati geometri, Lagrange avea trattati vari problemi relativi alla meccanica de' solidi e de' fluidi, creando una nuova scienza per queste come per tutte le altre quistioni di equilibrio e di moto: intendo parlare della Meccanica Analitica, opera cui anche oggidì si danno molte lodi, e viene chiamata la vera meccanica filosofica ma che nel fatto si riguarda poco più che un oggetto di erudizione. Avendo io avuta nella mia prima giovinezza particolare occasione di fare su quest'opera uno studio pertinace, erami formata un'idea così elevata della generalità e della forza de' suoi metodi, che giunsi a riputarli, in confronto dei metodi antecedentemente usati, un prodigio di invenzione non minore di quello del calcolo differenziale e integrale in confronto dell'analisi cartesiana: e pensai e scrissi essere impossibile che per l'innanzi ogni ricerca di meccanica razionale non si facesse per questa via. Esaminate in seguito le recenti memorie, e avendo notato come in esse non si faccia uso (se non forse qualche rara volta in maniera secondaria) dell'analisi che tanto mi avea colpito, credetti d'essermi ingannato, che cioè le nuove questioni di meccanica non si potessero assoggettare ai metodi della Meccanica Analitica. Provai però a convincermene anche per mezzo di un esperimento: e allora fu molta la mia sorpresa nell'accorgermi che in quella vece esse vi si accomodano egregiamente, e ne ricevono molta chiarezza: un andamento di dimostrazione che accontenta lo spirito: conferma in alcuni luoghi: cangiamento in alcuni altri: e quel che è più, aggiunta di nuovi teoremi. Ecco il motivo che mi determinò a pubblicare una serie di Memorie sull'enunciato argomento, per tentare di ridurre alla mia opinione qualche lettore: ma innanzi alle prove di fatto pensai mettere alcune riflessioni generali dirette a indicare, per quanto almeno è della mia capacità, il profondo di quella sapienza che trovasi nella maggior opera del sommo Geometra italiano.
I. La generalità dei metodi è ragione assai forte per indurci a preferirli ad altri più particolari. Nessuno leggerebbe di presente uno scritto in cui si proponesse di tirare le tangenti alle curve con alcuno dei metodi che precedettero il leibniziano,

nè farebbe buona accoglienza alla quadratura di uno spazio piano curvilineo conchiusa dietro ragionamenti simili a quelli con cui Archimede quadrò la parabola. Ora l'aver trovato nel calcolo delle variazioni quel punto altissimo in cui si uniscono tutte le questioni di meccanica, e possono in conseguenza essere tutte trattate di una maniera uniforme, è forse qualche cosa di meno grande che l'aver trovata la prima questione geometrica solubile in generale per mezzo della derivata, e la seconda per mezzo della primitiva dell'ordinata che riguardisi come funzione dell'ascissa?
II. Il metodo della M. A. non risulta (se ben si esamina) dalla traduzione in analisi di un solo e semplice principio meccanico, ver. gr. del principio del parallelogrammo, o del principio di D'Alembert: è un metodo che può dirsi l'elaborato di tutti i principi successivamente scoperti nella meditazione delle leggi della natura, e che però colla riunita potenza di tutti si fa strada alla soluzione de' problemi. È noto che un principio meccanico di massimo o minimo trovato da Eulero dietro la considerazione delle cause finali e sviluppato nel secondo supplemento al suo libro *Methodus inveniendi lineas curvas ecc.*, è quello da cui prese Lagrange le prime mosse per l'invenzione del suo metodo fondato sui calcolo delle variazioni.
III. Una questione di meccanica presenta sovente varie parti: i punti alle superficie dei corpi abbisognano di considerazioni particolari che non hanno egualmente luogo per quelli che sono nell'interno de' corpi stessi: e anche per linee individuate in queste superficie e per punti in queste linee possono darsi particolari circostanze. Con metodi meno generali le indicate diverse parti sono discusse successivamente: ma la M. A. le abbraccia tutte a una volta, perchè nella sua *equazione generalissima*, dietro un principio noto nel calcolo delle variazioni, si fanno separatamente nulle quantità opportunamente disposte sotto integrali triplicati, duplicati, e semplici: il che distribuisce in varie masse tutte le equazioni dietro le quali si analizza il moto o l'equilibrio compiutamente.
IV. All'utilità di darci il problema svolto e anatomizzato, per così dire, in tutti i particolari un'altra se ne aggiunge non meno importante, quella di farci vedere l'indipendenza in cui rimangono alcune delle indicate equazioni dai cambiamenti introdotti in alcune altre. Se, per esempio, si vogliono trasportare dal caso dell'equilibrio a quello del moto i teoremi fra le pressioni alle superficie dei corpi, si sente il bisogno di una dimostrazione. La M. A. vi supplisce colla semplice osservazione che il passaggio dall'equilibrio al moto introduce mutazione nella sola quantità sottoposta all'integrale triplicato, non alterando quelle che stanno sotto i duplicati, e che quindi le equazioni dedotte da questi ultimi restano le stesse. Come mai dopo veduta questa gran luce potremo ancora adattarci a' ripieghi che in qualche parte sono in urto colla natura della questione?
V. Ecco il maggior vantaggio del sistema della Meccanica Analitica. Esso ci fa mettere in equazione *fatti* di cui abbiamo idee chiare senza obbligarci a considerare le cagioni di cui abbiamo idee oscure: fatti certi invece di cagioni a esprimere l'azione delle quali si formano ipotesi dubbie e non troppo persuadenti. è desso un sistema che abbisogna appunto di quelle sole cognizioni a cui arriva la mente umana con sicurezza, e si astiene o può astenersi dal

pronunciare appunto dove non pare possibile mettere un fondo sodo ai nostri ragionamenti. Un sistema che assume pochi dati invece di un gran numero di elementi; un sistema in cui colla stessa fiducia si seguono i più vicini e i più lontani svolgimenti di calcolo, perché non vi si fanno da principio ommissioni di quantità insensibili, che lasciano qualche sospetto di errore non egualmente insensibile nel progresso. Convincersi di tutte queste proposizioni è il frutto di lungo studio sulla M. A. Soggiungerò qualche parola a schiarimento di alcuna di esse.

VI. L'azione delle forze interne attive o passive (secondo una nota distinzione di Lagrange) è qualche volta tale che possiamo farcene un concetto, ma il più sovente rimane alla corta nostra veduta torbida così da lasciarci tutto il dubbio che il magistero della natura sia ben diverso da quelle immagini manchevoli colle quali ci sforziamo di rappresentarcelo. Per un esempio: se trattisi del moto di un punto obbligato a stare sopra una superficie, possiamo rappresentarci con chiarezza la resistenza della superficie siccome una forza che opera normalmente alla superficie stessa, e stabilire con questa sola considerazione le equazioni generali del moto. Se trattasi invece di quelle forze che mantengono la continuità nelle masse in moto, io confesso che, almeno per me, il loro modo d'agire è sì inviluppato, che non posso accontentarmi alle maniere con cui vorrei immaginarmelo. Quando pertanto dietro alcuna di queste maniere io volessi stabilire le equazioni del movimento, non potrei attaccare fede ai risultati del mio calcolo: e molto più se facessi altresì delle supposizioni secondarie, e parecchie di quelle ommissioni accennate più sopra. Ma nella M. A. si contemplano gli effetti delle forze interne e non le forze stesse, vale a dire le equazioni di condizione che debbono essere soddisfatte, o certe funzioni che dalle forze sono fatte variare: questi effetti sono chiari anche nel secondo caso, e in tal modo, saltate tutte le difficoltà intorno alle azioni delle forze, si hanno le stesse equazioni sicure ed esatte che si avrebbero da una perspicua cognizione di esse azioni. Ecco il gran passo: si può poi, se si vuole, rivestire della rappresentazione delle forze i coefficienti indeterminati introdotti in maniera strumentale, e allora, determinati questi coefficienti a posteriori mediante le equazioni meccaniche, acquistare delle cognizioni intorno alle forze stesse. Seguendo un tal metodo nel primo dei due casi sopraccennati il risultato del calcolo si trova perfettamente d'accordo colla rappresentazione che ci eravamo fatta intorno all'intervento della forza passiva, e ciò non può che riuscire di molta soddisfazione. Nel secondo caso poi il risultato è d'accordo con quel tanto che vedevamo a priori: ed è poi un gran conforto il sapere ch'esso è sicuramente giusto anche dove i ragionamenti a priori erano deboli, anche dove entrando essenzialmente l'infinito non potevamo vedere al di là di poche congruenze, anche dove la punta della nostra intelligenza non poteva direttamente in nessuna guisa penetrare.

VII. Insisto su queste idee perché ne consegue, di qualunque valore esser possa, la mia opinione intorno a quella Meccanica fisica che si vuole adesso far sorgere a lato della Meccanica Analitica. Applaudo a questa nuova scienza: ma invece di vederla sorgere a lato della M. A., bramerei vedervela sorgere sopra: e mi

spiego. Quando le equazioni dell'equilibrio e del moto siano stabilite dietro principi inconcussi, sarà lecito il far delle ipotesi sulla costituzione interna dei corpi in modo di avere altrimenti le stesse equazioni; e allora quelle ipotesi, se non con sicurezza, almeno con probabilità potranno essere ricevute. Ciò anche servirà per determinare in qualche maniera certe quantità sulle quali l'analisi lagrangiana non pronuncia. Supponendo quindi i corpi come quelle ipotesi danno, potranno dedursi altre ed altre conseguenze che non avranno maggiore probabilità della ipotesi originaria: ma se poi in questo cammino ci sarà dato di avere altri punti di confronto colla natura nei quali non ci troviamo fuori di strada, l'ipotesi primaria acquisterà sempre maggiore consistenza. Non vorrei io però una meccanica fisica di cui le prime equazioni ragionate sopra supposizioni alquanto incerte non ottennessero se noti una lontana conferma, scendendo dal generale al particolare, per qualche corrispondenza con fenomeni osservati. La buona filosofia fatta esperta dalle aberrazioni di molti fra que' pensatori che fabbricarono sistemi intorno alle cose naturali, deduce dalla moltiplicità stessa e contrarietà delle loro opinioni, che non è retto quel metodo di filosofare il quale, senza sufficiente appoggio nel suo principio, ne ha uno soltanto nel suo fine. Se queste riflessioni sono giuste, ognun vede quanto interessi rimettere in credito e in pratica lo studio della M. A. la quale è la sola che a stabilire l'equazioni fondamentali abbisogna di pochi dati la cui verità non è disputabile.
VIII. Resta a sciogliere qualche difficoltà: la M. A. non è scienza al tutto perfetta: essa presenta alcuni passi mancanti e meno veri: essa conduce qualche volta a calcoli intrattabili. Gli ammiratori di Lagrange non vorranno pienamente ammettere queste asserzioni: ma quand'anche si ammettano, esse null'altro provano se non che a Lagrange come a Leibnitz mancò il tempo a riconoscere per intero la vastità di quel concepimento che si era formato nella sua mente, e riconosciutala, informarne altri a tutto agio. Leibnitz lasciò molto a fare ai suoi successori i quali compierono l'edificio di cui egli avea gettati i fondamenti ed erette molte parti: e i Rolle, i Lagny, i Nieuventyt che non vollero portar pietre a questo edificio certamente la sbagliarono. Tocca ai geometri successori di Lagrange a perfezionare la grand'opera ch'egli fondò e portò a tanta altezza: a rettificarne qualche luogo in cui egli pagò un lieve tributo all'umanità senza conseguenze che intacchino la sostanza del metodo, a spianarne qualche altro ove sono certe asprezze, a supplire alcune parti che tuttora si desiderano. E quanto alla malagevolezza e complicazione dei calcoli diremo: nulla è la fatica di un lungo calcolo, quando nel seguirlo sappiamo a non dubitarne che siamo molti uniti colla verità e colla verità giungeremo al fine: è gioja, è godimento in questa fatica sostenuta dall'aspettativa di un largo profitto. I grandi perfezionamenti poi introdotti nella scienza del calcolo dopo la morte di Lagrange valgono a superare alcune difficoltà a cui egli stesso erasi arrestato: ciò che rimane è un invito prezioso onde promuovere anche l'analisi col doppio scopo dell'invenzione e dell'applicazione.

Premesse queste riflessioni generali per fissare l'attenzione dei leggitori sull'eccellenza del metodo lagrangiano a cui intendo di attenermi: farò un brevissimo cenno di quella disposizione che penso dare alle seguenti memorie.

Comincerò da una sui corpi solidi rigidi nella quale si vedrà chiaro il modo con cui le nuove ricerche si attaccano alla M. A., e si troverà preparata l'analisi fondamentale che servir deve anche per quanto avrò a dire in appresso. Passerò nella seguente a parlare dei corpi estesi in generale: e quindi le teoriche saranno successivamente sviluppate secondo la concatenazione più naturale.

2.8 Il est bon de remarquer que cette valeur de $Dx\,Dy\,Dz$ est celle qu'on doit employer dans les intégrals triples relatives à x, y, z, lorsqu'on veut y substituer, à la place des variables x, y, z, des fonctions données d'autres variables a, b, c.

2.9 Osservisi la perfetta coincidenza di questo risultato con quello ottenuto dai due celebri geometri citati dal principio dell'introduzione dietro ragionamenti affatto diversi e nei due casi dell'equilibrio e del moto trattati separatamente. Raccomando di notare che nella mia analisi le A, B, C, D, E, F non sono pressioni che si esercitino sopra diversi piani, ma sono coefficienti, cui nel seguito attaccherò io pure una rappresentazione di forze secondo mi sembrerà più naturale: sono funzioni delle x, y, z, t di forma ancora incognita, ma di cui sappiamo che non cambia passando dall'una all'altra parte del corpo. Mi si può obbjettare che queste equazioni [...] sono state trovate coi metodi della M. A. nel solo caso dei sistemi solidi rigidi, laddove quelle dei due chiarissimi francesi si riferiscono anche a' solidi elastici e variabili. Rispondo che nella seguente memoria farò vedere come esse si generalizzano ad abbracciare tutti i casi contemplati dai citati Autori senza dipartirsi dagli andamenti analitici insegnati da Lagrange.

2.10 Lagrange est allé aussi loin qu'on puisse le concevoir, lorsqu'il a remplacé les liens physiques des corps par des équations entre les coordonnées de leurs différents points: c'est là ce qui constitute la *Mécanique analytique*; mais à côté de cette admirable conception, on pourrait maintenant élever la *Mécanique physique*, dont le principe unique serait de ramener tout aux actions moléculaires, qui transmettent d'un point à un autre l'action des forces données, et sont l'intermédiaire de leur équilibre.

2.11 L'usage que Lagrange a fait de ce calcul dans la *Mécanique Analytique* ne convient réellement qu'à des masses continues; et l'analyse d'après laquelle on étend les résultats trouvés de cette manière aux corps de la nature, doit être rejetè comme insuffisante.

2.12 [il] Sig. Poisson [...] vorrebbe ridurre tutto alle sole azioni molecolari. Io mi conformo a questo voto non ammettendo appunto oltre le forze esterne, che un'azione reciproca di attrazione e repulsione [...]. Non è già che io creda da abbandonarsi l'altra maniera usata da Lagrange, ché anzi io sono d'avviso che eziandio con essa si possano vantaggiosamente trattare molte moderne questioni, ed ho già pubblicato un saggio di un mio lavoro che può in parte provare questa mia asserzione.

2.13 Si guadagnarono alcuni nuovi teoremi, ma si perdette gran parte dei vantaggi e delle bellezze di un'analisi elaborata con lungo studio dai nostri maestri.

2.14 Mostrare come si sostenga ancora in gran parte l'analisi di D'Alembert, di Eulero e di Lagrange supponendo coi moderni la materia discontinua: conservare il tesoro di scienza trasmessoci dai nostri predecessori, e nondimeno progredire coi lumi del nostro secolo.

2.15 Disposizione ideale antecedente allo stato vero nella quale la materia del corpo stesso era contenuta in un parallelepipedo [...] e tutte le a non diversificano fra loro che di aumenti eguali ad α, le b di aumenti eguali a β, le c di aumenti eguali a γ [...].

2.16 [fatta salva] l'irregolarità voluta dalla discontinuità della materia, [...] ottengo una regolarità [...] necessaria pel meccanismo del calcolo quale è adoperato da Lagrange nella Meccanica analitica.

2.17 [...] l'espressione dell'azione molecolare può avere un valore sensibile pei punti estremamente vicini [...], l'azione molecolare è insensibile per distanze sensibili [...]. [...] il raggio della sfera di attività dell'azione molecolare, quantunque si estenda a un numero grandissimo di molecole, deve ancora considerarsi una quantità insensibile [...].

2.18 Scrissi più volte non parermi necessario il creare una nuova Meccanica, dipartendoci dai luminosi metodi della Meccanica analitica di Lagrange [...]. Però mi stettero e mi stanno anche attualmente contro autorità ben rispettabili, davanti alle quali io dovrei darmi per vinto [...]. Ma [...] credetti convenisse [...riunire] in questa Memoria i miei pensieri sull'argomento [...]. Perocchè non dissimulo accorgermi ora che ne' precedenti miei scritti alcune idee non furono esposte con sufficiente maturità: ve ne ha qualcuna troppo spinta, ve ne ha qualch'altra troppo timorosa: certe parti di quelle scritture potevano essere ommesse, [...] a più forte ragione quelle altre che [...] non mi sentirei più di ripetere [...].

2.19 Spero mettere in chiaro nella seguente Memoria che l'unico motivo pel quale la Meccanica Analitica parve restar addietro nella trattazione di alcuni problemi, fu che Lagrange nello scrivere dell'equilibrio e del moto di un corpo solido, non è disceso fino ad assegnare le equazioni spettanti a un solo punto qualunque di esso. Se questo avesse fatto, e lo potea benissimo senza uscire dai metodi insegnati nel suo libro, sarebbe giunto prontamente alle stesse equazioni cui arrivarono con molta fatica i Geometri francesi del nostro tempo, e che ora servono di base alle nuove teoriche. Però quello ch'egli non fece [...] può esser fatto da altri [...].

2.20 Abbiamo un teorema di analisi che ci somministra il mezzo di passare da un integrale finito definito a un integrale continuo parimenti definito.

2.21 [...] si vorrebbero trasmutare queste equazioni [...] in altre che non contenessero traccia delle a, b, c e non constassero che di quantità spettanti allo stato reale del corpo.

2.22 Le mentovate sei quantità in ambi i casi sono le espressioni analitiche contenenti l'effetto complessivo di tutte le azioni interne sopra il punto generico (p, q, r) ovvero (x, y, z).

2.23 Tal principio sta nel riferimento simultaneo di un qualunque sistema a due terne di assi ortogonali: esso può adoperarsi in due maniere e in entrambe produce grandiosi effetti. Si adopera in una prima maniera [...] a fine di dimostrare il principio delle velocità virtuali, e anche gli altri della conservazione del moto del centro di gravità, e delle aree. Invece di concepire in tal caso le δx, δy, δz dei diversi punti del sistema come velocità virtuali o spazietti infinitesimi descritti in virtù di quel moto fittizio (il quale fu poi altresì detto dopo Carnot un moto geometrico), è assai più naturale e non ha nulla di misterioso il ravvisarle quali aumenti che prendono le coordinate degli anzidetti punti quando il sistema si riferisce ad altri tre assi ortogonali vicinissimi ai primi, come se questi si fossero di pochissimo spostati. [...] allora si capisce chiaro come gli aumenti delle coordinate abbiano luogo senza alterazioni nelle azioni reciproche delle parti del sistema le une sulle altre.

2.24 Si un système de corps part d'une position donnée, avec un mouvement [velocità] arbitraire, mais tel qu'il eût été possible aussi de lui en faire prendre un autre tout-à-fait égal et directement opposé; chacun de ces mouvements sera nommé mouvement géométrique.

2.25 Il riferimento simultaneo del sistema a due terne di assi ortogonali giuoca poi efficacemente in un'altra maniera [...]. Qui s'intende parlare di quel metodo che lascia alle δx, δy, δz tutta la loro generalità e tratta le equazioni di condizione, introducendo moltiplicatori indeterminati. In tal caso la contemplazione delle due terne di assi giova per l'impianto delle dette equazioni di condizione, che altrimenti non si saprebbero assegnare in generale [...]. Un tal punto di vista parmi sfuggito a Lagrange e ad altri Geometri: a esso si riferisce quanto nella presente Memoria può essere più meritevole di attenzione.

2.26 Il n'est pas malaisé de déduire du Principe des vitesses virtuelles et de la généralisation thermodynamique de ce principe la conséquence suivant: Si un système est en équilibre lorsqu'il est assujetti à de certains liaisons, il demeura en équilibre lorsqu'on l'assujettira non seulement à ces liaisons mais encore à des nouvelles liaisons compatibles avec les premières.

2.27 On ajoutera donc cette intégrale $SF\delta ds$ à l'intégrale $SX\delta x + Y\delta y + Z\delta z$, qui exprime la somme des momens de toutes les forces extérieures qui agissent sur le fil [...], & égalent le tout à zéro, on aura l'équation générale de l'équilibre du fil à ressort. Or il est visible que cette équation sera de la même forme que celle [...] pour le cas d'un fil inextensible, & qu'en y changeant F en λ, les deux équations deviendront même identiques. On aura donc dans le cas présent les mêmes équations particulieres pour l'équilibre du fil qu'on a trouvées dans le cas de l'art. 31, en mettant seulement dans celle-ci F à la place de λ.

2.28 [Lagrange] nella sua M. A. [...] adottò un principio generale (§ 9. della Sez. II^a, e 6. della IV^a), mediante il quale l'espressione analitica dell'effetto di forze interne attive riesce affatto analoga a quella che risulta per le passive quando si hanno equazioni di condizione: il che si ottiene assumendo dei coefficienti indeterminati e moltiplicando con essi le variate di quelle stesse funzioni che rimangono costanti per corpi rigidi, o inestensibili, o liquidi. Se ci conformassimo a un tal metodo, potremmo a dirittura generalizzare i risultamenti ai quali siamo giunti nel capitolo precedente: io però preferisco astenermene, giacché la mia ammirazione pel grande Geometra non m'impedisce di riconoscere come in quel principio rimanga tuttavia alcun che di oscuro e di non dimostrato.

2.29 Infatti molte possono essere contemporaneamente le espressioni di quantità che le forze interne di un sistema tendono a far variare; quali di esse prenderemo, quali ommetteremo? Chi ci assicura che adoperando parecchie di tali funzioni soggette a mutamenti per l'azione delle forze interne, non facciamo ripetizioni inutili, esprimendo per mezzo di alcune un effetto già scritto con altre? E non potrebbe invece accadere che ommettessimo di quelle necessarie a introdursi affinché l'effetto complessivo delle forze interne venga espresso totalmente?

2.30 Circa la questione: *quali* sono le funzioni fatte variare dalle forze interne che si debbono usare a preferenza di altre, ho dimostrato che sono que' trinomj alle derivate. [...] Relativamente all'altra questione: *quante* poi debbano essere tali funzioni [...] ho risposto quante ce ne vogliono per risalire alle variate di que' trinomj poste uguali a zero.

2.31 Il concetto che Lagrange voleva ci formassimo delle forze, e che esponemmo nel prologo, è più generale di quello universalmente ammesso. S'intende facilmente da tutti essere la forza una causa che mediante la sua azione altera la grandezza di certe quantità. Nel caso più ovvio, avvicinando un corpo o un punto materiale ad un altro, cambia distanze, ossia fa variare lunghezze di linee rette: ma può invece far variare un angolo, una densità, ecc. In questi altri casi il modo di agire delle forze ci riesce oscuro, mentre ci par chiaro nel primo: ma forse la ragione di ciò è estrinseca alla natura delle forze. Per verità anche in quel primo caso non si capisce come faccia la forza a infondere la sua azione nel corpo sì da diminuirne od accrescerne la distanza da un altro corpo: nondimeno noi vediamo continuamente il fatto: l'osservazione giornaliera sopisce in noi la voglia di cercarne più in là. Se però sottilmente esaminando si trova che qui pure il modo di agire delle forze è misterioso, nessuna meraviglia ch'esso ci appaja oscuro negli altri casi. Voler ridurre in ogni caso, l'azione delle forze a quella che diminuisce una distanza, è impiccolire un concetto più vasto, è un non voler riconoscere che una classe particolare di forze. Generalmente parlando, a qual punto possono essere spinte le nostre cognizioni intorno alle cause che sottoponiamo a misura? forse a comprenderne l'intima natura, e il vero modo con cui agiscono? [...] Radunato tutto quanto vi è d'incognito nella unità di misura della stessa specie, noi diciamo di conoscere la quantità, lorché possiamo assegnarne i rapporti colla detta unità assunta originariamente arbitraria. Ora eziandio quando si concepiscono le forze alla maniera più generale di Lagrange, cioè siccome cause che fanno variare quantità talvolta diverse dalle linee, concorrono i dati necessari a poter dire che sappiamo misurarle: si ha tutto ciò che ragionevolmente ci è lecito di pretendere: se pare che ci manchi l'immagine con che rivestirne il concetto, è perché vogliamo colorirla come nel caso particolare delle forze che agiscono lungo le rette: un fondo incognito rimane sempre tanto in questi casi più generali, come in quello sì comune.

2.32 Le equazioni generali del moto di un punto qualunque (x, y, z) del corpo sono le (56) ove le L_1, L_2, ec. [...] si riducono a dipendere [...] [dal]la sola incognita $\psi(S)$ relativa all'azione molecolare. Ben è vero, che [...] le trovate equazioni si rapportano a quella composizione delle x, y, z in a, b, c che è ignota anzi inassegnabile; ma passiamo ora a vedere in qual modo, fermato il vantaggio di formole ottenute rigorosamente, si sormonta in quanto agli effetti l'accennata difficoltà.

2.33 [...es gibt] eine andere Auffassung des Prinzipes der virtuellen Verrückungen, die von vornherein nur die eigentlichen *Kräfte*, die Massenkräfte X, Y, Z und die Flachenkräfte $\bar{X}, \bar{Y}, \bar{Z}$ als gegeben betrachtet; es ist die folgende leichte Fortbildung der Formulierung von *G. Piola* : *Für das Gleichgewicht ist notwendig, dass die virtuelle Arbeit der angeführten Kräfte*

$$\int\int_{(V)}\int (X\delta x + Y\delta y + Z\delta z)dV + \int_{(S)}\int (\bar{X}\delta x + \bar{Y}\delta y + \bar{Z}\delta z)dS.$$

verschwindet für alle [...starren Bewegungen] *des ganzen Bereiches V* [...so dass] die *Komponenten der Spannungsdyade als Lagrangesche Faktoren gewisser Starrheitsbedingungen* erweisen.

2.34 Lorsque les pressions sont prises sur les planes légèrement obliques dans lesquels se sont changés les trois plans matériels primitivement rectangulaires et parallèles aux coordonnées, on a, pour les six composantes, les mêmes expressions, en fonction des dilatations et des glissements, que lorsque les déplacements sont très petits.

2.35 Ich werde die Coordinaten eines Punktes nach der Formänderung ξ, η, ζ nennen, die Coordinaten desselben Punktes vor derselben, x, y, z. Im natürlichen Zustande des Körpers denke ich mich durch den Punkt (x, y, z) drei Ebenen gelegt, parallel den Coordinaten-Ebenen; die Theile dieser Ebenen, welche unendlich nahe an den genannten Punkte liegen, gehen bei der Formänderung in Ebenen über, die mit den Coordinaten-Ebenen schiefe, endliche Winkel bilden, mit einander aber Winkel, die unendlich wenig von 90° verschieden sind. Die Drucke, die diese Ebenen nach der Formänderung auszuhalten haben, denke ich mich in Componenten nach den Coordinaten-Axen zerlegt, und nenne diese Componenten: $X_x, Y_x, Z_x, X_y, Y_y, Z_y, X_z, Y_z, Z_z$, in der Art, dass z. B. Y_x die y Componente des Druckes ist, den die Ebene auszuhalten hat, die von der Formänderung senkrecht zur x Axe war. Diese neun Drucke sind im Allgemeinem schief gegen die Ebenen gerichtet, gegen die sie wirken, und es sind nicht drei von ihnen drei anderen gleich, wie es bei unendlich kleinen Verschiebung der Fall ist. Stellt man die Bedingungen dafür auf, dass ein Theil des Körpers sich im Gleichgewichte befindet, der vor der Formänderung ein unendlich kleines Parallelepipedum ist, dessen Kanten parallel den Coordinaten-Axen sind, und die Längen dx, dy, dz haben, so kommt man zu den Gleichungen:

$$\left.\begin{aligned}\rho X &= \frac{\partial X_x}{\partial x} + \frac{\partial X_y}{\partial y} + \frac{\partial X_z}{\partial z}\\ \rho Y &= \frac{\partial Y_x}{\partial x} + \frac{\partial Y_y}{\partial y} + \frac{\partial Y_z}{\partial z}\\ \rho Z &= \frac{\partial Z_x}{\partial x} + \frac{\partial Z_y}{\partial y} + \frac{\partial Z_z}{\partial z}\end{aligned}\right\}\ldots \qquad (1)$$

wenn man mit ρ die Dichtigkeit des Körpers, mit X, Y, Z die Componenten der beschleunigenden Kraft bezeichnet, die auf den Körper im Punkte (ξ, η, ζ) wirkt. Man kommt zu diesen Gleichungen, indem man benützt, dass die Winkel und die Kanten des Parallelepipedums sich nur unendlich wenig geändert haben, übrigens aber dieselben Betrachtungen anstellt, durch die man bei unendlich kleinen Verschiebungen diese Gleichungen beweist.

2.36 Les distances mutuelles de points très-rapprochées ne varient que dans une petite proportion.

A.3 Quotations of Chap. 3

3.1 Delle ipotesi delle forze molecolari io conservo soltanto la prima parte, cioè ammetto soltanto che gli elementi dei corpi agiscano gli uni sugli altri nel senso della retta che li unisce e proporzionalmente al prodotto delle loro masse; il che porta a ammettere che le forze di coesione e di aderenza abbiano funzioni potenziali.

3.2 Le forze che agiscono secondo la legge di Newton sono quelle che emanano da ciascuno negli elementi infinitesimi di una data materia e che tendono a avvicinare oppure a allontanare tra loro questi elementi, in ragione diretta delle loro masse e in ragione inversa dei quadrati delle loro distanze.

3.3 Un'altra proprietà hanno queste forze che si deduce dal [...] [principio] fondamentale della Fisica moderna: il principio della conservazione delle forze; e che consiste nell'avere esse una funzione potenziale.

3.4 Per determinare le relazioni che debbono esistere tra le forze che agiscono sopra un corpo solido elastico omogeneo, e le deformazioni degli elementi dello stesso, affinché si abbia equilibrio, ci varremo del seguente principio di *Lagrange*: affinché un sistema, i cui moti virtuali siano invertibili, sia in equilibrio è necessario e sufficiente che il lavoro meccanico fatto dalle forze in un moto virtuale qualunque, sia uguale a zero.

3.5 Se in un corpo elastico omogeneo, due sistemi di spostamenti fanno rispettivamente equilibrio a due sistemi di forze, la somma dei prodotti delle componenti delle forze del primo sistema per le corrispondenti componenti degli spostamenti degli stessi punti nel secondo sistema è uguale alla somma dei prodotti delle componenti delle forze del secondo sistema per le componenti degli spostamenti nei medesimi punti del primo.

3.6 In questa memoria dimostro un teorema che, nella teorica delle forze elastiche dei corpi solidi, tiene il luogo che il teorema di Green ha nella teorica delle forze che agiscono secondo la legge di Newton, e quanto alle applicazioni mi limito a dedurne formule analoghe a quella di Green per le funzioni potenziali.

3.7 Il corso universitario, io l'ho compiuto (parte per leggerezza, parte per quell'indolenza che accompagna ordinariamente il malanimo cagionato dalle frequenti avversità casalinghe) seguendo il malvezzo di studiare quel tanto che basti per passare gli esami. Perdetti poi due anni in occupazioni affatto aliene dalle mie tendenze. Dopo questa dura prova, formai decisamente il proposito di rifarmi a studiare la matematica, e (questa è la sola cosa di cui sinceramente mi lodo) tolsi a studiare con tutta diligenza una dopo l'altra l'aritmetica, l'algebra, la geometria, la trigonometria, l'algebra superiore e il calcolo, come avrebbe fatto uno che avesse percorso tutt'altra Facoltà, che la matematica [...]. Ecco la mia suppellettile scientifica: sento che è molto scarsa. Sopratutto mi sta assai sul cuore d'essere *tamquam tabula rasa* delle dottrine spettanti al calcolo delle variazioni, ai lavori di Jacobi e di Abel, alle ricerche di Gauss sulle superficie, ecc.

3.8 Io sarei determinato di rifiutare l'offerta fattami dal Betti, per più ragioni. Prima di tutto per la necessità di mutare l'indirizzo dei miei studi, il che porta sempre con sè degli inconvenienti e dei perditempi, tanto più che parlandomi il Betti di studi preparatori da farsi in un Osservatorio, pare che le materie da trattarsi nella nuova cattedra non debbano essere puramente teoriche. In secondo luogo la cattedra di introduzione al calcolo mi piace di più e per la natura dell'argomento che ne forma l'oggetto, e per la maggior latitudine che lascia nella scelta degli studi. Finalmente mi spiacerebbe occupare un posto che l'opinione pubblica amerebbe meglio probabilmente affidare a un distinto cultore di studi affini, voglio dire al Codazzi; e che, anche prescindendo da ciò, potrebbe essere ambito da professori più provetti di me e già benemeriti dell'insegnamento. Quanto al vantaggio pecuniario che potrei avere dalla nomina a professore ordinario, esso non è che momentaneo, in quanto che io ho a sperare lo stesso risultato dopo un tirocinio più o meno lungo anche e nel posto che occupo adesso, e senza abbandonare l'università in cui ti ho a collega. Comunque sia, non ho voluto rispondere al Betti prima d'aver chiesto il tuo consiglio, che ti prego volermi far conoscere liberissimamente.

3.9 Siano q_1, q_2, q_3 le coordinate curvilinee d'un punto qualunque in uno spazio a tre dimensioni e sia:

$$ds^2 = Q_1^2 dq_1^2 + Q_2^2 dq_2^2 + Q_3^2 dq_3^2$$

l'espressione del quadrato d'un elemento lineare qualunque, in questo spazio. [...] dunque, ponendo

$$\delta\theta_1 = \frac{\partial \delta q_1}{\partial q_1} + \frac{\delta Q_1}{Q_1}, \qquad \delta\omega_1 = \frac{Q_2}{Q_3}\frac{\partial \delta q_2}{\partial q_3} + \frac{Q_3}{Q_2}\frac{\partial \delta q_3}{q_2},$$
$$\delta\theta_2 = \frac{\partial \delta q_2}{\partial q_2} + \frac{\delta Q_2}{Q_2}, \qquad \delta\omega_2 = \frac{Q_3}{Q_1}\frac{\partial \delta q_3}{\partial q_1} + \frac{Q_1}{Q_3}\frac{\partial \delta q_1}{q_3},$$
$$\delta\theta_3 = \frac{\partial \delta q_3}{\partial q_3} + \frac{\delta Q_3}{Q_3}, \qquad \delta\omega_3 = \frac{Q_1}{Q_2}\frac{\partial \delta q_1}{\partial q_2} + \frac{Q_2}{Q_1}\frac{\partial \delta q_2}{q_1}$$

si può scrivere

$$\frac{\delta ds}{ds} = \lambda_1^2 \delta\theta_1 + \lambda_2^2 \delta\theta_2 + \lambda_3^2 \delta\theta_3 + \lambda_2\lambda_3 \delta\omega_1 + \lambda_3\lambda_1 \delta\omega_2 + \lambda_1\lambda_2 \delta\omega_3$$

dove le tre quantità $\lambda_1, \lambda_2, \lambda_3$, definite da

$$\lambda_i = \frac{Q_i dQ_i}{ds}$$

sono i coseni direttori degli angoli che l'elemento lineare ds fa con le tre coordinate q_1, q_2, q_3 [...].

3.10 Ma quello che più importa di osservare, e che risulta all'evidenza dal processo qui tenuto per stabilire quelle equazioni, è che lo spazio al quale esse si riferiscono non è definito da altro che dall'espressione (1) dell'elemento lineare, senz'alcuna condizione per le funzioni Q_1, Q_2, Q_3. Quindi le equazioni (4), (4_a) posseggono una molto maggiore generalità che non le analoghe in coordinate cartesiane e, in particolare giova subito notare che esse sono indipendenti dal postulato d'Euclide.

3.11 Si ottiene così una deformazione, priva tanto di rotazione quanto di dilatazione, nella quale la forza e lo spostamento hanno in ogni punto la stessa (o la opposta) direzione e le grandezze costantemente proporzionali. Tale risultato, che non ha riscontro nello spazio euclideo, presenta una singolare analogia con certi concetti moderni sull'azione dei mezzi dielettrici.

3.12 Delle sei equazioni di condizione per le quantità $\alpha, \beta, \gamma, \lambda, \mu, \nu$ che sono citate nel §I, si dimostra ordinariamente la necessità, non già la sufficienza. Stimo perciò opportuno, stante l'importanza di queste equazioni rispetto allo scopo del presente lavoro, di aggiungere una deduzione delle medesime, la quale stabilisca chiaramente la proprietà loro di costituire le condizioni non solo necessarie, ma eziandio sufficienti, per l'esistenza delle tre componenti di spostamento u, v, w.

Si rammenti, dalla teoria generale delle deformazioni d'un mezzo continuo che, insieme colle citate componenti α, β, γ, λ, μ, ν intervengono altresì, con ufficio non meno essenziale, le tre quantità p, q, r definite dalle equazioni:

$$\frac{\partial w}{\partial y} - \frac{\partial v}{\partial z} = 2p, \qquad \frac{\partial u}{\partial z} - \frac{\partial w}{\partial z} = 2q, \qquad \frac{\partial v}{\partial x} - \frac{\partial u}{\partial y} = 2r \tag{a}$$

e rappresentanti le *componenti di rotazione* della particella circostante al punto (x, y, z). Ora dal sistema delle nove equazioni che si ottengono combinando le sei equazioni (2) del §I, colle precedenti tre (a), si possono ricavare i valori di tutte le derivate prime delle tre componenti di spostamento u, v, w, e questi valori sono i seguenti:

$$\begin{aligned}
&\frac{\partial u}{\partial x} = \alpha, && \frac{\partial u}{\partial y} = \frac{\nu}{2} - r, && \frac{\partial u}{\partial z} = \frac{\mu}{2} + q,\\
&\frac{\partial v}{\partial x} = \frac{\nu}{2} + r, && \frac{\partial v}{\partial y} = \beta, && \frac{\partial v}{\partial z} = \frac{\lambda}{2} - p,\\
&\frac{\partial w}{\partial x} = \frac{\mu}{2} - q, && \frac{\partial w}{\partial y} = \frac{\lambda}{2} + p, && \frac{\partial w}{\partial z} = \gamma.
\end{aligned} \tag{b}$$

Consideriamo le prime tre di queste equazioni, che forniscono i valori delle derivate prime della funzione u. Affinché, supposte date le quantità che entrano nei loro secondi membri, esista una funzione u soddisfacente a queste tre equazioni, è necessario e sufficiente che sieno soddisfatte tre note relazioni, le quali possono essere scritte come segue:

$$-\frac{\partial q}{\partial y} - \frac{\partial r}{\partial z} = \frac{1}{2}\left(\frac{\partial \mu}{\partial y} - \frac{\partial \nu}{\partial z}\right), \quad \frac{\partial q}{\partial x} = \frac{\partial \alpha}{\partial z} - \frac{1}{2}\frac{\partial \mu}{\partial x}, \quad \frac{\partial r}{\partial x} = \frac{1}{2}\frac{\partial \nu}{\partial x} - \frac{\partial \alpha}{\partial y}.$$

Da queste si deducono, colla permutazione ciclica, le due terne analoghe di condizioni necessarie e sufficienti per l'esistenza delle altre due funzioni v e w. Ma, eseguendo dapprima questa permutazione sulla sola prima delle tre precedenti condizioni, e sommando poscia membro a membro le tre equazioni così ottenute, si trova[3]

$$\frac{\partial p}{\partial x} + \frac{\partial q}{\partial y} + \frac{\partial r}{\partial z} = 0$$

talché la prima delle dianzi trovate tre condizioni può scriversi più semplicemente così:

---

[3] Original note by Beltrami: Questa relazione notissima risulta già dalle formale di definizione (a): ma, per lo scopo attuale, era necessario far constare che essa è inclusa nelle nove condizioni d'integrabilità di cui qui è parola.

$$\frac{\partial p}{\partial x} = \frac{1}{2}\left(\frac{\partial \mu}{\partial x} - \frac{\partial \nu}{\partial z}\right)$$

Per tal modo si ottiene il seguente sistema di relazioni differenziali fra le nove funzioni $\alpha, \beta, \gamma, \lambda, \mu, \nu, p, q, r$:

$$\begin{array}{lll} \frac{\partial p}{\partial x} = \frac{1}{2}\left(\frac{\partial \mu}{\partial x} - \frac{\partial \nu}{\partial z}\right) & \frac{\partial p}{\partial y} = \frac{1}{2}\frac{\partial \lambda}{\partial y} - \frac{\partial \beta}{\partial z} & \frac{\partial p}{\partial z} = \frac{\partial \gamma}{\partial y} - \frac{1}{2}\frac{\partial \lambda}{\partial z} \\ \frac{\partial q}{\partial x} = \frac{\partial \alpha}{\partial z} - \frac{1}{2}\frac{\partial \mu}{\partial x} & \frac{\partial q}{\partial y} = \frac{1}{2}\left(\frac{\partial \nu}{\partial z} - \frac{\partial \lambda}{\partial x}\right) & \frac{\partial q}{\partial z} = \frac{1}{2}\frac{\partial \mu}{\partial z} - \frac{\partial \gamma}{\partial x} \\ \frac{\partial r}{\partial x} = \frac{1}{2}\frac{\partial \nu}{\partial x} - \frac{\partial \alpha}{\partial y} & \frac{\partial r}{\partial y} = \frac{\partial \beta}{\partial x} - \frac{1}{2}\frac{\partial \nu}{\partial y} & \frac{\partial r}{\partial z} = \frac{1}{2}\left(\frac{\partial \lambda}{\partial x} - \frac{\partial \mu}{\partial y}\right) \end{array} \qquad (c)$$

Questo sistema d'equazioni contiene le condizioni necessarie e sufficienti per l'esistenzadi tre funzioni u, v, w soddisfacenti alle nove condizioni (b), ossia alle sei equazioni(2) del §2 e alle tre equazioni (a) di questa Nota.
Ciò posto, consideriamo come date soltanto le *sei componenti di deformazione,* $\alpha, \beta, \gamma, \lambda, \mu, \nu$. Se esistono tre funzioni u, v, w soddisfacenti alle equazioni (2) del §I, esistono certamente anche le tre funzioni p, q, r definite dalle equazioni (a) di questa Nota. Poiché dunque le derivate di queste tre ultime funzioni sono legate alle $\alpha, \beta, \gamma, \lambda, \mu, \nu$ dalle nove equazioni (c), bisogna che sieno soddisfatte le condizioni di integrabilità che risultano da queste ultime nove equazioni e che si riducono alle sei seguenti:

$$\begin{array}{ll} \frac{\partial^2 \beta}{\partial z^2} + \frac{\partial^2 \gamma}{\partial y^2} = \frac{\partial^2 \nu}{\partial y \partial z} & \frac{\partial^2 \alpha}{\partial y \partial z} = \frac{1}{2}\frac{\partial}{\partial x}\left(\frac{\partial \mu}{\partial y} + \frac{\partial \alpha}{\partial z} - \frac{\partial \nu}{\partial x}\right) \\ \frac{\partial^2 \gamma}{\partial x^2} + \frac{\partial^2 \alpha}{\partial z^2} = \frac{\partial^2 \mu}{\partial x \partial z} & \frac{\partial^2 \beta}{\partial x \partial z} = \frac{1}{2}\frac{\partial}{\partial y}\left(\frac{\partial \nu}{\partial x} + \frac{\partial \lambda}{\partial z} - \frac{\partial \mu}{\partial y}\right) \\ \frac{\partial^2 \beta}{\partial x^2} + \frac{\partial^2 \alpha}{\partial y^2} = \frac{\partial^2 \lambda}{\partial x \partial y} & \frac{\partial^2 \gamma}{\partial y \partial x} = \frac{1}{2}\frac{\partial}{\partial z}\left(\frac{\partial \nu}{\partial x} + \frac{\partial \mu}{\partial y} - \frac{\partial \lambda}{\partial z}\right) \end{array} \qquad (d)$$

le quali sono appunto quelle citate nel §I. Quando queste condizioni sono soddisfatte, esistono indubbiamente tre funzioni p, q, r soddisfacenti alle nove equazioni (c); ma si è già veduto che se queste nove equazioni son soddisfatte da nove funzioni $\alpha, \beta, \gamma, \lambda, \mu, \nu, p, q, r$ esistono tre funzioni u, v, w soddisfacenti alle condizioni (2) del §I eq. (a) della presente Nota: dunque le sei condizioni (d), evidentemente necessarie per l'esistenza di tre funzioni u, v, w soddisfacenti alle sole equazioni (2) del §I, sono anche sufficienti.

3.13 È però utile osservare che la sufficienza delle equazioni in discorso può essere stabilita in un modo del quale non può immaginarsi il più perentorio, cioè coll'integrazione diretta, la quale riesce facilissimamente come segue.

3.14 Per trattare i problemi del genere di quello che porta il nome di SAINT VENANT, giova poter disporre arbitrariamente di alcune delle sei componenti di deformazione. Esaminando questo punto alquanto più da vicino ho potuto convincermi che si possono assumere a arbitrio tre delle quantità a, b, e, f, g, h, purché non sieno quelle che si trovano già associate fra loro in una delle tre equazioni di condizione formanti la prima delle due terne testè ricordate. Per conseguenza, delle 20 terne che si possono formare colle sei componenti suddette, sono 17 quelle per una delle quali si può, in un determinato problema, fissare a arbitrio la forma di tutte tre le funzioni che la compongono.

3.15 Le sei componenti di pressione p_{xx}, p_{yy}, ... sono necessariamente soggette a certe condizioni, quando corrispondono a forze interne generate per pura deformazione; giacché esse devono in tal caso potersi esprimere, in un modo del tutto determinato (e dipendente dalla natura del corpo), per mezzo dello spostamento.

3.16 Queste ultime condizioni suppongono l'assenza di ogni forza esterna. Tralascio per brevità, di riportare le condizioni analoghe per il caso in cui questa forza esista e abbia le componenti X, Y, Z.

3.17 Nella versione francese della *Teoria dell'elasticità* di CLEBSCH, riveduta e commentata dall'illustre De Saint Venant, il quale ha recato con tale pubblicazione un nuovo e segnalato servigio agli studiosi di quell'importantissima teoria, si trova riassunto, in una Nota finale al §31 (pp. 252–282), il metodo già da lungo tempo proposto dallo stesso De Saint Venant per la ricerca dei limiti di resistenza dei corpi elastici. Questo metodo differisce da quello generalmente seguito, e accettato anche da CLEBSCH, per il principio sul quale esso si fonda e che consiste nell'assegnare un limite massimo alle dilatazioni anziché alle tensioni.

Per giustificare questo nuovo principio, De Saint Venant cita in particolare il caso semplicissimo d'un parallelepipedo rettangolo stirato, con una stessa forza unitaria, secondo uno, o secondo due, o secondo tutte tre le direzioni dei suoi assi di figura; e osserva che, mentre la tensione massima è, per ipotesi, la stessa in tutti tre i casi, la dilatazione massima è maggiore nel primo che nel secondo, e è parimente maggiore nel secondo che nel terzo, donde sembra ovvio il concludere che il pericolo di disgregazione sia maggiore nel primo caso che nel secondo e nel terzo.

Ora tale conclusione non mi pare così, legittima come per avventura potrebbe credersi a prima giunta. Lo stiramento di un corpo nel senso che diremo longitudinale è accompagnato, come è notissimo, da una contrazione in ogni senso trasversale, contrazione che è parzialmente impedita, o anche mutata in dilatazione, quando il corpo è sottoposto contemporaneamente a stiramenti trasversali; ne segue che la coesione molecolare è indebolita, nel senso longitudinale, più nel primo caso che nel secondo, ma è anche rinforzata, nel senso trasversale, più in quello che in questo, cosicché non è facile, né forse possibile, decidere a priori circa la prevalenza dell'un effetto sull'altro. Ma se non si può formulare alcuna precisa conclusione intorno a ciò, parmi tuttavia potersi ammettere come evidente, in base appunto all'esempio molto opportunamente

addotto da De Saint Venant, che la vera misura del cimento a cui è messa la coesione di un corpo elastico non debba essere desunta né dalla sola tensione massima, né dalla sola dilatazione massima, ma debba risultare, in un qualche modo, dall'insieme di tutte le tensioni, o di tutte le dilatazioni che regnano nell'intorno di ogni punto del corpo.
Ora queste tensioni e queste dilatazioni, rappresentate le une le altre da sei componenti distinte sono tra loro legate da relazioni lineari, le quali esprimonsi che le sei componenti di tensione sono le derivate rispetto alle sei componenti di deformazione, di un'unica funzione quadratica formata con queste seconde componenti; oppure che le sei componenti di deformazione sono le derivate rispetto alle sei componenti di tensione un'analoga funzione formata con queste ultime componenti. Quest'unica funzione che ha l'identico valore sotto le due diverse forme che essa prende nell'uno e nell'altro caso è il cosiddetto potenziale di elasticità e ha l'insigne proprietà di rappresentare l'energia riferita all'unità di volume che il corpo elastico possiede nell'intorno del punto che si considera, energia la quale è equivalente sia al lavoro che l'unità di volume del corpo può svolgere nel restituirsi dallo stato attuale allo stato naturale, sia al lavoro che hanno dovuto svolgere le forze esterne per condurre la detta unità di volume dallo stato naturale all'attuale suo stato di coazione elastica.
Dietro ciò mi pare evidente che la vera misura del cimento a cui è messa, in ogni punto del corpo, la coesione molecolare debba essere data dal valore che assume in quel punto il potenziale unitario d'elasticità e che a questo valore, anziché a quello di una tensione o di una dilatazione, si debba prescrivere un limite massimo, per preservare il corpo dal pericolo di disgregazione, limite naturalmente diverso, come nelle ordinarie teorie, secondo che si tratti di disgregazione prossima o di remota.
Questa conclusione, giustificata già di per se stessa dal significato dinamico del potenziale d'elasticità, è resa ancor più manifestamente plausibile da una proprietà analitica di questo potenziale, la quale deve certamente dipendere anch'essa dal suddetto significato, benché non ci sia ancora nota la dimostrazione rigorosa di tale dipendenza.
Voglio alludere alla proprietà che ha il detto potenziale d'essere una funzione quadratica essenzialmente positiva cioè una funzione che non si annulla se non quando tutte le sue sei variabili sieno nulle, e che si mantiene maggiore di zero per ogni altra sestupla di valori reali di queste variabili. In virtù di questa proprietà non si può imporre un limite al valore del potenziale d'elasticità senza imporre al tempo stesso un limite a quello di ciascuna componente, sia di tensione, sia di deformazione, cosicché l'uso del detto potenziale come misura della resistenza elastica non contraddice intrinsecamente ai criteri desunti sia dalla considerazione delle sole tensioni, sia da quella delle sole deformazioni. Praticamente poi il criterio desunto dal potenziale ha il grande vantaggio di non esigere la risoluzione preliminare di alcuna equazione e di ridursi alla discussione d'una formola che non può mai presentare ambiguità di segni.
[...] P.S. Dopo avere scritto quanto precede ho riconosciuto con piacere che le obbiezioni da me sollevate contro i modi fin qui usati di stabilire le condizioni

di coesione erano state formulate, quasi negli stessi termini, dal compianto ing. Castigliano, alle p. 128 e sg. della Théorie de l'équilibre des systèmes élastiques. Mi è grato il pensare che il dotto ingegnere: il quale aveva riconosciuto tutta l'importanza del concetto di potenziale elastico, avrebbe probabilmente approvata la mia proposta di fondare sovr'esso anche la deduzione delle condizioni anzidette.

3.18 Le variazioni $\delta x, \delta y, \delta z$ sono funzioni monodrome delle variabili u, v. Per l'inestensibilità della superficie, queste variazioni devono soddisfare alle tre condizioni:

$$\delta E = 0, \qquad \delta F = 0, \qquad \delta G = 0, \; (2)$$

dove

$$\begin{aligned} \frac{1}{2}\delta E &= \sum \frac{\partial x}{\partial u}\frac{\partial \delta x}{\partial u}, \\ \delta F &= \sum \left(\frac{\partial x}{\partial u}\frac{\partial \delta x}{\partial v} + \frac{\partial x}{\partial v}\frac{\partial \delta x}{\partial u}\right), \quad (2_a) \\ \frac{1}{2}\delta G &= \sum \frac{\partial x}{\partial v}\frac{\partial \delta x}{\partial v}. \end{aligned}$$

In virtù del principio di LAGRANGE l'equazione generale dell'equilibrio è dunque la seguente:

$$\int (X\delta x + Y\delta y + Z\delta z)\, d\sigma + \int (X_s\, \delta x + Y_s\, \delta y + Z_s\, \delta z)\, ds + \frac{1}{2}\int (\lambda\delta E + 2\mu\delta F + \nu\delta G)\, \frac{d\sigma}{H} = 0$$

dove λ, μ, ν sono tre moltiplicatori, funzioni di u e di v (il divisore $2H$ è stato introdotto, nell'ultimo integrale, per comodo dei calcoli successivi).

3.19 Un pezzo qualunque di superficie flessibile e inestensibile è mantenuto in equilibrio da una forza normale dovunque alla superficie stessa e proporzionale alla curvatura media locale. La tensione costante del contorno si trasmette equabilmente in ogni punto della superficie.

3.20 Un pezzo qualunque di superficie flessibile e inestensibile è mantenuto in equilibrio da una tensione costante e normale lungo il contorno e da una forza normale dovunque alla superficie stessa e proporzionale alla misura di curvatura locale, e da una tensione lungo il contorno, diretta secondo la tangente coniugata al contorno stesso e avente la componente normale proporzionale alla curvatura del contorno. Le linee normali sono le linee di curvatura della superficie, quelle di tensione tangenziale sono le linee asintotiche della superficie stessa.

3.21 I concetti e i metodi fondamentali di Green e di Gauss avevano aperto la via maestra per la integrazione generale della equazione di Laplace, base della teoria del potenziale; scopo del Betti fu di trasportare gli stessi metodi, prima nel campo della scienza dell'equilibrio elastico, poi in quella del calore. Coi lavori di Betti [...] si inaugura una nuova e lunga serie di ricerche schiettamente italiane sulla integrazione delle equazioni dell'elasticità, tanto che può dirsi che, se Galileo per il primo adombrò i problemi dell'equilibrio dei corpi elastici, fu merito dei geometri italiani, a più di due secoli di distanza, di aver largamente contribuito a svolgere la teoria generale di quelle equazioni nelle quali Navier aveva rappresentato e, per dir così, racchiuso tutto il meccanismo del fenomeno.

3.22 La Memoria, veramente mirabile, del Betti sulle equazioni della elasticità gettò su queste un fascio di luce nuova, inattesa, e preparò, specialmente in Italia, una fioritura di lavori quale poche altre memorie possono vantarsi di aver prodotto. Il suo teorema di reciprocità dovette sembrare una rivelazione. Con mezzi semplicissimi dava già una folla di risultati e permetteva di penetrare addentro nelle proprietà analitiche delle equazioni di cui si tratta.

3.23 Ill.mo Signor Rettore
della R. università di Roma
Sono note le mie idee politiche per quanto esse risultino esclusivamente dalla mia condotta nell'ambito parlamentare, la quale è tuttavia insindacabile in forza dell'articolo 51 dello Statuto fondamentale del Regno. La S.V. comprenderà quindi come io non possa in coscienza aderire all'invito da lei rivoltomi con lettera 18 corrente relativo al giuramento dei professori.
Con osservanza della S.V.
Vito Volterra

3.24 Il peut exister des cas dans lesquels un corps élastique tout en n'étant sujet à aucune action extérieure, c'est-à-dire sans être sujet ni aux forces extérieures agissant sur se points intérieurs, ni aux forces extérieures agissant sur sa surface, peut cependant ne pas se trouver à l'état naturel, mais être dans un état de tension qui varie d'une façon continue et régulière d'un point à l'autre.

3.25 Un corps élastique qui occupe un espace simplement connexe et dont la déformation est régulière peut tojours être amené à son état naturel à l'aide de déplacements finis, continus et monodromes de se points.
Au contraire nous pouvons dire:
Si un corps élastique occupe un espace multiplement connexe et si sa déformation est régulière, les déplacements des ses points ne sont pas nécessairement monodromes.

3.26 *Les distorsions*

1. Dans le Chapitre précédent j'ai montré que les corps élastiques occupant des espaces plusieurs fois connexes peuvent se trouver dans des états d'équilibre bien différents de ceux qu'on a quand les corps élastiques occupent des espaces

simplement connexes. Dans ces nouveaux états d'équilibre on a une déformation intérieure régulière du corps, sans toutefois que celui-ci soit sollicité par des forces extérieures.
Imaginons qu'on mène les coupures qui rendent simplement connexe l'espace occupé par le corps. A chacune d'elles correspondent six constantes que nous avons appelées les constantes de la coupure. Il est facile d'établir la signification mécanique de ces constantes au moyen des formules (III) du Chapitre précédent.
En effet, exécutons matériellement les coupures suivant les dites sections et laissons le corps reprendre son état naturel. Si, en reprenant cet état, certaines parties du corps viennent à se superposer entre elles, supprimons les parties excédents. Alors les formules (III) déjà rappelées nous montrent que les parcelles placées des deux côtés d'une même section et qui, avant la coupure, étaient en contact subissent, par le fait même de la coupure, un déplacement résultant d'une translation et d'une rotation égales pour tous les couples de parcelles adjacentes à une même section.
En prenant l'origine pour centre de réduction, les trois composantes de la translation et les trois composantes de la rotation, suivant les axes coordonnée, sont les six caractéristiques de la coupure.
Réciproquement, si le corps élastique multiplement connexe est pris à l'état naturel, on pourra, pour l'amener à l'état de tension, exécuter l'opération inverse, c'est-à-dire le sectionner afin de le rendre simplement connexe, déplacer ensuite les deux parties de chaque coupure, l'une par rapport à l'autre, de manière que les déplacements relatifs des différents couples de parcelles (qui adhéraient entre elles et que la coupure a séparées) soient résultantes des translations et des rotations égales; rétablir enfin la connexion et la continuité suivant chaque coupure, en retranchant ou en ajoutant la matière nécessaire et en ressoudant les parties entre elles. L'ensemble de ces opérations relatives à chaque coupure peut s'appeler une *distorsion* du corps et les six constantes de chaque coupure peuvent s'appeler les *caractéristiques de la distorsion*. Dans un corps élastique multiplement connexe, dont la déformation est régulière et qui a subi un certain nombre de distorsions, l'inspection de la déformation ne peut en aucun manière révéler les endroits où les coupures et les distorsions qui s'ensuivent se sont produites, et cela en vertu de la régularité elle-même. On peut dire en outre que les six caractéristiques de chaque distorsion ne sont pas des éléments dépendant du lieu où la coupure a été exécutée. En effet, le même procédé qui nous a servi à établir les formules (III) prouve que, si l'on prend dans le corps deux coupures qu'on peut transformer l'une dans l'autre par une déformation continue, les constantes relatives à l'une des coupures sont égales aux constantes relatives à l'autre: Il s'ensuit que les caractéristiques d'une distorsion ne sont pas des éléments spécifiques de chaque coupure, mais qu'elles dépendent exclusivement de la nature géométrique de l'espace occupé par le corps et de la déformation régulière à laquelle il est assujetti. Le nombre des distorsions indépendantes auxquels un corps élastique peut être soumis est évidemment égal à l'ordre de la connexion de l'espace occupé par le corps

moins 1. En conformité de ce que nous avons trouvé, deux coupures qu'on peut par une déformation continue transformer l'une dans l'autre s'appellent *équivalentes*. Nous dirons aussi qu'une distorsion est connue quand les caractéristiques et la coupure relative ou une autre coupure équivalente seront données.

2. Cela posé, deux questions se présentent naturellement, à savoir:

1° A des distorsions arbitrairement choisies correspondra-t-il toujours un état d'équilibre et une déformation régulière du corps si l'on suppose nulles les actions extérieures?

2° Les distorsions étant connues, quel est cet état de déformation? Pour relier ces problèmes à d'autres déjà connus nous démontrerons le théorème suivant: *Si dans chaque corps élastique isotrope plusieurs fois connexe on prend un ensemble arbitraire de distorsions, on pourra calculer un nombre infini de déformations régulières du corps qui correspondent à ces distorsions et qui sont équilibrées par des forces extérieures superficielles (que nous indiquons avec T) ayant la résultante nulle et le moment nul par rapport à un axe quelconque.*

Dés lors, pour reconnaitre si dans un corps isotrope les distorsions données correspondent à un état d'équilibre, les forces extérieures étant nulles, il suffira de voir si les forces extérieures T changées de signe et appliquées au contour du corps, quand celui-ci n'est sujet à aucune distorsion, déterminent un état de déformation régulière équilibrant les forces elles-mêmes. Si l'on peut calculer effectivement cet état de déformation, le problème concernant l'équilibre du corps soumis aux distorsions données sera résolu.

En effet, appelons Γ la déformation relative aux distorsions données et aux forces extérieures T trouvées, qui agissent sur la surface, et Γ' la déformation déterminée par ces forces extérieures changées de signe quand le corps ne subit aucune distorsion. La déformation Γ'' qui résulte de Γ et Γ' correspondra aux distorsions données et aux forces extérieures nulles. Les questions sont ainsi ramenées à voir si la déformation Γ' existe et à la trouver. Elles se réduisent donc à des problèmes d'élasticité où les distorsions ne paraissent pas, c'est-à-dire à des problèmes ordinaires d'élasticité.

Mais les forces extérieures T, agissant sur la surface, en vertu du théorème énoncé sont telles que si le corps est rigide elles s'équilibrent; il s'ensuit qu'elles satisfont aux conditions fondamentales nécessaires pour l'existence de la déformation Γ'.

Or tout dernièrement on a beaucoup avancé par des méthodes nouvelles dans l'étude du théorème d'existence pour les questions d'élasticité, c'est pourquoi on peut dire que, sauf certaines conditions relatives à la forme géométrique de l'espace occupé par le corps élastique (conditions que nous ne préciserons pas ici), Γ' et Γ'' existeront toujours.

Ces réserves faites, on pourra donc répondre affirmativement à la première question dans le cas des corps isotropes. La seconde question posée est relative au cas où le corps n'est pas sujet aux actions extérieures; mais elle peut se généraliser et l'on peut supposer les distorsions données et le corps sollicité par des forces extérieures déterminées. Alors, si le corps est isotrope, il suffit pour la résolution du problème de superposer à la déformation Γ déterminée par les distorsions et par les forces extérieures T, la déformation déterminée par les forces extérieures données et par les forces extérieures −T qui agissent sur la surface dans l'hypothèse que les distorsions manquent.

3.27 Per certi sistemi di equazioni a derivate parziali di 2° ordine, che chiamo simmetrici, e che comprendono, come caso particolare, le equazioni della elasticità.

3.28 Le soluzioni semplici ora definite possono essere considerate come rappresentanti ciascuna una deformazione speciale del corpo; le *L* poi rappresentano le componenti della pressione prodotta da tale deformazione sopra gli elementi superficiali. Perciò le [leggi delle soluzioni semplici] [...] esprimono che le forze superficiali sono dirette come gli spostamenti e a essi proporzionali.

3.29 Mi propongo di dimostrare come per le funzioni che rappresentano gli integrali delle equazioni dell'elasticità, nel caso della isotropia e dell'equilibrio, si possa stabilire una teoria analoga sotto molti rapporti alla teoria delle funzioni potenziali, e che ne costituisce in certo modo una estensione.

3.30 Ma è ovvio pensare che, oltre questi casi, altri ne esistono, offertici dall'osservazione. Noi possiamo immaginare che i lembi del taglio, oltre che spostati rigidamente fra loro, subiscano delle leggere deformazioni, come quando in un taglio prodotto in un corpo anulare incastriamo una sottile lente a faccie curve. E inoltre noi possiamo immaginare incastrati, o estratti, in corpi semplicemente connessi, come una sfera o un ellissoide, dei sottili corpi lentiformi e avere così in essi dello tensioni elastiche prodotte parimenti in assenza di forze esterne.

Ora è lecito domandare: questi fatti di così ovvia osservazione, già considerati anche dal Weingarten, non sono suscettibili di uno studio che abbia qualche analogia con quello delle distorsioni del Volterra? oppure la teoria della elasticità nella sua forma attuale non è ancora in grado di attaccare questi problemi? Per rispondere a tali domande è necessario prendere in esame e discutere i punti di partenza della teoria.

L'ipotesi fondamentale che sta a base delle considerazioni del Weingarten, è che le sei componenti delle tensioni interne variino con continuità da punto a punto, dopo ristabilita nel corpo la continuità, materiale, così che esso possa considerarsi nelle stesse condizioni statiche di un corpo compatto. Da ciò segue che anche le sei caratteristiche della deformazione (che sono funzioni lineari indipendenti delle tensioni) debbono godere delle stesse proprietà di continuità. Le ipotesi del Volterra sono più restrittive. Egli ammette:

1° la continuità delle caratteristiche della deformazione (da cui segue la continuità delle tensioni);

2° la continuità delle loro derivate prime e seconde.

Ora, per l'estensione che noi abbiamo di mira, nulla vieta di lasciare da parte questa seconda ipotesi, per la quale non è evidente una assoluta necessità meccanica. Ritorneremo così alle ipotesi del Weingarten, e potremo proporci di cercare se esistono deformazioni che soddisfacciano a queste ipotesi e non a quelle di Volterra. Per semplicità e chiarezza di linguaggio chiamerò distorsioni di Volterra le deformazioni che soddisfanno alle precedenti condizioni 1 e 2; distorsioni di Weingarten quelle che soddisfanno solamente alla 1. Non mi propongo qui di risolvere in modo generale la quistione enunciata, ma mostrerò con un esempio, che è però di una notevole generalità, che esistono distorsioni di Weingarten che non sono distorsioni di Volterra.

3.31 [è un termine aggiuntivo] che si può considerare come l'espressione dell'energia delle reazioni che lo spazio, rigido nella propria costituzione geometrica, oppone alla materia elastica che lo riempie, supponendo questa *inerte* nel senso che, obbligata a deformarsi nel detto spazio, essa tende a farlo come se lo spazio stesso fosse euclideo. L'ulteriore svolgimento della teoria dei mezzi elastici negli spazi curvi permetterà forse di rispondere alla domanda di Clifford: *se non potrebbe darsi che noi consideriamo come variazioni fisiche certi effetti realmente dovuti a cambiamenti della curvatura del nostro spazio; in altre parole, se alcune delle cause, che noi chiamiamo fisiche, e forse tutte, non fossero per avventura dovute alla costituzione geometrica dello spazio nel quale viviamo.*

A.4 Quotations of Chap. 4

4.1 Ora la Legge sulla pubblica istruzione del 13 novembre 1859 che creò la Scuola degli Ingegneri di Torino, non è priva di ambiguità su questo punto. I successivi Regolamenti parlano invece chiaramente del conferimento del diploma di *Ingegnere Laureato.*
Tale denominazione si trova nel Decreto Reale 11 ottobre 1863, nonché nei Regolamenti 17 ottobre 1860 e 11 ottobre 1866. Ma, per ciò che li riguarda, potrebbesi ancora sospettare che il titolo di Laurea debbasi ascrivere agli studi precedenti di Matematica, allora considerati come completi, sebbene limitati a un biennio.
Invece il Regolamento 14 novembre 1867 si esprime in modo che non ammette più dubbi dichiarando testualmente:
"La Scuola conferisce diplomi di Laurea di Ingegneri civili, meccanici, agricoli, metallurgici, chimici e architetti civili". Tale netta designazione manca nuovamente nei Regolamenti successivi, i quali talvolta evitano persino la parola diploma, evidentemente perché la omissione del titolo di Laurea non sia avvertita. Invece l'attuale, si direbbe con ostentazione, ripete a sazietà che il titolo conferito agli Ingegneri è un semplice diploma. Si volle insomma muovere un primo passo decisivo in quell'indirizzo che la Relazione della Commissione Reale per la riforma degli studi superiori tende a stabilire in modo generale attribuendo alle vecchie Facoltà universitarie la esclusività dell'insegnamento scientifico e per conseguenza il diritto al conferimento della dignità dottorale, e limitando l'ufficio di tutti gli studi di applicazione, (presa questa parola nel

senso più largo di abilitazione all'esercizio professionale) al conferimento di un qualsiasi diploma soltanto.
Strano indirizzo, oggi, quando la tecnica dell'Ingegnere, sciogliendosi dai limiti circoscritti del passato, si è svolta con una ampiezza impreveduta di studi fra le applicazioni più disparate, diventando non una sola ma un complesso di scienze, le quali rinnovarono per intero parecchi dei suoi vecchi Capitoli: quando questa tecnica immedesimandosi con la vita sociale in tutte le sue manifestazioni è diventata il primo suo fattore si è imposta in tutte le Amministrazioni pubbliche, è assurta al grado di mezzo insostituibile per il conseguimento di qualsiasi progresso.
Strano indirizzo, o non piuttosto giustificato, appunto da questo primato inatteso, che minaccia quelli stabiliti su vecchie tradizioni?

4.2 Nel 1866 fu fondata a Torino la Società degli Ingegneri e degli Industriali, sotto la presidenza di Pietro Paleocapa. Due anni dopo, a Milano, fu ricostituito l'antico Collegio degli Ingegneri e Architetti, che era stato fondato nel 1606 e soppresso nel 1797 in base alle disposizioni della Costituzione Cisalpina; primo presidente ne fu Luigi Tatti. Nel 1868, sotto l'impulso di Prospero Richelmy a Torino e Francesco Brioschi a Milano, direttori l'uno della Scuola di Applicazione, l'altro dell'istituto Tecnico Superiore, entrambe le Associazioni iniziarono la pubblicazione di Atti. L'esempio di Torino e Milano fu poi seguito negli anni settanta in altre città italiane fra cui Roma ove nel 1871, acquisita la libertà di associazione, si costituì un primo Circolo Tecnico, successivamente trasformato in Collegio degli Ingegneri e Architetti (1876).
Fra i periodici più autorevoli del tempo vi erano, quali organi di informazione sui lavori pubblici, il *Giornale del Genio Civile*, il *Monitore delle Strade Ferrate* (Torino, 1868) e il *Giornale dei lavori Pubblici e delle Strade Ferrate* (Roma, 1874); quali riviste tecnico-scientifiche *Il Politecnico* di Milano, che aveva ripreso le pubblicazioni nel 1866 con la direzione di Brioschi e *L'ingegneria Civile e le Arti Industriali* fondata a Torino nel 1875 da Giovanni Saccheri, già professore di disegno nella Scuola di applicazione. A questi due ultimi periodici collaborò in particolare Castigliano, con ripetuti interventi a partire dal 1876.

4.3 Quando nascono a Torino la Scuola di applicazione per gli ingegneri nel 1859, e il Regio Museo industriale nel 1862, il Piemonte sta attraversando un periodo di evoluzione verso profonde trasformazioni sociali ed economiche.
I problemi legati all'unità, finalmente raggiunta, hanno particolare effetto in Piemonte, soprattutto su Torino: da un lato, si avvia la trasformazione della città da capitale di un piccolo stato regionale ad importante centro urbano di una grande Nazione; dall'altro, si assiste al passaggio da una economia principalmente di guerra—anche se in uno stato relativamente all'avanguardia nel panorama europeo—ad un'altra di normalità, ma nell'ambito di una sopravvenuta situazione nazionale, caratterizzata da vaste aree di sottosviluppo e con problemi di integrazione sovraregionale.

L'ingegnere torinese nasce, quindi, in un clima di grande evoluzione, anche culturale.Nella prima metà del'Ottocento, Torino non si può ancora definire compiutamente una città industriale, ma essa è già un importante polo di innovazione tecnica e tecnologica.
Le attività produttive, a parte i Mulini della città e le officine dell'Arsenale, si svolgono comunque in un quadro sostanzialmente artigianale, anche se i primi segni tangibili di un cambiamento si scorgono nelle Esposizioni Nazionali di Arti e Meştieri che, sin dal 1827, trovano la loro sede istituzionale nel cortile del Castello del Valentino.
Le iniziative tecniche e produttive appena richiamate si inseriscono nello sforzo di internazionalizzazione (rivolto ovviamente in primis verso il resto d'Italia) e di *europeizzazione* ante litteram che il piccolo regno sabaudo ha perseguito come politica nel corso degli anni '40 e '50 del secolo XIX.
[...]
In questo clima fortemente attento alla formazione tecnica e professionale nasce ben presto l'esigenza di formare un ingegnere capace di gestire l'innovazione e primo attore della nuova società industriale.
Il 13 novembre 1859 il Governo del Regno di Sardegna promulga la Legge "Sul riordinamento dell'Istruzione Pubblica", nota con il nome del suo estensore, Gabrio Casati. La Legge Casati crea un ordinamento efficace, il cui indirizzo strutturale rimarrà immutato fino alla Riforma Gentile e pone le premesse ideologiche e le scelte pedagogico-didattiche del nuovo stato italiano. La Legge Casati dà l'avvio agli studi tecnici, triennali, ponendo le basi alla nuova struttura scolastica italiana, a partire dall'istruzione superiore, sino a quella elementare. È inoltre stabilito il nuovo ordinamento degli studi di ingegneria, che divide in due stadi la carriera scolastica degli ingegneri, lasciando alle università un primo stadio teorico o di preparazione, creando nuove scuole per il secondo stadio di scienze applicate.

4.4 Kalendis decembribus a MDCCCLXXIII Victorio Emmanuele II rege ant scialoia studiis administrandis praef quae fuerant can lateran aedes mathematicis disciplinis et artibus in doctrinae lucem vocatae patuere.

4.5 Queste pagine—in cui ho raccolte le lezioni da me impartite quest'anno agli allievi del Politecnico di Torino—rispecchiano fedelmente la concezione didattica a cui io ispiro tutto il mio insegnamento; il quale si propone, deliberatamente, finalità di alta cultura, e, solo subordinatamente, di preparazione professionale.
La scelta degli argomenti è stata fatta con quest'unica preoccupazione: di offrire allo studioso l'occasione di conoscere i principii fondamentali, di approfondirne il significato e la portata, di vedere come si possa su di essi costruire un corpo razionale di dottrine, e come questo possa poi venire, di volta in volta, utilizzato per risolvere problemi concreti.
Gli argomenti che meglio si prestano a tale scopo sono stati sviluppati a fondo. Altri, per se stessi non meno importanti, ma sotto questo punto di vista meno suggestivi, sono stati in tutto o in parte trascurati.

Il lettore non troverà qui la solita raccolta di soluzioni fatte, da applicare—a proposito od a sproposito—a tutti i problemi che la pratica tecnica gli potrà presentare. Ma potrà imparare a analizzare e a risolvere ciascuno di quei problemi, rendendosi conto del valore delle ipotesi su cui la soluzione si fonda e del grado di approssimazione ch'essa comporta.

4.6 Le nombre des équations d'équilibre pour les n points sera $3n$; si p est celui des équations qui doivent subsister entre les forces extérieures, indépendamment des tensions, pour qu'il y ait équilibre, le nombre des équations qui contiennent effectivement les tensions se réduira à $3n - p$. Ainsi, lorsque m sera $>3n - p$, les équations précédentes ne suffiront pas pour déterminer toutes les tensions. Il en sera de même quand le système contiendra un certain nombre de points fixes. Cette indétermination signifie qu'il y a une infinité de valeurs des tensions qui, combinées avec les forces extérieures données sont aptes à tenir le système en équilibre. Les valeurs des tensions effectives dépendent de l'élasticité respective des liens, et lorsque celle-ci est déterminée, il doit en être de même des tensions.

4.7 Lorsqu'un système élastique se met en équilibre sous l'action de forces extérieures, le travail développé par l'effet des tensions ou des compressions des liens qui unissent les divers points du système est un minimum.

4.8 Puisque, dans le cas que nous considérons, les tensions peuvent varier sans que l'équilibre cesse d'exister, on devra admettre que ces variations s'effectuent indépendamment de tout travail des forces extérieures; elles sont toujours accompagnées d'allongements ou d'accourcissements dans les divers liens correspondants, ce qui donne lieu, dans chacun d'eux, à un développement de travail. Les variations de longueur des liens doivent être supposées très petites pour que les positions respectives des divers points du système ne soient pas sensiblement altérées. Mais, puisque pendant ce petit mouvement intérieur l'équilibre continue à exister et que le travail des forces extérieures est nul, il s'ensuit que le travail total élémentaire des tensions ainsi développé est également nul.

Pour exprimer cette conséquence, soient T la tension d'un lien quelconque, δl la variation élémentaire de la longueur de ce lien; le travail développé par suite de la variation de tension correspondante sera $T\delta l$, et par conséquent, pour l'ensemble du système, on aura

$$\sum T\delta l = 0 \tag{A.5}$$

Soit l l'extension on l'accourcissement qu'a primitivement éprouvé le lien sous l'action de la tension T, on a, indépendamment du signe,

$$T = \epsilon l \tag{A.6}$$

ou ϵ est un coefficient que j'appellerai coefficient d'élasticité, e qui est fonction du module d'élasticité, de la section et de la longueur du lien.
Le travail développé pour produire cette variation de longueur l sera égal à $1/2\epsilon l^2$, et par suite le travail totale du système sera égal a $1/2 \sum \epsilon l^2$.
Mais en vertu des équations (1) et (2) on a:

$$\sum T\delta l = \sum \epsilon l \delta l = \delta \frac{1}{2} \sum \epsilon l^2 = 0 \tag{A.7}$$

Ce qui est la démonstration du principe énoncé auquel on peut encore parvenir par d'autre considérations. Il est également possible de l'exprimer d'une autre manière, car on a

$$\sum T\delta l = \sum \frac{1}{\epsilon} T\delta T = \delta \frac{1}{2} \sum \frac{1}{\epsilon} T^2 \tag{A.8}$$

4.9 Data una di quelle disposizioni d'equilibrio, se si suppone che il sistema passi gradatamente a un'altra vicinissima, il complesso delle forze esterne (X, Y, Z) non dovrà cessare di essere in equilibrio per ognuna di queste disposizioni, indipendentemente dalle forze interne; e siccome questo stato di equilibrio non dipende soltanto dalle intensità e direzione rispettiva delle forze, ma anche dalle posizioni de' punti di applicazione, ne segue che ogni nodo deve mantenersi costantemente nella stessa posizione, malgrado le variazioni che possono succedere nelle tensioni de' legami che vi corrispondono.

4.10 Sembra che il mio scritto venisse generalmente accolto con favore dagli scienziati che più si erano occupati di quell'argomento, né fu da essi messa in dubbio la esattezza del metodo da me proposto, fuorché dal sig. Emilio SABBIA il quale, in un opuscolo intitolato: Errore del principio di elasticità formolato dal signor L. Federigo MENABREA, Cenno critico di Emilio SABBIA, Torino 1869, impugna, con particolare vivacità, la verità di quel principio [...].
Percorrendo lo scritto del sig. SABBIA credei di scorgere l'equivoco in cui egli era incorso; e non avrei tardato a rispondere alle sue critiche, se altre cure assai più gravi non mi avessero allora trattenuto. Restituito a maggiore libertà, io mi accingeva a tal lavoro, quando mi fu comunicato uno scritto del valente cultore delle scienze matematiche il sig. Comm. ADOLFO PARODI, Ispettore generale de' lavori marittimi, che ha precisamente per oggetto l'opuscolo del sig. SABBIA. Egli così nitidamente ribatte gli appunti del sig. SABBIA che non saprei come meglio difendere il mio teorema che valendomi delle considerazioni stesse svolte dall'insigne autore.
[...] Non sarà neppure discaro all'Accademia di avere sott'occhio due nuove dimostrazioni dell'equazione di elasticità date l'una dal signor BERTRAND e l'altra dal sig. YVON VILLARCEAU, ambidue Membri dell'Istituto di Francia, i quali nelle pregievoli lettere delle quali trasmetto gli estratti presentano la quistione sotto punti di vista che io direi nuovi, e che conducono ai medesimi risultati [...].

[...La mia dimostrazione] venne giudicata, come si rileverà da uno degli scritti qui uniti, rigorosa abbastanza, e che ha almeno il pregio della semplicità e della chiarezza.

4.11 Sebbene in coincidenza de' risultati ottenuti dalla applicazione del principio di elasticità, con quelli ricavati da altri metodi speciali e non contestati fosse nella mia seconda memoria confermata da moltiplici esempi, e dovesse indurre a ammettere che il principio e il metodo che ne derivava erano esatti, tuttavia l'uno e l'altro furono per parte di alcuni, oggetto di aspre e strane denegazioni, mentre parecchi fra i più eminenti matematici di nostra epoca accolsero il principio con maggiore benevolenza. Non ostante le opposizioni fatte, le applicazioni del principio di elasticità si sono propagate e hanno vieppiù confermato l' esattezza, la semplicità e la generalità del metodo che ne deriva. Siccome questo racchiude sostanzialmente in sé tutti gli altri, credo di fare cosa utile cercando di togliere, circa la esattezza del medesimo, ogni dubbio che possa tuttora rimanere nelle menti più scrupolose in fatto di rigore matematico.

4.12 Quando un sistema elastico, suscettibile di uno stato neutro generale, si trova in equilibrio con forze esteriori, tra i diversi modi, in cui le tensioni si potrebbero immaginare distribuite sui legami in guisa di equilibrio contro dette forze, il modo, in cui esse sono effettivamente distribuite, soddisfa alla condizione, che il lavoro totale concentrato per le forze interiori è un *minimo*.

4.13 In un sistema elastico qualunque pervenuto in equilibrio sotto l'azione di forze esteriori tra le diverse posizioni che i punti mobili avrebbero potuto prendere, quelle, che presero effettivamente, soddisfano alla condizione che il lavoro totale sviluppato dalle forze interiori nei reciproci loro spostamenti è minimo.

4.14 Maintenant, *si* l'on imagine que le travail L^a reste constant [...], malgré la variation possible du travail des forces f, on aura aussi:

$$L^a + L^i + \delta L^i = 0.$$

d'où

$$\delta L^i = \sum f \delta \Delta \rho = 0.$$

4.15 En suivant la démonstration et traduisant en langage ordinaire le conséquences de l'équation [...], on est conduit à l'énoncé suivant qui n'offre plus aucune ambiguité.

La somme des quarrés des tensions, divisés respectivement par le coefficient d'élasticité du lien correspondant est un minimum; c'est-à-dire que cette somme est moindre que pour tout autre système de tensions capable d'assurer l'équilibre, lorsqu'on néglige les conditions relatives a l'extensibilité des liens. Permettez-moi, Monsieur, de vous soumettre en second lieu une démonstration fort simple de votre équation [...].

Soit l la longueur de l'un des liens, λ son allongement dans la position d'équilibre T sa tension égale à $\epsilon\lambda$, $T + \Delta T$ la tension du même lien à une autre solution des équations d'équilibre, lorsque les liens sont supposes inextensibles; les forces ΔT, si elles étaient seules, se feraient équilibre sur le système, puisque les forces T et les forces $T + \Delta T$, font, par hypothèse, équilibre aux mêmes forces extérieures (le système est celui dont le liens extensibles ont disparu). La somme des moments virtuels des forces ΔT est donc nulle pour tous les déplacements compatibles avec les liaisons autres que l'inextensibilité des liens. Mais, un de ces déplacements est celui qui se produit réellement et dans lequel le lien l s'allonge de λ égal à T/ϵ, on a par conséquent

$$\sum \frac{T\Delta T}{\epsilon} = 0$$

C'est précisément l'équation [...] dont le principe d'élasticité est la traduction immédiate.

4.16 Non tralasciai nelle varie occasioni anzi ricordate di esporre la genesi di quella teoria che ebbe origine, per quanto mi consta, in una memoria del Sig. *Vène* uffiziale superiore del Genio Francese, il quale fin dal 1818 e quindi nel 1836 (*Mémoire sur les lois que suivent les pressions*) enunziava il seguente teorema per il caso speciale di pressioni esercitate da pesi sopra punti d'appoggio omogenei: *La somme des Quarrés des poids doit être un minimum*. Di questo nuovo principio si faceva cenno nel *Bulletin des Sciences Mathématiques de FERUSSAC tome neuvième pag.* 7 in un articolo firmato S. In un'altro articolo che fa seguito al precedente, nello stesso torno pag. 10 e firmato A. C. il principio anzidetto venne esteso al caso di punti di appoggio non omogenei e a quello di pressioni prodotte sopra i punti d'appoggio per mezzo di spranghe rigide. L'Autore A. C. di quell'articolo si supponeva essere Augustin Cauchy; ma ulteriormente desso venne con maggiore probabilità attribuito al S. *A. Cournot*. – *Pagani* trattava il caso speciale di cordoni elastici fissi rispettivamente in una delle loro estremità e riuniti nell'altra in un nodo al quale era applicata una forza. Il Mossotti trattò nella sua *Meccanica* gli argomenti precedenti.

4.17 Ces pressions [...] sont des grandeurs hétérogènes aux forces par lesquelles sont engendrées [...].
La détermination des pressions doit être considérée comme une autre branche de la dynamique ou de la science des effets des forces; branche qui pourrait prendre le nom de dynamique latente [...].
S'il s'agit d'un système ayant plusieurs points par des obstacles fixes, chaque obstacle subira une pression proportionnelle à la droite infiniment petite que le point correspondant décrirait pendant l'élément du temps.

4.18 Ces pressions, prises en sens contraires, pourront être considérées comme des forces appliquées au système, et qui le maintiennent en équilibre, abstraction faite des obstacles.

4.19

$$F\delta f + F'\delta f' + \cdots - (P\delta p + P'\delta p' + \ldots) = 0$$

formule qui donnera les relations de l'équilibre, après qu'on aura réduit, au plus petit nombre possible, les variations indépendants, en tenant compte des liaisons propres du système, mais non pas de celles qui résultent de la présence des obstacles, maintenant remplacées par les forces P, P'[...]

4.20 Quand on a regard à la présence de ces obstacles pour réduire le nombre des variations, il vient simplement:

$$F\ \delta f\ + F'\ \delta f'\ +\ \ldots\ =\ 0;$$

donc aussi, dans le même cas:

$$P\ \delta p\ + P'\ \delta p'\ +\ \ldots\ =\ 0$$

ce qui résulte immédiatement de ce que le deux systèmes (F) et (P) sont équivalents.

4.21

$$p\ \delta p\ + p'\ \delta p'\ +\ \ldots\ =\ 0,$$

relation en vertu de laquelle la somme des quantités p^2, p'^2, etc., ou, par l'hypothèse, celle des carrés des pressions P^2, P'^2, etc. est un *minimum*; car il est facile de s'assurer que le case du *maximum* ne peut avoir lieu ici.

4.22 Par conséquence, les équations qui complètent, dans tous les cas, le nombre de celles qui sont nécessaires pour l'entière détermination des pressions, résultent de la condition que la somme des carrés de ces pressions soit un minimum.

4.23 Dès l'année 1857 j'avais fait connaître à l'Académie des Sciences de Turin l'énoncé de ce nouveau principe; puis en 1858 (séance du 31 mai) j'en avais fait l'objet d'une communication a l'Institut de France (Académie des Sciences). Dans la démonstration que j'en donnai je m'appuyais sur la considérations de la transmission du travail dans les corps. Quoique, selon moi, celle démonstrations fût suffisamment rigoureuse, elle parut à quelques géomètres trop subtile pour être acceptée sans contestation. D'un autre côté la signification des équations déduites de ce théorème n'était pas suffisamment indiqué. C'est pourquoi j'ai cru devoir reprendre cette étude qui a été plus d'une fois interrompue par suite des événements auxquels ma position m'à appelé à prendre part. Je présent aujourd'hui ces nouvelles recherches qui ont eu pour résultat de me conduire à une démonstration tout-à-fait simple et rigoureuse [...].

4.24 Pour donner à la question de la distribution de tension toute l'étendue qu'elle comporte sous le rapport physique, il faudrait tenir compte des phénomènes de *thermodynamique* qui se manifestent dans l'acte de changement de forme du corps ou système élastique; mais je considère le corps au moment où l'équilibre est établi entre les forces *intérieures* et extérieures, en supposant que la température n'a pas varié. Alors on peut admettre que le travail développé se résume dans celui qui se trouve concentré à l'*état latent* dans le système élastique par l'effet des forces extérieures.

4.25

$$\sum \epsilon_{pq}\lambda_{pq}\delta\lambda_{pq} = \sum \frac{1}{\epsilon_{pq}} T_{pq}\delta T_{pq} = 0.$$

qui est l'*équation d'élasticité*, de la quelle on conclut le théorème que nous avons énoncé au commencement de ce Mémoire, savoir que: *Lorsqu'un système élastique se met en équilibre sous l'action de forces extérieur, le travail intérieur, développé dan les changement de forme qui en dérive, est un* minimum.

4.26 Uguagliando a zero il coefficiente di ciascuna variazione si ha:

$$\lambda_{mn} = (A_n - A_m)\cos\phi_{mn} + (B_n - B_m)\cos\theta_{mn} + (C_n - C_m)\cos\psi_{mn} = 0$$

Paragonando queste espressioni di λ con quelle (3) si vedrà che sono identiche prendendo per valori de' coefficienti indeterminati $A_m = \alpha_m$; $B_m = \beta_m$, $C_m = \gamma_m$...ecc. Così tali espressioni condurranno agli identici risultati già ottenuti precedentemente. In tal modo resta dimostrata la esattezza del metodo dedotto dal principio di elasticità e è perciò confermato il principio medesimo.

4.27 Il Cav. *Rombaux* ingegnere capo delle ferrovie Romane, annunzia la pubblicazione sulla tettoja di *Arezzo*, di una memoria dalla quale egli prende argomento per trattare colla massima ampiezza, la quistione del riparto delle tensioni e delle pressioni de' sistemi elastici. Egli per ragione di semplicità, si vale del *principio di elasticità*, e con numerosi esempi analitici e numerici, dimostra la coincidenza de' risultati che se ne deducono, con quelli ottenuti con altri metodi.

4.28 Nel metodo della flessibilità si ammette che uno degli appoggi sia cedevole, quindi mediante le equazioni delle curve di flessione si calcola la espressione analitica della freccia che vi si manifesta, e ponendola poi uguale a zero si ottiene una equazione di flessibilità che esprime la condizione a cui deve soddisfare la reazione per rimettere il suo punto di applicazione nello stato di un' appoggio fisso. Secondo il principio di elasticità, allorché il prisma trovasi equilibrato sotto l'azione delle forze esterne, il lavoro molecolare sviluppatosi un minimo, e quindi la sua derivata per rapporto alla reazione predetta deve essere nulla: donde risulta una equazione di elasticità alla quale deve soddisfare la reazione stessa per conseguire il minimo lavoro. Nei due modi di procedere le equazioni di flessibilità e di elasticità completano le equazioni di equilibrio e fanno cessare l'indeterminazione.

4.29 [...] ne segue che le equazioni utili per determinare queste tensioni si riducono a $3n - 6$ e non bastano in generale a determinare tutte le incognite, se non quando il numero delle verghe sia uguale a $3n - 6$.

4.30 Con questa formula si possono esprimere le tensioni di tutte le verghe in funzione degli spostamenti dei vertici parallelamente agli assi: questi spostamenti sarebbero $3n$, se tutti i vertici potessero muoversi, ma a cagione delle condizioni a cui abbiamo assoggettato i tre vertici V_1, V_2, V_3, si ha $\xi_1 = 0$, $\eta_1 = 0$, $\zeta_1 = 0$; $\eta_2 = 0$, $\zeta_2 = 0$; $\zeta_3 = 0$, onde gli spostamenti incogniti si riducono a $3n - 6$.

4.31 Se determino le tensioni T_{pq} in modo che rendano minima l'espressione $\sum T_{pq}^2 \epsilon_{pq}$, supponendo che tra quelle tensioni debbano aver luogo le equazioni [1], nelle quali però si considerano costanti tutte le forze esterne X_p, Y_p, Z_p, e tutti gli angoli α_{pq}, β_{pq}, γ_{pq}, i valori delle tensioni che così si ottengono, coincidono con quelli ottenuti con il metodo degli spostamenti.

4.32 Uguagliando ora a zero i coefficienti dei differenziali di tutte le tensioni si otterranno tante equazioni quante sono queste tensioni, e aggiungendovi le $3n - 6$ equazioni (1) si avranno tante equazioni quante bastano a determinare tutte le tensioni e i $3n - 6$ moltiplicatori.

4.33 7. TEOREMA. – Consideriamo un sistema elastico formato di parti soggette a torsione, flessione o scorrimento trasversale, e di verghe congiunte a snodo con quelle parti e fra loro: io dico che se questo sistema viene sottoposto all'azione di forze esterne cosicché esso si deformi, le tensioni delle verghe dopo la deformazione sono quelle, che rendono minima l'espressione del lavoro molecolare del sistema, tenendo conto delle equazioni, che si hanno fra queste tensioni, e supponendo costanti le direzioni delle verghe e delle forze esterne.

4.34 Intanto se fra le equazioni [9] si considerano quelle, che contengono le tensioni delle verghe, le quali non sono congiunte per alcun estremo colle parti flessibili del sistema, si riconosce che esse son precisamente quelle, che si otterrebbero col metodo degli spostamenti per esprimere quelle tensioni, intendendo solo che in generale A, B, C rappresentino gli spostamenti del vertice V parallelamente agli assi: i tre vertici V_1, V_2, V_3 dei quali il primo è posto nell'origine delle coordinate, il secondo sull'asse delle x e il terzo nel piano delle xy, suppongo sian di quelli in cui concorrono soltanto verghe congiunte a snodo.
Ci resta solo a dimostrare che anche quelle fra le equazioni [1], le quali contengono le tensioni delle verghe, che con un estremo si congiungono alle parti flessibili del sistema, coincidono colle equazioni fornite dal metodo degli spostamenti.

4.35 Ma il lavoro delle forze esterne dev'essere uguale al lavoro interno o molecolare, e questo è indipendente dalle legge colla quale sono venute crescendo le forze esterne; dunque la formola [10] esprime il lavoro molecolare della deformazione, qualunque sia la legge colla quale hanno variato le forze, che l'hanno prodotta.

4.36 Ma abbiamo veduto che si esprime anche colla formola

$$\frac{dF}{dP} = dP + \frac{dF}{dQ}dQ + \frac{dF}{dR}dR + \cdots \frac{dF}{dT_1}dT_1 + \frac{dF}{dT_2}dT_2 + \cdots ;$$

dunque queste due espressioni, dovendo essere identiche qualunque siano i valori dei differenziali $dP, dQ, dR, \ldots, dT_1, dT_2, \ldots$ bisognerà che sia

$$\frac{dF}{dP} = p, \ \frac{dF}{dQ} = q, \ \frac{dF}{dR} = r, \ldots \frac{dF}{dT_1} = t_1, \ \frac{dF}{dT_2} = t_2, \ldots$$

4.37 Vedesi che anche quelle fra le equazioni [9], che contengono le tensioni $T_1, T, \ldots$ coincidono pienamente con quelle ottenute col metodo degli spostamenti.

4.38 13. *Applicazione a una trave sostenuta in più di due punti.* – Suppongo la trave orizzontale, rettilinea, omogenea, di azione costante, simmetrica rispetto al piano verticale che passa pel suo asse, e caricata di un peso uniformemente distribuito su ciascuna parte contenuta tra due appoggi successivi.
È chiaro che i valori dei momenti inflettenti per le sezioni in corrispondenza degli appoggi, sono funzioni dei pesi distribuiti sul solido e delle pressioni o reazioni degli appoggi; ora tenendo conto delle due equazioni dateci dalla statica tra i valori di queste reazioni, vedesi che tante di esse rimangono a determinarsi quanti sono gli appoggi, meno due, ossia tante quanti sono i momenti inflettenti sugli appoggi, poiché i momenti inflettenti sugli appoggi estremi sono nulli. Donde segue, che le reazioni degli appoggi si possono esprimere in funzione dei momenti inflettenti relativi agli appoggi medesimi, e perciò possiamo prendere per incognite questi momenti.
Queste incognite si debbono determinare colla condizione che il lavoro molecolare della trave sia un minimo; io trascuro il lavoro proveniente dallo scorrimento trasversale, onde il differenziale del lavoro molecolare di tutta la trave, riesce uguale alla somma di tante espressioni analoghe alla [15], quante sono le parti in cui la trave è divisa dagli appoggi, ossia le travate, avvertendo solo che per l'estrema travata di destra l'espressione [15] si riduce al solo primo termine, perciò $dm = 0$, e per l'estrema di sinistra si riduce al secondo termine, perché $dM = 0$.
Afflnché il lavoro molecolare sia un minimo, bisogna determinare i momenti inflettenti incogniti, uguagliando a zero i coefficienti dei differenziali di tutti questi momenti. Ora il differenziale del momento inflettente relativo all'appoggio B, non può entrare che in uno dei termini che provengono dal lavoro della travata AB e in uno di quelli che provengono dal lavoro della travata BC; cosicché chiamando a e a' le lunghezze di queste due travate, p e p' i pesi uniformemente distribuiti su di esse, m, m', m'' i momenti inflettenti relativi ai tre appoggi A, B, C; E il coefficiente di elasticità della trave e I il momento d'inerzia della sezione, i due termini che nell'espressione differenziale del

lavoro molecolare contengono il differenziale dm' sono:

$$\frac{a}{2EI}\left(\frac{m+2m'}{3}-\frac{1}{12}pa^2\right)dm'; \quad \frac{a'}{2EI}\left(\frac{2m'+m''}{3}-\frac{1}{12}p'a'^2\right)dm'.$$

Dunque uguagliando a zero il coefficiente di dm', si ottiene

$$am'+2(a+a')m'+m''a'-\frac{1}{4}(pa^3+p'a'^3)=0.$$

è questa appunto l'equazione dovuta a Clapeyron.

4.39 [La nuova dimostrazione] però pare non essere stata giudicata più rigorosa della prima, perché non ostante la grande bellezza e la evidente utilità del teorema del minimo lavoro, nessuno, ch'io sappia, credette di poterne trarre partito prima dell'anno 1872, in cui l'Ing. Giovanni Sacheri lesse alla Società degli Ingegneri e industriali di Torino una sua Memoria, nella quale si provò a applicare quel teorema [...]. Però di questa memoria non mi occorre parlare perché, contenendo solo un esempio numerico, non fece punto progredire la dimostrazione del teorema.

4.40 *10. Utilità del teorema del minimo lavoro.* – In pratica non avviene quasi mai che si adoperino dei sistemi elastici semplicemente articolati, cioè dei sistemi composti soltanto di verghe elastiche congiunte a snodo: invece sono continuamente adoperati dei sistemi che chiamerò *misti*, composti di travi rinforzate da saette o tiranti, cioè da verghe elastiche congiunte a snodo colle travi in diversi punti della loro lunghezza, e fra loro. Affinché dunque un teorema intorno ai sistemi elastici abbia un'utilità pratica, bisogna che esso sia applicabile ai sistemi misti. Questo pregio ha appunto il teorema del minimo lavoro, e è solo per ciò, che io mi sono adoperato, quanto ho potuto, a dimostrarne l'esattezza e l'utilità. Siccome però le sue proprietà riguardo ai sistemi semplicemente articolati si mantengono anche per quelli misti, come dimostrerò fra poco, dirò fin d'ora alcuni vantaggi che esso presenta su altri metodi nel calcolo dei sistemi articolati.

4.41 [...] se lo stato del sistema dopo la deformazione si può far dipendere da un piccolo numero di quantità legate fra loro da alcune equazioni di condizione, e se il lavoro molecolare del sistema nella deformazione si esprime per mezzo di quelle sole quantità, si otterranno i valori delle medesime considerandole come variabili legate alle equazioni di condizione, e cercando il sistema dei loro valori, che rende minima l'espressione del lavoro molecolare.

4.42 [...] Se di un sistema articolato deformato da date forze si sa esprimere il lavoro molecolare di una parte contenuta entro una certa superficie S in funzione delle tensioni delle verghe che congiungono questa parte alla rimanente, si otterranno le tensioni di queste verghe e di quelle esterne alla superficie S esprimendo che il lavoro molecolare di tutto il sistema è un minimo, tenuto conto [solo] *delle equazioni di equilibrio intorno a tutti i vertici esterni alla superficie S.*

4.43 Ciò posto, i due nuovi teoremi sono i seguenti:
1° Se per un sistema elastico qualunque il lavoro di deformazione espresso in funzione delle forze esterne si differenzia rispetto a una di queste forze, la derivata, che si ottiene, esprime lo spostamento del punto d'applicazione della forza proiettato sulla sua direzione.
2° Se la medesima espressione del lavoro di deformazione si differenzia rispetto al momento di una coppia, la derivata, che si ottiene, esprime la rotazione della linea, che congiunge i punti d'applicazione delle due forze della coppia.
Questi teoremi, la cui importanza è evidente, sono veri soltanto se le deformazioni sono piccolissime, per modo che le potenze degli spostamenti e delle rotazioni superiori alla prima siano trascurabili rispetto a questa. Essi possono riunirsi in un solo, ch'io chiamerò *teorema delle derivate del lavoro di deformazione* o più brevemente *teorema delle derivate del lavoro*.
3. Si vedrà in seguito che esso basta per risolvere tutte le questioni, che si presentano nella pratica intorno all'equilibrio dei sistemi elastici. Si vedrà pure che esso contiene come applicazione o meglio come semplice osservazione il *teorema del minimo lavoro delle deformazioni elastiche* o *principio d'elasticità*, che il Generale MENABREA ha pel primo enunciato in tutta la sua generalità nel 1857 e 1858 alle Accademie delle scienze di Torino e Parigi, e intorno al quale ha presentato nel 1868 un'altra Memoria all'Accademia delle scienze di Torino.

4.44 17. *Teorema delle derivate del lavoro di deformazione*—Se il lavoro di deformazione di un sistema articolato si esprime in funzione delle forze esterne, la sua derivata rispetto a una qualunque di queste ci dà lo spostamento del punto d'applicazione della medesima proiettato sulla sua direzione.

4.45 [...] se il lavoro di deformazione del sistema [articolato] si differenzia rispetto al momento M della coppia considerata, la derivata che si ottiene esprime l'angolo, di cui ha rotato intorno all'asse della coppia la retta, che congiunge i punti d'applicazione delle due forze della medesima coppia.

4.46 Io mi propongo di far vedere che anche per queste due classi di sistemi sono veri sia il *teorema delle derivate del lavoro di deformazione*, sia quello *del minimo lavoro* [...].
Per dare queste dimostrazioni io invocherò il *principio della conservazione delle energie*: io non avrei bisogno di farlo, se si ammettesse che quando un corpo elastico si deforma, l'azione, che si sviluppa tra due molecole vicine, è diretta secondo la linea, che ne congiunge i centri. Quest'ipotesi è stata ammessa finora, e alcuni autori insigni come LAMÉ e BARRÉ DE SAINT VENANT continuano a ammetterla, perché difatti è difficile farsi un'idea chiara d'un altro modo d'azione.
Siccome però il celebre astronomo GREEN nella sua *Teoria della luce* ha ammesso che l'azione tra due molecole possa aver luogo in una direzione diversa dalla retta, che ne congiunge i centri, ma tale però che abbia luogo il principio della conservazione delle energie, io procurerò di far vedere che

i nuovi teoremi sono veri indipendentemente dalla direzione in cui ha luogo l'azione tra le molecole dei corpi.

4.47 È molto importante persuadersi bene del rigore di questo ragionamento [relativo alle forze centrali], perché è assai probabile che il caso ora considerato sia quello, che ha luogo in natura. Ma per non introdurre nelle nostre ricerche alcuna restrizione, che non sia assolutamente necessaria, riferiremo qui il ragionamento di GREEN modificato dal signor BARRÉ DE SAINT VENANT, per dimostrare il teorema enunciato, qualunque sia la direzione dell'azione tra le molecole. Supponiamo che la deformazione di un corpo abbia luogo in un vaso impermeabile al calore, e che dopo aver fatto crescere, secondo una data legge, le forze esterne da zero sino ai loro valori finali, si facciano decrescere di nuovo sino a zero secondo un'altra legge che non sia esattamente inversa della prima. Poiché il corpo è in un vaso impermeabile al calore non può aver ricevuto calore, né può averne ceduto; e d'altra parte poiché il corpo ha ripreso il suo stato primitivo esso conterrà alla fine la stessa quantità di calore, che conteneva al principio.

Se dunque il lavoro fatto dalle forze esterne nel periodo del loro incremento non fosse esattamente uguale a quello raccolto nel periodo del loro decremento, si avrebbe una produzione o un consumo di lavoro, che non sarebbero compensati da un'equivalente quantità di calore consumata o prodotta. Il che è contro il *principio della conservazione delle energie.*

4.48 Dunque, poiché a cagione del disco perfettamente rigido la sezione resta piana nella deformazione del sistema, [...vale] il teorema seguente, importantissimo per la teoria della resistenza dei solidi.

Le derivate del lavoro di deformazione rispetto alle tre forze X, Y, Z e ai momenti delle tre coppie sopra definite esprimono gli spostamenti del centro della sezione parallelamente alle direzioni delle forze, e le tre rotazioni della sezione medesima intorno ai suoi assi principali di inerzia e alla sua perpendicolare condotta pel centro.

4.49 Avendogli io chiesto: Perché nel tuo libro non hai ritenuto preferibile come più generale l'ipotesi del Lamé? A ciò egli risposemi: E a cosa avrebbe giovato il secondo coefficiente? abbiano noi per la generalità dei corpi solidi delle serie esperienze che ne abbiano stabilito il valore?

La via tenuta dal Castigliano nel suo trattato maggiore non è sempre, rigorosamente parlando, quella che si direbbe via maestra e credo che a ciò fare egli sia stato indotto da una giustissima ragione. Lo scendere dal generale al particolare è il pregio precipuo delle opere che si indirizzano a menti in cui sono già mature le idee sul soggetto di cui si tratta; non è già la via migliore per un libro che deve servire per i dotti e a un tempo per chi ha brama di apprendere. E perciò che il nostro Autore premette la trattazione dei sistemi articolati in cui i solidi sono considerati soggetti a forze di trazione o compressione uniformi per tutta la loro sezione retta. Parte dunque da un caso semplicissimo per ascendere alle azioni reciproche d'una molecola colle sue vicine e ogni volta dimostra i principj della sovrapposizione degli effetti e del teorema delle derivate del lavoro. Dopo questa preparazione che ha abituato il lettore poco a poco a rendersi famigliari certe idee, egli ascende alla teoria generale del parallelepipedo elementare e stabilite le equazioni generali egli le applica a numerosi casi di flessione e torsione di solidi di forma svariata. Poscia egli passa alla parte delle applicazioni approssimative giustificando le ordinarie formole del trave e preparando i materiali per una rapida applicazione del suo teorema.

4.50 PREFACE

Cet ouvrage contient la *théorie de l'équilibre des systèmes élastiques exposée suivant une méthode nouvelle*, fondé sur quelques théorèmes qui sont tout-à-fait nouveaux, ou encore peu connus.

Comme faisant partie de cette théorie, on y trouvera la *théorie mathématique de l'équilibre des corps solides*, considérée particulièrement sous le point de vue de la résistance des matériaux.

Nous croyons que le moment est arrivé d'introduire dans l'enseignement cette manière rationnelle de présenter la résistance des matériaux, on abandonnant ainsi les méthodes anciennes que l'illustre LAMÉ a justement définies comme *mi-analytiques et mi-empiriques, ne servant qu'à masquer les abords de la veritable science*.

Nous donnerons maintenant quelques renseignements historiques sur la découverte des théorèmes dont on fait un usage presque continue dans tout le cours de cet ouvrage.

Ces théorèmes sont les trois suivants:

1° des dérivées du travail, première partie;

2° id. id. deuxième partie;

3° du moindre travail.

Le premier avait été déjà employé par le célèbre astronome anglais GREEN, mais seulement dans une question particulier, et n'avait point été énoncé et démontré d'une manière générale, ansi que nous le faisons dans le présent ouvrage.

Le second est le réciproque du premier, et nous croyons qu'il a été énoncé et démontré pour la première fois, en 1873, dans notre dissertation pour obtenir le diplôme d'Ingénieur à Turin: nous y avons donné ensuite plus d'étendue dans notre mémoire intitulé *Nuova teoria intorno all'equilibrio dei sistemi elastici*, publié dans les Actes de l'Académie des sciences de Turin en 1875.

Le troisième théorème peut être regardé comme un corollaire du second; mais de même que dans quelques autres questions de *maxima et minima*, il a été, pour ainsi dire, presenti plusieurs année avant la découverte du théorème principal.

[...] Voici maintenant quelques renseignements sur la redaction de notre travail. Comme notre but n'est pas seulement d'exposer une théorie, mais encore de faire apprécier ses avantages de breveté et de simplicité dans les applications pratiques, nous avons résolu, suivant la nouvelle méthode, non seulement la plupart des problèmes généraux qu'on traite dans les cours sur la résistance

des matériaux, mais nous avons encore ajouté plusieurs examples numériques pour le calcul dee systèmes élastiques les plus importantes.
[...] Quant aux calculs, nous ferons remarquer qu'il ne sont guère plus longs que dans les méthodes ordinairement suivies; et que, d'ailleurs, on pourra presque toujours les abréger sensiblement en négligeant quelques termes, qui influent peu sur le résultat.

4.51 11. *Théorème des dérivées da travail de déformation.*
Première Partie - Si l'on exprime le travail de déformation d'un système articulé, en fonction des déplacements relatifs des forces extérieures appliquées à ses sommets, on obtient une formule, dont les dérivées, par rapport à ces déplacements, donnent la valeur des forces correspondantes.
Seconde Partie - Si l'on exprime, au contraire, le travail de deformation d'un système articulé en fonction des forces extérieures on obtient une formule, dont les dérivées, par rapport à ces forces, donnent les déplacements relatifs de leurs points d'application.

4.52 Pour la seconde partie, observons que le travail de déformation du système, dû aux accroissements dR_p des forces extérieures doit être aussi représenté par la différentielle de la formule (15), qui est

$$\frac{1}{2}\sum R_p dr_p + \frac{1}{2}\sum r_p dR_p:$$

on a donc l'équation

$$\sum R_p dr_p = \frac{1}{2}\sum R_p dr_p + \frac{1}{2}\sum r_p dR_p,$$

d'où on tire

$$\sum R_p dr_p = \sum r_p dR_p;$$

et comme le premier membre de cette équation représente le travail de déformation du système pour les accroissements dR_p des forces extérieures, il en résulte que le second le représente aussi.
Or, si l'on appelle L le travail de déformation du système, dû aux forces R_p, il est évident que le travail infiniment petit dû aux accroissements dR_p sera représenté par la formule

$$\sum \frac{dL}{dR_p} dR_p$$

Cette formule devant être identique avec l'autre $\sum r_p dR_p$, il s'ensuit qu'on devra avoir pour chaque force

$$\frac{dL}{dR_p} = r_p$$

ce qui démontre la seconde partie du théorème.

4.53 1. Les résultantes $\mathscr{X}, \mathscr{Y}, \mathscr{Z}$ et les moments résultants M_x, M_y, M_z sont les dérivées du travail de déformation du système par rapport aux déplacements ξ_0, η_0, ζ_0, et aux rotations $\theta_x, \theta_y, \theta_z$.
2. Les trois déplacements ξ_0, η_0, ζ_0, et le trois rotations $\theta_x, \theta_y, \theta_z$, sont les dérivées du travail de déformation du système par rapport aux résultantes $\mathscr{X}, \mathscr{Y}, \mathscr{Z}$ et aux moments résultants M_x, M_y, M_z.

4.54 4. Travail de déformation du parallélépipède très-petit.
Dans le parallélépipède élémentaire dont les arêtes sont Δx, Δy, Δz, considérons la petite droite r joignant deux molécules très-rapprochées. Dans la déformation du corps, cette droite croît à partir de la longueur initiale jusqu'à la valeur $r(1 + \partial_r)$ et la tension entre les deux molécules croît proportionnellement à la dilatation, en sorte que quand la droite aura la longueur $r + \rho$, ρ étant une quantité plus petite que $r\partial_r$, la tension entre les deux molécules sera $\epsilon\rho$, en appelant ϵ un coefficient constant pour chaque couple de molécules, mais différent pour les divers couples.
Le travail de déformation de la droite r sera

$$\int_0^{r\partial_r} \epsilon\rho d\rho = \frac{1}{2}\epsilon r^2 \partial_r^2.$$

c'est-à-dire, en substituant à $\partial^2 r$ sa valeur donnée par la formule (8),

$$\tfrac{1}{2}\epsilon r^2(\partial_x \cos^2\alpha + \partial_y \cos^2\beta + \partial_z \cos^2\gamma + g_{yz}\cos\beta\cos\gamma + g_{xz}\cos\gamma\cos\alpha + g_{xy}\cos\alpha\cos\beta).^2$$

où l'on doit observer qu'en développant le carré, et réunissant les termes contenant les mêmes produits des cosinus $\cos\alpha$, $\cos\beta$, $\cos\gamma$ les termes distincts se réduisent à quinze. Pour avoir le travail de déformation de tout le parallélépipède, il faut additionner les expressions analogues à celle-ci pour tous les couples moléculaires qu'il contient.

4.55 Arrestiamoci alquanto a considerare quale sia dal punto di vista scientifico, la novità, la portata e la utilità di questo teorema delle derivate del lavoro e dell'altro, che si può dire gemello, del minimo lavoro.
Or bene questi teoremi se bene si considerano dal punto di vista della teoria generale non costituiscono enunciati essenzialmente nuovi. Già Legendre aveva dimostrato che data una funzione ϕ di n variabili x, si può formare colle sue derivate parziali una funzione ψ le di cui derivate parziali sono rispettivamente eguali alle variabili x. Era anche stato riconosciuto che se la ϕ è funzione

quadratica, risulta $\phi = \psi$. Più tardi l'illustre matematico inglese Giorgio Green da considerazioni sull'impossibilità del moto perpetuo fu condotto a stabilire che il lavoro di un sistema elastico era rappresentato da un potenziale degli spostamenti, e ciò nelle due celebri memorie sulla luce del 1839.
Era quindi completamente noto il substrato analitico che esprime la proprietà dei due teoremi di cui discorro; non credo però che siano mai stati formalmente enunciati forse perché in fondo non occorrevano al progresso della teoria generale, la quale colle considerazioni degli spostamenti viene a far uso delle stesse formole a cui quei due teoremi conducono.

4.56 Si faccia ora

$$\lambda = F_1 f_1 + F_2 f_2 + \cdots - L.$$

Differenziando si ha

$$d\lambda = f_1 dF_1 + f_2 dF_2 + \cdots + (F_1 df_1 + F_2 df_2 + \cdots - dL).$$

ma essendo nulla la quantità tra parentesi, resta:

$$d\lambda = f_1 dF_1 + f_2 dF_2 + \cdots$$

Considerando quindi λ come funzione delle forze F, si ha:

$$\frac{d\lambda}{dF_1} = f_1, \quad \frac{d\lambda}{dF_2} = f_2 \quad \text{ecc.}$$

4.57 Supporrò, che nella superficie di mutuo contatto dei diversi pezzi attrito non si sviluppi, o almeno sia trascurabile: se questo avviene, essi non sopporteranno che sforzi diretti nel senso della loro lunghezza. Ma debbo tosto soggiungere, che i sistemi articolati (senz'attrito) non esistono in realtà: esse sono mere astrazioni; contuttociò la loro teoria non è scevra di applicazioni pratiche, in tutti quei casi almeno, in cui gli effetti della flessione possono trascurarsi.

4.58 Anche i legami da cui il sistema è astretto per causa dele aste, che ne collegano i vertici, si possono esprimere mediante equazioni: [...] come dalle equazioni (1) derivano le forze rappresentate dalle espressioni (2), così dalle equazioni (3) deriveranno altre forze, le quali non saranno altro, che le tensioni delle aste: ma tra queste due specie di forze corre [...] la differenza che intercede tra le forze esterne e le forze interne: epperò tra le forze provenienti dalle equazioni (3), se ve ne è una A, ve ne esiste ancora un'altra $-A$.

4.59 Le formole ora trovate ci fornirebbero [...] la figura della trave deformata, [...] la variazione degli angoli, ed il lavoro sviluppato nella deformazione dalle forze esterne. Ma io lascierò qui tale argomento e passerò a discorrere [...] della distribuzione delle tensioni e delle pressioni, nei casi, nei quali, la statica dei corpi rigidi gettandoci nell'indeterminazione, è mestieri aver ricorso alle leggi dell'elasticità.

4.60 Si potrebbe dimandare quale sia il motivo pel quale coi metodi precedenti il problema della distribuzione delle tensioni e delle pressioni in un sistema elastico articolato si sia potuto risolvere ed in modo abbastanza spedito con tutto il rigore della teoria matematica dell'elasticità, mentre tanti altri problemi rimangono ancora insoluti per le troppe difficoltà che presentano. Ciò è dovuto al fatto, che nel caso ora considerato sono conosciute le leggi degli spostamenti dei diversi punti del sistema; imperocchè questo è il problema generale della teoria dell'elasticità: "Date le forze, che sollecitano un corpo, trovare gli spostamenti paralleli a tre assi, che fanno subire ad una molecola qualunque di essa. Quando tali spostamenti sieno conosciuti, come si è accennato al n° 8, sarà facilissimo trovare l'espressione delle forze elastiche provocate in ogni suo punto". Ma tutta la difficoltà versa appunto nel trovare la legge di questi spostamenti. La natura dei sistemi qualche volta indica a priori quale sia questa legge: un esempio l'abbiamo nei sistemi elastici articolati: in questi casi altro più non resta a fare, che a trovare la loro grandezza, conoscendo le forze estrinseche. Ma non saranno mai le ipotesi che ci guideranno alla verace conoscenza degli spostamenti; d'altronde quando sembra probabile un certo modo di deformazione, è agevole verificare se esso sia o no possibile; basta provare, se con una tale supposizione le equazioni dell'equilibrio interno restino soddisfatte. Io dico questo, perchè le teorie sulla resistenza dei materiali, come sono ordinariamente esposte, riposano su una ipotesi particolare intorno alla legge degli spostamenti; ipotesi che non è quasi mai verificata, come ora procurerò di dimostrare.

4.61 [...le quali] non possono sussistere altro che nel caso in cui a, b, c sieno funzioni lineari di z e p, q, r costanti; caso particolarissimo, e che non abbraccia certamente tutti quelli che si presentano in pratica. Ma [...] non mi fermerò ulteriormente su tal quistione. Essa può forse avere la sua utilità, quella almeno di far vedere una volta di più che non si arrecheranno mai perfezionamenti notevoli e di qualche vantaggio alle teorie della resistenza dei materiali senza svincolarsi da molte delle ipotesi (gratuite) da cui esse partono.

4.62 che devono essere verificate, perchè il problema sia possibile: epperò tale possibilità dipende essenzialmente dalla forma delle funzioni F, vedremo poi che dipende altresì dalla natura degli sforzi subiti dalle diverse aste. Questo non basta: nelle equazioni (25) alle variazioni delle coordinate si sostituiscano le loro espressioni per mezzo degli allungamenti delle diverse aste; ed a questi poi le tensioni corrispondenti. Si introduca quindi la condizione di uniforme resistenza: dopo d'aver eliminato le sei variazioni che vi entrano ancora, rimarranno m condizioni indipendenti dalle reazioni dei vincoli, che dovrebbero essere soddisfatte, perchè il problema fosse possibile: ma non lo saranno generalmente potendo le funzioni f_i essere qualunque. Concludiamo dunque, che *un sistema elastico articolato non si può ridurre ad essere di ugual resistenza, se il numero delle equazioni di condizione sia superiore a sei.*

4.63 Se il numero di tali equazioni non è superiore a sei, bisognerà vedere se le equazioni (28) reggano o non. Nel secondo caso potremo dire che è impossibile il determinare le sezioni delle diverse aste così da formare un sistema di egual resistenza: nel primo caso invece [...] questo sarà possibile [...] in un numero di modi k volte infinito; imperocchè nelle equazioni di equilibrio ponendo per T_{ij} il suo valore $TE_{ij}\sigma_{ij}$ ne seguiranno $3n - 6$ equazioni tra le aree delle sezioni rette delle $3n - 6 + k$ aste del sistema: ma intanto scegliendo ad arbitrio k di tali sezioni le prefate equazioni ne daranno sempre in valor determinato per le $3n - 6$ altre, e a ciascuna di tali sezioni si possono attribuire arbitrariamente infiniti valori diversi.

4.64 La quale ci dice, che *le variazioni di lunghezza dei diversi pezzi sono indipendenti dal modo, con cui si scelgono le k sezioni arbitrarie.*

4.65 Il lavoro delle forze esterne, e quindi anche quello delle forze molecolari nella deformazione non dipendono per nulla dal modo con cui venne fatta la scelta di quelle k sezioni.

4.66 Ora L e T non cambiano col variare la scelta delle k sezioni, il secondo membro dovrà dunque rimanere costante, comunque tal scelta venga fatta. Epperò possiamo dire, che in questo caso *la somma dei prodotti dei volumi delle singole sbarre pel rispettivo coefficiente di elasticità è indipendente dal modo con cui venne fatta la scelta delle k sezioni rimaste arbitrarie.* Che se poi il coefficiente E_{ij} è lo stesso per ogni sbarra posiamo ancora soggiungere, che, comunque si faccia tal scelta, *il peso della materia impiegata sarà sempre lo stesso.*

4.67 Il sistema si può rendere di egual resistenza ed in un numero di maniere semplicemente infinito, se tutte e dieci le aste sopportano sforzi della stessa natura.

4.68 Il numero dei punti essendo n, quello delle aste $n(n-1)/2$, bisogna fare alcune distinzioni: o si possono formare $(n^2 - 7n + 12)/2$ gruppi di cinque punti, pei quali dieci aste che li collegano sopportano sforzi della stessa natura (la quale peraltro può cambiare da un gruppo all'altro) e allora il sistema si può ancora ridurre ad essere di egual resistenza, e questo in $(n^2 - 7n + 12)/2$ infinite maniere differenti; o ciò è impossibile, ed allora, meno casi specialissimi, non si potrà ridurre il sistema ad essere di egual resistenza. Tuttavia, se sia possibile il formare alcuni gruppi di cinque punti, che soddisfino alle summenzionate condizioni, si possono foggiare le aste che li formano, cosicchè costituiscono un complesso di egual resistenza, e ciò in tante infinite maniere differenti, quanti sono questi gruppi.

4.69 [...] l'artifizio della soluzione consiste nel far dipendere la ricerca delle pressioni e delle tensioni incognite dalla ricerca di $3n-6$ altre quantità tante quante sono le equazioni di equilibrio fra loro indipendenti: cosa [...] sempre possibile. Nè questo artifizio è applicabile soltanto al mio problema, ma sì a ben altre quistioni più generali [...]: è noto infatti che la conoscenza delle forze molecolari destate in un corpo dipende da quella di sei funzioni legate fra loro da tre equazioni alle derivate parziali, equazioni che non sarebbero sufficienti a determinarle, se le sei funzioni in discorso non si potessero esprimere mercè tre altre soltanto. La natura poi di queste tre funzioni resta sempre determinata dal concetto che altri si fa sull'origine delle forze molecolari: nel caso delle forze elastiche queste tre funzioni sono gli spostamenti paralleli a tre assi di una molecola qualunque del corpo.

4.70 Mettendo poi nelle equazioni di equilibrio al posto delle tensioni le loro espressioni forniteci dalle (22) esse verranno a non contenere più che le variazioni delle coordinate, le quali potrannosi in tal modo determinare: una volta conosciuti i loro valori mercè le equazioni (22) si calcoleranno quelli delle tensioni.

4.71 Consideriamo il caso, in cui il sistema debba soddisfare a certe condizioni geometriche, il caso cioè in cui esista un certo numero di *equazioni alla superficie*, alle quali sieno obbligate le coordinate dei vertici del sistema (supporremo però che o non vi siano dei punti fissi, o, quando ve ne sono, si verifichino altresì le condizioni indicate nel n° 3). Sieno $m + 6$ queste condizioni: quando $m = 0$ non si ha alcuna difficoltà e questo argomento venne già discusso nel citato n° 3; se $m > 0$ le regole ivi enunziate non sono più sufficienti. Ma intorno a ciò osserveremo che le equazioni alla superficie, dovendo sempre sussistere qualunque sia il valore, che le coordinate vengano ad ottenere durante la deformazione, differenziate saranno pur soddisfatte sostituendo alle variazioni delle coordinate quelle effettive, che esse han subito sotto l'azione delle forze esterne. Ciò posto si ricavino con uno dei metodi precedenti le tensioni in funzione delle forze esterne e delle $m + 6$ reazioni dei vincoli: si esprimano le variazioni delle coordinate per mezzo di queste tensioni e si sostituiscano tali espressioni nelle $m+6$ equazioni di condizione differenziate: avremo così $m + 6$ equazioni tra le reazioni dei vincoli e sei variazioni delle coordinate, imperocchè per mezzo delle tensioni non si possono esprimere, che i valori di $3n - 6$ delle variazioni, e nel nostro caso tutte quante le variazioni sono determinate e niuna arbitraria. Ma combinando le equazioni di equilibrio se ne ricavano sei tra le forze esterne e le reazioni dei vincoli che congiunte colle prime $m + 6$ fanno $m + 12$ equazioni tra $m + 6$ reazioni e sei variazioni di coordinate, tante cioè quante sono le incognite del problema. Però si possono avere $m + 6$ equazioni tra le sole reazioni dei vincoli eliminando tra le prime $m + 6$ le sei variazioni delle coordinate.

4.72 Esse ci permetteranno dunque di ricavare nove delle tensioni in funzione della decima. Dopo ciò, facendo uso dei risultati già ottenuti, si passerà a trovar le tensioni dei diversi pezzi del rettangolo precedente, e così via di mano in mano sino a che si sia giunto all'appoggio. Tutte le tensioni si potranno esprimere mediante quella decima, che era rimasta indeterminata nel calcolo relativo al primo rettangolo, ma il cui valore si potrà trovare poi al termine dell'operazione. Sostituendo quindi il suo valore così trovato nelle espressioni precedenti, tutte le tensioni dei diversi pezzi diverranno conosciute.

4.73 Questo esempio serve a far vedere con quanta semplicità si possa sciogliere il problema della distribuzione delle tensioni, facendolo dipendere dalla ricerca di tante quantità soltanto quante sono le equazioni di equilibrio.

A.5 Quotations of Chap. 5

5.1 Nos épures obtinrent plus de succès que nos méthodes. Notre publication fut suivie d'un grand nombre de Statiques élémentaires, dans lesquelles, tout en reproduisant nos épures les plus simples (le plus souvent sans y non changer), les auteurs s'efforçaient d'en donner des démonstrations analytiques.

5.2 On est convenu de réserver le nom de *Statique graphique* à toute une catégorie de recherches récentes, qui constituent un corps de doctrine désormais bien coordonné et qui, prises dans leur ensemble, sont caractérisées par la double condition de mettre en ouvre les procédés constructifs du Calcul linéaire ou graphique, et de reposer sur la relation fondamentale qui existe entre le polygone des forces et le polygone funiculaire.
Le domaine de la *Statique graphique* étant ainsi, non pas rigoureusement défini, mais indiqué, on convient de designer sous le nom de *Statique géométrique* l'ensemble des autres applications de la Géométrie, et plus particulièrement de la Géométrie ancienne, à la statique.

5.3 Ed è ben manifesto che dietro di ciò la *Statica grafica* debba sempre più separarsi da quella analitica, non già, per fine diverso, ma perciò che vengono man mano abbandonati dall'una quei soggetti che nell'altra riescono trattati in modo più semplice e a un tempo più generale.
É così difatti che l'argomento delle travature reticolari strettamente indeformabili è quasi del tutto soppresso nella Statica analitica e riservato alla grafica, mentre quella conserva ancora di dominio proprio lo studio dello travature a membri sovrabbondanti.
La Statica grafica va coltivata nel suo giusto indirizzo, informandone il metodo alla Geometria moderna. Ed è a deplorarsi che essa, comparsa per la prima volta nell'opera magistrale del CULMANN, già improntata alla Geometria proiettiva da cui si ebbe le più belle e eleganti fra le sue dimostrazioni, abbia poi avuto degli autori che vollero appoggiarla alle sole risorse della Geometria elementare.
Questi autori sono scusabili soltanto se hanno voluto indirizzarsi a chi non è in grado di conoscere la Geometria proiettiva. I loro trattati, pur non mancando di utilità, sono però al medesimo livello dei trattati di *Meccanica analitica* svolti col solo calcolo elementare.
Non pochi lavori di Statica che videro la luce in questi ultimi anni, con titoli che alludono all'applicazione del metodo grafico, sono realmente sviluppati in parte analiticamente e in parte graficamente, vale a dire con metodo che può dirsi misto. Non crediamo nell'avvenire di un tale procedimento, il quale non lascia

presumere alcun carattere di generalità e che può ritenersi accettabile soltanto in via transitoria o nelle ulteriori applicazioni della Statica alle costruzioni.

5.4 Io credo che la poca diffusione, che ha avuto finora tra gli ingegneri l'applicazione della Statica grafica, sia derivata principalmente dalla mancanza di un ordinato libro acconcio allo insegnamento di questa nuova scienza. E sarebbe per me gran soddisfazione se, essendo giusta la mia opinione, il mio libro potesse riparare a tale difetto. Poiché certamente la Statica grafica è di tale importanza per lo studio della scienza degl'ingegneri in esercizio, ch'è a desiderarsi il suo maggior divulgamento, il quale si verificherà senza dubbio. Forse a tale diffusione contribuirà pure una qualità del mio libro cioè, che a intenderlo non è mestieri la conoscenza della così detta nuova geometria. Ciò io non ho fatto di proposito ma è da se stesso avvenuto, e ho a compiacermene, giacché credo che così sia reso un utile servigio agli ingegneri, che prima non avevano occasione di familiarizzarsi coi nuovi metodi della geometria, potendo esse abbracciar subito lo studio della Statica grafica, senza la necessità di occuparsi di una scienza ausiliaria.

5.5 L'addition et la sottration linéaire ou graphique sont deux opérations tellement simples, qu'il nous suffirait presque de le mentionner ici pour mémoire: additionner deux grandeurs linéaires a et b, c'est les mettre bout à bout sur une ligne indéfinie cd; le total est la grandeur qui se trouve comprise entre les deux limites qui ne s'aboutent pas. La règle est la même quel que soit le nombre de termes à additionner entre eux.

5.6 **21**. Se due segmenti concorrenti sono individuati in grandezza senso e direzione, s'intende per segmento loro risultante quello di chiusa della spezzata formata con due segmenti che abbiano la direzione propria. Tale segmento risultante AC resta pure individuato in grandezza senso e direzione. Se dei due segmenti componenti sono date anche le linee, il segmento AC dicesi *equipollente*, come lo sono AB e BC dei due componenti.

5.7

Préface de l'auteur

Les premières applications systématiques des méthodes graphiques, à la détermination des dimensions des diverses parties des constructions, sont dues à Poncelet. C'est en effet à l'école d'application du génie et de l'artillerie, à Metz, que ces méthodes, dont les beaux travaux de Monge avaient en quelque sorte jeté les bases, furent pour la première fois professées par Poncelet, devant un auditoire formé d'anciens élèves de l'école polytechnique de Paris, la seule où les sciences graphiques fussent enseignées à cette époque.
Poncelet avait reconnu le premier, que ces méthodes, tout en étant beaucoup plus expéditives que les méthodes analytiques, offraient cependant une approximation plus que suffisante dans la pratique puisque, quoi que l'on fasse, il ne sera jamais possible d'obtenir dans un projet rapporté sur le papier, une exactitude supérieure à celle donnée par une épure graphique.

Ces méthodes, appliquées à la théorie des voûtes et des murs de soutènement, ont été publiées dans le *Mémorial de l'officier du génie* (tomes, XII et XIII, années 1835 et 1840).
Poncelet n'a cependant pas fait usage, pour déterminer les résultantes, du polygone funiculaire, dont l'emploi offre des ressources si précieuses à la statique graphique,[4] et il était réservé à son successeur a l'école de Metz, M. Michon, d'en faire le premier l'application à la détermination des centres de gravité des voussoirs, dans sa *Théorie des voûtes.*[5]
La géométrie de position, à laquelle Poncelet a fait faire tant de progrès, n'était cependant pas à cette époque suffisamment avancée pour qu'il fût possible de la substituer complètement à la géométrie ordinaire (*Géométrie des Masses*) dans le développement et la démonstration des épures. Aussi Poncelet recourait-il, aussi souvent que possible, à la géométrie ordinaire, et lorsque les méthodes élémentaires ne lui suffisaient plus pour ses démonstrations, il se bornait à traduire en épures les formules algébriques.
Nous devons faire remarquer, du reste, que le premier Traité de géométrie de position, dans lequel il soit fait complètement abstraction de l'idée de mesure, n'a été publié qu'en 1847, par G. de Staudt, professeur de mathématiques à Erlangen (*Die Geometrie der Lage*, Nürenberg, 1847).
Quand nous fûmes appelé, en 1855, lors de la création de l'école polytechnique de Zurich, à professer le cours de construction (comprenant les terrassements, la construction des ponts, des routes et des chemins de fer), nous fûmes obligé d'introduire dans notre enseignement les méthodes graphiques de Poncelet pour suppléer aux lacunes des cours de mécanique appliquée. Ce cours ne comprenait alors à Zurich que les méthodes analytiques; il en était de même, à cette époque, à l'école des ponts et chaussées de Paris, et c'est en vain que l'on chercherait dans le *Cours de résistance des matériaux* de M. Bresse , les épures de Poncelet et de M. Michon.
Cette introduction des théories de la Statique graphique dans les cours de construction, ne laissait pas que de présenter certains inconvénients, en retardant outre mesure la marche des études; nous obtînmes, en 1860, la création d'un cours d'hiver (à deux leçons par semaine) obligatoire pour les ingénieurs, dans lequel nous traitions ceux des problèmes de statique appliqués à la construction, qui étaient susceptibles de solutions graphiques, et don l'enseignement ne trouvait pas place, faute de temps, dans le cours de mécanique technique (alors professé par M. Zeuner).
Telle fut l'origine de la Statique graphique. Les cours de construction (ponts et chemins de fer) qui rentraient plus particulièrement dans notre spécialité, et celui de statique, se trouvant ainsi réunis dans un même enseignement, nous

---

[4] (*) Varignon en fait mention dans sa Nouvelle mécanique publiée en 1687. – *Note by Culmann.*

[5] (**) C'est par l'effet du hasard qu'en 1845 un cours autographié sans nom d'auteur, ayant pour titre: Instruction sur la stabilité des constructions, est tombé entre nos mains. Celui qui nous l'a remis l'attribuait a M. Michon. Ce cours contient six leçons sur la stabilité des voûtes et quatre sur celle des murs de revétement. – *Note by Culmann.*

fûmes fréquemment amené à donner aux élevés des explications complémentaires sur les parties qu'ils n'avaient pas parfaitement comprises. Dans ces circonstances nous avons toujours trouvé qu'il était bien plus facile de rappeler des théorèmes de géométrie de position, dont la démonstration pouvait se faire à l'aide des lignes mêmes de l'épure, quo de recourir à des calculs analytiques dont les développements exigeaient l'emploi d'une feuille de papier séparé.
C'est ainsi que nous fûmes amené, pour ainsi dire irrésistiblement, à remplacer autant que possible l'algèbre par la géométrie de position. Pendant les premières années, les connaissances des élèves, dans cette matière, laissaient, il est vrai, un peu à désirer; mais depuis qu'un cours spécial de géométrie de position professé par M. Fiedler (auquel la *Géométrie descriptive* de cet auteur avait déjà préparé les élèves), a été introduit dans le programme des études, nous n'avons plus éprouvé aucune difficulté dans notre enseignement.
C'est lorsque cet enseignement eut pris quelque développement, que nous avons publié la première édition de notre *Statique graphique*. (La première moitié a paru en 1864 et la deuxième en 1865.)
Nos épures obtinrent plus de succès que nos méthodes. Notre publication fut suite d'un grand nombre de Statiques élémentaires, dans lesquelles, tout en reproduisant nos épures les plus simples (le plus souvent sans y rien changer), les auteurs s'efforçaient d'en donner des démonstrations analytiques.
Nous estimons que la vérité n'est pas là; qu'on ne parviendra jamais à tracer les lignes d'une épure et à exécuter simultanément les opérations algébriques que comporte l'explication de cette épure, nià se bien pénétrer de la signification de chaque ligne et à se représenter les relations statiques, si l'on se borne à traduire une formule dont les développements ne sont plus présents à la mémoire.
Nous devons toutefois excepter du reproche que nous nous croyons en droit d'adresser à nos successeurs, les auteurs italiens, et en particulier Cremona qui a introduit la Statique graphique dans l'enseignement de l'école polytechnique de Milan. Ce savant, auquel les sciences graphiques doivent de beaux travaux dont nous avons profité, ne dédaignait pas d'enseigner lui-même à ses élèves la géométrie de position. Bien que Cremona ait aujourd'hui quitté Milan pour Rome, l'enseignement de la Statique graphique est continué à l'école polytechnique de Milan dans le même esprit.
Les explications qui précédent nous ont paru nécessaires à l'historique de la Statique graphique, il nous reste à indiquer, en quelques mots, l'ordre que nous avons suivi dans notre ouvrage.
Le premier chapitre de la première partie traite du *calcul par le trait*. Bien qui'il soit étranger à la Statique proprement dite, il est nécessaire que les élèves le connaissent, et comme il n'est pas enseigné dans les cours préparatoires, nous avons pensé qu'il était indispensable do faire connaître ces méthodes, qui sont empruntées aux auteurs français et surtout à Cousinéry. Au calcul par le trait nous avons ajouté la cubature des terrassements, le mouvement des terres, la théorie de la règle à calcul, les méthodes si ingénieuses de M. Lalanne (aujourd'hui inspecteur général des ponts et chaussées et directeur de l'école

des ponts et chaussées de Paris) sur les représentations graphiques et sur les carrés logarithmiques.
La deuxième partie traite de la composition et de la décomposition des forces en général.
La troisième partie est consacrée aux forces parallèles et à leurs moments du premier et du second ordre, dont les applications à la théorie de l'élasticité, qui forme la quatrième partie de l'ouvrage, sont si nombreuses.

5.8 Nous avons essayé dans la deuxième édition, de joindre aussi brièvement que possible les solutions analytiques aux solutions purement géométriques. Les méthodes analytiques nouvelles ont le grand mérite de conduire directement au but et, en outre, de concorder avec les méthodes géométriques. Dans la plupart des cas nous avons pu déduire les formules des développements géométriques qui les précèdent. Ce mode de procéder a l'avantage de donner aux théorèmes une forme, qui, dans bien des cas, découle immédiatement des constructions géométriques, et, en outre, de laisser la choix, toutes les fois que nous donnons les deux solutions, entre la construction graphique et le calcul; dans la pratique c'est tantôt l'une des méthodes, tantôt l'autre qui conduit le plus rapidement au but. [...] Grâce à la méthode que nous avons suivie, nous avons montré à ceux qui cherchent à expliquer une épure analytiquement, comment il faut appliquer l'analyse pour faire ressortir l'identité des formules et des épures.

5.9 Les propriétés réciproques, que nous avons fait connaître jusqu'à présent entre le polygone funiculaire et le polygone des forces, et qui ont été indiquées pour la première fois par le professeur *Clerk Maxwell* dans le *Philosophical Magazine*, 1864, p. 250, se rapportent uniquement à des systèmes plans. Si on considère ces polygones comme les projections de polygones gauches, ces derniers peuvent être considérés de leur côté comme des formes réciproques d'un système focal. Cette théorie a été développée par Cremona dans son remarquable mémoire intitulé: *Le figure reciproche nella Statica grafica, Milano, Bernardoni*, 1872. Nous suivrons ici principalement ce dernier ouvrage.

5.10 Une force quelconque, sollicitant un arc, fait tourner l'extrémité de l'arc autour de son antipôle par rapport à l'ellipse centrale des $\Delta s/\epsilon\mathfrak{J}$; et la grandeur de la rotation est égale à la force, multipliée par le moment statique de ces $\Delta s/\epsilon\mathfrak{J}$ par rapport à sa direction.

5.11 Una reciprocità in un piano, dove due qualunque elementi omologhi si corrispondono in doppio modo (involutoriamente), ossia una reciprocità equivalente alla sua inversa, dicesi un sistema polare o una polarità; un punto e una retta che si corrispondono in una polarità piana si dicono polo e polare uno dell'altra.
La polarità in un piano può anche definirsi come una corrispondenza biunivoca fra i punti e le rette, tale che: se la retta corrispondente (polare) di un punto A passa per un punto B, la corrispondente (polare) di B passa per A.
Osservazione. Correlativamente (nello spazio) si può definire la polarità in una stella.

5.12 Mais s'il y avoit plusieurs poids suspendus en une mesme ligne, comme icy la ligne ABCDEF, ses poincts fermes extremes A, F, à laquelle sont sospendus 4 poids cognus, G, H, I, K; il est manifeste qu'on peut dire quel effort ils font à la corde, à chacune de ses parties AB, BC, CD, DE, EF: Car par example, produisant GB en haut vers L, & MN parallele à BC: Je dis BN donne BM, combien le poids G viendra l'effort quit est fait à AB.
Derechef BN donne MN, combien le poids G ce qui viendra sera l'effort qui est fait à BC.

5.13

THÉORÈME X.

I. Deux puissances quelconques K, L, dirigées à volonté, & appliquées en deux points quelconques C, D, d'une corde lâche et parfaitement flexible ACDB, attachée par le deux bouts à deux clous ou crochets A, B, demeurant encore en équilibre entr'elles, comme dans les Th. 8. 9. d'un point quelconque S soient faites SE, SF, SG parallèles aux trois côtez AC, CD, DB, du polygone ACDB que ces puissances font faire à cette corde; & d'un point F pris aussi à volonté sur SF, soient menées FE, FG, parallèles aux directions CK, DL, des puissances K, L, jusqu'à ce que deux lignes rencontrent SE, SG, en E, G. Cela fait, je dis qu'en ce cas d'équilibre les puissances K, L, seront entr'elles comme EF, FG, c'est-à-dire, K.L::EF.FG.
II. Réciproquement le corde ACDB étant donnée de position, c'est-à-dire, le polygone qu'elle forme étant donné, si d'un point S pris à volonté, on fait SE, SF, SG, parallèle aux trois côtez AC, CD, DB, de ce polygone; & que d'un point F pris aussi à volonté, on fait sur SF, on mene deux droites quelconques FE, FG, qui rencontrent SE, SG en E, G: deux puissances K, L, qui seroient entr'elles comme ces deux lignes FE, FG, & qui auraient leurs directions CK, DL, parallèles à ces mêmes lignes, chacune à chacune, retiendront la corde ACDB dans cette position donnée, y demeurant en équilibre entr'elles.

5.14 *Figure reciproche*. Proiettati questi due poliedri reciproci sopra un piano ortografico, a ogni lato della prima figura corrisponderà un lato parallelo nella seconda. Siccome poi a ρ spigoli formanti il contorno di una faccia dell'uno corrispondono i spigoli concorrenti nel vertice corrispondente dell'altro, così in proiezione a ρ lati concorrenti in un vertice corrisponderanno ρ lati paralleli formanti un poligono chiuso. Ogni spigolo in entrambi i poliedri è comune a due facce e congiunge due vertici; ogni faccia ha tre lati almeno e in ogni vertice concorrono almeno tre spigoli. Segue che nelle loro proiezioni ogni lato sarà comune a due poligoni e congiungerà due vertici e siccome ogni poligono avrà almeno tre lati, in ogni vertice concorreranno almeno tre lati. Gli elementi di un poliedro sono legati dalla relazione d'Eulero: $v + f = s + 2$ (1), dove v indica il numero dei vertici, f delle facce, s degli spigoli. Siccome ai v vertici nell'un poliedro corrispondono v facce nell'altro; alle f facce nell'uno, f vertici nell'altro, e agli s spigoli nell'uno gli s spigoli dell'altro, così la relazione (1) vale anche pel poliedro reciproco. Per le due figure ortografiche, se una consta di v vertici, f poligoni chiusi e s lati, l'altra consterà di f vertici, v poligoni chiusi e s lati. Se un poliedro ha un vertice all'infinito, l'altro ha una faccia perpendicolare al piano ortografico; onde, se una delle figure ortografiche ha un vertice all'infinito, il poligono corrispondente nell'altro si riduce a un segmento di retta su cui restano segnati dei punti corrispondenti al vertici della faccia di cui è proiezione. Le proiezioni ortografiche di due poliedri reciproci si dicono *figure reciproche*.

5.15 Due lati corrispondenti (r s), (r s)′ de' due poligoni si segano sopra una retta fissa che è parallela alla congiungente de' due poli O e O′. Questo teorema è fondamentale nei metodi di Culmann.

5.16 Sia dato un poligono piano di n lati 1, 2, 3, ..., $n-1$, n; e nello stesso piano siano inoltre date $n-1$ rette 1, 2, 3, ..., $n-1$ risp. parallele ai primi $n-1$ lati del poligono. Da un punto o polo, mobile nel piano (senz'alcuna restrizione), s'intendano projettati i vertici del poligono dato. Ora si immagini un poligono variabile di s lati, i primi $n-1$ vertici del quale 1, 2, 3, ..., $n-1$ debbano trovarsi ordinatamente nelle rette date omonime, mentre gli n lati (n . 1), (1 . 2), (2 . 3), ($n-1$. n) debbano essere paralleli ai raggi che dal polo projettano i vertici omonimi del poligono dato. Il punto di concorso di due lati qualsivogliano (r . $r+1$), (s . $s+1$) del poligono variabile cadrà in una retta determinata, parallela alla diagonale frai vertici (r . $r+1$), (s . $s+1$) del poligono dato.
Questo teorema, la cui dimostrazione per mezzo della sola geometria piana non pare ovvia, risulta invece a dirittura evidente, se si considerano le figure piane come projezioni ortografiche di poliedri reciproci.

5.17 Il diagramma reciproco di una travatura reticolare o diagramma di Cremona riunisce in una figura unica i poligoni di equilibrio di tutti i suoi nodi. In esso ogni segmento che misura lo sforzo S di un'asta, percorso una volta in un senso e una volta nell'altro, è lato comune ai due poligoni di equilibrio dei nodi estremi dell'asta; quindi compare una sola volta. Tralasciando la teoria generale, limitiamoci a indicare le reazioni che legano il diagramma reciproco allo schema della travatura, le sue principali proprietà e le regole pratiche per costruirlo.

5.18 Supposto che i nodi ai quali sono applicate le forze esterne si trovino tutti sul contorno dello schema della travatura, queste forze si dovranno prendere nell'ordine col quale sono incontrate da chi percorra il contorno suddetto. Quando non si seguano queste regole e le altre esposte più innanzi, si può ancora risolvere il problema della determinazione grafica degli sforzi interni, ma non si hanno più diagrammi reciproci, bensì figure più complicate o sconnesse, dove uno stesso segmento, non trovandosi al suo posto conveniente, dev'essere ripetuto o riportato per dar luogo alle costruzioni ulteriori, come accadeva nel vecchio metodo di costruire separatamente un poligono delle forze per ciascun nodo della travatura.

5.19 Questo metodo, che potrebbe dirsi statico, basta da sè solo alla determinazione grafica degli sforzi interni, al pari del metodo geometrico, esposto precedentemente, che si deduce dalla teoria delle figure reciproche e consiste nella costruzione successiva dei poligoni corrispondenti ai diversi nodi della travatura. Il metodo statico mi pare però meno semplice, e piuttosto può giovare in combinazione coll'altro, sopratutto per verificare l'esattezza delle operazioni grafiche, già eseguite.

5.20

Parte I. Geometria proiettiva

§ 1. Forme geometriche fondamentali.
§ 2. Sistemi armonici.
§ 3. Forme proiettive.
§ 4. Involuzioni.
§ 5. Generazione delle coniche.
§ 6. Poli e polari.
§ 7. Diametri delle coniche: ellisse, iperbole, parabola.
§ 8. Esercizi e costruzioni.
§ 9. Teorema di Desargues. Forme proiettive nelle coniche.
§ 10. Esercizi e costruzioni.
§ 11. Problemi di 2° grado.
§ 12. Fuochi delle coniche.
§13. Altri problemi e costruzioni.
§ 14. Coni e superficie gobbe di 2° grado.
§ 15. Esercizi.
§ 16. Proiettività delle forme geometriche fondamentali di seconda specie.
§ 17. Affinità e similitudine delle figure piane.
§ 18. Esercizi.
§ 19. Generazione delle superficie di 2° grado.
§ 20. Poli e piani polari rispetto a una superficie di 2° grado. Diametri, centro, assi.
§ 21. Proiettività delle forme geometriche fondamentali di 3^{a} specie.
§ Esercizi

Parte II. Calcolo grafico

§ 1. Addizione e sottrazione delle linee rette. Proprietà della somma vettoriale, poligono dei vettori. Sottrazione. Vettori paralleli.
§ 2. Moltiplicazione per uno scalare.
§ 3. Elevazione a potenza e estrazione di radice.
§ 4. Moltiplicazione di rette con rette.
§ 5. Trasformazione delle aree a contorno rettilineo.
§ 6. Tavole grafiche.
§ 7. Trasformazione delle figure circolari.
§ 8. Trasformazione delle figure curvilinee in generale.

§ 9. Teoria del planimetro.
§ 10. Cubatura di masse regolari di sterro e riporto.
§ 11. Cubatura di masse irregolari.
§ 12. Calcolo grafico dei movimenti di terra.

Parte III. Statica grafica

§ 1. Composizione delle forze applicate a un punto.
§ 2. Composizione di più forze, situate comunque in un piano.
§ 3. Corrispondenza proiettiva tra il poligono delle forze e il poligono funicolare.
§ 4. Esempi e casi particolari.
§ 5. Momenti di forze in un piano.
§ 6. Forze infinitamente piccole e infinitamente distanti.
§ 7. Equilibrio delle forze nel piano.
§ 8. Equilibrio delle forze nello spazio.
§ 9. Forze parallele in un piano.
§ 10. Centri di gravità.
§ 11. Momenti di inerzia.
§ 12. Ellissoide centrale.
§ 13. Ellissoide d'inerzia.
§ 14. Sistema di forze parallele le cui intensità siano proporzionali alle distanze dei punti d'applicazione da un piano.
§ 15. Ellissi d'inerzia.
§ 16. Sistema di forze parallele agente su una sezione piana.
§ 17. Costruzione dell'ellisse centrale e del nocciolo di una figura piana.
§ 18. Ellisse centrale e nocciolo di un profilo di rail.
§ 19. Ellisse centrale e nocciolo di un ferro a angolo.
§ 20. Distribuzione delle forze interne nelle sezioni di una travatura.
§ 21. Costruzione delle forze interne.

5.21 Col pubblicare ora per le stampe questo corso, d'indole scolastica, speriamo di rendere qualche servizio agli studiosi qualichesiansi di questa disciplina. Contiene numerosi esempi e problemi e moltissime figure, oltre a 1050. Ciò potrà sembrare soverchio per chi già si trova in possesso della materia, ma non per i principianti, a cui è dedicato particolarmente il libro.

5.22 Non intendiamo che sia da rifiutare in un libro di *Statica grafica* tutto quanto non sia informato alla Geometria pura. Vi hanno dei casi, in cui il metodo geometrico o non è giunto o non giungerà forse mai a sostituire quello analitico e ve ne ha altri poi, in cui sarebbe inespediente il metodo geometrico, presentandosi o meno generale o meno semplice di quello analitico.
Abbiamo voluto riportare per esempio nella prima parte il metodo geometrico di *Archimede* per determinare il baricentro di un segmento parabolico [...]. Ma ognun vede quanto più semplice sia in questo caso il metodo analitico.

5.23

Statica grafica. – Forze esterne

Prefazione alla parte seconda

CAPITOLO PRIMO.

FORZE CONCENTRATE.

§ 1. Nozioni preliminari.
§ 2. Composizione delle forze concorrenti.
§ 3. Composizione delle forze non concorrenti in un piano.
§ 4. Proprietà dei poligoni funicolari.
§ 5. Composizione delle forze non concorrenti in un piano col metodo del fascio funicolare; proprietà di questo
§ 6. Composizione e centro delle forze parallele nello spazio.
§ 7. Forze agenti per rotazione.
§ 8. Determinazione grafica del momento risultante di un sistema di forze in un piano.
§ 9. Decomposizione di una forza in altre compiane.
§ 10. Composizione delle coppie nello spazio.
§ 11. Composizione delle forze non concorrenti nello spazio.
§ 12. Altri due metodi per la composizione delle forze nello spazio.
§ 13. Asse centrale; sua determinazione; sue proprietà.
§ 14. Sistema polare individuato nello spazio da un sistema di forze.
§ 15. Interpretazione meccanica delle figure reciproche.
§ 16. Applicazione delle figure reciproche al disegno dei tetti.
§ 17. Decomposizione delle forze nello spazio.

CAPITOLO SECONDO.

FORZE RIPARTITE E EQUILIBRIO DEI CORPI SENZ'ATTRITO.

§ 1. Forze ripartite.
§ 2. Condizioni d'equilibrio di un corpo vincolato e reazioni dei vincoli.
§ 3. Sistemi di corpi in equilibrio.
§ 4. Problemi sull'equilibrio dei sistemi di corpi.
§ 5. Sistemi a equilibrio indifferente.

CAPITOLO TERZO.

EQUILIBRIO DEI CORPI APPOGGIATI CON ATTRITO.

§ 1. Attrito – Stabilità.
§ 2. Equilibrio di minima stabilità di un corpo.
§ 3. Sistemi di corpi appoggiati in equilibrio di minima stabilità.
§ 4. Attrito nelle catene.

CAPITOLO QUARTO.

STABILITà DEI CORPI APPOGGIATI.

§ 1. Rapporti di stabilità – Spinta dell'acqua. – Stabilità di una diga rispetto allo scorrimento, alla rotazione e alla compressione. – Diga a profilo triangolare. – Camini.
§ 2. Spinta delle materie semifluide e terre prive di coesione.
§ 3. Sistemi di corpi appoggiati per superfici estese formanti delle catene chiuse. – Volte. – Curva delle pressioni tangente a una linea data.
§ 4. Curve funicolari..

CAPITOLO QUINTO.

TRAVATURE RETICOLARI.

§ 1. Generazione delle travature reticolari strettamente indeformabili.
§ 2. Calcolazione delle travature indeformabili caricate ai nodi.
§ 3. Secondo problema. Metodo del diagramma per la calcolazione delle travature strettamente indeformabili.
§ 4. Travature applicate.
§ 5. Travature reticolari a membri caricati.
§ 6. Travature strettamente indeformabili con membri a più di due nodi.

CAPITOLO SESTO.

EFFETTI DELLE FORZE ESTERNE NELLE SEZIONI DEI SOLIDI.

§ 1. Forze esterne fisse.
§ 2. Diagrammi delle azioni componenti delle risultanti relative a tutte le sezioni del solido.
§ 3. Solidi a asse curvilineo.
§ 4. Travi orizzontali appoggiate agli estremi e soggette a carichi mobili.
§ 5. Diagramma delle forze taglianti in una sezione d'una trave percorsa direttamente da un carico ripartito uniformemente.
§ 6. Diagrammi dei momenti flettenti.
§ 7. Applicazione del poligono funicolare alla ricerca dei momenti massimi nelle sezioni di una trave percorsa da un sistema di carichi.
§ 8. Le travature reticolari indirettamente soggette a carichi mobili
§ 9. Travature reticolari a tre cerniere soggette a carichi mobili.

5.24 L'indirizzo che il CULMANN ha dato alla *Statica grafica* lascia presumere che il suo campo sia circoscritto nella meccanica applicata alle costruzioni civili. Ma il metodo grafico si presta non meno utilmente nello studio dell'equilibrio di minima stabilità dei corpi appoggiati. Con ciò la *Statica grafica* giova anche come corso preparatorio a quello di meccanica applicata alle macchine, in cui si va insinuando con ognor crescente vantaggio e al quale serve inoltre per tutto quanto comprende relativamente alle nozioni fondamentali sulla resistenza dei materiali.

5.25 Forma oggetto della seconda parte lo studio delle forze esterne, cioè delle azioni, non esclusa la gravità, che un corpo considerato isolatamente riceve da altri.
Nella Statica grafica si prendono raramente in considerazione gli angoli fra le forze, perché non entrano, come nella meccanica analitica, quali elementi necessarii per individuare le forze nello spazio.
Si fa un uso pure limitato dei momenti perché torna piú comodo e spedito l'operare sulle forze (segmenti) anziché sulle coppie (superficie).
L'attraente semplicità dei metodi che caratterizzano la statica grafica, permette di poter attaccare dopo poche nozioni diversi problemi che pel passato si solevano trattare soltanto in corso di ulteriori applicazioni.
Facciamo posto alle *forze ripartite* generalmente dimenticate negli ordinarii trattati di *Meccanica. Tutte le forze della natura sono composte d'elementi, sole forze che realmente esistano*, dice il *Belanger* nel suo *Corso di Meccanica* a pag. 37; *le altre sono delle concezioni della nostra mente che entrano nella scienza sotto il nome di somma o di risultante.*
Lo studio delle forze ripartite ammette direttamente alla statica dei corpi.
Consideriamo dei sistemi di corpi *deformabili, indeformabili* e *scioglibili.* I primi s'incontrano particolarmente nelle macchine e se ne considera l'*equilibrio di minima stabilità.* I sistemi indeformabili e scioglibili s'incontrano specialmente nelle costruzioni statiche. Di questi ultimi costituiti ordinariamente da catene di corpi appoggiati per superficie piane estese si considera la *stabilità* e dei sistemi indeformabili si studiano quelli strettamente indeformabili come *travature reticolari* avendo di mira particolarmente la determinazione delle *reazioni mutue* fra i corpi che le compongono.
Esempii e applicazioni produciamo sull'equilibrio delle catene di corpi appoggiati onde si acquisti facilità nel rilevare dove e come si trasmettono le pressioni, i corpi appoggiati di varia configurazione e in varie condizioni.
Lo studio delle azioni che esercitano forze in equilibrio sulle varie sezioni d'un solido cui sono applicate forma l'ultimo argomento di questa seconda parte. Esso ammette tosto allo studio delle forze interne che vengono trattate nella terza parte.

5.26 Un corpo non può da sé spostarsi se è in riposo o modificare il movimento che possiede senza l'intervento di una *causa* a esso esteriore. [...] Non se ne indaga l'origine; soltanto se ne *valuta l'effetto* [...]

5.27 Nella Statica si considerano dei corpi ideali che hanno tutte le loro dimensioni infinitesime, senza avere una forma determinata e che diconsi *elementi* o *punti materiali.* Si considerano inoltre a essi applicate delle forze di *grandezza finita* (ideali) le quali, essendo concentrate sopra un punto diconsi *forze concentrate.* [...]

Più punti di applicazione si dicono *rigidamente connessi* quando sieno collegati per modo che le loro distanze relative si conservino sempre invariabili, o quando facciano parte di un corpo indeformabile.
Quantunque nella Statica si considerino i corpi come materiali, pure da principio faremo astrazione del loro peso, cioè li riguarderemo come corpi *geometrici* o *nessi rigidi indefinitamente resistenti* dei punti d'applicazione che debbano formare un sistema di forma invariabile.

5.28 Culmann vuole fondare lo studio della Statica grafica su quello della Geometria di posizione, che ritiene necessario per lo svolgimento perfetto delle teorie della Statica grafica. Altri come Bauschinger, Levy, stimano sufficiente di seguire nella esposizione della Statica grafica metodi più semplici e elementari.
In questi due anni, durante i quali ho avuto l'onore d'insegnare la Statica grafica presso la Real Scuola d'applicazione per gl'Ingegneri di Torino, per le condizioni degli studi fatti dai miei uditori ho creduto bene di seguire le tracce di Bauschinger e di Levy. Ed ho redatto questo scritto, che mi sono deciso a pubblicare nella speranza che esso possa tornare utile a coloro i quali desiderino procedere per le vie più facili allo studio della statica grafica.

5.29 Espongo ora brevemente il programma della materia che ho cercato di sviluppare in questo scritto. Nel primo capitolo sono date alcune nozioni elementari di calcolo grafico che formano una introduzione utile allo studio della Statica grafica. Il secondo capitolo tratta delle proprietà geometriche dei poligoni funicolari. Nel terzo capitolo si definiscono le figure reciproche della Statica grafica, e se ne esaminano alcuni esempi dedotti dalla teoria geometrica dei poligoni funicolari. In questo capitolo viene enunciato un teorema relativo alle figure, le quali si possono riguardare come projezioni piane di poliedri. Tali figure ammettono sempre delle figure reciproche. La dimostrazione del teorema medesimo è data più tardi nel capitolo undicesimo mediante la teoria dei sistemi equivalenti di due forze nello spazio. Questo modo di considerare le figure reciproche come proiezioni piane di poliedri nello spazio è dovuto al chiarissimo Professore Cremona. Nel capitolo quarto si considerano i sistemi di forze concorrenti in unpunto. Il capitolo quinto tratta della composizione delle forzo giacenti in un piano mediante l'impiego del poligono delle forze e dei poligoni funicolari, dei quali diventano evidenti l'ufficio e l'importanza nella Statica grafica. Nel capitolo sesto si espone la teoria dei momenti delle forze e delle coppie giacenti in un piano, e si spiegano le costruzioni per ridurre i momenti ad una base. Nel capitolo settimo si risolvono diversi problemi sulla scomposizione delle forze in un piano. Nel capitolo ottavo si studiano vari problemi relativi all'equilibrio di un corpo soggetto a particolari condizioni e sotto l'azione di forze contenute tutte in un piano, come pure vari problemi relativi all'equilibrio dei poligoni articolati. Il capitolo nono si aggira intorno ai diagrammi degli sforzi di tensione e di compressione nei sistemi di sbarre. E si rende manifesta l'utilità delle figure reciproche per la descrizione dei diagrammi nel caso delle travature reticolari, delle quali si danno molti esempi ricavati dalle costruzioni. Si accenna ancora alla descrizione dei diagrammi degli sforzi di tensione e di compressione per altri sistemi di sbarre diversi dalle travature reticolari. Il capitolo decimo tratta dei diagrammi degli sforzi di taglio e dei momenti inflettenti per una trave orizzontale collocata so due appoggi e soggetta a carichi fissi o mobili. Il capitolo undicesimo è dedicato allo studio dei sistemi di forze nello spazio. Esso tratta della riduzione di un sistema di forze qualunque ad una forza ed una coppia, dei momenti delle forze rispetto ad un asse, della riduzione d'un sistema di forze qualunque ad un sistema equivalente di due forzo, della proprietà dei sistemi equivalenti di due forze nello spazio e dei poliedri reciproci. In questo capitolo si dà la dimostrazione del teorema enunciato nel capitolo terzo, e relativo alle figure che si possono riguardare corno projezioni piane di poliedri. Il capitolo dodicesimo tratta della determinazione del centro di un sistema di forze parallele. - Il capitolo tredicesimo ha per oggetto la determinazione dei centri di gravità delle linee, delle aree e dei volumi. - Il capitolo quattordicesimo si aggira intorno ai momenti di secondo ordine, e specialmente intorno ai momenti d'inerzia ed alla ellisse, d'inerzia d'un'area piana. In questo capitolo si determina ancora il centro d'un sistema di forze parallele.

Index

D. Capecchi and G. Ruta, *Strength of Materials and Theory of Elasticity in 19th Century Italy*, Advanced Structured Materials 52,
DOI 10.1007/978-3-319-05524-4

D

E

F

MIX
Papier aus verantwortungsvollen Quellen
Paper from responsible sources
FSC® C105338

If you have any concerns about our products,
you can contact us on
ProductSafety@springernature.com

In case Publisher is established outside the EU,
the EU authorized representative is:
Springer Nature Customer Service Center GmbH
Europaplatz 3, 69115 Heidelberg, Germany

Printed by Libri Plureos GmbH
in Hamburg, Germany